Introduction to Ordinary Differential Equations with *Mathematica*®

Solutions Manual

ALFRED GRAY   MICHAEL MEZZINO   MARK A. PINSKY

Introduction to Ordinary Differential Equations with *Mathematica*®

Solutions Manual

 **Web-Enhanced**

ISBN 978-0-387-98232-8   ISBN 978-1-4612-1736-7 (eBook)
DOI 10.1007/978-1-4612-1736-7

THE
ELECTRONIC
LIBRARY
OF
SCIENCE

TELOS, The Electronic Library of Science, is an imprint of Springer-Verlag New York with publishing facilities in Santa Clara, California. Its publishing program encompasses the natural and physical sciences, computer science, economics, mathematics, and engineering. All TELOS publications have a computational orientation to them, as TELOS' primary publishing strategy is to wed the traditional print medium with the emerging new electronic media in order to provide the reader with a truly interactive multimedia information environment. To achieve this, every TELOS publication delivered on paper has an associated electronic component. This can take the form of book/diskette combinations, book/CD-ROM packages, books delivered via networks, electronic journals, newsletters, plus a multitude of other exciting possibilities. Since TELOS is not committed to any one technology, any delivery medium can be considered.

The range of TELOS publications extends from research level reference works through textbook materials for the higher education audience, practical handbooks for working professionals, as well as more broadly accessible science, computer science, and high technology trade publications. Many TELOS publications are interdisciplinary in nature, and most are targeted for the individual buyer, which dictates that TELOS publications be priced accordingly.

Of the numerous definitions of the Greek word "telos," the one most representative of our publishing philosophy is "to turn," or "turning point." We perceive the establishment of the TELOS publishing program to be a significant step towards attaining a new plateau of high quality information packaging and dissemination in the interactive learning environment of the future. TELOS welcomes you to join us in the exploration and development of this frontier as a reader and user, an author, editor, consultant, strategic partner, or in whatever other capacity might be appropriate.

TELOS, The Electronic Library of Science
Springer-Verlag Publishers
3600 Pruneridge Avenue, Suite 200
Santa Clara, CA 95051

THE ELECTRONIC LIBRARY OF SCIENCE

TELOS Diskettes

Unless otherwise designated, computer diskettes packaged with TELOS publications are 3.5" high-density DOS-formatted diskettes. They may be read by any IBM-compatible computer running DOS or Windows. They may also be read by computers running NEXTSTEP, by most UNIX machines, and by Macintosh computers using a file exchange utility.

In those cases where the diskettes require the availability of specific software programs in order to run them, or to take full advantage of their capabilities, then the specific requirements regarding these software packages will be indicated.

TELOS CD-ROM Discs

For buyers of TELOS publications containing CD-ROM discs, or in those cases where the product is a stand-alone CD-ROM, it is always indicated on which specific platform, or platforms, the disc is designed to run. For example, Macintosh only; Windows only; cross-platform, and so forth.

TELOSpub.com (Online)

Interact with TELOS online via the Internet by setting your World-Wide-Web browser to the URL: http://www.telospub.com.

The TELOS Web site features new product information and updates, an online catalog and ordering, samples from our publications, information about TELOS, data-files related to and enhancements of our products, and a broad selection of other unique features. Presented in hypertext format with rich graphics, it's your best way to discover what's new at TELOS.

TELOS also maintains these additional Internet resources:

gopher://gopher.telospub.com
ftp://ftp.telospub.com

For up-to-date information regarding TELOS online services, send the one-line e-mail message:

send info to: *info@TELOSpub.com.*

PREFACE

The purpose of this companion volume to our text is to provide instructors (and eventually students) with some additional information to ease the learning process while further documenting the implementations of *Mathematica* and **ODE**.

In an ideal world this volume would not be necessary, since we have systematically worked to make the text unambiguous and directly useful, by providing in the text worked examples of every technique which is discussed at the theoretical level. However, in our teaching we have found that it is helpful to have further documentation of the various solution techniques introduced in the text.

The subject of differential equations is particularly well-suited to self-study, since one can always verify by hand calculation whether or not a given proposed solution is a bonafide solution of the differential equation and initial conditions. Accordingly, we have not reproduced the steps of the verification process in every case, rather content with the illustration of some basic cases of verification in the text. As we state there, students are strongly encouraged to verify that the proposed solution indeed satisfies the requisite equation and supplementary conditions.

With some due consideration, and at the possible risk of additional space, we have included the complete statement of every exercise whose solution is listed here. We feel strongly that the learning process begins with questions, to which there may be many different possible types of answers. With the statement of the exercise in front of the reader, there is no need to constantly refer to the text while verifying the solution of an exercise.

The level of detail of the solutions is quite varied as we move from section to section, again with some justification. In those cases where a solution requires the **ODE** code in order to find a symbolic or graphical solution, we have included the lines of code without further comment, following the detailed discussions of **ODE** and *Mathematica* in the main body of text. This applies in particular in chapters 2,4,10. In those sections which involve substitution of a specific equation type into a formula obtained in the text, there is little or no need for further documentation of the method of solution. This applies to the solutions of Chapters 3,8,9,15,16,20,21. At the other extreme, we have gone to great length in chapters 6,7,11,18,19 involving *applications* of differential equations, to give detailed documentations of all of the steps used in modelling a given physical application and its solution in terms of differential equations. This is traditionally the most challenging part of teaching a first course in differential equations, where students typically need the most assistance. This is especially important, since many students taking this course have not taken physics recently, and need to be reminded about conventions on units, force, mass, etc. Many of the solutions in these chapters contain graphs produced by *Mathematica*, reflecting our viewpoint that graphical output is instrumental in enhancing the learning process.

Another area in which we have endeavored to give a complete documentation concerns exercises which involve *mathematical proofs*. We feel that historically, the subject of differential equations has been instrumental in motivating many methods and viewpoints in mathematical analysis, and that some of these can be exhibited in a first course in differential equations. While the text itself contains a number of theorems with proofs, we have also used the exercises to augment this, especially in Chapter 3, Chapter 14, Chapter 15, Chapter 17 and Chapter 20 where issues of existence and uniqueness of solutions, stability of solutions, convergence of series, and related theoretical questions of analysis are treated. For those students who anticipate further work in mathematical analysis, this theoretical background from differential equations will provide a storehouse of interesting and non-trivial examples for motivation in further work. For example, in Chapter 20 the text contains the proof of the method of majorants for the homogeneous second-order equation, while the exercises and Solutions Manual contain the corresponding proof for the non-homogeneous second-order equation. By contrast, in Chapter 15, we have put all of the existence-uniqueness theory for linear systems into the exercises, in order to lighten the text.

Additional comments on this volume may be addressed to the authors, at any one of the following email addresses:

Alfred Gray gray@bianchi.umd.edu

Michael Mezzino mezzino@gauss.cl.uh.edu

Mark Pinsky pinsky@math.nwu.edu

SOLUTIONS

SOLUTIONS MANUAL

Chapter 1

Section 1.1, page 3

For each of the following differential equations, determine whether it is *ordinary* or *partial*, *linear* or *nonlinear* and find the *order*.

1. $y' + y^2 = \sin(t)$

Solution. There is only one independent variable. The highest derivative which appears is the first derivative, but there is a (quadratic) nonlinear term. Hence the equation is ordinary, nonlinear, first-order.

2. $y'' + 100y = \sin(t)$

Solution. There is only one independent variable. The highest derivative which appears is the second derivative and all terms are linear. Hence the equation is ordinary, linear, second-order.

3. $y'' + 4y = e^y$

Solution. There is only one independent variable. The highest derivative which appears is the second derivative but there is an (exponential) nonlinear term. Hence the equation is ordinary, nonlinear, second-order.

4. $\dfrac{\partial u}{\partial t} = \dfrac{\partial(u^2)}{\partial x}$

Solution. There are two independent variables, t and x. The highest derivative which appears is the first derivative, but there is a (quadratic) nonlinear term Hence the equation is partial, nonlinear, first-order.

5. $y' + t\, y = \sin(t)$

Solution. There is only one independent variable. The highest derivative which appears is the first derivative and all terms are linear. Hence the equation is ordinary, linear, first-order.

6. $y'' + y^2 = e^t$

Solution. There is only one independent variable. The highest derivative which appears is the second derivative and there is a (quadratic) nonlinear term. Hence the equation is ordinary, nonlinear, second-order.

7. $u_{tt} = u_{xx} + u_{yy}$

Solution. There are three independent variables, t, x and y. The highest derivative which appears is the second derivative and all terms are linear. Hence the equation is partial, linear, second-order.

8. $\dfrac{\partial u}{\partial t} = \dfrac{\partial^2 u}{\partial x^2}$

Solution. There are two independent variables t and x. The highest derivative which appears is the second derivative and all terms are linear. Hence the equation is partial, linear, second-order.

Section 1.3, page 15

Compute the following indefinite integrals

1. $\displaystyle\int t^{3/2} dt$

Solution. In general, the indefinite integral of the function t^n is

$$\frac{t^{n+1}}{(n+1)} + C$$

unless $n = -1$. Applying this formula to the case $n = 3/2$ gives

$$\int t^{3/2} dt = \frac{2t^{5/2}}{5} + C.$$

2. $\displaystyle\int t \sin(2t) dt$

Solution. We use the method of integration by parts with $u = t$ and $dv = \sin(2t)$. Thus

$$\int u\,dv = uv - \int v\,du$$

$$= t\left(\frac{-\cos(2t)}{2}\right) + \int \frac{\cos(2t)}{2}\,dt = \frac{-t\cos(2t)}{2} + \frac{\sin(2t)}{4} + C.$$

Hence $\int t\sin(2t)dt = \dfrac{\sin(2t)}{4} - \dfrac{t\cos(2t)}{2} + C.$

3. $\displaystyle\int \frac{dt}{t(t-1)}$

Solution. We use the method of partial fractions. Thus

$$\int \frac{dt}{t(t-1)} = \int \left(\frac{1}{t-1} - \frac{1}{t}\right)dt = \log|t-1| - \log|t| + C.$$

4. $\displaystyle\int t^2 e^{-t^3}\,dt$

Solution. The integrand is the derivative of the function $(-1/3)e^{-t^3}$. Hence the integral is $(-1/3)e^{-t^3} + C.$

5. $\displaystyle\int_1^2 t^{3/2}dt$

Solution. Using the indefinite integral found in Exercise 1, we have

$$\int_1^2 t^{3/2}dt = \frac{2t^{5/2}}{5}\Big|_1^2 = \frac{2}{5}(2^{5/2} - 1) = \frac{8\sqrt{2}}{5} - \frac{2}{5} \approx 1.862.$$

6. $\displaystyle\int_a^b t\sin(2t)dt$

Solution. Using the indefinite integral found in Exercise 2, we have

$$\int_a^b t\sin(2t)dt = \frac{a\cos(2a)}{2} - \frac{b\cos(2b)}{2} + \frac{\sin(2b)}{4} - \frac{\sin(2a)}{4}.$$

7. $\displaystyle\int_e^{2e} \frac{dt}{t(t-1)}$

Solution. Using the indefinite integral found in Exercise 3, we have

$$\int_{e}^{2e} \frac{dt}{t(t-1)} = \log|t-1| - \log|t| \Big|_{e}^{2e}$$

$$= \log(2e-1) - \log(e-1) - \log(2e) + 1$$

$$= \log(2e-1) - \log(e-1) - \log(2).$$

8. $\displaystyle\int_{0}^{\infty} t^2 e^{-t^3} dt$

Solution. Using the indefinite integral found in Exercise 4, we have

$$\int_{0}^{\infty} e^{-t^3} dt = -\frac{1}{3}e^{-t^3} \Big|_{0}^{\infty} = \frac{1}{3}.$$

9. Show that $y(x) = (1/c)\cosh(cx)$ is a solution of (1.8).

Solution. Computing the first derivative of $y(x)$, we obtain $y'(x) = \sinh(cx)$. Substituting this into the right-hand-side of (1.8), we obtain

$$c\sqrt{1 + \big(\sinh(cx)\big)^2} = c\,\cosh(cx) = y''(x).$$

Chapter 2

Section 2.2, page 28

1. Express the differential equation $y''(t) + t\,y'(t) + (1+t^2)y(t) = t/(1+t)$ using the "prime" notation. Do not evaluate the expression.

Solution.

```
y''[t] + t y'[t] + (1 + t^2)y[t]  ==  t/(1 + t)
```

2. Express the differential equation in Problem 1 using the operator **D**. Do not evaluate the expression.

Solution. Either

```
D[y[t],t,t] + t D[y[t],t] + (1 + t^2)y[t]  ==  t/(1 + t)
```

or

```
D[y[t],{t,2}] + t D[y[t],t] + (1 + t^2)y[t] == t/(1 + t)
```

3. Define the function $f(t) = 1 + t + \sin(t) + e^t - \sqrt{5 - t^2}$ and compute the exact and numerical values of $f(2)$.

Solution. The *Mathematica* definition of the function is

```
f[t_]:= 1 + t + Sin[t] + E^t - Sqrt[5 - t^2]
```

Its exact and numerical value are computed with **f[2]** and **f[2]//N**, giving

```
         2
2 + E  + Sin[2]
```

and

```
10.2984
```

4. Define the function $f(t, c) = \cos(t) + t^{3/2} + \dfrac{2}{3}c.$

Solution. The function is defined with the command

```
f[t_,c_]:= Cos[t] + t^(3/2) + (2/3)c
```

5. Compute the exact value of $\dfrac{1}{3 \cdot 5 \cdot 7 \cdot 9}$; then use **N** to find a numerical approximation.

Solution. The exact and numerical values of $(3 \cdot 5 \cdot 7 \cdot 9)^{-1}$ are computed with **1/(3 5 7 9)** and **1/(3 5 7 9)//N**, giving

```
 1
---
945
```

and

```
0.0010582
```

6. Compute the following derivatives:

$$\frac{d}{dt}\left(\frac{\log(\sin(t))}{\tan(t)}\right), \qquad \frac{d^2}{dt^2}\left(\frac{t^3 + 1}{t\,\sin(t)}\right).$$

Solution. The derivatives are computed with

```
D[Log[Sin[t]]/Tan[t],t]
```

and

```
D[(t^3 + 1)/(t Sin[t]),{t,2}]//Apart
```

giving

$$Cot[t]^2 - Csc[t]^2 Log[Sin[t]]$$

and

$$\frac{(2 + 2 t^3 + 2 t Cot[t] - 4 t^4 Cot[t] + t^2 Cot[t]^2 + t^5 Cot[t]^2) Csc[t]}{t^3}$$

$$+ \frac{(1 + t^3) Csc[t]^3}{t}$$

7. Compute the following integrals:

$$\int \left(t^3 e^{5t} \cos(t)\right)dt, \qquad \int_0^\pi (\sin(t))^7 dt.$$

Solution. The integrals are computed with

```
Integrate[t^3 E^(5 t)Cos[t],t]//Simplify
```

and

```
Integrate[(Sin[t])^7,{t,0,Pi}]
```

giving

$$(E^{5t} (-357 Cos[t] + 2145 t Cos[t] - 6084 t^2 Cos[t] +$$

$$10985 t^3 Cos[t] - 360 Sin[t] + 1443 t Sin[t] -$$

$$2535 t^2 Sin[t] + 2197 t^3 Sin[t])) / 57122$$

and

$$\frac{32}{35}$$

8. Verify the fundamental theorem of calculus for $g(t) = t\,e^t + \sec(t)$. [In other words, show that the derivative of the integral of $g(t)$ equals $g(t)$.]

Solution. We use

```
D[Integrate[t E^t + Sec[t],t],t]//Simplify
```

to get

```
 t
E  t + Sec[t]
```

9. Write the function f defined by $f(t) = t^3 + \tan(t)$ as a *Mathematica* function and compute the symbolic and numerical values of $f(\pi/4)$.

Solution. Since we already used the symbol **f** in Exercise 3, we first use **Remove[f]** or **Clear[f]** to clear it. Next, we define a new function **f** by either of the commands

```
f = Function[t,t^3 + Tan[t]]
```

or

```
f = (#^3 + Tan[#])&
```

Then the symbolic value of $f(\pi/4)$ is found with

```
f[Pi/4]
```

yielding

```
      3
    Pi
1 + ──
    64
```

The numerical value of $f(\pi/4)$ is found with

```
f[Pi/4]//N
```

giving

```
1.48447
```

Section 2.3, page 32

Do the following plots with *Mathematica*.

1. Plot $\left(\sin(t^2)\right)^2$ from -2π to 2π.
Solution.

```
Plot[Sin[t^2]^2,{t,-2Pi,2Pi}]
```

2. Use **ParametricPlot** to plot the following plane curves.
   a. The **ellipse** as $t \longrightarrow \left(\sin(t), 2\cos(t)\right)$, with $0 \le t \le 2\pi$.
   b. The **parabola** as $t \longrightarrow \left(4t, t^2\right)$, with $-2 \le t \le 2$.
   c. The **astroid** as $t \longrightarrow \left((\sin(t))^3, (\cos(t))^3\right)$, with $0 \le t \le 2\pi$.
Solution.

```
a. ParametricPlot[{Sin[t],2Cos[t]},{t,0,2Pi}]
b. ParametricPlot[{4t,t^2},{t,-2,2}]
c. ParametricPlot[{Sin[t]^3,Cos[t]^3},{t,0,2Pi}]
```

   a.

```
ParametricPlot[{Sin[t],2Cos[t]},{t,0,2Pi}]
```

   b.

```
ParametricPlot[{4t,t^2},{t,-2,2}]
```

   c.

```
ParametricPlot[{Sin[t]^3,Cos[t]^3},{t,0,2Pi}]
```

3. Viviani's curve is the space curve $t \longrightarrow \left(1 + \cos(t), \sin(t), 2\sin(t/2)\right)$. Use **ParametricPlot3D** to plot it from -2π to 2π.
Solution.

```
ParametricPlot3D[{1 + Cos[t],Sin[t],2Sin[t/2]},
{t,-2Pi,2Pi}]
```

4. Use **ContourPlot** to plot contours defined by x^2y^3. Use the ranges $-2\pi \le x \le 2\pi$ and $-2\pi \le y \le 2\pi$.
Solution.

```
ContourPlot[x^2 y^3,{x,-2Pi,2Pi},{y,-2Pi,2Pi},
PlotPoints->100]
```

Chapter 3

Section 3.1, page 38

Find the general solution of each of the following first-order differential equations.

1. $y' = (t^3 + 2)^3 t^2$

Solution. Using the binomial theorem, the equation is written

$$y' = t^2(t^9 + 6t^6 + 12t^3 + 8) = t^{11} + 6t^8 + 12t^5 + 8t^2,$$

which is integrated to yield

$$y = \frac{8t^3}{3} + 2t^6 + \frac{2t^9}{3} + \frac{t^{12}}{12} + C,$$

where C is a constant.

2. $y' = e^{3t}$

Solution. The indefinite integral of the exponential function e^{3t} is $e^{3t}/3$, so that the general solution of the equation is

$$y = \frac{e^{3t}}{3} + C,$$

where C is a constant.

3. $y' = t \sin(t)$

Solution. We integrate by parts with $u = t$ and $dv = \sin(t)\,dt$. Hence $v = -\cos(t)$ and

$$\int u\,dv = u\,v - \int v\,du = -t\,\cos(t) + \int \cos(t)\,dt;$$

hence the solution is

$$y = C - t\cos(t) + \sin(t),$$

where C is a constant.

4. $t\,y' = 2$

Solution. The equation is rewritten $y' = 2/t$, which is integrated to yield

$$y = C + 2\log(t),$$

where C is a constant.

Solve the following initial value problems.

5. $\begin{cases} y' = t^3 + 4t, \\ y(0) = 1 \end{cases}$

Solution. The general solution of the equation is obtained by integration, thus, $y(t) = t^4/4 + 2t^2 + C$. The arbitrary constant is determined from the initial conditions by $1 = y(0) = C$, hence

$$y = 1 + 2t^2 + \frac{t^4}{4}.$$

6. $\begin{cases} y' = \dfrac{\sin(t)}{\cos(t)}, \\ y(\pi/4) = 1 \end{cases}$

Solution. The general solution of the equation is obtained by integration as $y(t) = -\log(\cos(t)) + C$, which is valid as long as $\cos(t) > 0$. To determine C, we write

$$1 = y(\pi/4) = -\log(\cos(\pi/4) + C = \log(\sqrt{2}) + C,$$

hence $C = 1 - \log(\sqrt{2})$ with the solution

$$y = 1 - \frac{\log(2)}{2} - \log\left(\cos(t)\right).$$

7. $\begin{cases} y' = t\,e^{3t}, \\ y(1) = 0 \end{cases}$

Solution. The general solution is obtained by integration by parts with $u = t, dv = e^{3t}$. Hence

$$y(t) = \frac{t\,e^{3t}}{3} - \frac{1}{3}\int e^{3t}\,dt = \frac{t\,e^{3t}}{3} - \frac{e^{3t}}{9} + C.$$

The initial condition requires $0 = y(1) = e^3(-2/3) + C$ and thus

$$y = \frac{-2e^3}{9} + e^{3t}\left(\frac{t}{3} - \frac{1}{9}\right).$$

8. $\begin{cases} t\,y' = \log(t), \\ y(1) = 0 \end{cases}$

Solution. We write the equation as $y' = t^{-1}\log(t) = u'\,u$, which can be integrated as

$$y(t) = \frac{u^2}{2} + C = \frac{1}{2}(\log(t))^2 + C.$$

Applying the initial condition gives $0 = y(1) = C$, hence the solution is

$$y = \frac{(\log(t))^2}{2}.$$

Section 3.2, page 44

In each of Problems 1–10 find the general solution of the differential equation.

1. $y' + 4y = 2t + 3e^{3t}$

Solution. The integrating factor is e^{4t}, so that the equation can be integrated by writing

$$\frac{d}{dt}(y\,e^{4t}) = e^{4t}(2t + 3e^{3t}) = 2t\,e^{4t} + 3e^{7t}.$$

Thus

$$y\,e^{4t} = \int \left(2t\,e^{4t} + 3e^{7t}\right) dt = \frac{(4t-1)e^{4t}}{16} + \frac{3e^{7t}}{7} + C.$$

Hence the solution is $y(t) = \dfrac{t}{2} - \dfrac{1}{8} + \dfrac{3e^{3t}}{7} + Ce^{-4t}$.

2. $y' - 2y = t\,e^{3t}$

Solution. The integrating factor is e^{-2t}, so that the equation can be integrated by writing

$$\frac{d}{dt}(y\,e^{-2t}) = e^{-2t}(t\,e^{3t}) = t\,e^{t}.$$

Thus, $y\,e^{-2t} = (t-1)e^{t} + C$ with the solution $y(t) = t\,e^{3t} - e^{3t} + C\,e^{2t}$.

3. $y' + 2t\,y = 3t\,e^{-t^2}$

Solution. The integrating factor is e^{-t^2}, so that the equation can be integrated by writing

$$\frac{d}{dt}(y\,e^{t^2}) = 3t.$$

Thus, $y\,e^{t^2} = \frac{3t^2}{2}$ with the solution $y(t) = \dfrac{3t^2}{2e^{t^2}} + \dfrac{C}{e^{t^2}}$.

4. $y' + \dfrac{1}{t}y = e^{t^2}$

Solution. The integrating factor is $e^{\log(t)} = t$, so that the equation can be integrated by writing

$$\frac{d}{dt}(t\,y) = t\,e^{t^2}.$$

Thus, $t\,y = e^{t^2}/2 + C$ with the solution $y(t) = \dfrac{e^{t^2}}{2t} + \dfrac{C}{t}$.

5. $t\,y' + 5y = t^3$

Solution. First, we rewrite the equation as $y' + (5/t)y = t^2$. The integrating factor is $e^{5\log(t)} = t^5$, so that the equation can be integrated by writing

$$\frac{d}{dt}(t^5\,y) = t^7.$$

Thus, $t^5\,y = t^8/8 + C$ with the solution $y(t) = \dfrac{t^3}{8} + \dfrac{C}{t^5}$.

6. $(1+t^2)y' + 9y = 0$

Solution. First, we rewrite the equation as $y' + 9y/(1+t^2)y = 0$. The integrating factor is $e^{9\arctan(t)}$, so that the equation can be integrated by writing

$$\frac{d}{dt}(y\,e^{9\arctan(t)}) = 0.$$

Thus, $y\,e^{9\arctan(t)} = C$ with the solution $y(t) = C\,e^{-9\arctan(t)}$

7. $y' - y\,e^t = 2t\,e^{e^t}$

Solution. The integrating factor is e^{-e^t} so that the equation can be integrated by writing

$$\frac{d}{dt}(y\,e^{-e^t}) = 2t.$$

Thus, $y\,e^{-e^t} = t^2 + C$. with the solution $y(t) = t^2 e^{e^t} + C\,e^{e^t}$.

8. $y' + y = \sin(t)$

Solution. The integrating factor is e^t so that the equation can be integrated by writing

$$\frac{d}{dt}(y\,e^t) = e^t\,\sin(t).$$

Thus $y\,e^t = \dfrac{1}{2}e^t\,\sin(t) - \dfrac{1}{2}e^t\,\cos(t) + C$ with the solution

$$y(t) = \frac{\sin(t) - \cos(t)}{2} + C\,e^{-t}.$$

9. $y' + y \cot(t) = 5e^{\cos(t)}$

Solution. The integrating factor is $e^{\log(\sin(t))} = \sin(t)$ so that the equation can be integrated by writing

$$\frac{d}{dt}(y \sin(t)) = 5 \sin(t)e^{\cos(t)}.$$

Thus, $y \sin(t) = -5e^{\cos(t)} + C$ with the solution $y(t) = \dfrac{C}{\sin(t)} - \dfrac{5e^{\cos(t)}}{\sin(t)}$.

10. $y' + 2y \cos(t) = (\sin(t))^2 \cos(t)$

Solution. The integrating factor is $e^{2\sin(t)}$ so that the equation can be integrated by writing

$$\frac{d}{dt}(y\, e^{2\sin(t)}) = \sin(t)^2 \cos(t)e^{2\sin(t)}$$

Thus $y\, e^{2\sin(t)} = \dfrac{1 - 2\sin(t) + 2(\sin(t))^2}{4}e^{2\sin(t)+C}$ with the solution

$$y(t) = \frac{\left(1 - 2\sin(t) + 2(\sin(t))^2\right)}{4} + C\, e^{-2\sin(t)}.$$

In each of Problems 11–20 solve the initial value problem.

11. $\begin{cases} y' + 4y = 2t + 3e^{3t}, \\ y(0) = 5 \end{cases}$

Solution. We use the general solution of Exercise 1: $y(t) = \dfrac{t}{2} - \dfrac{1}{8} + \dfrac{3e^{3t}}{7} + C\, e^{-4t}$,

where the constant C is chosen to satisfy the initial condition. This requires

$$5 = y(0) = 0 - \frac{1}{8} + \frac{3}{7} + C,$$

which is solved to yield $C = 5 + (1/8) - (3/7) = 263/56$ with the solution

$$y(t) = \frac{t}{2} - \frac{1}{8} + \frac{3e^{3t}}{7} + \frac{263e^{-4t}}{56}.$$

12. $\begin{cases} y' - 2y = t\, e^{3t}, \\ y(3) = 1 \end{cases}$

Solution. We use the general solution of Exercise 2: $y(t) = t\, e^{3t} - e^{3t} + C\, e^{2t}$, where the constant C is chosen to satisfy the initial condition. This requires

$$1 = y(3) = 3\,e^9 - e^9 + C\,e^6,$$

which is solved to yield $C = e^{-6}(1 - 2e^9) = e^{-6} - 2e^3$, with the solution $y(t) = t\,e^{3t} - e^{3t} + (e^{-6} - 2e^3)e^{2t}$.

13. $\begin{cases} y' + 2t\,y = t^3 e^{-t^2}, \\ y(0) = 4 \end{cases}$

Solution. We use the general solution of Exercise 3: $y(t) = \dfrac{3t^2}{2e^{t^2}} + \dfrac{C}{e^{t^2}}$,

where the constant C is chosen to satisfy the initial condition. This requires $4 = y(0) = C$ with the solution $y(t) = \dfrac{t^4 e^{-t^2}}{4} + 4e^{-t^2}$.

14. $\begin{cases} y' + \dfrac{1}{t}y = e^{t^2}, \\ y(1) = 2 \end{cases}$

Solution. We use the general solution of Exercise 4: $y(t) = \dfrac{e^{t^2}}{2t} + \dfrac{C}{t}$,

where the constant C is chosen to satisfy the initial condition. This requires $2 = y(1) = e/2 + C$, so that $C = 2 - (e/2)$ with the solution $y(t) = \dfrac{e^{t^2}}{2t} + \dfrac{2}{t} - \dfrac{e}{2t}$.

15. $\begin{cases} t\,y' + 5y = t^3, \\ y(1) = 1 \end{cases}$

Solution. We use the general solution of Exercise 5: $y(t) = \dfrac{t^3}{8} + \dfrac{C}{t^5}$,

where the constant C is chosen to satisfy the initial condition. This requires $1 = y(1) = 1/8 + C$, so that $C = 7/8$, with the solution $y(t) = \dfrac{t^3}{8} + \dfrac{7}{8t^5}$.

16. $\begin{cases} (1+t^2)y' + 9y = 0, \\ y(3) = 1 \end{cases}$

Solution. We use the general solution of Exercise 6: $y(t) = C\,e^{-9\arctan(t)}$,

where the constant C is chosen to satisfy the initial condition. This requires $1 = y(3) = C \exp{-9\arctan(3)}$ so that $C = \exp\left(9\arctan(3)\right)$ with the solution $y(t) = \exp\left(9\arctan(3) - 9\arctan(t)\right)$.

17. $\begin{cases} y' - y e^t = 2t\, e^{e^t}, \\ y(0) = 1 \end{cases}$

Solution. We use the general solution of Exercise 7: $y(t) = t^2 e^{e^t} + C e^{e^t}$, where the constant C is chosen to satisfy the initial condition. This requires $1 = y(0) = C e$ so that $C = 1/e$, with the solution $y(t) = t^2 e^{e^t} + \dfrac{e^{e^t}}{e}$.

18. $\begin{cases} y' + \dfrac{3}{t} y = \dfrac{\sin(t)}{t^3}, \\ y(\pi) = 0 \end{cases}$

Solution. The integrating factor is t^3 so that the so that the equation can be integrated by writing

$$\frac{d}{dt}(t^3 y) = \sin(t).$$

Thus, $t^3 y = -\cos(t) + C$. The constant C is determined from the initial conditions as $0 = y(\pi) = 1 + C$, hence $C = -1$ with the solution $y(t) = -\dfrac{1 + \cos(t)}{t^3}$.

19. $\begin{cases} t\, y' + 2y = \cos(t), \\ y(\pi/2) = 0 \end{cases}$

Solution. First, write the equation as $y' + (2/t)y = \sec(t)/t$. The integrating factor is t^2, so that the so that the equation can be integrated by writing

$$\frac{d}{dt}(t^2 y) = t \cos(t).$$

Thus, $t^2 y = t \sin(t) - \cos(t) + C$ The constant C is determined from the initial conditions as $0 = y(\pi/2) = C$, with the solution $y(t) = \dfrac{\cos(t)}{t^2} + \dfrac{\sin(t)}{t} - \dfrac{\pi}{2t^2}$.

20. $\begin{cases} \cos(t)y' - \sin(t)y = 1, \\ y(0) = 1 \end{cases}$

Solution. First, write the equation in the form $y' - \tan(t)y = \cos(t)$. The integrating factor is $e^{\log \cos(t)} = \cos(t)$ so that the equation can be integrated by writing

$$\frac{d}{dt}(\cos(t)y) = 1.$$

Thus, $\cos(t) y = t + C$. The constant C is determined from the initial conditions as $1 = y(0) = C$, with the solution $y(t) = t \sec(t) + \sec(t)$.

21. Find the general solution of the differential equation

$$\frac{dy}{dt} = \frac{1}{e^y - t}. \qquad\qquad (S.1)$$

[Hint: Consider t as a function of y.]

Solution. Considering t as a function of y leads to the equation

$$\frac{dt}{dy} = e^y - t, \qquad\qquad (S.2)$$

which is a first-order linear equation in t as a function of y, equivalently written $t' + t = e^y$. An appropriate integrating factor for (S.2) is e^y, leading to

$$e^y(t + t') = e^{2y} \qquad \text{or} \qquad (t\,e^y)' = e^{2y}$$

We integrate this equation as

$$t\,e^y = \frac{e^{2y}}{2} + C. \qquad\qquad (S.3)$$

Solving (S.3) for t, we obtain

$$t = \frac{e^y}{2} + C e^{-y}. \qquad\qquad (S.4)$$

22. (Continuation of Problem 21) Solve the differential equation (S.1) with the initial condition $y(t_0) = Y_0$, where t_0 and Y_0 are such that $e^{Y_0} \neq t_0$.

Solution. In order to satisfy the initial condition we must have

$$t_0 = \frac{e^{Y_0}}{2} + \frac{C}{e^{Y_0}},$$

which is solved for C as

$$C = e^{Y_0}\left(t_0 - \frac{e^{Y_0}}{2}\right). \qquad\qquad (S.5)$$

We substitute (S.5) into (S.4), obtaining

$$t = \frac{e^y}{2} + e^{Y_0}\left(t_0 - \frac{e^{Y_0}}{2}\right)e^{-y} \qquad\qquad (S.6)$$

We multiply (S.6) by $2e^y$, obtaining

$$(e^y)^2 - 2t(e^y) + (2t_0 - e^{Y_0})e^{Y_0} = 0.$$

This equation can be solved for e^y by means of the quadratic formula as follows:

$$e^y = \frac{1}{2}\left(2t \pm \sqrt{(2t)^2 - 4(2t_0 - e^{Y_0})e^{Y_0}}\right)$$

$$= t \pm \sqrt{t^2 - 2t_0 e^{Y_0} + e^{2Y_0}}$$

$$= t \pm \sqrt{t^2 - t_0^2 + t_0^2 - 2t_0 e^{Y_0} + e^{2Y_0}}$$

$$= t \pm \sqrt{t^2 - t_0^2 + (t_0 - e^{Y_0})^2}.$$

In order to decide which sign to choose, we again invoke the initial condition; setting $t = t_0$, we have $e^{Y_0} = t_0 \pm |e^{Y_0} - t_0|$. If $e^{Y_0} > t_0$ we must choose the plus sign, whereas if $e^{Y_0} < t_0$ we must choose the minus sign; thus

$$y = \begin{cases} \log\left(t + \sqrt{t^2 - t_0^2 + (t_0 - e^{Y_0})^2}\right) & \text{for} \quad e^{Y_0} > t_0, \\[2mm] \log\left(t - \sqrt{t^2 - t_0^2 + (t_0 - e^{Y_0})^2}\right) & \text{for} \quad e^{Y_0} < t_0. \end{cases}$$

These solutions are valid on the interval where the argument of the logarithm is defined and remains positive.

23. (Continuation of Problem 21) Solve the differential equation (3.33) with the initial condition $y(t_0) = Y_0$, where t_0 and Y_0 are such that $e^{Y_0} = t_0$. Show that there are *two* solutions (3.33) satisfying $y(t_0) = Y_0$, but these solutions are only defined for $t^2 \geq t_0^2$.

Solution. If $e^{Y_0} = t_0$, then the initial conditions provide no new information and we have the ambiguity $e^y = t \pm \sqrt{t^2 - t_0^2}$, equivalently

$$y = \log\left(t \pm \sqrt{t^2 - t_0^2}\right).$$

These solutions are valid on the interval where the argument of the logarithm is defined and remains positive. In particular, the square root must be well-defined, so that we must have $t^2 - t_0^2 \geq 0$, or equivalently $t^2 \geq t_0^2$.

24. Prove that the following **superposition principle** holds for first-order linear equations: if y_1 is a solution of the equation

$$y_1' + p(t)y_1 = q_1(t) \qquad \text{(S.7)}$$

and y_2 is a solution of the equation

$$y_2' + p(t)y_2 = q_2(t), \qquad \text{(S.8)}$$

then $y = y_1 + y_2$ is a solution of the equation

$$y' + p(t)y = q_1(t) + q_2(t).$$

Solution. When we add (S.8) and (S.8), we get

$$y_1' + y_2' + p(t)(y_1 + y_2) = q_1(t) + q_2(t). \tag{S.9}$$

Now use the linearity of differentiation to convert (S.8) to

$$(y_1 + y_2)' + p(t)(y_1 + y_2) = q_1(t) + q_2(t).$$

25. Show that the following **subtraction principle** holds for first-order linear equations: if y_1 and y_2 are both solutions of the equation $y' + p(t)y = q(t)$, then the difference $y = y_1 - y_2$ is a solution of the differential equation $y' + p(t)y = 0$.

Solution. Since

$$y_1' + p(t)y_1 = q(t) \qquad \text{and} \qquad y_2' + p(t)y_2 = q(t),$$

we have

$$y_1' - y_2' + p(t)(y_1 - y_2) = 0. \tag{S.10}$$

The fact that $y_1' - y_2' = (y_1 - y_2)'$ converts (S.10) to

$$(y_1 - y_2)' + p(t)(y_1 - y_2) = 0.$$

26. Use Problem 24 to show that the solution of the initial value problem for

$$y' + p(t)y = q(t)$$

with $y(t_0) = Y_0$ can be written in the form $y = y_P + y_H$, where y_P is the particular solution of the equation with $y(t_0) = 0$ and y_H is the solution of the differential equation $y' + p(t)y = 0$ with $y(t_0) = Y_0$.

Solution. In Problem 25 let $y_1 = y$ and $y_2 = y_P$. The difference $y_1 - y_2$ must solve the homogeneous equation with initial value 0. In other words, $y - y_P = y_H$, where y_H is a solution of the homogeneous equation with $y_H(t_0) = Y_0 - Y_0 = 0$.

27. This problem gives a version of the method **variation of parameters** (see Section 9.4) which is suited to a first-order linear equation. Suppose that y_H is a solution of the differential equation $y' + p(t)y = 0$ and that y is a solution of the first-order linear equation $y' + p(t)y = q(t)$; define $z(t) = y/y_H$.

   a. Show that $z(t)$ satisfies the equation $z'(t) = q(t)/y_H(t)$.

b. Use part **a** to show that $y(t) = y_H(t) \int \left(\dfrac{q(t)}{y_H(t)} \right) dt.$

Solution. For part **a** we use the quotient rule for derivatives and the differential equations satisfied by the numerator and denominator. In detail,

$$
\begin{aligned}
z'(t) &= \frac{d}{dt} \left(\frac{y}{y_H} \right) = \frac{y_H y' - y y_H'}{y_H^2} \\
&= \frac{y_H(q(t) - p(t)y) - y(-p(t)y_H)}{y_H^2} = \frac{y_H q(t)}{y_H^2} = \frac{q(t)}{y_H(t)}.
\end{aligned}
$$

For part **b** we integrate the equation found in part **a**. Thus

$$
z(t) = \int \frac{q(t)}{y_H(t)} dt \quad \text{and} \quad y(t) = y_H(t)z(t) = y_H(t) \int \frac{q(t)}{y_H(t)} dt.
$$

28. This exercise illustrates the possibility of uniquely solving an equation by means of conditions at infinity.

    a. Find the general solution of the equation

$$
y' - y = \cos(t) - \sin(t). \tag{S.11}
$$

    b. Find the unique solution of (S.11) which remains bounded when t tends to ∞.

Solution. **a.** We multiply (S.11) by the integrating e^{-t}, obtaining

$$
\left(y e^{-t} \right)' = e^{-t} \left(\cos(t) - \sin(t) \right). \tag{S.12}
$$

We integrate (S.12), obtaining

$$
y e^{-t} = \int e^{-t} \left(\cos(t) - \sin(t) \right) dt. \tag{S.13}
$$

To evaluate the integral on the right-hand side of (S.13), we use the *Mathematica* command

```
Integrate[E^(-t)(Cos[t] - Sin[t]),t]
```

with the result

```
Sin[t]
——————
  t
 E
```

Therefore, $ye^{-t} = e^{-t}\sin(t) + C$, and we have the solution

$$y(t) = Ce^t + \sin(t), \tag{S.14}$$

where C is a constant.

b. The exponential function e^t is not bounded when $t \longrightarrow \infty$, but the sine function is bounded when $t \longrightarrow \infty$. Therefore, a solution $y(t)$ of the form (S.14) can be bounded when $t \longrightarrow \infty$ if and only if $C = 0$; hence the unique bounded solution of (S.11) is $y(t) = \sin(t)$.

29. Show that the general solution to the first-order linear equation

$$a(t)y'(t) + b(t)y(t) = c(t)$$

is given by the formula

$$y(t) = \exp\left(-\left(\int \frac{b(t)dt}{a(t)}\right)\right)\left(C + \int \exp\left(\int \frac{b(s)ds}{a(s)}\bigg|_{s=t}\right)\left(\frac{c(t)}{a(t)}\right)dt\right),$$

where C is a constant.

Solution. Follow the steps 1–4 on page 44.

Section 3.3, page 52

Find the general solution to each of the following separable equations.

1. $y' = \dfrac{5y}{t(y-3)}$

Solution. $\dfrac{e^y}{y^3} = Ct^5$

```
ODE[y' == 5y/(t(y - 3)),y,t,Method->Separable]
```

$$\left\{\frac{E^y}{y^3} == t^5\ C[1],\ y == 0\right\}$$

2. $y't^2 = y(y-3)$

Solution. $y = \dfrac{3}{1 - Ce^{-3/t}}$ and $y = 0$.

```
ODE[y't^2 == y(y - 3),y,t,Method->Separable]
```

$$
\{\{y \;\to\; \frac{3\;E^{3/t}}{E^{3/t} + C[1]}\}, \; \{y \;\to\; 0\}\}
$$

3. $t^3 y' = \sqrt{1 - y^2}$

Solution. $y = -\sin\left(\dfrac{1}{2t^2} + C\right)$

```
ODE[t^3 y' == Sqrt[1 - y^2],y,t,Method->Separable]
```

$$
\{\{y \;\to\; -\mathrm{Sin}[\frac{1}{2\,t^2} - C[1]]\}\}
$$

4. $e^{y^3} = \csc(t) y^2 y'$

Solution. $e^{-y^3} = 3\cos(t) + C$

```
ODE[E^(y^3) == Csc[t]y^2 y',y,t,Method->Separable,
    Form->Equation]
```

$$
\frac{-1}{3\,E^{y^3}} == C[1] - \mathrm{Cos}[t]
$$

5. $(1 + t + y + t\,y)y' = 1$

Solution. $\exp\left(y + \dfrac{y^2}{2}\right) = (1 + t)C$

```
ODE[(1 + t + y + t y)y' == 1,y,t,Method->Separable,
    Form->Equation]
```

$$
y + \frac{y^2}{2} == C[1] + \mathrm{Log}[1 + t]
$$

6. $\log(y\,y') = t$

Solution. $\dfrac{y^2}{2} = e^t + C$

```
ODE[Log[y'y]==t,y,t,Method->Separable]
```

$$\frac{y^2}{2} == E^t + C[1]$$

7. $(y \cos(y) + \sin(y))y' = t + \log(t)$

Solution. $y \sin(y) = -t + \dfrac{t^2}{2} + t \log(t) + C$

```
ODE[(y Cos[y] + Sin[y])y' == t + Log[t],y,t,
Method->Separable]
```

$$y \; Sin[y] \; == -t + \frac{t^2}{2} + C[1] + t \; Log[t]$$

8. $\dfrac{y'}{\sin(3t)} - t^3 e^{2y} = 0$

Solution. $-\dfrac{1}{2e^{2y}} = \dfrac{2t\cos(3t)}{9} - \dfrac{t^3\cos(3t)}{3} - \dfrac{2\sin(3t)}{27} + \dfrac{t^2\sin(3t)}{3} + C$

```
ODE[y'/Sin[3t] - t^3 E^(2y) == 0,y,t,
Method->Separable,Form->Equation]
```

$$\frac{-1}{2 E^{2 y}} == C[1] + \frac{2\; t\; Cos[3\; t]}{9} - \frac{t^3\; Cos[3\; t]}{3} - \frac{2\; Sin[3\; t]}{27} + \frac{t^2\; Sin[3\; t]}{3}$$

9. $\dfrac{\sin(y)y'}{t} = \dfrac{t\, e^t}{y}$

Solution. $-y\cos(y) + \sin(y) = (t^2 - 2t + 2)e^t + C$

```
ODE[Sin[y]y'/t == t E^t/y,y,t,Method->Separable]
```

$$-(y\; Cos[y]) + Sin[y] == E^t\; (2 - 2\; t + t^2) + C[1]$$

10. $y' = \left(\sin(t)\right)^2 \left(\sin(2y)\right)^2$

Solution. $-\cos(2y)\csc(y)\sec(y) = 2t - \sin(2t) + C$

```
ODE[y' == Sin[t]^2Sin[2y]^2,y,t,Method->Separable]
```

$$-(Cos[2\; y]\; Csc[y]\; Sec[y]) == 2\; t + C[1] - Sin[2\; t]$$

Solve the following initial value problems.

11. $\begin{cases} y' = t(1+y^2), \\ y(0) = 1 \end{cases}$

Solution. $y(t) = \tan\left(\dfrac{t^2}{2} + \dfrac{\pi}{4}\right)$

```
ODE[{y'==t(1 + y^2),y[0] == 1},y,t,
Method->Separable,PlotSolution->{{t,-2,2}}]
```

```
                   2
            Pi + 2 t
{{y -> Tan[---------]}}
               4
```

12. $\begin{cases} y\,y' = t^2, \\ y(1) = 3 \end{cases}$

Solution. $y(t) = \sqrt{\dfrac{2t^3 + 25}{3}}$

```
ODE[{y'y == t^2,y[1] == 3},y,t,
Method->Separable,PlotSolution->{{t,-2,2}}]
```

```
                    3
        Sqrt[25 + 2 t ]
{{y -> ---------------}}
          Sqrt[3]
```

13. $\begin{cases} y^2 y' = e^t, \\ y(1) = 4 \end{cases}$

Solution. $y(t) = \sqrt[3]{3e^t + 64 - 3e}$

```
ODE[{y'y^2 == E^t,y[1] == 4},y,t,
Method->Separable,PlotSolution->{{t,-1,1}}]
```

```
                       t 1/3
{{y -> (64 - 3 E + 3 E )   }}
```

14. $\begin{cases} t\,y\,y' = 1, \\ y(1) = 5 \end{cases}$

Solution. $y(t) = \sqrt{\log(t^2) + 25}$

```
ODE[{t y'y == 1,y[1] == 5},y,t,
Method->Separable,PlotSolution->{{t,-2,2}},
PostSolution->PowerExpand]
```

```
{{y -> Sqrt[25 + 2 Log[t]]}}
```

15. $\begin{cases} y' = -\dfrac{t}{y}, \\ y(1) = 1 \end{cases}$

Solution. $y(t) = \sqrt{2 - t^2}$

```
ODE[{y' == -t/y,y[1] == 1},y,t,
Method->Separable,PlotSolution->{{t,-2,2}}]
```

```
                         2
{{y -> Sqrt[2 - t ]}}
```

16. $\begin{cases} e^t y' = y, \\ y(0) = 4 \end{cases}$

Solution. $y(t) = 4 \exp(1 - e^{-t})$

```
ODE[{E^t y' == y,y[0] == 4},y,t,
Method->Separable,PlotSolution->{{t,-2,2}}]
```

```
                       -t
            1 - E
{{y -> 4 E        }}
```

17. $\begin{cases} t\,y' + \sin(y) = 0, \\ y(1) = \pi \end{cases}$

Solution. $y(t) = \pi$

```
Note: Only get the trivial solution.
ODE[{t y' + Sin[y] == 0,y[1 ]== Pi},y,t,
Method->Separable]
```

```
ODE[{t y'+Sin[y] == 0,y[1] == Pi},y,t,
Method->Trivial]
```

18. $\begin{cases} y' = -t\,y, \\ y(0) = \dfrac{1}{\sqrt{2\pi}} \end{cases}$

Solution. $y(t) = \dfrac{e^{-t^2/2}}{\sqrt{2\pi}}$

```
ODE[{y' == -t y,y[0] == 1/Sqrt[2Pi]},y,t,
Method->Separable,PlotSolution->{{t,-3,3}}]
```

$$\{\{y \to \frac{1}{E^{t^2/2}\ \mathrm{Sqrt[2\ Pi]}}\}\}$$

19. $\begin{cases} y\,y'\sqrt{1-t^2} = 1, \\ y(0) = 1 \end{cases}$

Solution. $y(t) = \sqrt{1 + 2\arcsin(t)}$

```
ODE[{y'y Sqrt[1 - t^2] == 1,y[0] == 1},y,t,
Method->Separable,PlotSolution->{{t,-1,1}}]
```

```
{{y -> Sqrt[1 + 2 ArcSin[t]]}}
```

20. $\begin{cases} y' = t^2 y, \\ y(0) = 4 \end{cases}$

Solution. $y(t) = 4e^{t^3/3}$

```
ODE[{y' == t^2 y,y[0] == 4},y,t,
Method->Separable,PlotSolution->{{t,-1,0.2}}]
```

```
{{y -> 4 (E^{t^3})^{1/3} }}
```

Use Theorem 3.2 to determine if the following equations are separable.

21. $y' = \dfrac{t\,y + t + y + 1}{t\,y}$

Solution.

```
f[t_,y_] = (t y + t + y + 1)/(t y)
SeparableQ[f][t,y]
```

```
True
```

22. $t(y^2 + 5y + t)\,dt + (t^2 - 1)\,dy = 0$

Solution.

```
f[t_,y_] = t(y^2+5y+t)/(1-t^2)
SeparableQ[f][t,y]
```

```
False
```

23. $y' = \cos(y)\big(\cos(t+y) + \cos(t-y)\big)$

   Solution.

```
f[t_,y_] = Cos[y](Cos[t + y] + Cos[t - y])
SeparableQ[f][t,y]
```

```
True
```

24. $e^{ty}(y^2 + 5y + 4)\,dt + (t^2 - 1)\,dy = 0$

   Solution.

```
f[t_,y_] = Exp[t y](y^2 + 5y + 4)/(1-t^2)
SeparableQ[f][t,y]
```

```
True
```

Section 3.4, page 61

Find the general solution of each of the following exact equations.

1. $(3y + t\,y^2) + (3t + t^2 y)y' = 0$

   Solution. Writing the equation in the form $(3y + t\,y^2)dt + (3t + t^2 y)dy = 0$, we see that $M_y - N_t = 0$; the potential function Φ must satisfy $\Phi_t = 3y + t\,y^2$, $\Phi_y = 3t + t^2 y$. The Φ_t equation requires that $\Phi(t, y) = 3t\,y + t^2 y^2/2 + F(y)$, whence the Φ_y equation requires further that $3t + t^2\,y + F'(y) = 3t + t^2\,y$, hence $F(y) = C$ and the solution is $3t\,y + t^2\,y^2/2 = C$, which can be written $\dfrac{t\,y(6 + t\,y)}{2} = C$

2. $(2y - e^y) + t(2 - e^y)y' = 0$

   Solution. Writing the equation in the form $(2y - e^y)dt + t(2 - e^y)dy = 0$, we see that $M_y - N_t = 0$; the potential function Φ must satisfy $\Phi_t = 2y - e^y$, $\Phi_y = 2t - t\,e^y$. The Φ_t equation requires that $\Phi(t, y) = 2t\,y - t\,e^y + F(y)$, whence the Φ_y equation requires further that $2t - t\,e^y + F'(y) = 2t - t\,e^y$, hence $F(y) = C$ and the solution is $t(2y - e^y) = C$.

3. $(t + y)y' + y = 0$

   Solution. Writing the equation in the form $y\,dt + (t + y)dy = 0$, we see that $M_y - N_t = 0$; the potential function Φ must satisfy $\Phi_t = y$, $\Phi_y = t + y$. The Φ_t equation requires that $\Phi(t, y) = t\,y + F(y)$, whence the Φ_y equation requires further that $t + F'(y) = t + y$, hence $F'(y) = y$, $F(y) = y^2/2 + C$ and the solution is $t\,y + \dfrac{y^2}{2} = C$.

4. $2t\,e^y + (t^2 - y^2 - 2y)y' = 0$

Solution. Writing the equation in the form $2t\,e^y\,dt + (t^2 - y^2 - 2y)e^y\,dy = 0$, we see that $M_y - N_t = 0$; the potential function Φ must satisfy $\Phi_t = 2t\,e^y$, $\Phi_y = (t^2 - y^2 - 2y)e^y$. The Φ_t equation requires that $\Phi(t, y) = t^2 e^y + F(y)$, whence the Φ_y equation requires further that $t^2 e^y + F'(y) = (t^2 - y^2 - 2y)e^y$, hence $F'(y) = -y^2 e^y - 2y e^y$, $F(y) = -y^2 e^y + C$ and the solution is $e^y(t^2 - y^2) = C$.

Find the general solution of each of the following equations by finding a suitable integrating factor.

5. $(y - t)y' + y = 0$

Solution. Writing the equation in the form $y\,dt + (y-t)\,dy = 0$, we see that $M_y - N_t = 1 - (-1) = 2$, so that $(M_y - N_t)/M = 2/y$ and we can find an integrating factor which only depends on y through the equation $\mu(y) = \exp \int (-2/y)dy = 1/y^2$. Multiplying the equation by $\mu(y)$ yields the exact equation $dt/y + (y - t)/y^2\,dy$. The potential function $\Phi(t, y)$ must satisfy $\Phi_t = 1/y$, $\Phi_y = (y - t)/y^2$. The Φ_t equation requires that $\Phi(t, y) = t/y + F(y)$, whence the Φ_y equation requires further that $F'(y) = 1/y$, $F(y) = \log(y) + C$. The solution is written

$$\begin{cases} \mu(y) = 1/y^2, \\ \log(y) + t/y = C \end{cases}$$

6. $(2y + y^2) - t\,y' = 0$

Solution. Writing the equation in the form $(2y + y^2)\,dt - t\,dy = 0$, we see that $M_y - N_t = 2 + 2y + 1 = 2y + 3$, so that $(M_y - N_t)/M = (2y + 1)/(2y + y^2)$ and we can find an integrating factor which only depends on y through the equation $\mu(y) = \exp \int (-(2y + 1)/(2y + y^2))dy = 1/y^{3/2}\sqrt{y+2}$. Multiplying the equation by $\mu(y)$ yields an exact equation. The solution is written

$$\begin{cases} \mu(y) = \dfrac{1}{y^{3/2}\sqrt{y+2}}, \\ y = \dfrac{-2t^2}{t^2 + C}, \quad y = 0 \end{cases}$$

7. $(2t^2 + 3t\,y - y^2) + (t^2 - t\,y)y' = 0$

Solution. Writing the equation in the form $(2t^2 + 3t\,y - y^2)\,dt + (t^2 - t\,y)\,dy = 0$, we see that $M_y - N_t = (3t - 2y) - (2t - y) = t - y$, so that $(M_y - N_t)/N = 1/t$ and we can find an integrating factor which only depends on t through the equation $\mu(t) = \exp \int (dt/t) = t$. Multiplying the equation by $\mu(t)$ yields an exact equation. The solution is written

$$\begin{cases} \mu(t) = t, \\ \dfrac{t^4}{2} + t^3 y - \dfrac{t^2 y^2}{2} = C \end{cases}$$

8. $y - t y' = 0$

Solution. Writing the equation in the form $y \, dt - t \, dy = 0$, we see that $M_y - N_t = 1 - (-1) = 2$, so that $(M_y - N_t)/N = -2/t$ and we can find an integrating factor which only depends on t through the equation $\mu(t) = \exp \int (-2dt/t) = t^{-2}$. Multiplying the equation by $\mu(t)$ yields an exact equation. The solution is written

$$\begin{cases} \mu(t) = 1/t^2, \\ y = C t \end{cases}$$

9. $(1+t)e^t + (y e^y - t e^t)y' = 0$

Solution. Writing the equation in the form $(1+t)e^t \, dt + (y e^y - t e^t)dy = 0$, we see that $M_y - N_t = (t+1)e^t$, so that $(M_y - N_t)/M = 1$ and we can find an integrating factor which only depends on y through the equation $\mu(y) = \exp \int (-dy) = e^{-y}$. Multiplying the equation by $\mu(y)$ yields an exact equation. The solution is written

$$\begin{cases} \mu(y) = e^{-y}, \\ \dfrac{y^2}{2} + t e^{t-y} = C \end{cases}$$

10. $2t y + (y^2 - t^2)y' = 0$

Solution. Writing the equation in the form $2t y \, dt + (y^2 - t^2)dy = 0$, we see that $M_y - N_t = 2t - (-2t) = 4t$, so that $(M_y - N_t)/N = 2/y$, and we can find an integrating factor which only depends on y through the equation $\mu(y) = \exp \int (-2dy/y) = 1/y^2$. Multiplying the equation by $\mu(y)$ yields an exact equation. The solution is written

$$\begin{cases} \mu(y) = \dfrac{1}{y^2}, \\ y + \dfrac{t^2}{y} = C, \quad y = 0 \end{cases}$$

Solve the following initial value problems.

11. $\begin{cases} 2t - y + (2y - t)y' = 0, \\ y(2) = 4 \end{cases}$

Solution. Writing the equation in the form $(2t - y) \, dt + (2y - t) \, dy = 0$, we see that $M_y - N_t = -1 + 1 = 0$, so that the equation is exact. The potential function $\Phi(t, y)$

must satisfy $\Phi_t = (2t - y)$, $\Phi_y = (2y - t)$, from which $\Phi(t, y) = t^2 - t y + F(y)$, $-t + F'(y) = 2y - t$, $F(y) = y^2 + C$, with the solution $t^2 - t y + y^2 = C$. The constant is determined from the initial condition as $C = 2^2 - 2 \times 4 + 4^2 = 12$. The solution is $t^2 - t y + y^2 = 12$.

12.
$$\begin{cases} e^t + y + (2y + t + y e^y)y' = 0, \\ y(0) = 1 \end{cases}$$

Solution. Writing the equation in the form $(e^t + y) dt + (2y + t + ye^y) dy = 0$, we see that $M_y - N_t = 1 - 1 = 0$, so that the equation is exact. The potential function $\Phi(t, y)$ must satisfy $\Phi_t = e^t + y$, $\Phi_y = 2y + t + ye^y$, from which $\Phi(t, y) = t y + e^t + F(y)$, $t + F'(y) = 2y + t + ye^y$, $F(y) = y^2 + ye^y - e^y + C$. with the solution $t y + e^t + y^2 + ye^y - e^y = C$. The constant is determined from the initial condition as $C = 1 + 1 + e - e = 2$. The solution is written $e^t + t y + y^2 + (y - 1)e^y = 2$

13.
$$\begin{cases} \dfrac{y\cos(t)}{t} + \dfrac{2y\sin(t)}{t^2} + \dfrac{\sin(t)y'}{t} = 0, \\ y(\pi/2) = 1 \end{cases}$$

Solution. Writing the equation in the form $(y\cos(t)/t + 2y\sin(t)/t^2) dt + (\sin(t)/y) dy = 0$, we see that $M_y - N_t = 3\sin(t)/t^2$, so that the $(M_y - N_t)/N = 3/t$ and we can find an integrating factor as $\mu(t) = \exp \int (3 dt/t) = t^3$. Multiplying by $\mu(t)$ produces an exact equation, for which the potential function $\Phi(t, y)$ must satisfy $\Phi_y = t^2 \sin(t)$, $\Phi_t = (2t \sin(t) + t^2 \cos(t)$, which is integrated to yield $y t^2 \sin(t) = C$. The initial condition further requires that $C = (\pi/2)^2$ and hence we have the solution

$$\begin{cases} \mu(t) = t^3, \\[2mm] y = \dfrac{\pi^2 \csc(t)}{4t^2} \end{cases}$$

14.
$$\begin{cases} y\cos(t)\sin(y) + \cos(y)\sin(t)y\, y' = 0, \\ y(\pi/4) = \pi/4 \end{cases}$$

Solution. Writing the equation in the form $y \cos(t) \sin(y) dt + y \cos(y) \sin(t) dy = 0$, we have $M_y - N_t = \cos(t)\sin(y)$ and $(M_y - N_t)/M = 1/y$, so that an integrating factor can be found in the form $\mu(y) = \exp(\int dy/y) = 1/y$, which leads to $\sin(y)\sin(t) = C$. The constant C is obtained from the initial conditions as $C = (1/\sqrt{2})^2 = 1/2$ and we have the solution

$$\begin{cases} \mu(y) = 1/y, \\[2mm] \sin(t)\sin(y) = \dfrac{1}{2} \end{cases}$$

15. Show that the differential equation $y f(t\,y)dt + t\,g(t\,y)dy = 0$ has an integrating factor of the form

$$\mu(t, y) = \frac{1}{t\,y(f - g)}.$$

Solution. In order to determine whether the indicated function is an integrating factor, we first compute the relevant partial derivatives:

$$\frac{\partial}{\partial y}\left(\frac{y\,f(t\,y)}{t\,y(f(t\,y) - g(t\,y))}\right) = \frac{\partial}{\partial y}\left(\frac{f(t\,y)}{t(f(t\,y) - g(t\,y))}\right)$$

$$= \frac{t(f(t\,y) - g(t\,y))t\,f'(t\,y) - f(t\,y)t^2(f'(t\,y) - g'(t\,y))}{t^2(f(t\,y) - g(t\,y))^2}$$

$$= \frac{t^2 f(t\,y)g'(t\,y) - t^2 f'(t\,y)g(t\,y)}{t^2(f(t\,y) - g(t\,y))^2}$$

$$= \frac{f(t\,y)g'(t\,y) - f'(t\,y)g(t\,y)}{(f(t\,y) - g(t\,y))^2},$$

and

$$\frac{\partial}{\partial t}\left(\frac{t\,g(t\,y)}{t\,y(f(t\,y) - g(t\,y))}\right) = \frac{\partial}{\partial t}\left(\frac{g(t\,y)}{y(f(t\,y) - g(t\,y))}\right)$$

$$= \frac{y(f(t\,y) - g(t\,y))y\,g'(t\,y) - g(t\,y)y^2(f'(t\,y) - g'(t\,y))}{y^2(f(t\,y) - g(t\,y))^2}$$

$$= \frac{y^2 f(t\,y)g'(t\,y) - y^2 f'(t\,y)g(t\,y)}{y^2(f(t\,y) - g(t\,y))^2}$$

$$= \frac{f(t\,y)g'(t\,y) - f'(t\,y)g(t\,y)}{(f(t\,y) - g(t\,y))^2}.$$

The equality of these two expressions proves that we have found an integrating factor.

16. Use Problem 15 to show that the differential equation

$$y(1 - t\,y)dt - t(1 + t\,y)dy = 0$$

has $\mu = 1/(2t\,y)$ as an integrating factor. This leads to the solution $y\,e^{t\,y} = C\,t$.

Solution. If we set $f(t\,y) = 1 - t\,y$ and $g(t\,y) = -(1 + t\,y)$, then our equation has the form $yf(t\,y)dt + tg(t\,y)dy = 0$. From Problem 15, we obtain an integrating factor of the form

$$\mu(t, y) = \frac{1}{t\,y(f - g)} = \frac{1}{2t\,y}.$$

17. Show that if the differential equation has the form $t^a y^b (p\,y\,dt + q\,t\,dy) + t^d y^e (r\,y\,dt + s\,t\,dy) = 0$, where a, b, d, e, p, q, r, s are constants, then an integrating factor of the form $t^k y^n$ for specific values of k and n can be found.

Solution. First, observe that $t^{\alpha-1} y^{\beta-1}$ is an integrating factor for $\alpha\,y\,dt + \beta\,t\,dy$. Indeed, $t^{\alpha-1} y^{\beta-1} (\alpha\,y\,dt + \beta\,t\,dy) = d(t^\alpha y^\beta)$. In fact, the more general expression $t^{\alpha-1} y^{\beta-1} F(t^\alpha y^\beta)$ is also an integrating factor. Hence, $t^{p-a-1} y^{q-b-1} F(t^p y^q)$ is an integrating factor for $t^a y^b (p\,y\,dt + q\,t\,dy)$ and $t^{r-d-1} y^{s-e-1} G(t^r y^s)$ is an integrating factor for $t^d y^e (r\,y\,dt + s\,t\,dy)$. Therefore, if F and G can be determined such that

$$t^{p-a-1} y^{q-b-1} F(t^p y^q) = t^{r-d-1} y^{s-e-1} G(t^r y^s),$$

an integrating factor for the original equation can be found. Let $F(z) = z^u$ and $G(z) = z^v$, then $t^k y^n$ will be an integrating factor provided

$$\begin{cases} k = (u+1)p - a - 1 = (v+1)q - b - 1, \\ n = (u+1)r - d - 1 = (v+1)s - e - 1. \end{cases} \tag{S.15}$$

If $ps - qr \neq 0$, these equations determine u and v as we can see by rewriting (S.15) as

$$\begin{cases} up - vq = a + q - p - b, \\ ur - vs = d + s - r - e. \end{cases}$$

After computing u and v, k and n follow immediately.

18. For which values of the constants a, b, c and f does the first-order differential equation

$$(at\,y + b\,y^2) + (ct^2 + f\,t\,y)y' = 0 \tag{S.16}$$

have an integrating factor that is a function $\mu(t)$ of t alone? Find a formula for μ.

Solution. Working directly from first principles, the condition for exactness is

$$\frac{\partial}{\partial y}\left(\mu(t)(at\,y + b\,y^2)\right) = \frac{\partial}{\partial t}\left(\mu(t)(ct^2 + f\,t\,y)\right),$$

or

$$\mu(t)(at + 2b\,y) = \mu'(t)(ct^2 + f\,t\,y) + \mu(t)(2ct + f\,y). \tag{S.17}$$

Both sides of (S.17) are linear functions of y. Equality can take place in (S.17) for all y if and only if corresponding coefficients are equal. Therefore

$$(a - 2c)\mu(t) - ct\,\mu'(t) = 0, \qquad (f - 2b)\mu(t) + f\,t\mu'(t) = 0. \qquad \text{(S.18)}$$

If any coefficient in (S.18) is nonzero, then the system of simultaneous linear equations (S.18) has a solution if and only if the determinant is zero, that is, if and only if

$$(a - 2c)f + c(f - 2b) = 0,$$

which simplifies to

$$f(a - c) - 2bc = 0. \qquad \text{(S.19)}$$

To obtain a formula for $\mu(t)$, we first consider the case $c \neq 0$. Then we have from the first equation of (S.18) that

$$\frac{\mu'(t)}{\mu(t)} = \frac{(a - 2c)}{ct},$$

for which a solution is $\mu(t) = t^r$ with $r = (a - 2c)/c$; direct substitution of this shows that the second equation of (S.18) is also satisfied, and we have found an integrating factor in case $c \neq 0$. In case $c = 0$, the condition (S.19) gives $a\,f = 0$, which means that either $a = 0$ or $f = 0$. If $a = 0$ but $f \neq 0$, then we can obtain an integrating factor from the second equation in (S.18) as $\mu(t) = t^r$ with $r = (2b - f)/f$; the first equation is automatically satisfied since $a = c = 0$. If $f = 0$ but $a \neq 0$, then the first equation of (S.18) is $a\mu(t) = 0$, which implies $\mu(t) \equiv 0$; hence there is no integrating factor in this case. Finally, if $c = 0$ and both a and f are zero, then the second equation in (S.18) is $2b\,\mu(t) = 0$, which gives $\mu(t) \equiv 0$ unless $b = 0$; hence there is no integrating factor in this case.

19. Suppose that μ is an integrating factor for $M\,dt + N\,dy = 0$ leading to the solution $\Phi(t, y) = C$, and that F is any function of Φ. Show that $\mu\,F(\Phi)$ is also an integrating factor for $M\,dt + N\,dy = 0$.

Solution. If $d\Phi = \mu M\,dt + \mu N\,dy = 0$, then

$$d \int F(\Phi)\,d\Phi = F(\Phi)\,d\Phi = \mu\,F(\Phi)M\,dt + \mu\,F(\Phi)N\,dy = 0.$$

Hence $\mu F(\Phi)$ is also an integrating factor for $M\,dt + N\,dy = 0$.

20. Show that if

$$\frac{M_y - N_t}{N\,y - M\,t} = f(z)$$

where $z = t\, y$, then

$$\mu = e^{\int f(z)\,dz}$$

is an integrating factor for $M\,dt + N\,dy = 0$.

Solution. If $\mu = e^{\int f(z)\,dz}$ with $z = t\, y$, then

$$\frac{\partial(\mu\,M)}{\partial y} - \frac{\partial(\mu\,N)}{\partial t} = \mu\,M_y + M\,\mu f(z)\,t - \big(\mu\,N_t + N\,\mu\,f(z)\,y\big),$$

which simplifies to

$$\frac{\partial(\mu\,M)}{\partial y} - \frac{\partial(\mu\,N)}{\partial t} = \mu\big((M_y - N_t) - f(z)(N\,y - M\,t)\big) = 0.$$

Hence, $e^{\int f(z)\,dz}$ with $z = t\, y$ is an integrating factor for $M\,dt + N\,dy = 0$.

Use the method of Exercise 20 to solve the following differential equations.

21. $(t\,y^2 + y)\,dt + t\,dy = 0$

   Solution. Letting $M = t\,y^2 + y$ and $N = t$, we see that

$$\frac{M_y - N_t}{N\,y - M\,t} = -\frac{2}{t\,y} = f(t\,y).$$

   Therefore, if we let $z = t\, y$, then

$$\mu(z) = e^{\int f(z)\,dz} = \frac{1}{(t\,y)^2}$$

   is an integrating factor for $M\,dt + N\,dy = 0$, which leads to the solution

$$\frac{t}{e^{1/(t\,y)}} = C.$$

   An additional solution (which can be found by inspection) is $y = 0$.

22. $(3y + 8t\,y^2)\,dt + (2t + 6y\,t^2)\,dy = 0$

   Solution. Letting $M = 3y + 8t\,y^2$ and $N = 2t + 6y\,t^2$, we see that

$$\frac{M_y - N_t}{N\,y - M\,t} = -\frac{1 + 4t\,y}{t\,y + 2(t\,y)^2} = f(t\,y)$$

Therefore, if we let $z = t\, y$, then

$$\mu(z) = e^{\int f(z)\, dz} = \frac{1}{t\, y + 2(t\, y)^2}.$$

is an integrating factor for $M\, dt + N\, dy = 0$, which leads to the solution

$$t^3\, y^2(1 + 2t\, y) = C.$$

23. $(y^3 - 2y\, t^2)\, dt + (2t\, y^2 - t^3)\, dy = 0$

Solution. Letting $M = y^3 - 2y\, t^2$ and $N = 2t\, y^2 - t^3$, we see that

$$\frac{M_y - N_t}{N\, y - M\, t} = \frac{1}{t\, y} = f(t\, y).$$

Therefore, if we let $z = t\, y$, then

$$\mu(z) = e^{\int f(z)\, dz} = t\, y$$

is an integrating factor for $M\, dt + N\, dy = 0$, which leads to the solution

$$t^2\, y^2(y^2 - t^2) = C.$$

24. $(y\, t^2 + y^2)\, dt - t^3\, dy = 0$

Solution. Letting $M = y\, t^2 + y^2$ and $N = -t^3$ we see that

$$\frac{M_y - N_t}{N\, y - M\, t} = -\frac{2}{t\, y} = f(t\, y)$$

Therefore, if we let $z = t\, y$, then

$$\mu(z) = e^{\int f(z)\, dz} = \frac{1}{(t\, y)^2}$$

is an integrating factor for $M\, dt + N\, dy = 0$, which leads to the solution

$$\frac{t}{y} - \frac{1}{t} = C.$$

An additional solution (which can be found by inspection) is $y = 0$.

Section 3.5, page 65

Find the general solution of each of the following homogeneous differential equations.

1. $y' = \dfrac{t-y}{t}$

Solution. The substitution $z = y/t$ transforms the equation into

$$t z' + z = 1 - z$$
$$t z' = 1 - 2z$$

so that we obtain the separable equation

$$\frac{dz}{1-2z} = \frac{dt}{t}.$$

We integrate both sides to get

$$1 - 2z = C t^{-2},$$

where C is a constant of integration. Thus, the general solution is

$$y = \frac{t}{2} + \frac{C}{t}.$$

2. $y' = \dfrac{2t+y}{t}$

Solution. The substitution $z = y/t$ transforms the equation into

$$t z' + z = 2 + z$$
$$t z' = 2$$

so that we obtain the separable equation

$$\frac{dz}{2} = \frac{dt}{t}.$$

We integrate both sides to get

$$z = 2 \log(t) + C,$$

where C is a constant of integration. Thus, the general solution is

$$y = 2t \log(t) + C t.$$

3. $y' = \dfrac{t^2 + 5y^2}{t\,y}$

Solution. The substitution $z = y/t$ transforms the equation into

$$t\,z' + z = \frac{1}{z} + 5z$$

$$t\,z' = \frac{1}{z} + 4z$$

so that we obtain the separable equation

$$\frac{z\,dz}{1 + 4z^2} = \frac{dt}{t}.$$

We integrate both sides to get

$$\frac{\log(1 + 4z^2)}{8} = \log(t) + C$$

$$1 + 4z^2 = C\,t^8,$$

where C is a constant of integration. Thus, the general solution is $y = \pm t \sqrt{\dfrac{C\,t^8 - 1}{4}}$.

4. $y' = \dfrac{t^2 - 3y^2}{t\,y}$

Solution. The substitution $z = y/t$ transforms the equation into

$$t\,z' + z = \frac{1}{z} - 3z$$

$$t\,z' = \frac{1}{z} - 4z$$

so that we obtain the separable equation

$$\frac{z\,dz}{1 - 4z^2} = \frac{dt}{t}.$$

We integrate both sides to get

$$\frac{\log(1 - 4z^2)}{8} = -\log(t) + C$$

$$4z^2 - 1 = C\,t^{-8},$$

where C is a constant of integration. Thus, the general solution is $y = \pm \sqrt{\dfrac{t^2}{4} + \dfrac{C}{t^6}}$.

5. $y' = \dfrac{4y^4 + t^4}{4t\,y^3}$

Solution. The substitution $z = y/t$ transforms the equation into

$$t\,z' + z = \frac{1}{4z^3} + z$$

$$t\,z' = \frac{1}{4z^3}$$

so that we obtain the separable equation

$$4z^3\,dz = \frac{dt}{t}.$$

We integrate both sides to get

$$z^4 = \log(t) + C$$

where C is a constant of integration. Thus, the general solution is $y^4 = t^4(\log(t) + C)$.

6. $y' = \dfrac{6y^4 + t^4}{4t\,y^3}$

Solution. The substitution $z = y/t$ transforms the equation into

$$t\,z' + z = \frac{6z}{4} + \frac{1}{4z^3}$$

$$t\,z' = \frac{z}{2} + \frac{1}{4z^3}$$

so that we obtain the separable equation

$$\frac{4z^3\,dz}{1 + 2z^4} = \frac{dt}{t}.$$

We integrate both sides to get

$$\frac{1}{2}\log(1 + 2z^4) = \log(t) + C$$

$$1 + 2z^4 = C\,t^2$$

where C is a constant of integration. Thus, the general solution is $t^4 + 2y^4 = C\,t^6$

7. $y' = \dfrac{6yt}{2t^2 - y^2}$

Solution. The substitution $z = y/t$ transforms the equation into

$$t z' + z = \frac{6z}{2 - z^2}$$

$$t z' = \frac{4z + z^3}{2 - z^2}$$

so that we obtain the separable equation

$$\frac{2 - z^2}{4z + z^3} = \frac{dt}{t}.$$

This can be integrated by the method of partial fractions, when we write

$$\frac{2 - z^2}{4z + z^3} = \frac{1}{2z} - \frac{3z/2}{z^2 + 4}$$

We integrate both sides to get

$$\frac{1}{2} \log(z) - \frac{3}{4} \log(z^2 + 4) = \log(t) + C$$

$$\frac{z^2}{(z^2 + 4)^3} = C t^4$$

where C is a constant of integration. Thus, the general solution is $\dfrac{y^2}{(y^2 + 4t^2)^3} = C.$

8. $y' = \dfrac{y + \sqrt{y^2 + t^2}}{t}$

Solution. The substitution $z = y/t$ transforms the equation into

$$t z' + z = z + \sqrt{z^2 + 1}$$

$$t z' = \sqrt{z^2 + 1}$$

so that we obtain the separable equation

$$\frac{dz}{\sqrt{z^2 + 1}} = \frac{dt}{t}.$$

This can be integrated by

$$\sinh^{-1}(z) = \log(t) + C$$

where C is a constant of integration. Thus, the general solution is $y = t \sinh(\log(t) + C) = C t^2 + 1/C$

9. $y'(t^2 y^2 - t^4) = t^2 y^2 - y^4$

Solution. The substitution $z = y/t$ transforms the equation into

$$t z' + z = \frac{z^2 - z^4}{z^2 - 1} = -z^2$$

$$t z' = -z - z^2$$

so that we obtain the separable equation

$$\frac{dz}{z + z^2} = -\frac{dt}{t}.$$

This can be integrated by partial fractions as

$$\log(z) - \log(1 + z) = -\log(t) + C$$

$$\frac{z + 1}{z} = Ct$$

where C is a constant of integration. Thus, the general solution is

$$y = \frac{t}{Ct - 1}.$$

10. $y' = \dfrac{4t\, y + 8y^2}{4t^2 + y^2}$

Solution. The substitution $z = y/t$ transforms the equation into

$$t z' + z = \frac{4z + 8z^2}{4 + z^2}$$

$$t z' = \frac{8z^2 - z^3}{4 + z^2}$$

so that we obtain the separable equation

$$\frac{4 + z^2}{8z^2 - z^3}\, dz = \frac{dt}{t}.$$

This can be integrated by partial fractions by writing

$$\frac{4+z^2}{8z^2-z^3} = \frac{1}{16z} + \frac{17}{16}\frac{1}{8-z} + \frac{1}{2z^2}$$

$$\frac{\log(z)}{16} - \frac{17\log(8-z)}{16} - \frac{1}{2z} = \log(t) + C$$

where C is a constant of integration. Thus, the general solution is $z^{1/16}(8-z)^{-17/16}e^{-1/2z} = C\,t$ where $y = t\,z$.

Solve the following initial value problems.

11. $\begin{cases} y' = \dfrac{4t\,y + 8y^2}{4t^2 + y^2}, \\[2mm] y(1) = 2 \end{cases}$

Solution. We can use the general solution found in 10. Thus $z^{1/16}(8-z)^{-17/16}e^{-1/2z} = C\,t$. The initial condition translates into $z(1) = y(1)/1 = 2$, so that $C = 2^{1/16}\,6^{-17/16}e^{-1/4}$ and the solution y can be obtained from the implicit relation

$$\frac{y}{t(8-y/t)^{17}} = \frac{t^{16}\,e^{8t/y}}{6^{16}\,e^4}$$

12. $\begin{cases} y' = \dfrac{y}{t} + \dfrac{t}{y}, \\[2mm] y(4) = 0 \end{cases}$

Solution. The substitution $z = y/t$ transforms the equation into

$$t\,z' + z = z + \frac{1}{z}$$

$$t\,z' = \frac{1}{z}$$

so that we obtain the separable equation

$$z\,dz = \frac{dt}{t}.$$

This can be integrated by writing

$$\frac{z^2}{2} = \log(t) + C,$$

where C is a constant of integration. The initial condition gives $0 = \log(4) + C$, thus, $z^2 = 2(\log(t) - \log(4)) = 2\log(t/4)$, $z = \sqrt{2\log(t/4)}$ and the solutions $y = \pm t\sqrt{2\log(t/4)}$.

13. $\begin{cases} y' = \dfrac{y - \sqrt{t^2 + y^2}}{t}, \\[3mm] y(1) = 0 \end{cases}$

Solution. The substitution $z = y/t$ transforms the equation into

$$t z' + z = z - \sqrt{1 + z^2}$$
$$t z' = \sqrt{1 + z^2}$$

so that we obtain the separable equation

$$\frac{dz}{\sqrt{1 + z^2}} = \frac{dt}{t}.$$

This can be integrated by writing

$$\sinh^{-1}(z) = \log(t) + C,$$

where C is a constant of integration. The initial condition gives $0 = 0 + C$, thus, $z = \sinh(\log(t)) = t/2 - 1/2t$. Thus, the solution is $y = \dfrac{1 - t^2}{2}$.

14. $\begin{cases} y' = \dfrac{t\,y - y^2}{t^2}, \\[3mm] y(1) = 1 \end{cases}$

Solution. The substitution $z = y/t$ transforms the equation into

$$t z' + z = z - z^2$$
$$t z' = -z^2$$

so that we obtain the separable equation

$$\frac{dz}{z^2} = -\frac{dt}{t}.$$

This can be integrated by writing

$$\frac{-1}{z} = -\log(t) + C,$$

where C is a constant of integration. The initial condition gives $-1 = 0 + C$, thus, $z = \sinh(\log(t)) = t/2 - 1/2t$. Thus, $1/z = 1 + \log(t)$ and the solution is

$$y = \frac{t}{1 + \log(t)}$$

15. $\begin{cases} y' = \dfrac{t+y}{t}, \\ y(1) = 1 \end{cases}$

Solution. The substitution $z = y/t$ transforms the equation into

$$t\,z' + z = 1 + z$$
$$t\,z' = 1$$

so that we obtain the separable equation

$$dz = \frac{dt}{t}.$$

This can be integrated by writing

$$z = \log(t) + C,$$

where C is a constant of integration. The initial condition gives $1 = 0 + C$, thus, $z = 1 + \log(t)) = t/2 - 1/2t$ and the solution is $y = t\big(1 + \log(t)\big)$

16. $\begin{cases} y' = \dfrac{y+2t}{2y-t}, \\ y(1) = 1 \end{cases}$

Solution. The substitution $z = y/t$ transforms the equation into

$$t\,z' + z = \frac{z+2}{2z-1}$$

$$t\,z' = \frac{2 + 2z - 2z^2}{2z - 1}$$

so that we obtain the separable equation

$$\frac{2z - 1}{2 + 2z - 2z^2} = \frac{dt}{t}.$$

This can be integrated directly by writing

$$\frac{1}{2}\log(2 + 2z - 2z^2) = -\log(t) + C$$

$$2 + 2z - 2z^2 = C/t^2,$$

where C is a constant of integration. The initial condition gives $C = 2$. When we solve this quadratic equation for z and set $y = tz$, the solution is $y = \dfrac{t + \sqrt{5t^2 - 4}}{2}$

17. $$\begin{cases} y' = \dfrac{t^2 + 2t\,y - y^2}{t^2 - 2t\,y - y^2}, \\ y(1) = 1 \end{cases}$$

The substitution $z = y/t$ transforms the equation into

$$t z' + z = \frac{1 + 2z - z^2}{1 - 2z - z^2}$$

$$t z' = \frac{1 + z + z^2 + z^3}{1 - 2z - z^2}$$

so that we obtain the separable equation

$$\frac{1 - 2z - z^2}{1 + z + z^2 + z^3} = \frac{dt}{t}.$$

The solution is $y = \dfrac{1 + \sqrt{1 + 4t - 4t^2}}{2}$

18. $$\begin{cases} y' = \dfrac{y(y^2 + 3t^2)}{2t^3}, \\ y(1) = 1 \end{cases}$$

Solution. The substitution $z = y/t$ transforms the equation into

$$t z' + z = \frac{z^3 + 3z}{2}$$

$$t z' = \frac{z^3 + z}{2}$$

so that we obtain the separable equation

$$\frac{dz}{z^3 + z} = \frac{dt}{2t}.$$

which can be solved by partial fractions; thus

$$\frac{1}{z^3 + z} = \frac{1}{z} - \frac{z}{z^2 + 1}$$

$$\log(z) - \frac{1}{2}\log(z^2 + 1) = \frac{1}{2}\log(t) + C$$

$$\frac{z}{\sqrt{1 + z^2}} = C t^{1/2}$$

where C is a constant of integration, which may be obtained form the initial conditions as $C = 1/\sqrt{2}$. Solving for z yields $z = \sqrt{t/(2-t)}$ and the solution $y = \dfrac{t^{3/2}}{\sqrt{2-t}}$.

19. Consider a first-order differential equation of the form

$$\frac{dy}{dt} = f\left(\frac{a_1 t + b_1 y + c_1}{a_2 t + b_2 y + c_2}\right), \tag{S.20}$$

where $a_1, b_1, c_1, a_2, b_2, c_2$ are constants with a_2, b_2, c_2 not all zero.

a. Show that the transformation

$$\tilde{t} = t - h, \qquad \tilde{y} = y - k$$

puts (S.20) in the form

$$\frac{d\tilde{y}}{d\tilde{t}} = f\left(\frac{a_1\tilde{t} + b_1\tilde{y} + a_1 h + b_1 k + c_1}{a_2\tilde{t} + b_2\tilde{y} + a_2 h + b_2 k + c_2}\right). \tag{S.21}$$

b. Show that if $a_1 b_2 - a_2 b_1 \neq 0$, then h and k can be determined so that

$$a_1 h + b_1 k + c_1 = 0 \qquad \text{and} \qquad a_2 h + b_2 k + c_2 = 0,$$

yielding a differential equation which is homogeneous in $\tilde{t}$ and $\tilde{y}$.

Solution. **a.** We compute

$$a_1 t + b_1 y + c_1 = a_1(\tilde{t} + h) + b_1(\tilde{y} + k) + c_1 = a_1\tilde{t} + b_1\tilde{y} + (a_1 h + b_1 k + c_1)$$

and

$$a_2 t + b_2 y + c_2 = a_2(\tilde{t} + h) + b_2(\tilde{y} + k) + c_2 = a_2\tilde{t} + b_2\tilde{y} + (a_2 h + b_2 k + c_2).$$

On the other hand, since $d\tilde{t}/dt = 1$ and we have

$$\frac{dy}{dt} = \frac{d}{dt}(\tilde{y} + k) = \frac{d\tilde{y}}{dt} = \frac{d\tilde{y}}{d\tilde{t}}\frac{d\tilde{t}}{dt} = \frac{d\tilde{y}}{d\tilde{t}}.$$

Thus, we get (S.21).

b. In order to achieve the desired form, we must have

$$\begin{cases} a_1 h + b_1 k = -c_1, \\ a_2 h + b_2 k = -c_2. \end{cases} \tag{S.22}$$

Then (S.22) is a system of simultaneous linear equations for the unknowns h and k. There will be a unique solution in case the determinant of (S.22) is nonzero, in other words when $a_1 b_2 - a_2 b_1 \neq 0$.

Use Exercise 19 to solve the following equations.

20. $y' = \dfrac{t-y}{t+y+2}$

Solution. In this case we have $h - k = 0$ and $h + k + 2 = 0$, which has the unique solution $h = k = -1$. Making the substitutions of Exercise 19, we have

$$\frac{d\tilde{y}}{d\tilde{t}} = \frac{(\tilde{t}-1)-(\tilde{y}-1)}{(\tilde{t}-1)+(\tilde{y}-1)+2} = \frac{\tilde{t}-\tilde{y}}{\tilde{t}+\tilde{y}}$$

Using the technique for homogeneous equations, we write $\tilde{y} = \tilde{t}\, z$; thus

$$\frac{d}{d\tilde{t}}\tilde{y} = \frac{d}{dt}(\tilde{t}\, z) = \tilde{t}\, z' + z = \frac{1-z}{1+z}$$

and

$$\tilde{t}\frac{dz}{d\tilde{t}} = \frac{1-z-z(1+z)}{1+z} = \frac{1-2z-z^2}{1+z},$$

which is the separable equation

$$\tilde{t}\frac{dz}{d\tilde{t}} = \frac{1-2z-z^2}{1+z}, \qquad \text{or} \qquad \frac{1+z}{1-2z-z^2}\, dz = \frac{d\tilde{t}}{\tilde{t}}.$$

It can be integrated in the form

$$\frac{-\log(1-2z-z^2)}{2} = \log|\tilde{t}| + C.$$

We then make the substitutions $\tilde{y} = \tilde{t}\, z$, $\tilde{y} = y + 1$ and $\tilde{t} = t + 1$, obtaining

$$y = -2 - t \pm \sqrt{2 + 4t + 2t^2} + C.$$

21. $y' = \dfrac{t+y+1}{t-y}$

Solution. In this case we have $h + k + 1 = 0$ and $h - k = 0$, which has the unique solution $h = k = -1/2$. Making the substitutions of Exercise 19, we have

$$\frac{d\tilde{y}}{d\tilde{t}} = \frac{(\tilde{t}-1/2)+(\tilde{y}-1/2)+1}{(\tilde{t}-1/2)(\tilde{y}-1/2)} = \frac{\tilde{t}+\tilde{y}}{\tilde{t}-\tilde{y}}$$

Using the technique for homogeneous equations, we write $\tilde{y} = \tilde{t}\, z$; thus

$$\frac{d}{d\tilde{t}}\tilde{y} = \frac{d}{dt}(\tilde{t}\, z) = \tilde{t}\, z' + z = \frac{1+z}{1-z}$$

and

$$\tilde{t}\frac{dz}{d\tilde{t}} = \frac{1+z-z(1-z)}{1-z} = \frac{1+z^2}{1-z},$$

which is the separable equation

$$\tilde{t}\frac{dz}{d\tilde{t}} = \frac{1+z^2}{1-z}, \qquad \text{or} \qquad \frac{1-z}{1+z^2}\,dz = \frac{d\tilde{t}}{\tilde{t}}.$$

This can be integrated in the form

$$\left|\arctan(z) - \frac{1}{2}\right|\log(1+z^2) = \log|\tilde{t}| + C,$$

and then make the substitutions $\tilde{y} = \tilde{t}\,z$, $\tilde{y} = y + 1/2$ and $\tilde{t} = t + 1/2$. Thus, we get

$$\frac{\exp\left(\arctan\left(\dfrac{1/2+y}{1/2+t}\right)\right)}{\sqrt{1+\left(\dfrac{1/2+y}{1/2+t}\right)^2}} = (1/2+t)C_1,$$

where C_1 is a constant.

22. $y' = (y-t)^2$

 Solution. Let $z = y - t$, then

$$\frac{dy}{dt} = \frac{dz}{dt} + 1,$$

and the original equation can be rewritten as

$$\frac{dz}{dt} = z^2 - 1,$$

which is a separable equation. It can be integrated to yield

$$z = \frac{1 - e^{2tC}}{1 + e^{2tC}}.$$

We then make the substitution $z = y - t$ to obtain

$$y = t + \frac{1 - e^{2tC}}{1 + e^{2tC}}.$$

23. $y' = \dfrac{1}{t+y}$

Solution. Let $z = t + y$, then

$$\frac{dy}{dt} = \frac{dz}{dt} - 1,$$

and the original equation can be rewritten as

$$\frac{dz}{dt} = \frac{1+z}{z},$$

which is a separable equation. It can be integrated to yield

$$z - \log(1+z) = \log|t| + C.$$

We then make the substitution $z = t + y$ to obtain

$$y - \log(1 + t + y) = C.$$

24. Consider the differential equation $y' = f(t, y)$.

    a. Prove that if $f(t, y)$ is generalized homogeneous of degree k, then

$$t \frac{\partial(tf/y)}{\partial t} = -k \left(y \frac{\partial(tf/y)}{\partial y} \right),$$

    b. Prove that if $f(t, y)$ is generalized homogeneous of degree k, then the transformation $y = z\, t^k$ applied to $y' = f(t, y)$ leads to a separable equation in t and z.

Solution. **a.** Suppose that $f(t, y) = t^{k-1} f(y/t^k)$; then

$$\frac{\partial(tf/y)}{\partial t} = \frac{k \left(t^k f(y/t^k) - y\, f'(y/t^k) \right)}{t\, y}$$

and

$$\frac{\partial(tf/y)}{\partial y} = \frac{- \left(t^k f(y/t^k) - y\, f'(y/t^k) \right)}{y^2}.$$

Hence,

$$t \frac{\partial(tf/y)}{\partial t} = \frac{k \left(t^k f(y/t^k) - y\, f'(y/t^k) \right)}{y} = -k\, y \frac{\partial(tf/y)}{\partial y}.$$

    b. Let $y(t) = z(t)\, t^k$; then $y'(t) = k\, t^{k-1} z(t) + t^k z'(t) = t^{k-1} f(z)$. Therefore, $t\, z'(t) = f(z) - k\, z$, a separable equation.

Show that the following differential equations are generalized homogeneous and solve them by the method of Exercise 24.

25. $y' = (t^4 y^2 + 1)/t^3$

  Solution. To check generalized homogeneity, we compute

$$t\frac{\partial(tf/y)}{\partial t} = -\frac{2}{y t^2} + 2y t^2 = 2y\frac{\partial(tf/y)}{\partial y}.$$

We want $y = z/t^2$, where $k = -2$. Therefore, the original equation becomes

$$t z' = z^2 + 2z + 1,$$

which after separating and integrating becomes

$$z(t) = -\frac{\log(t) + 1 + C}{\log(t) + C}.$$

Letting $y = z/t^2$ finally yields

$$y(t) = -\frac{\log(t) + 1 + C}{t^2(\log(t) + C)}.$$

26. $y' = (t y + 1)/t^2$

  Solution. To check generalized homogeneity, we compute

$$t\frac{\partial(tf/y)}{\partial t} = -\frac{1}{t y} = y\frac{\partial(tf/y)}{\partial y}.$$

We want $y = z/t$, where $k = -1$. Therefore, the original equation becomes

$$t z' = 2z + 1,$$

which after separating and integrating becomes

$$z(t) = C t^2 - \frac{1}{2}.$$

Letting $y = z/t$ finally yields

$$y(t) = C t - \frac{1}{2t}.$$

27. $y' = (y t^2 + 1)/t^3$

Solution. To check generalized homogeneity, we compute

$$t \frac{\partial(tf/y)}{\partial t} = -\frac{2}{t^2 y} = 2y \frac{\partial(tf/y)}{\partial y}.$$

We want $y = z/t^2$, where $k = -2$. Therefore, the original equation becomes

$$t z' = 3z + 1,$$

which after separating and integrating becomes

$$z(t) = C t^3 - \frac{1}{3}.$$

Letting $y = z/t$ finally yields

$$y(t) = C t - \frac{1}{3t^2}.$$

28. $y' = (t^2 y^2 + 1)/t^2$

Solution. To check generalized homogeneity, we compute

$$t \frac{\partial(tf/y)}{\partial t} = -\frac{1}{t y} + t y = y \frac{\partial(tf/y)}{\partial y}$$

hence; we want $y = z/t$, where $k = -1$. Therefore, the original equation becomes

$$t z' = z^2 + z + 1,$$

which after separating and integrating becomes

$$\frac{2}{\sqrt{3}} \arctan\left(\frac{1 + 2z(t)}{\sqrt{3}}\right) = C + \log(t).$$

Letting $y = z/t$ finally yields

$$y = \frac{-1 + \sqrt{3} \tan\left(\dfrac{\sqrt{3}\,(C + \log(t))}{2}\right)}{2t}.$$

Section 3.6, page 70

Find the general solution of each of the following Bernoulli equations.

1. $y' + y = y^4 \sin(t)$

   Solution.   The equation is already in the leading-coefficient-unity form (step 1), from which we recognize the exponent $n = 4$, hence the equation of step 2 is $u' - 3u = -3\sin(t)$, for which the integrating factor is e^{-3t}, so that

   $$u\,e^{-3t} = -3\int e^{-3t}\sin(t)\,dt = \frac{3}{10}\cos(t)e^{-3t} + \frac{9}{10}\sin(t)e^{-3t} + C.$$

   When we multiply by e^{3t} and form $y = u^{-1/3}$, we find the solution

   $$y = \left(C\,e^{3t} + \frac{3}{10}\cos(t) + \frac{9}{10}\sin(t)\right)^{-1/3}.$$

2. $t^3 y' - 2t\,y = y^3$

   Solution. The leading-coefficient-unity form (step 1) is $y' - (2/t^2)y = y^3/t^3$, from which we recognize $n = 3$ and the equation of step 2 is $u' + (4/t^2)u = -2/t^3$, for which the integrating factor is $e^{-4/t}$ so that

   $$u\,e^{-4/t} = -2\int \frac{e^{-4/t}}{t^3}\,dt = -e^{-4/t}\left(\frac{1}{2t} + \frac{1}{8}\right) + C.$$

   When we multiply by $e^{4/t}$ and form $y = u^{-1/2}$, we find the solutions

   $$y = \pm\frac{1}{\sqrt{-1/8 - 1/2t + C\,e^{4/t}}}.$$

3. $t\,y' + 2y = \dfrac{e^t}{y}$

   Solution. The leading-coefficient-unity form (step 1) is $y' + (2/t)y = y^{-1}(e^t/t)$, from which we recognize $n = -1$ and the equation of step 2 is $u' + (4/t)u = 2(e^t/t)$, for which the integrating factor is t^4 so that

   $$u\,t^4 = 2\int t^3 e^t\,dt = (2t^3 - 6t^2 + 12t - 12)e^t + C.$$

   When we multiply by t^{-4} and form $y = u^{1/2}$, we find the solutions

   $$y = \pm\sqrt{\frac{C + e^t(-12 + 12t - 6t^2 + 2t^3)}{t^4}}.$$

4. $t\,y' + 2y = \dfrac{t}{y^3}$

Solution. The leading-coefficient-unity form (step 1) is $y' + (2/t)y = y^{-3}$, from which we recognize $n = -3$ and the equation of step 2 is $u' + (8/t)u = 4$, for which the integrating factor is t^8 so that

$$u\,t^8 = \int 4\,t^8\,dt = \frac{4t^9}{9} + C.$$

When we multiply by t^{-8} and form $y = u^{1/4}$, we find the solutions

$$y = \pm\left(\frac{4t}{9} + \frac{C}{t^8}\right)^{1/4}.$$

Solve the following initial value problems.

5. $\begin{cases} y' + \dfrac{y}{4t} = -e^{\sqrt{t}}y^3, \\[2mm] y(1) = 1 \end{cases}$

Solution. The equation is already in leading-coefficient-unity form (step 1), from which we recognize $n = 3$ and the equation of step 2 is $u' - (1/2t)u = 2e^{\sqrt{t}}$, for which the integrating factor is $t^{-1/2}$ so that

$$u\,t^{-1/2} = 2\int t^{-1/2}e^{\sqrt{t}}\,dt = 4e^{\sqrt{t}} + C.$$

The constant is determined from the initial conditions by noting that $u(1) = 1$, thus, $1 = 4e + C$. When we multiply by $t^{1/2}$ and form $y = u^{1/2}$, we find the solution

$$y = \frac{1}{t^{1/4}(1 - 4e + 4e^{\sqrt{t}})^{1/2}}.$$

6. $\begin{cases} y' - 2y\tan(t) + y^2(\tan(t))^4 = 0, \\[2mm] y(0) = 2 \end{cases}$

Solution. The equation is already in leading-coefficient-unity form (step 1), from which we recognize $n = 2$ and the equation of step 2 is $u' + 2\tan(t)u = -\tan(t)^4$, for which the integrating factor is $\sec(t)^2$ so that

$$u(\sec(t))^2 = \int \sec(t)^2\tan(t)^4\,dt = \frac{(\tan(t))^5}{5} + C.$$

The constant is determined from the initial conditions by noting that $u(0) = 1/2$, thus, $C = 1/2$. When we multiply by $\cos(t)^2$ and form $y = u^{-1}$, we find the solution

$$y = \frac{10\sec(t)^2}{5 + 2(\tan(t))^5}.$$

7. $\begin{cases} y' - \dfrac{t\,y}{2} = t\,y^5, \\ y(1) = 1 \end{cases}$

Solution. The equation is already in leading-coefficient-unity form (step 1), from which we recognize $n = 5$ and the equation of step 2 is $u' + 2t\,u = -4t$, for which the integrating factor is e^{t^2} so that

$$u\,e^{t^2} = -\int 4t\,e^{t^2}\,dt = -2\,e^{t^2} + C.$$

The constant is determined from the initial conditions by noting that $u(1) = 1$; thus, $e = -2e + C$ and $C = 3e$. When we multiply by e^{-t^2} and form $y = u^{-1/4}$, we find the solution $y = \left(-2 + 3e^{1-t^2}\right)^{-1/4}$.

8. $\begin{cases} y' + \dfrac{y}{t} = t^3 y^3, \\ y(2) = 1 \end{cases}$

Solution. The equation is already in leading-coefficient-unity form (step 1), from which we recognize $n = 3$ and the equation of step 2 is $u' - (2/t)\,u = -2t^3$, for which the integrating factor is t^{-2} so that

$$u\,t^{-2} = -2\int t\,dt = -t^2 + C.$$

The constant is determined from the initial conditions by noting that $u(2) = 1$, thus, $2^{-2} = -4 + C$, $C = 17/4$. When we multiply by t^2 and form $y = u^{-1/2}$, we find the solution $y = \dfrac{2}{t\sqrt{17 - 4t^2}}$

9. $\begin{cases} 2y' - \dfrac{y}{t} + \cos(t)y^3 = 0, \\ y(2) = 1 \end{cases}$

Solution. The leading-coefficient-unity form (step 1) is $y' - (1/2t)y = -\cos(t)y^3/2$, from which we recognize $n = 3$ and the equation of step 2 is $u' + (1/t)u = \cos(t)$, for which the integrating factor is t so that

$$ut = \int t \cos(t)\, dt = t \sin(t) + \cos(t) + C.$$

The constant is determined from the initial conditions by noting that $u(2) = 1$, thus, $C = 2 - \cos(2) - 2\sin(2)$. When we multiply by t^{-1} and form $y = u^{-1/2}$, we find the solution $y = \dfrac{\sqrt{t}}{\sqrt{2 - \cos(2) + \cos(t) - 2\sin(2) + t\sin(t)}}$

10. $\begin{cases} t\,y' + y = \dfrac{1}{y^2}, \\ y(1) = 2 \end{cases}$

Solution. The leading-coefficient-unity form (step 1) is $y' + (1/t)y = (1/t)y^{-2}$, from which we recognize $n = -2$ and the equation of step 2 is $u' + (3/t)u = 3/t$, for which the integrating factor is t^3 so that

$$u\,t^3 = \int 3t^2\, dt = t^3 + C.$$

The constant is determined from the initial conditions by noting that $u(1) = 8$; thus, $8 = 1 + C$ and $C = 7$.

When we multiply by t^{-3} and form $y = u^{1/3}$, we find the solution $y = \dfrac{\sqrt[3]{t^3 + 7}}{t}$

Chapter 4

Section 4.4, page 90

Use **ODE** with the option **Method->FirstOrderLinear** to find the general solution of each of the following differential equations:

1. $y' - t y = t$

Solution.

```
ODE[y' - t y == t,y,t,Method->FirstOrderLinear]

                    2
                 t /2
{{y[t] -> -1 + E     C[1]}}
```

2. $t\,y' + y = t\log(t)$

Solution.

```
ODE[t y' + y == t Log[t],y,t,Method->FirstOrderLinear]
```

$$\{\{y[t] \rightarrow \frac{-t}{4} + \frac{C[1]}{t} + \frac{t\;Log[t]}{2}\}\}$$

3. $t\,y' + y = t\sin(t^2)$

Solution.

```
ODE[t y' + y == t Sin[t^2],y,t,Method->FirstOrderLinear]
```

$$\{\{y[t] \rightarrow \frac{2\;C[1] - Cos[t^2]}{2\;t}\}\}$$

4. $t^2 y' + y - 2t\,y = t^2$

Solution.

```
ODE[t^2 y' + y - 2t y == t^2,y,t,
Method->FirstOrderLinear]
```

$$\{\{y[t] \rightarrow t^2\;(1 + E^{1/t}\;C[1])\}\}$$

5. $y' + \dfrac{y}{t\log(t)} = \dfrac{1}{t}$

Solution.

```
ODE[y' +  y/(t Log[t]) == 1/t,y,t,
Method->FirstOrderLinear]
```

$$\{\{y[t] \rightarrow \frac{C[1]}{Log[t]} + \frac{Log[t]}{2}\}\}$$

6. $y' + y = t^5\cos(t)^2$

Solution.

```
ODE[y' + y == t^5 Cos[t]^2,y,t,Method->FirstOrderLinear]
```

$$\{\{y[t] \rightarrow -60 + 60\ t - 30\ t^2 + 10\ t^3 - \frac{5\ t^4}{2} + \frac{t^5}{2} + \frac{C[1]}{E^t} -$$

$$\frac{1404\ Cos[2\ t]}{3125} + \frac{492\ t\ Cos\{2\ t\}}{625} + \frac{42\ t^2\ Cos[2\ t]}{125} -$$

$$\frac{22\ t^3\ Cos[2\ t]}{25} + \frac{3\ t^4\ Cos[2\ t]}{10} + \frac{t^5\ Cos[2\ t]}{10} - \frac{528\ Sin[2\ t]}{3125} -$$

$$\frac{456\ t\ Sin[2\ t]}{625} + \frac{144\ t^2\ Sin[2\ t]}{125} - \frac{4\ t^3\ Sin[2\ t]}{25} -$$

$$\frac{2\ t^4\ Sin[2\ t]}{5} + \frac{t^5\ Sin[2\ t]}{5}\}\}$$

7. $y' - \dfrac{3y}{t^2} = \dfrac{1}{t^2}$

Solution.

```
ODE[y' - (3/t^2)y == 1/t^2,y,t,Method->FirstOrderLinear]
```

$$\{\{y[t] \rightarrow -(\frac{1}{3}) + \frac{C[1]}{E^{3/t}}\}\}$$

8. $y' - \dfrac{2y}{t} = t\log(t)$

Solution.

```
ODE[y' - (2/t)y == t Log[t],y,t,
Method->FirstOrderLinear]
```

$$\{\{y[t] \rightarrow t^2\ C[1] + \frac{t^2\ Log[t]^2}{2}\}\}$$

9. $t\,y' + 2y = \cos(t)^2$

Solution.

```
ODE[t y' + 2 y == Cos[t]^2,y,t,Method->FirstOrderLinear]
```

$$\{\{y[t] \to \frac{1}{4} + \frac{C[1]}{t^2} + \frac{\cos[2\ t]}{8\ t^2} + \frac{\sin[2\ t]}{4\ t}\}\}$$

10. $(1+t^2)^3 y' + 4t\,y = \dfrac{t}{(1+t^2)^2}$

Solution.

    ODE[(1 + t^2)^3 y' + 4 t y == t(1 + t^2)^(-2),y,t,
    Method->FirstOrderLinear]

$$\{\{y[t] \to \frac{1}{4} + \frac{1}{4 + 8\ t^2 + 4\ t^4} + E^{1/(1 + 2\ t^2 + t^4)}\ C[1]\}\}$$

Use **ODE** (or **DSolve**) to solve the following initial value problems:

11. $\begin{cases} t\,y' + y = t^4 + t^3, \\[4pt] y(1) = 1/2 \end{cases}$

Solution.

    ODE[{t y' + y == t^4 + t^3,y[1] == 1/2},y,t,
    Method->FirstOrderLinear]

$$\{\{y \to \frac{1}{20\ t} + \frac{t^3}{4} + \frac{t^4}{5}\}\}$$

12. $\begin{cases} 2y + t\,y' = e^t, \\[4pt] y(1) = 1 \end{cases}$

Solution.

    ODE[{t y' + 2y == E^t,y[1] == 1},y,t,
    Method->FirstOrderLinear]

$$\frac{3}{}\ \frac{4}{1}\ \frac{1}{t}\ \frac{t}{} \quad y[t] \to \frac{}{} + \frac{}{} + \frac{}{} - 20\ t\ 4\ 5$$

13. $\begin{cases} y\cot(t) + y' = 2\csc(t), \\[4pt] y(\pi/2) = 1 \end{cases}$

Solution.

```
ODE[{y' + Cot[t]y == 2Csc[t],y[Pi/2] == 1},y,t,
Method->FirstOrderLinear]
```

```
{{y[t] -> Csc[t] - Pi Csc[t] + 2 t Csc[t]}}
```

14. $\begin{cases} 2y + t\,y'(t) = 2\sin(t), \\ y(\pi) = 1/\pi \end{cases}$

Solution.

```
ODE[{t y' + 2y == 2Sin[t],y[Pi] == 1/Pi},y,t,
Method->FirstOrderLinear]
```

$$\{\{y[t] \to -(\frac{Pi}{2}) - \frac{2\ Cos[t]}{t} + \frac{2\ Sin[t]}{2}\}\}$$

15. $\begin{cases} y'\cot(t) + y = 4\sin(t), \\ y(-\pi) = 0 \end{cases}$

Solution.

```
ODE[{y'Cot[t] + y == 4Sin[t],y[-Pi] == 0},y,t,
Method->FirstOrderLinear]
```

```
{{y[t] -> -4 Pi Cos[t] - 4 t Cos[t] + 4 Sin[t]}}
```

16. $\begin{cases} 2(1+t)y + t(2+t)y' = 1 + 3t^2, \\ y(-1) = 1 \end{cases}$

Solution.

```
ODE[{t(2 + t)y'+2(1 + t)y == 1 + 3t^2,y[-1] == 1},y,t,
Method->FirstOrderLinear]
```

$$\{\{y[t] \to \frac{1}{2\ t + t^2} + \frac{t}{2\ t + t^2} + \frac{t^3}{2\ t + t^2}\}\}$$

17. $\begin{cases} y'\cos(t) - y\sin(t) = t^3 e^{t^2}, \\ y(0) = 1 \end{cases}$

Solution.

```
ODE[{Cos[t] y' - Sin[t] y == t^3 E^(t^2),
     y[0] == 1},y,t,Method->FirstOrderLinear]
```

$$\{\{y[t] \to \frac{3\ \text{Sec}[t]}{2} - \frac{E^{t^2}\ \text{Sec}[t]}{2} + \frac{E^{t^2}\ t^2\ \text{Sec}[t]}{2}\}\}$$

18.
$$\begin{cases} 3.1y' + \log(t)y = 0, \\ y(1.2) = 2.3 \end{cases}$$

Solution.

```
ODE[{3.1 y' + Log[t]y == 0,y[1.2] == 2.3},y,t,
Method->FirstOrderLinear]
```

$$\{\{y \to \frac{1.68\ (E)^{t\ 0.323}}{t^{0.323\ t}}\}\}$$

19.
$$\begin{cases} 0.05y' + t y = t^2, \\ y(0.1) = 0 \end{cases}$$

Solution.

```
ODE[{0.05 t^2 y' + t y == t^2,y[0.1] == 0},y,t,
Method->FirstOrderLinear]
```

```
{{y[t] -> 0.952381 t}}
```

20.
$$\begin{cases} y' - 0.5t y = e^{0.25t^2}, \\ y(0) = 1 \end{cases}$$

Solution.

```
ODE[{y' - 0.5t y == E^(0.25t^2),y[0] == 1},y,t,
Method->FirstOrderLinear]
```

$$\{\{y \to E^{0.25\ t^2.}\ (1. + t)\}\}$$

Plot the solutions to the following initial value problems:

21. $y' + y = t \sin(t)$, $y(0) = 0$.   Use the range $-2\pi < t < 2\pi$.

Solution.

```
ODE[{y' + y == t Sin[t],y[0] == 0},y,t,
Method->FirstOrderLinear,
PlotSolution->{{t,-2Pi,2Pi}}]
```

$$\{\{y[t] \rightarrow \frac{-1}{2\,E^t} + \frac{Cos[t]}{2} - \frac{t\,Cos[t]}{2} + \frac{t\,Sin[t]}{2}\}\}$$

22. $y'\sin(t) + y\cos(t) = \sin(t)^2 e^t$, $y(\pi/2) = 0$.   Use the range $-\pi < t < \pi$.

Solution.

```
ODE[{Sin[t] y' + Cos[t]y == Sin[t]^2E^t,
y[Pi/2] == 0},y,t,
Method->FirstOrderLinear,
PlotSolution->{{t,-2Pi,2Pi}},
SuppressAsymptotes->True]
```

$$\{\{y[t] \rightarrow \frac{-3\,E^{Pi/2}\,Csc[t]}{5} + \frac{E^t\,Csc[t]}{2} - \frac{E^t\,Cos[2\,t]\,Csc[t]}{10} - $$

$$\frac{E^t\,Csc[t]\,Sin[2\,t]}{5}\}\}$$

23. $y' - \dfrac{2y}{t} = t$, $y(1) = 1$.   Use the range $0.01 < t < 1$.

Solution.

```
ODE[{y' - (2/t)y == t,y[1] == 1},y,t,
Method->FirstOrderLinear,
PlotSolution->{{t,0.01,1}}]
```

$$\{\{y[t] \rightarrow t^2 + t^2\,Log[t]\}\}$$

24. Use **FirstOrderLinear** to find the general solution of each of the differential equations 1–10.

Solution.

    1. **FirstOrderLinear[1&,-#&,#&,C][t]**

    2. **FirstOrderLinear[#&,1&,#Log[#]&,C][t]**

    3. **FirstOrderLinear[#&,1&,#Sin[#^2]&,C][t]**

    4. **FirstOrderLinear[#^2&,(1-2#)&,#^2&,C][t]**

5. `FirstOrderLinear[1&,1/(#Log[#])&,(1/#)&,C][t]`

6. `FirstOrderLinear[1&,1&,#^5Cos[#]^2&,C][t]`

7. `FirstOrderLinear[1&,(-3/#^2)&,(1/#^2)&,C][t]`

8. `FirstOrderLinear[1&,(-2/#)&,#Log[#]&,C][t]`

9. `FirstOrderLinear[#&,2&,Cos[#]^2&,C][t]`

10. `FirstOrderLinear[(1+#^2)^3&,4#&,#/(1+#^2)^2&,C][t]`

25. Use `FirstOrderLinear` to find the general solution of each of the differential equations 1–10 of Section 3.2.

Solution.

1. `FirstOrderLinear[1&,4&,2#+3Exp[3#]&,C][t]`

2. `FirstOrderLinear[1&,-2&,#Exp[3#]&,C][t]`

3. `FirstOrderLinear[1&,2#&,3#Exp[-#^2]&,C][t]`

4. `FirstOrderLinear[1&,1/#&,Exp[#^2]&,C][t]`

5. `FirstOrderLinear[#&,5&,#^3&,C][t]`

6. `FirstOrderLinear[(1+#^2)&,9&,0&,C][t]`

7. `FirstOrderLinear[1&,-Exp[#]&,2#Exp[Exp[#]]&,C][t]`

8. `FirstOrderLinear[1&,1&,Sin[#]&,C][t]`

9. `FirstOrderLinear[1&,Cot[#]&,5Exp[Cos[#]]&,C][t]`

10. `FirstOrderLinear[1&,2Cos[#]&,Sin[#]^2Cos[#]&,C][t]`

Section 4.5, page 94

Use **ODE** with the option **Method->Separable** to find the general solution of each of the following differential equations.

1. $\sin(t)(\sin(y))^2 - (\cos(t))^2 y' = 0$

Solution.

```
ODE[Sin[t] Sin[y[t]]^2 - Cos[t]^2 y'[t] == 0,y[t],t,
Method->Separable]
```

2. $y^3 + t^6 y' = 0$

Solution.

```
ODE[y[t]^3 + t^6 y'[t] == 0,y[t],t,
Method->Separable,Form->Equation]
```

3. $1 + \sqrt{a^2 - t^2} y' = 0$

Solution.

```
ODE[1 + Sqrt[a^2 - t^2] y'[t] == 0,y[t],t,
Method->Separable]
```

4. $2y(t^2 - t - 1) + (t^3 - t)y' = 0$

Solution.

```
ODE[2y[t](t^2 - t - 1) + (t^3 - t)y'[t] == 0,y[t],t,
Method->Separable,Form->Equation]
```

5. $y^2 + 2y + 5 - y' = 0$

Solution.

```
ODE[y[t]^2 + 2y[t] + 5 - y'[t] == 0,y[t],t,
Method->Separable,Form->Equation]
```

6. $y^2 \cos(\sqrt{t}) - 2\sqrt{t}\, e^{1/t} y' = 0$

Solution.

```
ODE[y[t]^2Cos[Sqrt[t]] - 2Sqrt[t]E^(1/y[t])y'[t] == 0,
y[t],t,Method->Separable,Form->Equation]
```

7. $y + (1 + t^2)\arctan(t)y' = 0$

Solution.

```
ODE[y[t] + (1 + t^2)ArcTan[t]y'[t] == 0,y[t],t,
Method->Separable,Form->Equation]
```

8. $y' = (y^3 + 8)t^3 \sin(t)$

Solution.

```
ODE[y' == (y^3 + 8)t^3 Sin[t],y,t,
Method->Separable]
```

Section 4.6, page 96

Use **ODE** with the option **Method->IntegratingFactor** to find the general solution of each of the following differential equations. Also, determine which equations are solvable with the option **Method->FirstOrderExact**.

1. $t^3 + t\, y^4 + 2y^3 \dfrac{dy}{dt} = 0$

Solution.

```
ODE[t^3 + t y^4 + 2y^3y' == 0,y,t,
Method->IntegratingFactor,Form->Equation]
```

2. $t^2 + y^2 + 2t\,y\dfrac{dy}{dt} = 0$

Solution.

```
ODE[t^2 + y^2 + 2t y y' == 0,y,t,
Method->IntegratingFactor,Form->Equation]
```

3. $2t\,y + (y^2 - 3t^2)\dfrac{dy}{dt} = 0$

Solution.

```
ODE[2t y + (y^2 - 3t^2)y' == 0,y,t,
Method->IntegratingFactor,Form->Equation]
```

4. $1 - 2t\,y^2 + 2t\,y(1 - t - t\,y^2)\dfrac{dy}{dt} = 0$

Solution.

```
ODE[1 - 2t y^2 + 2t y(1 - t - t y^2)y' == 0,y,t,
Method->IntegratingFactor,Form->Equation]
```

5. $y(1 + t\,y) + t(1 - t\,y)\dfrac{dy}{dt} = 0$

Solution.

```
ODE[y(1 + t y) +t(1 - t y)y' == 0,y,t,
Method->IntegratingFactor,Form->Equation]
```

6. $t\left(y\dfrac{dy}{dt} - 1\right) - (t\,y^2 - 2t) = 0$

Solution.

```
ODE[t(y y' - 1) - (t y^2 - 2t) == 0,y,t,
Method->IntegratingFactor,Form->Equation]
```

7. $y^2 e^{ty^2} + 1 + (2t\,y\,e^{ty^2} - 3y^2)\dfrac{dy}{dt} = 0$

Solution.

```
ODE[y^2 E^(t y^2) + 1 + (2t y E^(t y^2) - 3y^2)y' == 0,
y,t,Method->FirstOrderExact]
```

8. $4t^3y^3 - 2y + \left(3t^4y^2 - 2t + 2y + \dfrac{2}{y}\right)\dfrac{dy}{dt} = 0$

Solution.

```
ODE[4t^3y^3 - 2y + (3t^4y^2 - 2t + 2y + 2/y)y' == 0,
y,t,Method->FirstOrderExact]
```

9. $2t\sin(t^2) + 3\cos(y) - 3t\sin(y)\dfrac{dy}{dt} = 0$

Solution.

```
ODE[2t Sin[t^2] + 3Cos[y] - 3t Sin[y]y' == 0,y,t,
Method->FirstOrderExact]
```

10. $y^2 - 3 - \dfrac{y}{t^2 + ty} + \left(\dfrac{1}{t+y} + 2y + 2y\right)\dfrac{dy}{dt} = 0$

Solution.

```
ODE[y^2-3 - y/(t^2 + t y) + (1/(t+y) + 2t y + 2y)y' == 0,
y,t,Method->FirstOrderExact]
```

Section 4.7, page 99

Use **ODE** with the option **Method->FirstOrderHomogeneous** to find the general solution of each of the following differential equations.

1. $y' = \dfrac{y^2}{t^2 - ty}$

Solution.

```
ODE[y' - y^2/(t^2 - t y) == 0,y,t,
Method->FirstOrderHomogeneous]
```

2. $y' = \dfrac{2ty}{3t^2 + y^2}$

Solution.

```
ODE[y' == 2t y/(3t^2 + y^2),y,t,
Method->FirstOrderHomogeneous,Form->Equation]
```

3. $y' = -\dfrac{15t + 11y}{9t + 5y}$

Solution.

```
ODE[y' == -(15t + 11y)/(9t + 5y),y,t,
Method->FirstOrderHomogeneous,Form->Equation]
```

4. $y' = \dfrac{y-t}{y+t}$

Solution.

```
ODE[y' == (y - t)/(y + t),y,t,
Method->FirstOrderHomogeneous,Form->Equation]
```

5. $t\,y' - y = \sqrt{t^2 + y^2}$

Solution.

```
ODE[t y' - y == Sqrt[t^2 + y^2],y,t,
Method->FirstOrderHomogeneous]
```

6. $y' = -\dfrac{y(2t^3 - y^3)}{t(2y^3 - t^3)}$

Solution.

```
ODE[y' == -(y/t)(2t^3 - y^3)/(2y^3 - t^3),y,t,
Method->FirstOrderHomogeneous,Form->Equation]
```

7. $y' = \dfrac{y - t - t\,\cos(y/t)}{t}$

Solution.

```
ODE[y' == (y - t - t Cos[y/t])/t,y,t,
Method->FirstOrderHomogeneous,Form->LogEquation]
```

8. $y' = \dfrac{y + t\,e^{y/t}}{t}$

Solution.

```
ODE[y' == (y + t Exp[y/t])/t,y,t,
Method->FirstOrderHomogeneous,Form->LogEquation]
```

9. $y' = \dfrac{y - 2t\,\tanh(y/t)}{t}$

Solution.

```
ODE[y' == (y - 2t Tanh[y/t])/t,y,t,
Method->FirstOrderHomogeneous,Form->Equation]
```

10. $y' = -\dfrac{t^2 + t\,y + y^2}{t^2}$

Solution.

```
ODE[y' == -(t^2 + t y + y^2)/t^2,y,t,
Method->FirstOrderHomogeneous,Form->LogEquation]
```

Section 4.8, page 101

Use **ODE** with the option **Method->Bernoulli** to find the general solution of each of the following differential equations.

1. $y' - y = t\,e^{5t}y^3$

Solution.

```
ODE[y' - y == t E^(5t)y^3,y,t,
Method->Bernoulli,Form->Equation,
PostSolution->{ExpandAll,PowerExpand}]
```

2. $y' + 5y = t\,\sin(t)e^t y^5$

Solution.

```
ODE[y' + 5y == t Sin[t]E^t y^5,y,t,
Method->Bernoulli,Form->Equation]
```

3. $y' + \dfrac{y}{t-3} = 5(t-3)y^{3/2}$

Solution.

```
ODE[y' + y/(t - 3) == 5(t - 3)y^(3/2),y,t,
Method->Bernoulli]
```

4. $t^3 y' - 2t\,y = \dfrac{t}{y^{10}}$

Solution.

```
ODE[t^3 y' - 2t y == t y^-10,y,t,
Method->Bernoulli]
```

Use **ODE** with the option **Method->Bernoulli** to solve each of the following initial value problems.

5. $\begin{cases} t\,y' = t\,y^2 + (1+t)y, \\ y(1) = 1 \end{cases}$

Solution.

```
ODE[{t y' == t y^2 + (1 + t)y,y[1] == 1},y,t,
Method->Bernoulli]
```

6. $\begin{cases} 3(1+t^2)y' = 2t\,y(y^3 - 1), \\ y(0) = 2 \end{cases}$

Solution.

```
ODE[{3(1 + t^2)y' == 2t y(y^3 - 1),y[0] == 2},y,t,
Method->Bernoulli]
```

7. $\begin{cases} y' + \tan(2t)y = y^2 \cos(2t)\sin(2t), \\ y(0) = 1 \end{cases}$

Solution.

```
ODE[{y' + Tan[2t]y == y^2 Cos[2t]Sin[2t],y[0] == 1},y,t,
Method->Bernoulli,Form->Equation]
```

8. $\begin{cases} y' = \dfrac{y^2 + 2t\,y}{t^2}, \\ y(1) = 2 \end{cases}$

Solution.

```
ODE[{y' == (y^2 + 2t y)/t^2,y[1] == 2},y,t,
Method->Bernoulli]
```

Section 4.9, page 106

Use **ODE** with the option **Method->Clairaut** to find both the general solution and the singular solution of each of the following differential equations.

1. $y = t\,y' + (1 + y'^2)^{1/2}$

Solution.

```
ODE[y == t y' + (1 + y'^2)^(1/2),y,t,
Method->Clairaut,Form->Equation]
```

2. $y = t\,y' + (y')^3$

Solution.

```
ODE[y == t y' + y'^3,y,t,
Method->Clairaut,Form->Equation]
```

3. $y = t\,y' + \sin(y')$

Solution.

```
ODE[y == t y' + Sin[y'],y,t,
Method->Clairaut,Form->Equation]
```

4. $y = t\,y' + 3 - 1/y'$

Solution.

```
ODE[y == t y' + 3-1/y',y,t,
Method->Clairaut,Form->Equation]
```

5. $y = t\,y' + \exp(y')$

Solution.

```
ODE[y == t y' + E^y',y,t,
Method->Clairaut,Form->Equation]
```

6. $y = t\,y' + \log(y')$

Solution.

```
ODE[y == t y' + Log[y'],y,t,
Method->Clairaut,Form->Equation]
```

7. $y = t\,y' + (y')^4/4$

Solution.

```
ODE[y == t y' + y'^4/4,y,t,
Method->Clairaut,Form->Equation]
```

8. $y = t\,y' + 2\exp(y')$

Solution.

```
ODE[y == t y' + 2E^y',y,t,
Method->Clairaut,Form->Equation]
```

Use **ODE** with the option **Method->Lagrange** to find both the general solution and the singular solution of each of the following differential equations.

9. $y = 2t\,y' + \log(y')$

Solution.

```
ODE[y == 2t y' + Log[y'],y,t,
Method->Lagrange]
```

10. $y = t(1 + y') + y'^2$

Solution.

```
ODE[y == t(1 + y') + y'^2,y,t,
Method->Lagrange]
```

11. $y = (3/2)t\,y' + \exp(y')$

Solution.

```
ODE[y == (3/2)t y' + Exp[y'],y,t,
Method->Lagrange]
```

12. $y = 2t\,y' + \cos(y')^2$

Solution.

```
ODE[y == 2t y' + Cos[y']^2,y,t,
Method->Lagrange]
```

Section 4.10, page 111

Use **ODE** to solve each of the following initial value problems. Plot the solutions over appropriate intervals.

1. $\begin{cases} y' = \sin(t^2), \\ y(0) = 0 \end{cases}$

Solution.

```
ODE[{y' == Sin[t^2],y[0] == 0},y,t,
Method->Separable,
PlotSolution->{{t,-3Pi,3Pi},PlotPoints->100}]
```

2. $\begin{cases} y' = t^2\cos(t^2), \\ y(0) = 0 \end{cases}$

Solution.

```
ODE[{y' == t^2Cos[t^2],y[0] == 0},y,t,
Method->Separable,
PlotSolution->{{t,-Pi,3Pi},PlotPoints->100}]
```

3. $\begin{cases} y' = ty + y^5, \\ y(0) = 4 \end{cases}$

Solution.

```
ODE[{y' ==  t y + y^5,y[0] == 4},y,t,
Method->Bernoulli,PlotSolution->{{t,-Pi,3Pi},
PlotPoints->100}]
```

4. $\begin{cases} y' = 1 - t^2 + y^2, \\ y(0) = 1 \end{cases}$

Solution.

```
ODE[{y' == 1 - t^2 + y^2,y[0] == 1},y,t,
Method->Riccati,KnownSolution->t,
PlotSolution->{{t,-2Pi,2Pi}}]
```

5. Prove Equation (4.30).

Solution. Let $u = z/\sqrt{2}$, then

$$\text{erf(t)} = \frac{2}{\sqrt{\pi}} \int_0^t e^{-u^2} du = \frac{2}{\sqrt{\pi}} \int_{-\infty}^t e^{-u^2} du - 1 = \sqrt{\frac{2}{\pi}} \int_{-\infty}^{t\sqrt{2}} e^{-z^2/2} dz - 1.$$

Using

$$\Phi(t) = \frac{1}{\sqrt{2\pi}} \int_{-\infty}^t e^{-w^2/2} dw,$$

we obtain

$$\text{erf(t)} = 2\Phi\left(t\sqrt{2}\right) - 1.$$

6. Show that $\text{erfi}(t) = -i\,\text{erf}(i\,t)$.

Solution. Let $u = -i\,z$, then

$$\text{erfi(t)} = -i\,\frac{2}{\sqrt{\pi}} \int_0^{it} e^{-z^2} dz = -i\,\text{erf}(i\,t).$$

Section 4.11, page 113

Use **ODE** to define solutions to the following initial value problems. Then plot the solutions over an appropriate interval.

1.
$$\begin{cases} y' - t\,y = \sin(t), \\ y(0) = 0 \end{cases}$$

Solution.

```
y1[t_] = ODE[{y' - t y == Sin[t],y[0] == 0},y,t,
             Method->FirstOrderLinear,Form->Explicit]
         Plot[y1[t],{t,-1,1}];
```

2.
$$\begin{cases} y' = \sin(t^3), \\ y(0) = 0 \end{cases}$$

Solution.

```
y2[t_] = ODE[{y' == Sin[t^3]^2,y[0] == 0},y,t,
             Method->FirstOrderLinear,Form->Explicit]
         Plot[y2[t],{t,-0.5,0.5},PlotRange->All];
```

3.
$$\begin{cases} y' = \sin(e^t), \\ y(0) = 0 \end{cases}$$

Solution.

```
y3[t_] = ODE[{y' == Sin[E^t],y[0] == 0},y,t,
             Method->FirstOrderLinear,Form->Explicit]
         Plot[y3[t],{t,0,3Pi/2},PlotRange->All,
         PlotPoints->100];
```

4.
$$\begin{cases} y' = e^{\cos(t)}, \\ y(0) = 1 \end{cases}$$

Solution.

```
y4[t_] = ODE[{y' == Exp[Cos[t]],y[0] == 1},y,t,
             Method->FirstOrderLinear,Form->Explicit]
         Plot[y4[t],{t,-2Pi,2Pi},PlotRange->All];
```

5.
$$\begin{cases} y' + t\,y = \dfrac{\sin(t)}{t}, \\ y(1) = 0 \end{cases}$$

Solution.

```
y5[t_] = ODE[{y' + t y == Sin[t]/t,y[1] == 0},y,t,
           Method->FirstOrderLinear,Form->Explicit]
Plot[y5[t],{t,-2Pi,2Pi},PlotRange->All];
```

6. $\begin{cases} y' + y = e^{-t^2}, \\ y(0) = 1 \end{cases}$

Solution.

```
y6[t_] = ODE[{y' + y == Exp[-t^2],y[0] == 1},y,t,
           Method->FirstOrderLinear,Form->Explicit]
Plot[y6[t],{t,-2,6},PlotRange->All];
```

7. $\begin{cases} y' = e^{-t^2}\cos(t), \\ y(0) = 1 \end{cases}$

Solution.

```
y7[t_] = ODE[{y' == Exp[-t^2] Cos[t],y[0] == 1},y,t,
           Method->FirstOrderLinear,Form->Explicit]
Plot[y7[t],{t,-Pi,Pi},PlotRange->All];
```

8. $\begin{cases} y' + t y = \sin(e^{-t})y^2, \\ y(0) = 1 \end{cases}$

Solution.

```
y8[t_] = ODE[{y' + t y == Sin[Exp[-t]] y^2,y[0] == 1},
           y,t,Method->Bernoulli,Form->Explicit]
Plot[y8[t],{t,-2Pi,2Pi},PlotRange->All];
```

Section 4.12, page 116

In each of the Exercises 1–6 find the general solution to the Riccati equation given a known solution y_K. Work each problem by hand and with *Mathematica*. (Exercise 6 involves the function **Erfi**.)

1. $\begin{cases} y' = 6 + 5y + y^2, \\ y_K = -2 \end{cases}$

Solution. $y(t) = \dfrac{2 - 3C\,e^t}{C\,e^t - 1}$

2. $\begin{cases} y' = 9 + 6y + y^2, \\ y_K = -3 \end{cases}$

Solution. $y(t) = \dfrac{-1 - 3t - 3C}{t + C}$

3. $\begin{cases} y' = -2 - y + y^2, \\ y_K = 2 \end{cases}$

Solution. $y(t) = -\dfrac{e^{3t} - 6C}{e^{3t} + 3C}$

4. $\begin{cases} y' = e^{2t} + (1 + 2e^t)y + y^2, \\ y_K = -e^t \end{cases}$

Solution. $y(t) = -\dfrac{e^t(1 + e^t + C)}{e^t + C}$

5. $\begin{cases} y' = \sec(t)^2 - \tan(t)y + y^2, \\ y_K = \tan(t) \end{cases}$

Solution. $y(t) = \tan(t) - \dfrac{\sec(t)}{C - \log\left(\dfrac{\cos(t)}{1 + \sin(t)}\right)}$

6. $\begin{cases} y' = 1 - t - y + t\,y^2, \\ y_K = 1 \end{cases}$

Solution. $y(t) = \dfrac{-2e^{1/4 - t + t^2} + 4e^{1/4}C + \sqrt{\pi}\,\mathrm{erfi}((2t - 1)/2)}{2e^{1/4 - t + t^2} + 4e^{1/4}C + \sqrt{\pi}\,\mathrm{erfi}((2t - 1)/2)}$

In each of the Exercises 7–10 solve the initial value problems, both by hand and with *Mathematica*.

7. $\begin{cases} y'e^{-t} + y^2 - 2y\,e^t = 1 - e^{2t}, \\ y(0) = 2, \quad y_K = e^t \end{cases}$

Solution. $y(t) = e^t + e^{-t}$

```
ODE[{y' E^-t+y^2-2y E^t == 1-E^(2t),y[0] == 2},
y,t,Method->Riccati,KnownSolution->E^t]
```

8. $$\begin{cases} t^2 y' = t^2 y^2 + ty + 1, \\ y(1) = 1, \quad y_K = -1/t \end{cases}$$

Solution. $y(t) = \dfrac{1 + 2\log(t)}{t - 2t\log(t)}$

```
ODE[{t^2 y' == t^2 y^2 + t y + 1,y[1] == 1},y[t],t,
Method->Riccati,KnownSolution->-1/t]
```

9. $$\begin{cases} y' = e^{2t} + (1 + 2e^t)y + y^2, \\ y(0) = 1, \quad y_K = -e^t \end{cases}$$

Solution. $y(t) = \dfrac{e^t(1 - 2e^t)}{2e^t - 3}$

```
ODE[{y' == Exp[2t] + (1 + 2Exp[t])y + y^2,y[0] == 1},
y,t,Method->Riccati,KnownSolution->-Exp[t]]
```

10. $$\begin{cases} y' = -4/t^2 - y/t + y^2, \\ y(1) = 1, \quad y_K = 2/t \end{cases}$$

Solution. $y(t) = -2\dfrac{t^4 - 3}{t^5 + 3t}$

```
ODE[{y' == -4/t^2 -y/t + y^2,y[1] == 1},y,t,
Method->Riccati,KnownSolution->2/t]
```

11. Suppose that $y(t)$ is a solution of the Riccati equation $y' = p(t) + q(t)y + r(t)y^2$ with $r(t) \neq 0$. Define $w(t) = \exp\left(-\int r(t)y(t)dt\right)$. Show that $y = -w'/(r(t)w)$ and that $w(t)$ satisfies the second-order linear equation

$$w'' - \left(q(t) + \frac{r'(t)}{r(t)}\right)w' + p(t)r(t)w = 0.$$

Solution. We have

$$w' = -r(t)y \exp\left(-\int r(t)y(t)dt\right) = -r(t)y\, w,$$

so that

$$y = -\frac{w'}{r(t)w} \quad \text{and} \quad y' = -\frac{w''}{r(t)w} + \frac{w'^2}{r(t)w^2} + \frac{w'r'}{r(t)^2 w}.$$

Hence $y' = p(t) + q(t)y + r(t)y^2$ implies that

$$-\frac{w''}{r(t)w} + \frac{w'^2}{r(t)w^2} + \frac{w'r'}{r(t)^2w} = p(t) - \frac{q(t)w'}{r(t)w} + \frac{w'^2}{r(t)w^2}. \qquad \text{(S.23)}$$

The w'^2 terms cancel in (S.23). Thus when we multiply both sides of (S.23) by $-r(t)w$, we get

$$w'' - \left(q(t) + \frac{r'(t)}{r(t)}\right)w' + p(t)r(t)w = 0.$$

12. Suppose that $w(t)$ is a solution of the linear second-order equation $w'' + P(t)w' + Q(t)w = 0$, and that $w(t) \neq 0$. Define $y(t) = w'(t)/w(t)$ and show that $y(t)$ is a solution of the Riccati equation $y' = -Q(t) - P(t)y - y^2$.

Solution. We have

$$w' = y\,w \qquad \text{and} \qquad w'' = y'w + y\,w' = y'w + y^2w.$$

Hence $w'' + P(t)w' + Q(t)w = 0$ implies that

$$y'w + y^2w + P(t)y\,w + Q(t)\,w = 0.$$

When we divide this equation by w, we get $y' = -Q(t) - P(t)y - y^2$.

Chapter 5

Section 5.1, page 126

For each of the following equations, find the region in the (t, y) plane where the existence and uniqueness theorem (Theorem 5.1) guarantees the existence of a unique solution through any specified initial point.

1. $y' = t/y$

   Solution. The function $f(t, y) = t/y$ is continuous with a continuous derivative except at points where $y = 0$, that is, the t-axis. Therefore, we have a unique solution passing through all points except those on the t-axis.

2. $y' = \dfrac{3t + 2y}{2t - 5y}$

   Solution. The function $f(t, y) = (3t + 2y)/(2t - 5y)$ is continuous with a continuous derivative except at points where $2t - 5y = 0$. Therefore, we have a unique solution passing through all points except those on the line $2t - 5y = 0$.

3. $y' = (t^2 + y^2 - 4)^{1/3}$

Solution. The function $f(t, y) = (t^2 + y^2 - 4)^{1/3}$ is continuous with a continuous derivative except at points where $t^2 + y^2 - 4 = 0$. Therefore, we have a unique solution passing through all points except those on the circle $t^2 + y^2 = 4$.

4. $y' = \dfrac{\log|(t+1)y|}{y - t^3}$

Solution. The function

$$f(t, y) = \frac{\log|(t+1)y|}{(y - t^3)}$$

is continuous with a continuous derivative except at points where the denominator is zero or where the logarithm is undefined; therefore we have a unique solution passing through all points except those on the curve $y - t^3 = 0$, on the line $t = -1$, or on line $y = 0$.

5. For which values of α can one expect to have a unique solution to the equation $y' = |y|^\alpha$ for any specified initial point in the (t, y) plane?

Solution. The function $f(y) = |y|^\alpha$ is continuous everywhere, but for $y > 0$ the derivative is $f'(y) = \alpha|y|^{\alpha-1}$, which is not continuous when $y \longrightarrow 0$ unless $\alpha \geq 1$. Therefore, we can expect a unique solution only if $\alpha \geq 1$.

6. Use ODE to find and plot three solutions to the initial value Problem (5.7).

Solution.

```
ODE[y' == y^(1/3),y,t,
Method->Separable,PostSolution->{Simplify,PowerExpand,N}]
```

7. Show that the solution $y(t)$ to the initial value problem

$$\begin{cases} y' = \dfrac{\sin(y)}{\cosh(t)}, \\ y(0) = \pi/2 \end{cases}$$

satisfies $0 < y(t) < \pi$ for all t. Use ODE to find and plot the solution.

Solution. The function $\sin(y)$ is positive for $0 < y < \pi$ and vanishes at $y = 0$ and at $y = \pi$. Hence there are two critical solutions $y_1(t) \equiv 0$ and $y_2(t) \equiv \pi$. If another solution satisfied $y(t) = 0$ for some t, then we would have $y(t) = 0$ for all t, by uniqueness. Similarly, if $y(t) = \pi$ for some t, then we must have $y(t0 = \pi$ for all t. A suitable code using ODE is

```
prob7[t_] = ODE[{y'== Sin[y]/Cosh[t],y[0] == Pi/2},y,
            {t,0,10},MaxSteps->2000,Form->Explicit,
            Method->NDSolve,
            PlotSolution->{{t,0,10},PlotRange->All}];
```

We note from the graphics that the solution converges when $t \longrightarrow \infty$ to a limiting value about 2.7, which is strictly less than π. A more accurate estimate can be obtained with.

```
prob7[10]
```

2.73162

Section 5.2, page 130

For each of the following differential equations, determine whether or not the hypotheses of the theorem on global existence (Theorem 5.3) are satisfied, by finding appropriate functions $p(t)$ and $q(t)$.

1. $y' = t + y^2$

Solution. Global existence does not hold. Proof: If $y^2 \leq |y|p(t) + q(t)$, then divide by y^2 and let $y \longrightarrow \infty$ to obtain the contradiction $1 \leq 0$.

2. $y' = \log(1 + y^2)$

Solution. Global existence holds. For example, let $p(t) = 1$ and $q(t) = 0$. (Examine the graph of $\log(1 + y^2)$ against the graph of y, or note that

$$\frac{d(\log(1 + y^2))}{dy} = \frac{2y}{1 + y^2} \leq 1,$$

so that $\log(1 + y^2) \leq \int_0^y 1\, ds = y$ for $y \geq 0$.)

3. $y' = \cos(t^2 + y)$

Solution. Global existence holds. For example, let $p(t) = 0$ and $q(t) = 1$.

4. $y' = t \sin(t)$

Solution. Global existence holds. For example, let $p(t) = 0$ and $q(t) = t \sin(t)$.

5. $y' = \sinh(y)$

Solution. Global existence does not hold. Proof: If $e^y \leq |y|p(t) + q(t)$, then divide by e^y and let $y \longrightarrow \infty$ to obtain the contradiction $1 \leq 0$.

6. $y' = \tanh(y)$

Solution. Global existence holds. For example, let $p(t) = 0$ and $q(t) = 1$.

7. $y' = \sinh(t)\sin(y) + e^{t^2}$

Solution. Global existence holds. For example, let $p(t) = \sinh(t)$ and $q(t) = e^{t^2}$.

8. $y' = t\log(t^2 + y^2)$

Solution. Global existence holds. For example, let $p(t) = 1$ and $q(t) = t\log(t^2)$. Proof: From Problem 3 we have

$$\log(t^2 + y^2) = \log(t^2) + \log(1 + y^2/t^2) \le \log(t^2) + |y/t|.$$

Now multiply this equation by t.

9. $y' = e^{-y^2}$

Solution. Global existence holds. For example, let $p(t) = 0$ and $q(t) = 1$.

10. $y' = y e^{-y^2}$

Solution. Global existence holds. For example, let $p(t) = 1$ and $q(t) = 0$.

For each of the following initial value problems, show explicitly that an explosion occurs by solving the equation by separation of variables and finding the explosion time T.

11. $\begin{cases} y' = y^4, \\ y(0) = 1 \end{cases}$

Solution. The differential equation is separable; it is solved by writing $dy/y^4 = dt$ and integrating as $y^{-3}/3 = -t + C$. Setting $t = 0$ gives the value $C = 1$ with the solution $y(t) = (1 - 3t)^{-1/3}$. Explosion occurs when the denominator is zero; therefore $T = 1/3$.

12. $\begin{cases} y' = 4 + y^2, \\ y(0) = 0 \end{cases}$

Solution. The differential equation is separable; it is solved by writing $dy/(4+y^2) = dt$ and integrating as $(1/2)\arctan(y/2) = t + C$. Setting $t = 0$ gives the value $C = 0$ with the solution $y(t) = 2\tan(2t)$. Explosion occurs when the tangent becomes infinite at $\pi/2$; hence $T = \pi/4$.

13. $\begin{cases} y' = y^2 + 3y + 2, \\ y(0) = 0 \end{cases}$

Solution. The differential equation is separable; it is solved by writing $dy/(y + 1)(y + 2) = dt$ and using the method of partial fractions. This gives the solution $y(t) = -2 + \dfrac{2}{2 - e^t}$. Explosion occurs when the denominator is zero; thus, $2 - e^T = 0$, or $T = \log(2)$.

14. $\begin{cases} y' = e^y, \\ y(0) = 0 \end{cases}$

Solution. The differential equation is separable; it is solved by writing $dy/e^y = dt$ and integrating to obtain $e^{-y} = t + C$. The initial condition gives $C = 1$; hence the solution $y(t) = \log(1/(1-t))$. Explosion occurs when the argument of the logarithm reaches zero; hence $T = 1$.

Section 5.3, page 136

Use the method of Picard iteration to find approximations to the solutions of the following equations.

1. $\begin{cases} y' = 2y, \\ y(0) = 1 \end{cases}$

Solution. Beginning with the trial solution $y_0(t) = 1$, we find

$y_0(t) = 1,$
$y_1(t) = 1 + 2t,$
$y_2(t) = 1 + 2t + 2t^2,$
$y_3(t) = 1 + 2t + 2t^2 + 4t^3/3$

2. $\begin{cases} y' = y^2, \\ y(0) = 1 \end{cases}$

Solution. Beginning with the trial solution $y_0(t) = 1$, we find

$y_0(t) = 1,$
$y_1(t) = 1 + t,$
$y_2(t) = 1 + t + t^2 + t^3/3,$
$y_3(t) = 1 + t + t^2 + t^3 + 2t^4/3 + t^5/3 + t^6/9 + t^7/63$

3. $\begin{cases} y' = 1 + 3y, \\ y(0) = 2 \end{cases}$

Solution. Beginning with the trial solution $y_0(t) = 2$, we find

$$y_0(t) = 2,$$
$$y_1(t) = 2 + 7t,$$
$$y_2(t) = 2 + 7t + 21t^2/2,$$
$$y_3(t) = 2 + 7t + 21t^2/2 + 21t^3/2$$

4. $$\begin{cases} y' = -2t\,y, \\ y(0) = 1 \end{cases}$$

Solution. Beginning with the trial solution $y_0(t) = 1$, we find

$$y_0(t) = 1,$$
$$y_1(t) = 1 - t^2,$$
$$y_2(t) = 1 - t^2 + t^4/2,$$
$$y_3(t) = 1 - t^2 + t^4/2 + t^8/24 - t^{10}/12$$

5. $$\begin{cases} y' = -y^3/2, \\ y(0) = 1 \end{cases}$$

Solution. Beginning with the trial solution $y_0(t) = 1$, we find

$$y_0(t) = 1,$$
$$y_1(t) = 1 - t/2,$$
$$y_2(t) = 1 - t/2 + 3t^2/8 - t^3/8 + t^4/64$$

6. $$\begin{cases} y' = 1 + y, \\ y(1) = 0 \end{cases}$$

Solution. Beginning with the trial solution $y_0(t) = 0$, we find

$$y_0(t) = 0,$$
$$y_1(t) = -1 + t,$$
$$y_2(t) = -1/2 + t^2/2,$$
$$y_3(t) = -2/3 + t/2 + t^3/6$$

7. $$\begin{cases} y' = 1 + 4y + 2y^2, \\ y(0) = 0 \end{cases}$$

Solution. Beginning with the trial solution $y_0(t) = 0$, we find

$$y_0(t) = 0,$$
$$y_1(t) = t,$$
$$y_2(t) = t + 2t^2 + 2t^3/3,$$
$$y_3(t) = t + 2t^2 + 10t^3/3 + 8t^4/3 + 32t^5/15 + 8t^6/9 + 8t^7/63$$

8. $\begin{cases} y' = 1 - t\,y^2, \\ y(0) = 0 \end{cases}$

Solution. Beginning with the trial solution $y_0(t) = 0$, we find

$y_0(t) = 0,$
$y_1(t) = t,$
$y_2(t) = t - t^4/4,$
$y_3(t) = t - t^4/4 + t^7/14 - t^{10}/160$

Section 5.5, page 144

Use **ODE** with the options **PlotField** and **Method->None** to plot the direction fields together for the following differential equations.

1. $y' = \dfrac{y+t}{y-t}$

Solution.

```
ODE[y' == (y + t)/(y -t),y,t,Method->None,
PlotField->{{{t,-2,2},{y,-2,2}}}]
```

2. $y' = \sin\left(\dfrac{y+t}{y-t}\right)$

Solution.

```
ODE[y' == Sin[(y + t)/(y -t)],y,t,Method->None,
PlotField->{{{t,-2,2},{y,-2,2}}}]
```

3. $y' + y = t\,y^3$

Solution.

```
ODE[y' + y==t y^3,y,t,Method->None,
PlotField->{{{t,-2,2},{y,-2,2}}}]
```

4. $y' = e^{2t} + (1+2e^t)y + y^2$

Solution.

```
ODE[y' == Exp[2t] + (1+2Exp[t])y +  y^2,y,t,
Method->None,
PlotField->{{{t,-Pi,Pi},{y,-2,2}}}]
```

5. $y' = \sin(t)/t$

Solution.

```
ODE[{y'==Sin[t]/t,y[0]==0},y,t,Method->None,
PlotField->{{{t,-Pi,Pi},{y,-2,2}}}]
```

6. $y' = t + y^2$

Solution.

```
ODE[y' == t+y^2,y,t,Method->None,
 PlotField->{{{t,-2,2},{y,-2,2}}}]
```

Use **ODE** with the option **PlotField** to plot the direction fields for the following differential equations. Indicate several possible solution curves.

7. $y' = y + y^3$

Solution.

```
ODE[{y' == y + y^3,y[1]==1},y,t,
PlotSolution->{{t,-2,2},
PlotStyle->{{RGBColor[1,0,0],AbsoluteThickness[2]}}},
PlotField->{{{t,-2,2},{y,-2,2}},
PlotRange->{{-2,2},{-2,2}}},
SuperimposePlots->True]
```

8. $y' = 1 - y/t$

Solution.

```
ODE[{y' == 1 - y/t,y[1]==0},y,t,
PlotField->{{{t,-2,2},{y,-2,2}},
PlotRange->{{-2,2},{-2,2}}},
PlotSolution->{{t,-2,2},
PlotStyle->{{RGBColor[1,0,0],AbsoluteThickness[2]}}},
SuperimposePlots->True]
```

9. $y' = \sin(t)/y$

Solution.

```
ODE[{y' == Sin[t]/y,y[0]==1},y,t,
PlotField->{{{t,-2Pi,2Pi},{y,-2Pi,2Pi}}},
PlotSolution->{{t,-2Pi,2Pi},
PlotStyle->{{RGBColor[1,0,0],AbsoluteThickness[2]}}},
SuperimposePlots->True]
```

10. $y' = t/y$

Solution.

```
ODE[{y' == t/y,y[0]==1},y,t,
PlotField->{{{t,-2,2},{y,-2,2}}},
PlotSolution->{{t,-2,2}},
PlotStyle->{{RGBColor[1,0,0],AbsoluteThickness[2]}},
SuperimposePlots->True]
```

11. $y' = \cos(t)/\sin(y)$

Solution.

```
ODE[{y' == Cos[t]/Sin[y],y[0]==Pi/2},y,t,
PlotField->{{{t,-2Pi,2Pi},{y,-2Pi,2Pi}}},
PlotSolution->{{t,-2Pi,2Pi}},
PlotStyle->{{RGBColor[1,0,0],AbsoluteThickness[2]}},
SuperimposePlots->True]
```

Solution.

12. $y' = e^{\sin(t)} \cos(t)$

Solution.

```
ODE[{y' == Exp[Sin[t]]Cos[t],
y[0]==0},y,t,Method->Separable,
PlotField->{{{t,-2Pi,2Pi},{y,-4,4}}},
PlotSolution->{{t,-2Pi,2Pi}},
PlotStyle->{{RGBColor[1,0,0],AbsoluteThickness[2]}},
SuperimposePlots->True]
```

13. $y' = y - \cos\big((\pi/2)t\big)$

Solution.

```
ODE[{y' == y - Cos[Pi/2 t],y[0]==0},y,t,
Method->FirstOrderLinear,
PlotField->{{{t,-2,2},{y,-2,2}}},
PlotSolution->{{t,-2,2}},
PlotStyle->{{RGBColor[1,0,0],AbsoluteThickness[2]}},
SuperimposePlots->True]
```

14. $y' = t\,e^{y}$

Solution.

```
ODE[{y' == t Exp[y],y[0]==0},y,t,Method->Separable,
PlotField->{{{t,-2,2},{y,-2,2}}},
PlotSolution->{{t,-1.3,1.3}},
PlotStyle->{{RGBColor[1,0,0],AbsoluteThickness[2]}},
SuperimposePlots->True]
```

Section 5.6, page 153

Use **ODE** with the option **PlotField** to plot the direction fields of the following differential equations and estimate the critical values. If possible, find the exact locus of critical values for each problem.

1. $y' = 2y - e^{-y}$

Solution.

```
ODE[y' == 2y - Exp[-y],y,t,Method->None,
PlotField->{{{t,-2,2},{y,-2,2}}}]
FindRoot[2y - Exp[-y] == 0,{y,0}]
```

2. $y' = y^4 + 3y^2 + y - 7$

Solution.

```
ODE[y' == y^4 + 3y^2 + y -7,y,t,Method->None,
PlotField->{{{t,-2,2},{y,-2,2}}}]
FindRoot[y^4 + 3y^2 + y -7 == 0,{y,-2}]
FindRoot[y^4 + 3y^2 + y -7 == 0,{y,2}]
```

3. $y' = y^2 - y - \cos(y)$

Solution.

```
ODE[y' == y^2 - y - Cos[y],y,t,Method->None,
PlotField->{{{t,-2,2},{y,-2,2}}}]
FindRoot[y^2 - y - Cos[y] == 0,{y,-2}]
FindRoot[y^2 - y - Cos[y] == 0,{y,2}]
```

4. $y' = e^{-y^2} - y^4$

Solution.

```
ODE[y' == Exp[-y^2] - y^4,y,t,Method->None,
PlotField->{{{t,-2,2},{y,-2,2}}}]
FindRoot[Exp[-y^2] - y^4 == 0,{y,-1}]
FindRoot[Exp[-y^2] - y^4 == 0,{y,1}]
```

Find the critical points of each of the following equations and classify each of the critical points as stable, unstable or semistable. Plot the auxiliary graph and some of the solutions. If possible use **ODE** to solve the equation.

5. $y' = y - 3y^2$
   Solution. $y = 0$ is unstable and $y = 1/3$ is stable.

6. $y' = y - y^3$
   Solution. $y = 1$ and $y = -1$ are stable and $y = 0$ is unstable.

7. $y' = 1 - \cos(y)$
   Solution. The critical points are of the form $y = 2n\pi$ for integers n and they are all semistable.

8. $y' = \sin(y^2)$
   Solution. The critical points are of the form $y = \pm\sqrt{n\pi}$ for nonnegative integers n. $y = 0$ is semistable. If n is even and $n \neq 0$, then $y = \sqrt{n\pi}$ is unstable and $y = -\sqrt{n\pi}$ is stable. If n is odd, then $y = \sqrt{n\pi}$ is stable and $y = -\sqrt{n\pi}$ is unstable.

9. $y' = y^2(6 - y)(4 + y)$
   Solution. $y = 0$ is semistable, $y = -4$ is unstable and $y = 6$ is stable.

10. $y' = y^2$
    Solution. $y = 0$ is semistable.

11. $y' = y^3$
    Solution. $y = 0$ is unstable.

12. $y' = y^{2n+1}$, $n = 2, 3, 4, \ldots$
    Solution. $y = 0$ is unstable.

13. $y' = y^{2n}$, $n = 2, 3, 4, \ldots$
    Solution. $y = 0$ is semistable.

14. $y' = \sinh(y)$
    Solution. $y = 0$ is unstable.

15. $y' = y(e^y - 1)$
    Solution. $y = 0$ is semistable.

16. $y' = -y^{2n+1}$, $n = 2, 3, 4, \ldots$

Solution. $y = 0$ is stable.

17. $y' = -y^{2n}$, $n = 2, 3, 4, \ldots$

Solution. $y = 0$ is semistable.

Chapter 6

Section 6.1, page 165

1. A colony of bacteria doubles in 4 hours and 20 minutes. Assume an exponential growth law. Find the growth rate k.

   Solution. In this problem we are given that the doubling time is

   $$T_d = 4\ 1/3 \text{ hours} = 15600 \text{ seconds.}$$

   Hence the growth rate is given by

   $$k = \frac{0.693}{15600} = 0.000044 \text{ second}^{-1}.$$

2. Assuming an exponential growth law, compute the growth rates of the countries listed in Example 6.2.

3. For each of the following population models with constant growth determine and solve the initial value problem that describes it.

   a. The growth rate is $k = 5$ and $P(2) = 6$.

   b. The population doubles in one unit of time and $P(0) = 3$

   c. The growth rate is $k = 0.0135/$year and $P(1993) = 3 \times 10^9$.

   Solution. a. The differential equation to be solved is $P'(t) = 5P(t)$ with the initial condition $P(2) = 6$. The general solution of this differential equation is $P(t) = C\,e^{5t}$. The initial condition gives the equation $6 = C\,e^{10}$, which is solved to obtain $C = 2e^{-10}$. Thus, the solution is $P(t) = 6e^{-10}e^{5t} = 6e^{5t-10}$.

   b. The general solution of the equation is $P(t) = C\,e^{kt}$. If the population doubles in one unit of time, then we must have $e^k = 2$. The initial condition gives $3 = C\,e^0 = C$ and so we have $P(t) = 3 \times 2^t$.

   c. The differential equation is $P'(t) = k\,P(t)$ with $k = 0.0135$. The solution is $P(t) = 3 \times 10^9 e^{0.0135(t-1993)}$.

4. A population with constant growth rate k is found to have the values $P_0 = 151$ million when $t_0 = 1950$ and $P_1 = 203$ million when $t_1 = 1970$. Find the growth rate.

   Solution.
   $$k = \frac{\log(203/151)}{1970 - 1950} = \frac{\log(1.34437)}{20} = 0.0147963 \text{ year}^{-1}$$

5. A report by the 1994 UN Conference on World Population estimates that: the world population will increase from 5.7 billion in 1994 to 8.3 billion in 2025. Assuming zero migration, estimate the growth rate k.

Solution. The growth rate in this case is obtained as

$$k = \frac{\log(8.3/5.7)}{2025 - 1994} \approx \frac{\log(1.456)}{31} \approx 0.0121/\text{year}.$$

6. According to the 1994 UN Conference on World Population, it is estimated that the U.S. population will increase from 260.75 million in 1994 to 322 million in 2025. Assuming zero migration, estimate the growth rate k.

Solution. The growth rate in this case is obtained as

$$k = \frac{\log(322/260.75)}{2025 - 1994} \approx \frac{\log(1.235)}{31} \approx 0.00750 \text{ year}^{-1}.$$

7. According to the 1994 UN Conference on World Population, it is estimated that the population of China will increase from 1.192 billion in 1994 to 1.504 billion in 2025. Assuming zero migration, estimate the growth rate k.

Solution. The growth rate in this case is obtained as

$$k = \frac{\log(1.504/1.192)}{2025 - 1994} \approx \frac{\log(1.26174)}{31} \approx 0.007578/\text{year}.$$

8. An environmental organization states that in the next 6 seconds 24 people will be added to the population of the Earth. Assuming a constant growth rate, no migration and a current population of 3 billion, compute:

    a. the increase of the Earth's population in one hour;

    b. the increase of the Earth's population in one day;

    c. the increase of the Earth's population in one year;

    d. the time necessary for the Earth's population to double in size.

Solution. **a.** The population satisfies the relation $P(t) = P_0 e^{kt}$. Measuring time in seconds, from the data given we have

$$P(6) = 3,000,000,024 = 3,000,000,000\, e^{6k},$$

from which

$$e^{6k} = 1.000000008 = 1 + 8 \times 10^{-9}, \ k = (1/6) \times 8 \times 10^{-9}.$$

Thus, after one hour we have

$$P(3600) = P_0 e^{3600k} = P_0 \left(e^{6k}\right)^{600}$$
$$= 3 \times 10^9 (1.000000008)^{600} = 3,000,014,400$$

to the nearest integer, so that the increase in population in one hour is 14,400.

b. From $e^k = 1.000000008$, after one day we have

$$P(24 \times 3600) = P_0 e^{24 \times 3600k} = P_0 \left(e^{6k}\right)^{24 \times 600}$$
$$= 3 \times 10^9 (1.000000008)^{14400} = 3,000,345,600$$

to the nearest integer, so that the increase in population in one day is 345,600.

c. From $e^k = 1.000000008$ after one year we have

$$P(365 \times 24 \times 3600) = P_0 e^{365 \times 24 \times 3600k} = P_0 \left(e^{6k}\right)^{365 \times 24 \times 600}$$
$$= 3 \times 10^9 (1.000000008)^{525600} = 3,128,833,616$$

to the nearest integer, so that the increase in population in one year is 128,833,616.

d. From Lemma 6.1 the time for the population to double is given by

$$T_d = \frac{\log(2)}{k} = \frac{0.693}{1.333 \times 10^{-9}} = 0.520 \times 10^9 \text{ seconds} = 16.48 \text{ years}$$

9. Solve Exercise 8 if we assume a *linear* rate of growth, that is, $P(t) = P_0 + rt$ with 24 people added every 6 seconds. Find **a** the increase in one hour, **b** the increase in one day, **c** the increase in one year and **c** the time necessary to double the population.

Solution. **a.** If we assume a purely linear rate of growth, then the population must satisfy $P(t) = P_0 + rt$; the number r can be computed from the data as

$$r = \frac{P(6) - P_0}{6} = \frac{24}{6} = 4.$$

Therefore, at the end of 1 hour we have

$$P(3600) = P_0 + 3600r = 3 \times 10^9 + 3600 \times 4 = 3,000,014,400,$$

or an increase of $14,400$.

b. At the end of one day the population must satisfy

$$P(3600 \times 24) = P_0 + 3600 \times 24r = 3 \times 10^9 + 3600 \times 24 \times 4 = 3,000,345,600$$

or an increase of 345, 600.

    c. At the end of one year the population must satisfy

$$P(3600 \times 24 \times 365) = P_0 + 3600 \times 24r = 3 \times 10^9 + 3600 \times 24 \times 365 \times 4$$

$$= 3, 126, 144, 000,$$

or an increase of 126, 144, 000.

    d. In this model the time T_d for the population to double is obtained by solving $P(T_d) = 2P_0$, or equivalently

$$2P_0 = P_0 + rT_d;$$

thus

$$T_d = \frac{P_0}{r} = \frac{3 \times 10^9}{r} = 7.5 \times 10^8 \text{ seconds} = 23.78 \text{ years.}$$

10. Find the population size $P(t)$ for all $t > 0$, given a growth rate of $k = 0.03$, a migration rate of $r = 0.15$ and an initial population size of $P(0) = 2.4 \times 10^8$.

Solution. From the solution formula (6.13) we have

$$P(t) = 5(e^{0.03t} - 1) + (2.4 \times 10^8)e^{0.03t}.$$

Alternatively **ODE** can be used as follows:

```
ODE[{P' == (3/100)P + 15/100,P[0] == 24 10^7},P,t]
```

The result is

```
                         (3 t)/100
{{P -> -5 + 240000005 E            }}
```

11. Find the population $P(t)$ for all $t > 0$, given that growth rate is $k = 2$, the migration rate is $r = 3$ and the population at $t = 1$ is $P(1) = 10^{23}$.

Solution. From the solution formula (6.13) we have

$$P(t) = -\frac{3}{2} + \left(10^{23} + \frac{3}{2}\right)e^{2(t-1)}.$$

Alternatively **ODE** can be used as follows:

```
ODE[{P' == 2.0P + 3.0,P[1] == 10^23},P,t]
```

The result is

```
                   22  2. t
{{P -> -1.5 + 1.35335 10   E    }}
```

12. Solve the following initial value problem involving growth with migration:

$$\begin{cases} 4P' = P + 1, \\ P(0) = 16. \end{cases}$$

Solution. From the solution formula (6.13) we have $P(t) = 17e^{t/4} - 1$

Alternatively we use

ODE[{4P' == P + 1,P[0] == 16},P,t]

to get

```
                t/4
{{P -> -1 + 17 E    }}
```

13. A population with constant growth rate and migration rate is found to have the values $P(1800) = 5.31$ million, $P(1810) = 7.24$ million and $P(1820) = 9.64$ million. Find the growth and migration rates.

Solution. The growth rate is determined from

$$e^{10k} = (9.64 - 7.24/7.24 - 5.31) = 1.24,$$

so that

$$k = \log(1.24)/10 = 0.02$$

The migration rate is determined from

$$\frac{r}{k} = \frac{((7.24)^2 - (9.64)(5.31))}{(9.64 + 5.31 - 2(7.24))} 2.62,$$

and $r = 2.62 \times 0.02 = 0.05$

14. Prove Lemma 6.3.

Solution. (i) If the population doubles from $P(t_1)$ at $t = t_1$ to $2P(t_1)$ at $t = t_1 + T$, then from (6.12) we must have

$$2P(t_1) + \frac{r}{k} = \left(P(t_1) + \frac{r}{k}\right)e^{kT},$$

or

$$e^{kT} = \frac{2P(t_1) + r/k}{P(t_1) + r/k} = 2 - \frac{r/k}{P(t_1) + r/k} = 2 - \frac{r}{k\,P(t_1) + r}.$$

Taking the logarithm and dividing by T gives (i). To prove (ii), we note that in case $r = 0$ we have $T = \log(2)/k$. Since $r > 0$ and $k > 0$ we have $r/(kP(t_1) + r) > 0$, and since the logarithm function is increasing, $T = \log(2)/k$ is greater than the value obtained in case $r > 0$, proving (ii). For (iii), we note that by increasing the value of $P(t_1)$ we decrease the value of $r/(k\,P(t_1) + r)$, which *increases* the argument of the logarithm, hence increases the value of T.

15. Derive the inequalities (6.23).

   Proof. From (6.16) and the fact that both $P_2 - P_1$ and $P_1 - P_0$ are positive, we have

$$k > 0 \iff \frac{P_2 - P_1}{P_1 - P_0} > 1$$

$$\iff P_2 - P_1 > P_1 - P_0$$

$$\iff P_2 + P_0 > 2P_1.$$

Thus, the second inequality of (6.23) is equivalent to $k > 0$. Furthermore, it follows that k and $P_2 - 2P_1 + P_0$ always have the same sign. Now (6.22) is equivalent to

$$r = \left(P_1^2 - P_0 P_2\right)\frac{k}{P_2 - 2P_1 + P_0}.$$

Hence $r > 0$ is equivalent to the first inequality of (6.23).

Section 6.2, page 169

Solve the following population models with time-dependent growth and/or migration:

1.
$$\begin{cases} P' = 4 + \cos(t), \\ P(0) = 3 \end{cases}$$

   Solution. In this case we have $k(t) = 0$ and so the solution is simply obtained by integration:

$$P(t) = P(0) + \int_0^t \left(4 + \cos(s)\right) ds = 3 + 4t + \sin(t).$$

The solution can be found and plotted using **ODE**:

```
ODE[{P' == 4 + Cos[t],P[0] == 3},P,t,
Method->FirstOrderLinear,
PlotSolution->{{t,0,4Pi}}]
```

2. $\begin{cases} P' = 4P + \cos(t), \\ P(0) = 3 \end{cases}$

Solution. This initial value problem can be solved directly using the integrating factor e^{-4t} as follows. First, we write

$$\left(e^{-4t} P(t)\right)' = e^{-4t} \cos(t);$$

we integrate this equation as

$$e^{-4t} P(t) = 3 + \frac{-4e^{-4t} \cos(t) + 4 + e^{-4t} \sin(t)}{17}.$$

Thus, the solution is

$$P(t) = \frac{\sin(t)}{17} - \frac{4 \cos(t)}{17} + \frac{55e^{-4t}}{17}.$$

The solution can be found and plotted using **ODE**:

```
ODE[{P' == 4P + Cos[t],P[0] == 3},P,t,
Method->FirstOrderLinear,
PlotSolution->{{t,0,4Pi}}]
```

3. $\begin{cases} P' = 4\cos(t) + \cos(t)P, \\ P(0) = 3 \end{cases}$

Solution. This initial value problem is solved by the integrating factor $e^{-\sin(t)}$ to obtain $\left(P(t)e^{-\sin(t)}\right)' = 4e^{-\sin(t)} \cos(t)$; we integrate this equation as

$$P(t)e^{-\sin(t)} = 3 + \int_0^t 4e^{-\sin(s)} \cos(s)ds = 3 + 4(1 - e^{-\sin(t)}).$$

Thus, the solution is

$$P(t) = 3e^{\sin(t)} + 4(e^{\sin(t)} - 1) = 7e^{\sin(t)} - 4.$$

The solution can be found using **ODE**:

```
ODE[{P' == 4Cos[t] + Cos[t]P,P[0] == 3},P,t,
Method->FirstOrderLinear,
PlotSolution->{{t,0,4Pi}}]
```

4. $\begin{cases} P' = 5 - \sin(t) + 3\cos(t), \\ P(1) = 4 \end{cases}$

Solution. The solution is obtained by direct integration as

$$P(t) = 5t + \cos(t) + 3\sin(t) - 1 - \cos(1) - 3\sin(1)$$

Alternatively, the following *Mathematica* command can be used:

```
ODE[{P' == 5 - Sin[t]  +3 Cos[t],P[1] == 4},P,t,
Method->FirstOrderLinear,
PlotSolution->{{t,0,4Pi}}]
```

5. Suppose that the growth rate and migration rate are given by

$$k = 0.02 \quad \text{and} \quad r(t) = 3 \times 10^4 + 2 \times 10^4 \cos(t),$$

and that the initial population is $P_0 = 7 \times 10^5$. Find the population at $t = 1, 2, 3, 4$ and compare with the corresponding model of constant migration with $k = 0.02$ and $r = 3 \times 10^4$.

6. Suppose that the growth rate and migration rate are given by $k(t) = k$ and $r(t) = r_0 + r_1 t$, where k, r_0 and r_1 are constants.

    a. Solve the resulting differential equation (6.24) with $P(0) = P_0$.

    b. Assume the following census data for the 1960 US population:

$$k = 0.014, \quad r_0 = 2.5 \times 10^5, \quad r_1 = 5.2 \times 10^3, \quad P_0 = 1.8 \times 10^8.$$

Use the solution of **a** to compute the population in the years 1970, 1980, 1990, 2000 (corresponding to $t = 10, 20, 30, 40$.)

Section 6.3, page 176

1. Consider the logistic equation $dP/dt = P(6 - P)$ with the initial condition $P(0) = P_0$.

    a. Find the critical points and classify their stability.

    b. Solve the equation.

    c. If $P_0 = 1$, find the time t for which $P(t) = 5$.

    d. Plot representative solutions of $dP/dt = P(6 - P)$.

Solution. **a.** The function $f(P) = P(6 - P)$ is zero precisely when $P = 0$ or $P = 6$, and we have

$$f'(0) = 6 > 0 \qquad \text{and} \qquad f'(6) = 6 - 12 = -6 < 0.$$

Therefore, Theorem 5.5 implies that 0 is an unstable critical point and 6 is a stable critical point.

    b. The equation $dP/dt = P(6 - P)$ can be solved as in Example 6.6, to yield

$$P(t) = \frac{6P_0}{(6 - P_0)e^{-6t} + P_0}.$$

Alternatively, the command

```
ODE[{P' == P(6 - P),P[0] == P0},P,t,Method->Bernoulli]
```

yields

```
                6 t
         6 E       P0
{{P -> ---------------}}
               6 t
       6 - P0 + E   P0
```

    c. We must solve the equation

$$5 = \frac{6}{5e^{-6t} + 1}$$

for t. The solution is

$$t = \frac{\log(25)}{6} \approx 0.536. \tag{S.24}$$

Note that (S.24) also follows directly from (6.39). We can also use *Mathematica* to obtain (S.24) as follows. First, we solve the initial value problem with

```
prob1[t_] = ODE[{P' == 6P - P^2,P[0] == 1},P,t,
              Method->Bernoulli,Form->Explicit]
```

```
     6 t
  6 E
---------
     6 t
 5 + E
```

Then

```
FindRoot[prob1[t] - 5,{t,1}]
```

yields

```
{t -> 0.536479}
```

d. We use

```
ODE[{P' == 6P - P^2,P[0] == P0},P,t,
Method->Bernoulli,Parameters->{{P0,1,6,1}},
PlotSolution->{{t,0,1.5},
PlotRange->{Automatic,{0,6}}}]
```

to get

```
          6 t
     6 E      P0
{{P -> ——————————————}}
              6 t
     6 - P0 + E    P0
```

and the plot

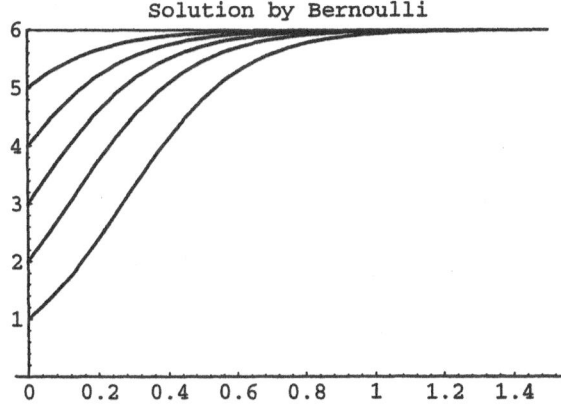

e. The command

```
ODE[{P' == 6P - P^2,P[0] == 1},P,t,Method->AllSymbolic]
```

shows that $dP/dt = P(6 - P)$ can be solved by the methods **Bernoulli**, **Separable** and **DSolve**.

2. The Gompertz model of population growth is described by the equation

$$\frac{dP}{dt} = P(k - b \log(P)), \qquad (S.25)$$

where k and b are positive constants.

    a. Find the critical points of (S.25) and discuss their stability.

    b. Show that if $0 < P(0) < e^{k/b}$, then $P(t)$ tends to $e^{k/b}$ when $t \longrightarrow \infty$.

    c. Solve the differential equation (S.25) as a separable equation.

    d. Plot representative solutions of (S.25) in the case that $k = 1$ and $b = 4$.

Solution. **a.** and **b.** The critical points are 0 and $e^{k/b}$. The auxiliary graph shows that $P(k - b \log(P))$ is positive for $0 < P < e^{k/b}$ and negative for $P < 0$ or $P > e^{k/b}$. Hence 0 is an unstable critical point and $e^{k/b}$ is a stable critical point. Note that Theorem 5.5 cannot be applied to the critical point 0 since the derivative of $P \longmapsto P(k - b \log(P))$ is infinite at $P = 0$.

The auxiliary graph is obtained with the command

```
Plot[P(1 - 4 Log[P]),{P,0.01,2}]
```

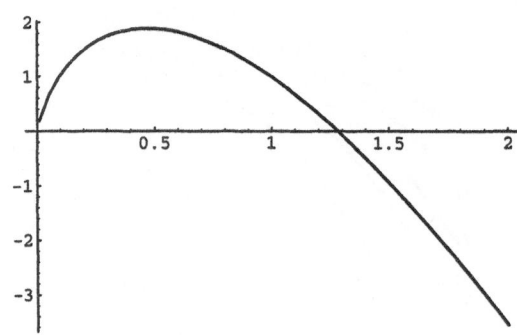

c. and **d.** The command

```
ODE[{P' == P(1 - 4 Log[P]),P[0] == P0},
P,t,Method->Separable,
PostSolution->{PowerExpand,ExpandAll},
Parameters->{{P0,0,E^(1/4),0.2}},
PlotSolution->{{t,0,1.5}}]
```

yields

```
                        b t       -(b t)
        k/b - k/(b E   )   E
{{P -> E                     P0           }}
```

and the plot

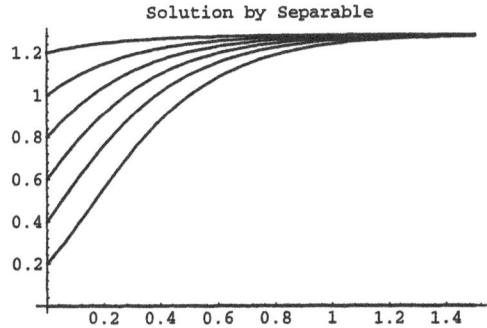

Solution by Separable

3. Suppose that the logistic equation (6.31) is used to model the natural growth of halibut in the Pacific Ocean. Let $P(t)$ be the total mass at time t of halibut measured in kilograms. Assume that $k = 0.71/$year, $b = 88.2 \times 10^{-9}$ and $P(0) = 0.25\, k/b$.

   a. Find the value of P after two years.

   b. Find the time t for which $P(t) = 0.75\, k/b$.

   c. Plot the predicted total mass from 0 years to 10 years.

Solution. **a.** First, we note that

```
N[0.25 0.71/(88.2 10^-9)]
```

yields

```
            6
2.01247 10
```

telling us that $P(0) = 2.01247 \times 10^6$. The solution of the logistic equation can be defined in *Mathematica* by

```
Logistic[{k_,b_},{t0_,P0_}][t_]:=
k P0 /((k - b P0)E^(-k t + k t0) + b P0)
```

Then

```
Logistic[{0.71,8.82 10^-9},{0,2.01247 10^6}][t]
```

yields the general formula

$$\frac{1.42885\ 10^{6}\ E^{0.71\ t}}{0.69225 + 0.01775\ E^{0.71\ t}}$$

In particular

Logistic[{0.71,8.82 10^-9},{0,2.01247 10^6}][2]

gives

$$7.72034\ 10^{6}$$

Thus $P(2) = 7,720,340$ kilograms.

    b. We use (6.39) to compute

$$t = \frac{1}{k}\log\left(\frac{k\,(P(0))^{-1} - b}{k\,(P(t))^{-1} - b}\right) = \frac{1}{k}\log\left(\frac{k\,(k/(4b))^{-1} - b}{k\,(3k/(4b))^{-1} - b}\right)$$

$$= \frac{\log(9)}{k} = \frac{\log(9)}{0.71} \approx 3.09 \text{ years.}$$

    c. We use

Plot[Evaluate[Logistic[
    {0.71,8.82 10^-9},{0,2.01247 10^6}][t]],{t,0,10}]

to obtain the plot

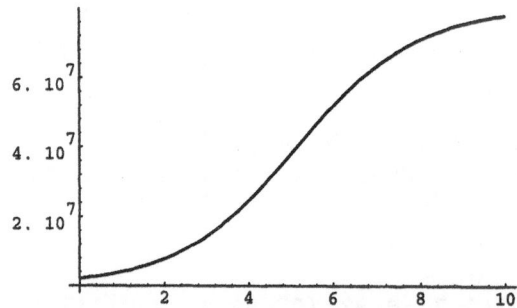

4. Consider the solution (6.38) of the logistic equation with $k > 0$ and $P(t_0) = P_0$, where $0 < P_0 < k/b$.

    a. Show that $P(t) < P_0 e^{k(t-t_0)}$ for any fixed $t > t_0$.

    b. Show that if $k > 0$ is fixed and $b \longrightarrow 0$, then $P(t) \longrightarrow P_0 e^{k(t-t_0)}$ for any fixed $t > t_0$.

Solution. **a.** Since $e^{-k(t-t_0)} \leq 1$, we have

$$\left(k - b\,P_0\right)e^{-k(t-t_0)} + b\,P_0 = k\,e^{-k(t-t_0)} + b\,P_0(1 - e^{-k(t-t_0)}) \geq k\,e^{-k(t-t_0)}.$$

Therefore

$$P(t) = \frac{k\,P_0}{\left(k - b\,P_0\right)e^{-k(t-t_0)} + b\,P_0} \leq \frac{k\,P_0}{k\,e^{-k(t-t_0)}} = P_0 e^{k(t-t_0)}.$$

b. Since $(k - b\,P_0)e^{-k(t-t_0)} + b\,P_0 \longrightarrow k\,e^{-k(t-t_0)}$ as $b \longrightarrow 0$, we have

$$\lim_{b \to 0} P(t) = \lim_{b \to 0} \frac{k\,P_0}{\left(k - b\,P_0\right)e^{-k(t-t_0)} + b\,P_0} = \frac{k\,P_0}{k\,e^{-k(t-t_0)}} = P_0 e^{k(t-t_0)}.$$

5. Suppose that the logistic equation (6.31) is modified to

$$\frac{dP}{dt} = -P\left(k - b\,P\right), \tag{S.26}$$

where $k > 0$. (This models exponential decay in the presence of a *threshold*.) Assume that $P(t_0) = P_0$.

    a. Find the critical points and discuss their stability.

    b. Show that if $0 \leq P_0 < k/b$, then $P(t)$ tends to zero when $t \longrightarrow \infty$.

    c. Show that if $P_0 > k/b$, then $P(t)$ becomes infinite at some positive time.

Solution. **a.** The critical points of (S.26) are clearly 0 and k/b. Furthermore, 0 is stable and k/b is unstable.

    b. The solution of (S.26) is

$$P(t) = \frac{-k\,P_0}{\left(-k + b\,P_0\right)e^{k(t-t_0)} - b\,P_0}$$

$$= \frac{k\,P_0}{\left(k - b\,P_0\right)e^{k(t-t_0)} + b\,P_0}. \tag{S.27}$$

If $0 \leq P_0 < k/b$, then both terms in the denominator of (S.27) are positive and the first tends to ∞ as $t \longrightarrow \infty$. Hence $P(t)$ tends to zero.

c. If $P_0 > k/b$, then the first term in the denominator of (S.27) is negative, while the second term is positive, Thus, the denominator is zero when

$$e^{k(t-t_0)} = \frac{-b\,P_0}{k - b\,P_0}.$$

6. Suppose that the logistic equation is further modified to

$$\frac{dP}{dt} = -P\bigl(k - b\,P\bigr)\bigl(k - c\,P\bigr),$$

where k, b and c are positive constants with $b > c$. (This models exponential decay in the presence of a *double threshold*). Assume that $P(t_0) = P_0$.

    a. Find the critical points and discuss their stability.

    b. Show that if $0 \le P(t_0) < k/b$, then $P(t)$ tends to zero when $t \longrightarrow \infty$.

    c. Show that if $P(t_0) > k/b$, then $P(t)$ tends to b/c when $t \longrightarrow \infty$.

Solution. 0 is a stable critical point, k/b is an unstable critical point and k/c is a stable critical point. In every case the solution tends to the nearest stable critical solution, either $P \equiv 0$ or $P \equiv k/c$, depending on the initial condition.

7. Suppose that the logistic model is modified to

$$\frac{dP}{dt} = P\bigl(k - b\,P\bigr)^2. \tag{S.28}$$

where k and b are positive constants. Assume that $P(t_0) = P_0$.

    a. Find the critical points of (S.28) and classify their stability.

    b. Solve the differential equation (S.28) implicitly.

    c. Find the time t for which $P(t) = 0.9k/b$, if $P_0 = 0.1k/b$.

Solution. **a.** 0 is an unstable critical point and k/b is a semistable critical point.

    b. The implicit solution to the separable equation (S.28) may be found using

```
ODE[{P' == P(k - b P)^2,P[t0] == P0},P,t,
Method->Separable,Form->LogEquation]
```

yielding

```
    1           Log[P]     Log[k - b P]
----------   +  ------  -  ------------   ==
 2                2             2
k   - b k P       k             k

       1                     Log[P0]     Log[k - b P0]
----------   + t - t0 +  ------  -  -------------
 2                           2             2
k   - b k P0                k             k
```

Thus, the implicit solution to (S.28) is

$$\frac{1}{k^2} \log\left(\frac{P(t)}{k - b P(t)}\right) + \frac{1}{k^2 - k b P(t)}$$

$$= t - t_0 + \frac{1}{k^2} \log\left(\frac{P_0}{k - b P_0}\right) + \frac{1}{k^2 - k b P_0}. \qquad (S.29)$$

c. We solve (S.29) for t as follows:

$$\frac{1}{k^2} \log\left(\frac{9}{b}\right) + \frac{10}{k^2} = t + \frac{1}{k^2} \log\left(\frac{1}{9b}\right) + \frac{10}{9k^2}$$

or

$$t = \frac{1}{k^2}\left(\frac{80}{9} + 2\log(9)\right) \approx \frac{13.2833}{k^2}.$$

8. Show that (6.48) implies (6.47).

Solution. Equation (6.48) implies that

$$(k - b P_0)e^{-k\mathbf{L}_d} + b P_0 = \frac{k}{2}.$$

When we solve this equation for $e^{k\mathbf{L}_d}$, we get

$$e^{k\mathbf{L}_d} = \frac{2(k - 2b P_0)}{(k - b P_0)} = \log\left(2 + \frac{2b P_0}{k - 2b P_0}\right). \qquad (S.30)$$

Then (S.30) implies

$$\mathbf{L}_d(P_0) = \mathbf{L}_d = \frac{1}{k} \log\left(2 + \frac{2b P_0}{k - 2b P_0}\right).$$

9. Find the doubling time $\mathbf{L}_d(P_0)$ for the logistic model in the case that $k = 3.06$, $b = 0.02$ and $P_0 = 25.4$.

Solution. From formula (6.47) we have

$$\mathbf{L}_d(25.4) = \frac{1}{3.06} \log\left(2 + \frac{2(0.02)(25.4)}{3.06 - 2(0.02)(25.4)}\right) = 0.089744.$$

10. Suppose that we have three equally spaced, times $t_0 < t_1 < t_2$, that is, $t_2 - t_1 = t_1 = t_0$. We denote the corresponding population values by P_0, P_1 and P_2. Prove that

$$\frac{1}{P_1} < \frac{1}{2}\left(\frac{1}{P_0} + \frac{1}{P_2}\right) \qquad \text{and} \qquad \frac{1}{P_1} < \sqrt{\frac{1}{P_0 P_2}}. \qquad (S.31)$$

Solution. From (6.40) we have

$$\frac{P_2^{-1} - P_1^{-1}}{P_1^{-1} - P_0^{-1}} < 1. \tag{S.32}$$

Since $P_0 < P_1 < P_2$, both numerator and denominator on the left-hand side of (S.32) are negative. When we multiply out (S.32) respecting the signs, we find that

$$\frac{1}{P_1} - \frac{1}{P_0} < \frac{1}{P_2} - \frac{1}{P_1},$$

which is equivalent to the first inequality of (S.31). To prove the second, we use (6.41). The first inequality of (S.31) says that the numerator on the right-hand side of (6.41) is negative. But the k/b is positive; therefore the denominator on the right-hand side of (6.41) must also be negative, which means that

$$\frac{1}{P_1^2} - \frac{1}{P_0}\frac{1}{P_2} < 0,$$

proving the second inequality of (S.31).

11. A certain contagious virus is known to die out if it is present in sufficiently small quantitities, but will multiply without bound if enough is present. Assume the model of Exercise 5 with the values $k = 0.14$ and $b = 0.42$. Determine the critical level below which the virus will die out.

   Solution. Equation (S.26) has two critical points, 0 and k/b. The first of these is stable and the second is unstable. For this problem $k/b = 0.33$. If the initial amount of virus is below 0.33, then the virus will die out.

Section 6.4, page 183

1. Consider the (subthreshold) harvesting model governed by

$$\frac{dP}{dt} = P(4 - P) - 3.$$

   Find the extinction time, given that $0 < P(0) < 1$.

   Solution. We write the equation as $dP/dt = P(4 - P) - 3 = (3 - P)(P - 1)$. The extinction time is obtained as a definite integral by writing

$$\frac{dP}{(3 - P)(P - 1)} = dt$$

to obtain

$$T = \int_0^{P(0)} \frac{dP}{(3-P)(1-P)} = \frac{1}{2}\log\left(\frac{3-P(0)}{3(1-P(0))}\right).$$

2. Consider the (threshold) harvesting model governed by

$$\frac{dP}{dt} = P(4-P) - 4.$$

Find extinction time, given that $0 < P(0) < 2$.

Solution. We write the equation as $dP/dt = -(P-2)^2$. The extinction time is obtained as a definite integral by writing

$$T = \int_0^{P(0)} \frac{dP}{(2-P)^2} = \frac{P(0)}{2(2-P(0))}.$$

3. Consider the (subthreshold) harvesting model governed by

$$\frac{dP}{dt} = P(6-P) - 5.$$

   a. Find the critical points and classify their stability.
   b. Solve the equation.

Solution. a. We write the equation as

$$dP/dt = f(P) = (5-P)(P-1),$$

so that the critical points are $P = 1$ and $P = 5$; then $f'(1) = 6 - 2 \times 1 > 0$, so that $P = 1$ is an unstable critical point, whereas $f'(5) = 6 - 2 \times 5 < 0$ and $P = 5$ is a stable critical point.

   b. The solution $P(t)$ is given by the *Mathematica* command

```
ODE[{P' == -(P - 1)(P - 5),P[0] == P0},P,t,
Method->Separable,Form->Explicit,
PostSolution->{Apart}]
```

and the output

```
        4 (-5 + P0)
5 + ─────────────────────
      4 t           4 t
    5 - E    - P0 + E    P0
```

4. Consider the (threshold) harvesting model governed by

$$\frac{dP}{dt} = P(6 - P) - 9.$$

    a. Find the critical points and classify their stability.

    b. Solve the equation.

    c. Find the time t so that $P(t) = 0$ if $P(0) < 3$.

    d. For which values of $P(0) = P_0$ do we have $P(t) = 0$ for some t? What happens to $P(t)$ for other values of $P(0)$?

5. Consider the (superthreshold) harvesting model with

$$\frac{dP}{dt} = P(6 - P) - 13.$$

    a. Show that there are no critical points.

    b. Solve the equation.

    c. Find the time t for which $P(t) = 0$, given that $P(0) = P_0$.

Solution. $P = 3$ is a semistable critical point.

The solution is $P(t) = 3 + \dfrac{P(0) - 3}{1 + t(P(0) - 3)}$. If $P(0) < 3$ then $P(t) = 0$, when $t = \dfrac{P(0)}{3(3 - P(0))}$. If $P(0) > 3$, then $P(t)$ is never zero and $P(t) \longrightarrow 3$ when $t \longrightarrow \infty$.

6. Suppose that the harvesting model is modified to

$$\frac{dP}{dt} = P(k - b\,P) - h\,P,$$

where $0 < h < k$ are constants.

    a. Find the critical points P_-, P_+ and classify their stability.

    b. Show that $P(t) > 0$ for all t if $P(0) > 0$.

Section 6.5, page 192

Plot the actual population together with the exponential and logistic approximations for the populations of each of the following states.

1. Maryland. Use the data

```
marylandpopulation=
{{1700,0.03},{1725,0.07},{1750,0.14},{1775,0.23},
{1800,0.34},{1825,0.44},{1850,0.58},{1875,0.83},
{1900,1.19},{1925,1.50},{1950,2.34},{1975,4.10}}
```

2. Texas. Use the data

```
texaspopulation=
{{1850,0.21},{1875,1.71},{1900,3.05},
{1925,4.85},{1950,7.71},{1975,12.24}}
```

3. Illinois. Use the data

```
illinoispopulation=
{{1850,0.85},{1875,2.95},{1900,4.82},
{1925,6.67},{1950,8.71},{1975,11.15}}
```

Section 6.6, page 199

Solve the following temperature equalization models and find the relaxation time:

1. $\begin{cases} T'(t) = -3(T - 68), \\ T(0) = 40 \end{cases}$

Solution. We apply the general solution formula (6.63) with $T(0) = 40$, $k = 3$ and $T_e = 68$ with the result

$$T(t) = 68 - 28e^{-3t}.$$

The relaxation time is $\tau = 1/3$. The relevant *Mathematica* command is

```
ODE[{T' == -3(T - 68),T[0] == 40},T,t,
Method->FirstOrderLinear,
PlotSolution->{{t,0,1}}]
```

2. $\begin{cases} T'(t) = -k(T - T_e), \\ T(t_0) = 0, \qquad (k,\, T_e \text{ constant}) \end{cases}$

Solution.

We must solve the initial value problem

$$\begin{cases} \dfrac{d}{dt}\left(e^{kt}T(t)\right) = e^{kt}k\,T_e, \\ T(t_0) = T_0. \end{cases} \tag{S.33}$$

The general solution of (S.33) is

$$T(t) = T_0 e^{-k(t-t_0)} + T_e(1 - e^{-k(t-t_0)}). \qquad (S.34)$$

We are given $T(t_0) = T_0 = 0$, so (S.34) reduces to

$$T(t) = T_e(1 - e^{-k(t-t_0)}).$$

The relaxation time is $\tau = 1/k$. The relevant *Mathematica* command is

```
ODE[{T' == -k(T - Te),T[t0] == 0},T,t,
Method->FirstOrderLinear]
```

3. $$\begin{cases} T'(t) = -4(T - 68), \\ \lim_{t \to \infty} e^{4t} T'(t) = -3 \end{cases}$$

Solution. From (6.63) we compute $T'(t) = -k\,T(0)e^{-kt} + k\,T_e e^{-kt}$; thus

$$e^{kt} T'(t) = -k\,T(0) + k\,T_e.$$

From the data of the problem we have $k = 4$ and $T_e = 68$. Hence the limiting condition is

$$-3 = \lim_{t \to \infty} e^{4t} T'(t) = -4T(0) + 4 \times 68 = -4T(0) + 272,$$

which is solved to yield $T(0) = (272 + 3)/4 = 68 + (3/4)$, and the solution

$$T(t) = 68 + \frac{3e^{-4t}}{4}.$$

The relaxation time is $\tau = 1/4$.

4. $$\begin{cases} T'(t) + 4T = 100, \\ T(0) = -5 \end{cases}$$

Solution. We write this in the standard form (6.61) as $T'(t) = -4(T - 25)$, so that we may apply (6.63) with $k = 4$, $T_e = 25$ and $T(0) = -5$ to obtain the solution

$$T(t) = 25 - 30\,e^{-4t}.$$

The relaxation time is $\tau = 1/4$.

```
ODE[{T' + 4T == 100,T[0] == -5},T,t,
Method->FirstOrderLinear]
```

5. A cold glass of water at 35°F reaches a temperature of 42°F after 5 minutes in a room of temperature 68°F. Determine the relaxation time. Find the temperature of the water after 15 minutes after its temperature was 35°F.

Solution. From (6.64) we have

$$e^{-5k} = \frac{42 - 68}{35 - 68} = \frac{26}{33} = 0.7879,$$

so that

$$5k = -\log(0.7879) = 0.2384.$$

hence $k = 0.2384/5 = 0.0477$, and the relaxation time is $\tau = 1/(0.0477) = 20.9722$ minutes. To find the temperature after 15 minutes we use (6.63) to write

$$T(15) = 68 + (35 - 68)e^{-15k} = 68 - 33(.7879)^3$$
$$= 68 - 16.1396 = 51.86°F$$

6. A cup of coffee initially at 200°F cools to 175°F in 3 minutes in a room of temperature 70°F. Find the relaxation time. When will the temperature of the coffee reach 112°F?

Solution. From (6.64) we have

$$e^{-3k} = \frac{175 - 70}{200 - 70} = \frac{105}{130} = 0.8077$$

or

$$3k = -\log(0.8077) = 0.2136.$$

Hence $k = (0.2136)/3 = 0.0712$ and the relaxation time $\tau = 1/(0.0712) = 14.0466$ minutes. To find the time necessary to reach 112°F we use (6.63) to write

$$\frac{112 - 70}{200 - 70} = e^{-kt}$$

and the solution

$$t = -(1/k)\log\left(\frac{112 - 70}{200 - 70}\right) = -\frac{\log(0.3231)}{0.0712} = 15.8708 \text{ minutes.}$$

7. A building is cooled to a temperature of 75°F on a 90°F day, at which time the air conditioner is turned off. Assuming a relaxation time of 4 hours and that the air conditioner is not turned back on, find the temperature at the end of 2 hours. When will the temperature reach 85°F?

Solution. The temperature is obtained from (6.64) as

$$T(2) = 90 + (75 - 90)e^{-2/4} = 90 - 15e^{-0.5} = 90 - 15 \times 0.6065 = 80.902°F$$

To find the time necessary to reach 85°F, we solve (6.64) for t:

$$e^{-t/2} = \frac{85 - 90}{75 - 90} = 0.3333, \qquad t = -2\log(0.3333) = 2.1972 \text{ hours}$$

8. An oven, whose maximum temperature is 400°F, can reach 380°F in 15 minutes. A potato must be cooked for 1 hour between 350°F to 400°F. How long does it take to cook a potato if the oven and the potato are initially at room temperature of 75°F? What can be said if the oven is preheated to 350°F?

Solution. Since the initial temperature is assumed to be 75°F, and the final temperature of the oven is 400°F, the temperature equalization model yields the following initial value problem:

$$\begin{cases} T' = -k(T - 400), \\ T(0) = 75. \end{cases}$$

This first-order initial value problem is easily integrated to give

$$T(t, k) = 400 - 325e^{-kt}. \qquad (S.35)$$

We can solve (S.35) when we incorporate the additional information that the oven temperature will be 380°F after 15 minutes. Therefore,

$$380 = 400 - 325e^{-k15}$$

or

$$k = -\frac{1}{15}\log\left(\frac{20}{325}\right) \approx 0.185873/\text{minute}. \qquad (S.36)$$

We can now determine the time required for the oven to reach 350°F. Using the value of k given by (S.36), we get

$$350 = 400 - 325 \exp(\log(20/325)^{1/15} t). \qquad (S.37)$$

When we solve (S.37) for t, we obtain

$$t = \frac{\log(50/325)}{\log(20/325)^{1/15}} \approx 10.07 \text{ minutes}.$$

Preheating the oven saves 10 minutes.

9. Experimental data suggests that a baked potato will undergo a 10% drop in temperature 20 minutes after it is placed in a room whose temperature is 75°F. If the human palette cannot tolerate food temperatures exceeding 125°F, how long must a person wait before eating the potato?

Solution.

Since the initial temperature is assumed to be 400°F and the final temperature is the ambient temperature of the room or 75°F, the temperature equalization model yields the following initial value problem:

$$\begin{cases} T' = -k(T - 75), \\ T(0) = 400. \end{cases}$$

This first-order initial value problem is easily integrated to give

$$T(t, k) = 75 + 325e^{-kt}. \tag{S.38}$$

We can solve (S.38) for k when we incorporate the additional information that the temperature of the potato will be 160°F = 40% of 400°F after 10 minutes. Therefore,

$$k = -\frac{1}{10} \log(85/325) \approx 0.134117/\text{minute}.$$

We may now determine the time required for the potato to cool to 125°F. Using the value of k given by (S.36), we get

$$125 = 75 + 325 \exp(\log(85/325)^{1/10} t),$$

When we solve (S.37) for t, we obtain

$$t = \frac{\log(50/325)}{\log(85/325)^{1/10}} \approx 13.9564 \text{ minutes.}$$

10. A barn has no heating or cooling and is subject to the daily variation of external temperature, which varies from 40°F to 80°F according to the formula

$$T_e(t) = 60 + 20 \cos\left(\frac{\pi t}{12}\right),$$

where t is measured in hours. Assuming a relaxation time of 3 hours, find the maximum and minimum steady-state temperatures. If the maximum external temperature occurs at 2 PM, when does the maximum temperature of the barn occur?

Solution. We apply (6.71) with the values $k = 1/3$, $A_0 = 60$, $A_1 = 20$ and $\omega = \pi/12 = 0.2616$. This yields

$$T_{max} = 60 + \frac{20 \times (1/3)}{\sqrt{(0.2616)^2 + (0.3333)^2}} = 78.42°F,$$

or

$$T_{max} = 60 - \frac{20 \times (1/3)}{\sqrt{(0.2616)^2 + (0.3333)^2}} = 41.57°F$$

To find the time of occurrence, we solve $\tan(\omega T) = \omega/k = 0.2616/0.3333 = 0.7848$, to obtain

$$T = \frac{\arctan(\omega/k)}{\omega} = \frac{\arctan(0.8748)}{0.2616} = \frac{0.6654}{0.2616} = 2.5438 \text{ hours}$$

If the maximum external temperature occurs at 2:00 PM, then the maximum temperature inside the barn occurs at 4:33 PM.

11. Two friends sit down for a cup of coffee. The first person pours the cream into her coffee immediately. The second person waits t seconds before pouring cream into her coffee. Which cup of coffee is hotter at this time, the pre-mixed cup or the just-mixed cup? [Hint: Assume that the coffee and cream both have the same relaxation time and that when mixed the new temperature instantaneously becomes the weighted average of the two temperatures.]

Solution. From (6.63) we have in general for either cup

$$T(t) = T_e + \left(T(0) - T_e\right)e^{-kt}, \tag{S.39}$$

where T_e is room temperature. Assuming M_1 ounces of coffee at temperature T_1 and M_2 ounces of cream at temperature T_2, the mixture assumes the temperature $(M_1 T_1 + M_2 T_2)/(M_1 + M_2)$, so that after t seconds (S.39) shows that the temperature of the pre-mixed cup is

$$T_{pre-mixed}(t) = T_e + \left(\frac{M_1 T_1 + M_2 T_2}{M_1 + M_2} - T_e\right)e^{-kt}. \tag{S.40}$$

On the other hand, for the person who first lets her coffee cool, (S.39) shows that its temperature after t seconds is

$$T_{black\ coffee}(t) = T_e + (T_1 - T_e)e^{-kt},$$

whereas the temperature of the cream is

$$T_{cream}(t) = T_e + (T_2 - T_e)e^{-kt},$$

and the weighted average of these two temperatures is

$$T_{\text{just mixed}}(t) = \frac{M_1 T_{\text{black coffee}}(t) + M_2 T_{\text{cream}}(t)}{M_1 + M_2}$$

$$= \frac{M_1\left(T_e + (T_1 - T_e)e^{-kt}\right) + M_2\left(T_e + (T_2 - T_e)e^{-kt}\right)}{M_1 + M_2}. \qquad \text{(S.41)}$$

We rewrite (S.41), keeping separate track of the coefficients of T_e and e^{-kt} to obtain

$$T_{\text{just mixed}}(t) = T_e + \frac{M_1(T_1 - T_e) + M_2(T_2 - T_e)}{M_1 + M_2}e^{-kt}$$

$$= T_e + \left(\frac{M_1 T_1 + M_2 T_2}{M_1 + M_2} - T_e\right)e^{-kt},$$

which is the same as (S.40). Therefore, both the pre-mixed cup and the just-mixed cup have the same temperature.

12. **a.** It is required to find the time of a person's death, given measurements of the temperature of the corpse at two later times. Assuming that $t_1 < t_2$ and we are given $T(t_1) = T_1$ and $T(t_2) = T_2 < T_1$, find the time $t < t_1$ for which $T(t) = 98.6°F$.

b. Illustrate the calculations of part **a** in the case that $t_1 = 1$ hours, $t_2 = 2$ hours, $T_1 = 95°F$, $T_2 = 92°F$ and room temperature is assumed to be $T_e = 68°F$.

Solution. We first find the value of k from the given data. To do this, we apply (6.64) twice:

$$T_1 = T(t_1) = T_e + (T_0 - T_e)e^{-kt_1},$$

$$T_2 = T(t_2) = T_e + (T_0 - T_e)e^{-kt_2}.$$

Subtracting T_e from both equations and dividing one by the other gives

$$\frac{T_1 - T_e}{T_2 - T_e} = e^{-k(t_1 - t_2)},$$

from which

$$k(t_2 - t_1) = \log\left(\frac{T_1 - T_e}{T_2 - T_e}\right). \qquad \text{(S.42)}$$

Equation (S.42) gives the required value of k. To find the time t for which $T(t) = 98.6$, we again use (6.64) to write

$$e^{-k(t - t_1)} = \frac{T(t) - T_e}{T_1 - T_e} = \frac{98.6 - T_e}{T_1 - T_e},$$

which can be solved for $t - t_1$ as

$$k(t - t_1) = \log\left(\frac{T_1 - T_e}{98.6 - T_e}\right). \tag{S.43}$$

The value of k obtained in (S.42) can be inserted into (S.43) to solve for the desired value of t.

b. In case $t_1 = 1$, $t_2 = 2$, $T_1 = 95$, $T_2 = 92$ and $T_e = 68$, Equation (S.42) becomes

$$k = \log\left(\frac{95 - 68}{92 - 68}\right) = \log(1.1250) = 0.1177$$

and (S.43) gives

$$0.1177(t - t_1) = \log\left(\frac{95 - 68}{98.6 - 68}\right) = \log(0.8824) = 0.1177,$$

so that $t = t_1 - 0.1252/0.1177 = t_1 - 1.0637$; the death occurred approximately one hour and 4 minutes before the first temperature reading.

Chapter 7

Section 7.1, page 212

1. Suppose that a particle moves in a constant force field F_0 with no frictional forces. Find the velocity $v(t)$ and the position $y(t)$ for all later times t.

   Solution. Since the force is constant, by Newton's Second Law (7.1) the motion of the particle is governed by the differential equation $m\,y'' = F_0$, equivalently $y'' = v' = F_0/m$. The initial conditions are $y(0) = Y_0$ and $y'(0) = v(0) = V_0$. Integrating once, we obtain

$$y'(t) = v(t) = V_0 + \int_0^t v'(s)\,ds = V_0 + \frac{F_0 t}{m}.$$

A second integration gives $y(t) = Y_0 + V_0\,t + \dfrac{F_0 t^2}{2m}$.

2. Suppose that a particle moves in a constant force field with a restoring force proportional to the negative of the velocity, and suppose that the initial velocity V_0 is positive.

   a. Find formulas for the velocity $v(t)$ and the position $y(t)$.

   b. Show that $v(t) < \hat{v}(t)$ and $y(t) < \hat{y}(t)$, where $\hat{v}(t)$ and $\hat{y}(t)$ are obtained from Exercise 1.

   Solution. (i) In this case the force is given by $F = F_0 - k\,v$ with $k > 0$, so that the differential equation of motion is

$$m\,y'' = F_0 - k\,y'. \tag{S.44}$$

We consider (S.44) to be a first-order equation in y'. The initial conditions are $y(0) = Y_0$ and $y'(0) = v(0) = V_0$. Integrating (S.44) once with the integrating factor $e^{kt/m}$ gives

$$v' + \frac{k\,v}{m} = \frac{F_0}{m}, \qquad \text{or} \qquad (v\,e^{kt/m})' = \frac{F_0}{m}e^{kt/m},$$

which we integrate as

$$v(t)e^{kt/m} - V_0 = \int_0^t \frac{F_0}{m}e^{ks/m}\,ds = \frac{F_0\,m}{m\,k}(e^{kt/m} - 1).$$

Hence

$$v(t) = V_0 e^{-kt/m} + \frac{F_0}{k}(1 - e^{-kt/m}).$$

Integrating once more gives

$$y(t) = Y_0 + \int_0^t v(s)\, ds = Y_0 + V_0 \frac{m}{k}(1 - e^{-kt/m}) + \frac{F_0}{k}\left(t - \frac{m}{k}(1 - e^{-kt/m})\right).$$

(ii) For $x > 0$ we have $1 - x < e^{-x} < 1$. (The negative exponential function is less than one for positive arguments, but lies above its tangent line at $(0,1)$). Therefore, $e^{-kt/m} < 1$ and $1 - e^{-kt/m} < kt/m$ for all $t > 0$. Hence

$$v(t) < V_0 + \frac{F_0}{m}t = \hat{v}(t).$$

To obtain the corresponding inequality for $y(t)$, note that for $x > 0$ we have

$$1 - e^{-x} = \int_0^x e^{-t}\, dt > \int_0^x (1 - t)\, dt = x - \frac{x^2}{2}.$$

Thus

$$x - (1 - e^{-x}) < x^2/2 \qquad\qquad\qquad (\text{S.45})$$

for $x > 0$. Applying (S.45) to the formula for $y(t)$ with $x = kt/m$, we have

$$t - \frac{m}{k}(1 - e^{-kt/m}) < \frac{kt^2}{2m},$$

$$y(t) < Y_0 + t V_0 + \frac{F_0 t^2}{2m} = \hat{y}(t).$$

3. Suppose that a fully equipped parachutist weighing 175 pounds falls from a height of 6000 feet and opens the parachute after 12 seconds of free fall. Assume an air resistance of $0.8v$ with the parachute closed and an air resistance of $10v$ with the parachute open, where v denotes the velocity.

    a. Find the speed of fall and the distance fallen when the parachute opens.

    b. What is the limiting velocity after the parachute opens?

    c. What is the total time of descent?

Solution. This problem is solved in two separate parts. Before the parachute opens the motion is governed by the differential equation $m\, y'' = m\, g - k\, v$, where $m\, g = 175$ and $k = 0.8$, with the initial conditions $y(0) = 0$, and $y'(0) = 0$. After the parachute opens the motion is governed by a differential equation of the same form with $m\, g = 175$ and $k = 10$. The new initial conditions are obtained from the position and velocity found in the first part at $t = 12$.

a. Before the parachute opens we have $0 \le t \le 12$ and $m\, v' = m\, g - k\, v$; this is solved as in Exercise 2 to obtain

$$v(t) = \frac{m\, g}{k}(1 - e^{-kt/m})$$

$$= (175/0.8)(1 - e^{-0.8 \times 32t/175})$$

$$= 218.75(1 - e^{-0.1463\, t}),$$

and

$$y(t) = \frac{m\, g}{k}(t - (m/k)(1 - e^{-kt/m})$$

$$= 40(t - 6.836(1 - e^{-0.1463\, t}).$$

Hence

$$v(12) = 218.75(1 - e^{-1.7556}) \approx 180.95 \text{ feet/second,}$$

and

$$y(12) = 218.75(12 - 6.836(1 - e^{-1.7556}) \approx 1388 \text{ feet.}$$

b. After the parachute opens we have $t \ge 12$, and we must solve $m\, v' + k\, v = m\, g$ with $k = 10$ and the initial condition $v(12) = 180.95$ feet/second.

$$v(t) = v(12)e^{-k/m(t-12)} + \frac{m\, g}{k}\bigl((1 - e^{-k(t-12)/m}\bigr),$$

so that the limiting velocity is

$$\lim_{t \to \infty} v(t) = \frac{m\, g}{k} = 175/10 = 17.5 \text{ feet/second.}$$

To find the total time of fall, we note that for $k\, t/m$ very large the exponential terms can be neglected in comparison with the other terms. The exact solution is

$$y(t) = y(12) + v(12)(m/k)(1 - e^{-k(t-12)/m})$$

$$+ \frac{m\, g}{k}(t - 12 - (m/k)(1 - e^{-k(t-12)/m})).$$

We have

$$y(t) \approx y(12) + v(12)(m/k) + \frac{m\, g}{k}(t - 12 - m/k)$$

The total time of fall T is obtained by setting $y(T) = 6000$; thus,

$$6000 = y(12) + v(12)(m/k)(1 - e^{-k(T-12)/m})$$
$$+ \frac{m\,g}{k}(T - 12 - (m/k)(1 - e^{-k(T-12)/m}))$$

$$\approx y(12) + v(12)(m/k) + \frac{m\,g}{k}(t - 12 - m/k).$$

Thus

$$6000 - y(12) - v(12)(m/k) \approx (m\,g/k)(T - 12 - \frac{m}{k})$$

and

$$T - 12 - m/k = (k/(m\,g))\big(6000 - y(12) - v(12)(m/k)\big).$$

Recalling that $k = 10$, we have

$$\frac{m\,g}{k} = 17.5, \qquad m/k = 175/32/10 = 0.5468$$

and

$$v(12)m/k = 180.95 \times 0.5468 = 98.9435.$$

Thus, we get

$$T \approx 12 + 0.5468 + (6000 - 1388 - 98.94)/17.5$$

$$\approx 12 + 0.5468 + 257.886$$

$$\approx 270.443 \text{ seconds.}$$

4. A particle moves in a force field which varies sinusoidally with time according to

$$F(t) = F_0 + F_1 \cos(\omega t - \alpha),$$

where F_0, F_1 and α are constants. Assume further that there is a restoring force which is negatively proportional to the velocity. Find the velocity $v(t)$ and the position $y(t)$. Show that $\lim_{t \to \infty} v(t)$ does not exist, but the limit of $y(t)/t$ when $t \longrightarrow \infty$ does exist and find it.

Solution. The motion is governed by the differential equation

$$m\frac{dv}{dt} = F_0 + F_1 \cos(\omega t - \alpha) - k\,v,$$

or equivalently

$$\frac{dv}{dt} + \frac{k}{m}v = \frac{1}{m}(F_0 + F_1 \cos(\omega t - \alpha)).$$

This can be solved with the integrating factor $e^{kt/m}$ as

$$(v(t)e^{kt/m})' = \frac{1}{m}e^{kt/m}(F_0 + F_1\cos(\omega t - \alpha))$$

$$v(t)e^{kt/m} = V_0 + \frac{1}{m}\int_0^t e^{ks/m}(F_0 + F_1\cos(\omega s - \alpha))\,ds$$

$$= V_0 + \frac{F_0}{k}(e^{kt/m} - 1)$$

$$+ \frac{F_1/m}{(k/m)^2 + \omega^2}\left(\frac{k}{m}(e^{kt/m}\cos(\omega t - \alpha) - 1) + e^{kt/m}\omega\sin(\omega t - \alpha)\right)$$

$$v(t) = V_0 e^{-kt/m} + \frac{F_0}{k}(1 - e^{-kt/m}) +$$

$$\frac{F_1/m}{(k/m)^2 + \omega^2}\left(\frac{k}{m}\cos(\omega t - \alpha) - \frac{k}{m}e^{-kt/m} + \omega\sin(\omega t - \alpha)\right).$$

The position $y(t)$ is given by a further integration as

$$y(t) = Y_0 + \int_0^t v(s)\,ds$$

$$= Y_0 + \frac{V_0 m}{k}(1 - e^{-kt/m}) + \frac{F_0}{k}(t - (m/k)(1 - e^{-kt/m}))$$

$$+ \frac{F_1/m}{(k/m)^2 + \omega^2}\left((k/m\omega)\sin(\omega t - \alpha) - (1 - e^{-kt/m}) + (1 - \cos(\omega t - \alpha))\right)$$

When $t \longrightarrow \infty$ the terms $\sin(\omega t - \alpha)$ and $\cos(\omega t - \alpha)$ do not have limits, therefore $\lim_{t\to\infty} v(t)$ does not exist. To obtain the limiting average velocity, we note that

$$\lim_{t\to\infty}\frac{\cos(\omega t - \alpha)}{t} = \lim_{t\to\infty}\frac{\sin(\omega t - \alpha)}{t} = \lim_{t\to\infty}\frac{1}{t} = \lim_{t\to\infty}\frac{e^{-kt/m}}{t} = 0,$$

so that the limit is obtained as the coefficient of t, namely

$$\lim_{t\to\infty}\frac{y(t)}{t} = \frac{F_0}{k}.$$

5. Consider the model upward motion of a body of mass m travelling in the presence of air resistance. Assume the motion is described by the nonlinear differential equation

$$m\frac{dv}{dt} = -mg - kv^2, \qquad v > 0 \qquad\qquad (S.46)$$

where m, g and k are positive constants.

    a. Solve the given differential equation with $v(0) = V_0 > 0$.

    b. If $v(0) = V_0 > 0$, find the time t for which $v(t) = 0$.

    c. If $v(0) = V_0 > 0$, find the total height to which the body travels. [Hint: Use (7.11).]

Solution. **a.** We use **ODE** to solve (S.46), as follows:

```
ODE[{m v'/(m g  + k v^2)  == -1,v[0]==V0}, v,t,
Method->Separable,Form->Explicit]
```

$$
-\left(\frac{\text{Sqrt[g] Sqrt[m] Tan}\left[\dfrac{\text{Sqrt[g] Sqrt[k] t}}{\text{Sqrt[m]}} - \text{ArcTan}\left[\dfrac{\text{Sqrt[k] V0}}{\text{Sqrt[g] Sqrt[m]}}\right]\right]}{\text{Sqrt[k]}}\right)
$$

Thus, the solution is

$$
v(t) = \frac{1}{\sqrt{k/(m\,g)}} \frac{V_0\sqrt{k/(m\,g)} - \tan(t\sqrt{kg/m})}{1 - V_0\sqrt{k/(m\,g)}\tan(t\sqrt{kg/m})}. \tag{S.47}
$$

This solution is only valid as long as $v(t) > 0$.

    b. To find the time T at which $v(T) = 0$, we solve the equation (S.47), noting that the numerator must be zero in this case; thus

$$
V_0\sqrt{k/(m\,g)} - \tan(T\sqrt{kg/m}) = 0,
$$

or

$$
T = \sqrt{m/kg}\ \arctan(V_0\sqrt{k/(m\,g)}).
$$

    c. To obtain the total height y_{max}, we write the equation (S.46) as

$$
\frac{dy}{dv} = -\frac{m\,v}{m\,g + k\,v^2}. \tag{S.48}
$$

Equation (S.48) can be solved with **ODE** using the command

```
xxx = ODE[{y'== -m v/(m g  + k v^2),y[V0]  == 0},y,v,
       Method->Separable]
```

which results in

```
                   2 m/(2 k)
           (g m + k V0 )
{{y -> Log[-------------------]}}
                   2 m/(2 k)
           (g m + k v )
```

When $y = 0$ we have $v = V_0$, and when $y = y_{max}$ we have $v = 0$. Moreover,

xxx /. v->V0

yields

$$Log[\frac{(g\ m\ +\ k\ V0\ )^{2\ m/(2\ k)}}{(g\ m)^{m/(2\ k)}}]$$

Hence

$$y_{max} = \frac{m}{2k} \log(k\,v^2 + m\,g)\Big|_0^{V_0} = \frac{m}{2k} \log\left(1 + \frac{k\,V_0^2}{m\,g}\right).$$

6. Consider the model downward motion of a body of mass m travelling in the presence of air resistance. Assume the motion is described by the nonlinear differential equation

$$m\frac{dv}{dt} = -m\,g + k\,v^2, \qquad v < 0 \qquad\qquad \text{(S.49)}$$

where m, g and k are positive constants.

    a. Solve the given differential equation with $v(0) = V_0 < 0$.

    b. Find the limit of $v(t)$ when $t \longrightarrow \infty$

Solution. **a.** We write (S.49) in the form

$$\frac{dv}{dt} = -g - \frac{k}{m}v^2.$$

This differential equation is a separable equation which can be solved by writing

$$\frac{dv}{g - (k/m)v^2} = -dt$$

and applying the method of partial fractions. Thus

$$\frac{1}{2\sqrt{g}} \left(\frac{1}{\sqrt{g} + v\sqrt{k/m}} + \frac{1}{\sqrt{g} - v\sqrt{k/m}}\right) dv = -dt.$$

When we integrate from 0 to t, each of the integrals on the left can be expressed in terms of a logarithm, whereas the right-hand side is simply $-t$; thus, we obtain

$$\frac{1}{2}\sqrt{m/kg}\left(\log\left|\sqrt{g} + v\sqrt{k/m}\right| - \log\left|\sqrt{g} - v\sqrt{k/m}\right|\right)\Big|_{V_0}^{v(t)} = -t,$$

or

$$\log\left|\frac{\sqrt{g}+v(t)\sqrt{k/m}}{\sqrt{g}-v(t)\sqrt{k/m}}\right| = \log\left|\frac{\sqrt{g}+V_0\sqrt{k/m}}{\sqrt{g}-V_0\sqrt{k/m}}\right| - 2t\sqrt{kg/m}.$$

Taking the exponentials of both sides yields

$$\frac{\sqrt{g}+v(t)\sqrt{k/m}}{\sqrt{g}-v(t)\sqrt{k/m}} = e^{-2t\sqrt{kg/m}}\left(\frac{\sqrt{g}+V_0\sqrt{k/m}}{\sqrt{g}-V_0\sqrt{k/m}}\right).$$

This fraction is solved for $v(t)$ to obtain

$$v(t)\sqrt{k/m} = \frac{\sqrt{g}(\sqrt{g}+V_0\sqrt{k/m})e^{-2t\sqrt{kg/m}} - \sqrt{g}(\sqrt{g}-V_0\sqrt{k/m})}{(\sqrt{g}+V_0\sqrt{k/m})e^{-2t\sqrt{kg/m}} + (\sqrt{g}-V_0\sqrt{k/m})}$$

b. When $t \longrightarrow \infty$ the exponential terms tend to zero, and we have

$$\sqrt{\frac{k}{m}}\lim_{t\to\infty} v(t) = \frac{-\sqrt{g}(\sqrt{g}-V_0\sqrt{k/m})}{(\sqrt{g}-V_0\sqrt{k/m})} = -\sqrt{g}.$$

Thus

$$\lim_{t\to\infty} v(t) = -\sqrt{\frac{mg}{k}}.$$

7. The models described in Exercises 5 and 6 can be subsumed in the form

$$m\frac{dv}{dt} + k\,v|v| = -m\,g \tag{S.50}$$

a. Write (S.50) in the form $dv/dt = F(v)$.

b. Show that $F(v) < 0$ for $v > -\sqrt{mg/k}$, while $F(v) > 0$ for $v > \sqrt{mg/k}$.

c. Using the theory developed in Section 5.6, show that (S.50) has exactly one critical point which is stable.

d. Use this information to sketch the graph of the solution of (S.50) for $t \geq 0$

e. Use this information to sketch the graph of the solution of (7.19) for $t \geq 0$ in the following cases: (i) $V_0 > 0$; (ii) $V_0 = 0$; (iii) $-\sqrt{mg/k} < V_0 < 0$; (iv) $V_0 < -\sqrt{mg/k}$.

Solution. **a.** In this case $F(v) = -g - k\,v|v|/m$.

b. We consider separately three cases. If $v < -\sqrt{m\,g/k}$, then $|v| = -v$ and $v^2 > m\,g/k$, so that

$$
\begin{aligned}
F(v) &= -g - k\,v|v|/m \\
&= -g + k\,v^2/m \\
&> -g + k\frac{(m\,g/k)}{m} \\
&= -g + g \\
&= 0.
\end{aligned}
$$

Similarly, if $0 > v > -\sqrt{m\,g/k}$ then $v^2 < m\,g/k$ and

$$
\begin{aligned}
F(v) &= -g + k\,v^2/m \\
&< -g + k\frac{(m\,g/k)}{m} \\
&= -g + g \\
&= 0.
\end{aligned}
$$

Finally, if $v \geq 0$ then

$$
\begin{aligned}
F(v) &= -g - -k\,v|v|/m \\
&= -g - k\,v^2/m \\
&\leq -g \\
&< 0.
\end{aligned}
$$

c. From part **b.** we see that $F(v) \neq 0$ unless $v = v_{\text{crit}} = -\sqrt{m\,g/k} < 0$, in which case $F'(v) = k\,v/m < 0$, hence the critical point v_{crit} is stable.

8. It is observed that a raindrop measuring one sixteenth of an inch in diameter falls with a limiting velocity of 15 miles per hour. Assuming a model of the form (7.4), find the frictional constant k in this model. (Recall that the density of water is 62.4 pounds/foot^3.)

Solution. Newton's second law for this model is written

$$
m\frac{dv}{dt} + k\,v = -m\,g.
$$

We eliminate the time variable by using the transformations (7.12) and (7.13), obtaining

$$
m\,v\frac{dv}{dy} + k\,v = -m\,g,
$$

which is the separable equation

$$\frac{dv}{dy} = -\frac{(m\,g + k\,v)}{m\,v}$$

or equivalently

$$\frac{m\,v\,dv}{m\,g + k\,v} = -dy.$$

The left-hand side can be integrated by the method of partial fractions to obtain

$$\frac{m}{k}\left(v - \frac{m\,g}{k}\log|m\,g + k\,v|\right) = -y + C.$$

The constant of integration is obtained by noting that when $y = 0$ we have $v = V_0$, thus

$$C = \frac{m}{k}\left(V_0 - \frac{m\,g}{k}\log|m\,g + k\,v_0|\right).$$

At the maximum height y_{max} we must have $v = 0$, hence

$$y_{max} = C + \frac{m^2 g}{k^2}\log(m\,g) = \frac{m\,V_0}{k} + \frac{m^2 g}{k^2}\log\left(\frac{1}{1 + k\,v_0/m\,g}\right).$$

b. In this case we have

$$y_{max} = \frac{(0.25)(20)}{1/30} + \left(\frac{0.25}{1/30}\right)^2 (9.8)\log\left(\frac{1}{1 + 20(1/30)/(9.8 \times 0.25)}\right)$$

$$= 150 - (56.26)(9.8)(0.24) = 17.7 \text{ meters}.$$

c. To find the time at which the ball hits the ground, we rewrite the equation from part **a** as

$$\frac{m}{k}\left(v - \frac{m\,g}{k}\log|m\,g + k\,v|\right) = -y + \frac{m}{k}\left(V_0 - \frac{m\,g}{k}\log|m\,g + k\,v_0|\right).$$

We must solve for v, given $y = 0$, $V_0 = 20$, $m = 0.25$, $k = 1/30$, $g = 9.8$. Since we are not interested in the solution $v = 20$, we start the search at a negative value. This is done by typing

```
FindRoot[{v-(0.25)(9.8)30  Log[(0.25)(9.8) + v/30] ==
          20-(0.25)(9.8)30  Log[(0.25)(9.8) + 20/30]},
          {v,-10}]
```

to which *Mathematica* responds with

```
{v -> -16.9217}
```

Hence the object reaches the ground at a speed of less than 17 meters/second.

9. A ball of mass m is thrown upward with initial velocity V_0 and experiences a frictional force proportional to the velocity, with proportionality constant k.

   a. Find the formula for the height y as a function of velocity.

   b. Find the maximum height y_{max} attained by the ball.

   c. Illustrate the result of part b in case $m = 0.25$ kilograms, $V_0 = 20$ meters/second, $k = 1/30$ kilograms/second.

   d. Using the values in part c, find the speed at which the ball hits the ground. [Hint: The solution will involve a transcendental equation, which can be solved using *Mathematica*'s command **FindRoot**.]

   e. Using the values from part c, plot the height as a function of the velocity.

Solution. **a.** Newton's second law for this model is written

$$m\frac{dv}{dt} + kv = -mg.$$

We eliminate the time variable by using the transformations (7.12) and (7.13) to obtain

$$mv\frac{dv}{dy} + kv = -mg,$$

or equivalently

$$\frac{mvdv}{mg + kv} = -dy. \tag{S.51}$$

We can integrate (S.51) by the method of partial fractions to obtain

$$\frac{m}{k}\left(v - \frac{mg}{k}\log(mg + kv)\right) = -y + C. \tag{S.52}$$

The constant of integration in (S.52) is found by noting that $v = V_0$ when $y = 0$. Thus

$$C = \frac{m}{k}\left(V_0 - \frac{mg}{k}\log(mg + kV_0)\right),$$

and from (S.52) we get

$$y = \frac{m}{k}\left(v - \frac{mg}{k}\log(mg + kv)\right) - \frac{m}{k}\left(V_0 - \frac{mg}{k}\log(mg + kV_0)\right). \tag{S.53}$$

This formula can also be obtained with *Mathematica* in two steps. First, we put

```
xxx = ODE[{m v v'/(m g + k v) == -1,v[0] == V0},v,y,
Method->Separable,Form->Equation]
```

whose output is

```
k v - g m Log[g m + k v]        y    k V0 - g m Log[g m + k V0]
------------------------- == -(-) + -------------------------
           2                    m               2
          k                                     k
```

Then from

```
zzz = Solve[xxx,y]
```

we get

```
       m (k v - k V0 - g m Log[g m + k v] + g m Log[g m + k V0])
{{y -> -(-----------------------------------------------------------)}}
                                  2
                                 k
```

b. At the maximum height y_{max} we must have $v = 0$; hence (S.53) implies that

$$y_{max} = -\frac{m V_0}{k} - \frac{m^2 g}{k^2} \log\left(\frac{m g}{m g + k V_0}\right).$$

The relevant *Mathematica* command

```
zzz /. {y->ymax,v ->0}
```

leads to

```
          m (-(k V0) - g m Log[g m] + g m Log[g m + k V0])
{{ymax -> -(-------------------------------------------------)}}
                              2
                             k
```

c. In this case we have

$$y_{max} = \frac{0.25 \times 20}{1/30} - \left(\frac{0.25}{1/30}\right)^2 \times 9.8 \times \log\left(\frac{9.8 \times 0.25}{9.8 \times 0.25 + (1/30)20}\right)$$

$$= 150 - (56.25 \times 9.8 \times 0.24) = 17.3273 \text{ meters.}$$

The same result can be obtained using

```
y /. zzz /. {m->0.25,k->1/30,V0->20,g->9.8,v->0}
```

with the output

```
{17.3273}
```

d. To find the time at which the ball hits the ground, we must solve (S.53) for v, given that $y = 0$, $V_0 = 20$, $m = 0.25$, $k = 1/30$ and $g = 9.8$. A numerical solution is called for, so we use

```
FindRoot[Evaluate[xxx /.
{y->0,m->0.25,k->1/30,V0->20,g->9.8}],{v,-10}]
```

to which *Mathematica* responds with

```
{v -> -16.9217}
```

Hence the object reaches the ground at a speed of slightly less than 17 meters/second.

e. We define the height function in *Mathematica* by

```
height[v_]:=
y /. zzz /. {m->0.25,k->1/30,V0->20,g->9.8}
```

Then typing **height[v]** shows us the explicit formula:

$$\{-225. \ (2.11841 + \frac{v}{30} - 2.45 \ Log[2.45 + \frac{v}{30}])\}$$

Part **d** suggests that we plot up to 17, so we use

```
Plot[height[v]//Evaluate,{v,-20,22}]
```

obtaining the plot

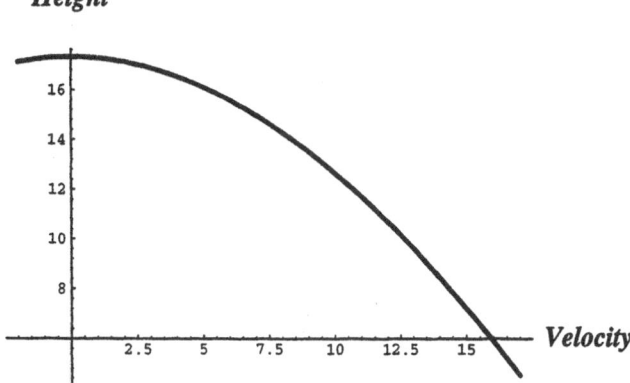

Height as a function of velocity

Section 7.2, page 217

1. Compute the acceleration due to gravity and escape velocity for each planet and the moon.

Solution.

Planet	Acceleration due to Gravity	Escape Velocity
Mercury	$g_{Mercury}$ = 3.70 meters/second²	$V_{Mercury}$ = 4.25 kilometers/second
Venus	g_{Venus} = 8.76 meters/second²	V_{Venus} = 10.3 kilometers/second
Earth	g_{Earth} = 9.80 meters/second²	V_{Earth} = 11.2 kilometers/second
Mars	g_{Mars} = 3.72 meters/second²	V_{Mars} = 5.03 kilometers/second
Jupiter	$g_{Jupiter}$ = 24.8 meters/second²	$V_{Jupiter}$ = 59.6 kilometers/second
Saturn	g_{Saturn} = 10.4 meters/second²	V_{Saturn} = 35.5 kilometers/second
Uranus	g_{Uranus} = 8.87 meters/second²	V_{Uranus} = 21.3 kilometers/second
Neptune	$g_{Neptune}$ = 11.1 meters/second²	$V_{Neptune}$ = 23.5 kilometers/second
Pluto	g_{Pluto} = .651 meters/second²	V_{Pluto} = 1.22 kilometers/second
Moon	g_{Moon} = 1.62 meters/second²	V_{Moon} = 2.38 kilometers/second

2. Suppose that a projectile is sent into space from Earth at the escape velocity $V_{Earth} = \sqrt{2g\,R}$. Show that $y(t)/t$ tends to zero when $t \longrightarrow \infty$, but that the limit

$$\lim_{t \to \infty} \frac{y(t)}{t^{2/3}}$$

exists and find its value.

Solution. In this case we can apply the result from formula (7.30)

$$y(t) = \left(\frac{3\sqrt{2g}}{2} R\,t + R^{3/2} \right)^{2/3} - R$$

When we divide by $t^{2/3}$ the second term tends to zero when $t \to \infty$. For the first term we have

$$\frac{\left((3\sqrt{2g}/2)\mathbf{R}\,t + \mathbf{R}^{3/2}\right)^{2/3}}{t^{3/2}} = \left(\frac{3\sqrt{2g}}{2}\mathbf{R} + \frac{\mathbf{R}^{2/3}}{t}\right)^{2/3}$$

The term in the parenthesis tends to $(3\sqrt{2g}/2)\mathbf{R}$ when $t \to \infty$, therefore the limit is

$$\lim_{t \to \infty} \frac{y(t)}{t^{2/3}} = \left(\frac{3\sqrt{2g}}{2}\mathbf{R}\right)^{2/3}.$$

3. Suppose that a projectile is sent into space from Earth at an initial velocity less than the escape velocity: $V_0 < V_{\text{Earth}}$. Find a formula for the maximum distance reached by the projectile. [Hint: At the maximum the projectile must have zero velocity.]

Solution. We can use the energy equation expressed in (7.26) as

$$v^2 = V_0^2 - 2g\,\mathbf{R} + \frac{2g\,\mathbf{R}^2}{\mathbf{R}+y}.$$

At the maximum distance we have $v = 0$; thus

$$V_0^2 - 2g\,\mathbf{R} + \frac{2g\,\mathbf{R}^2}{\mathbf{R}+y} = 0 \qquad\qquad\qquad \text{(S.54)}$$

When we solve (S.54) for y, we obtain

$$y = \frac{\mathbf{R}V_0^2}{2\mathbf{R}g - V_0^2}.$$

4. Suppose that a projectile is sent against a *constant* gravitational field $y'' = -g$ with initial velocity V_0. Find the maximum distance reached by the projectile and show that it is strictly greater than the maximum distance found in Exercise 3.

In a constant gravitational field starting from $y(0) = 0$, $y'(0) = V_0$ we have $y'' = -g$, $v = y' = V_0 - g\,t$, $y = V_0 t - g\,t^2/2$. At the maximum distance we have $v = 0$, thus, $t = V_0/g$ and $y = V_0(V_0/g) - g(V_0/g)^2/2 = V_0^2/2g$. To compare this with the result of Problem 3, note that $2\mathbf{R}g - V_0^2 < 2\mathbf{R}g$, thus,

$$\frac{\mathbf{R}V_0^2}{2g\mathbf{R} - V_0^2} > \frac{\mathbf{R}V_0^2}{2\mathbf{R}g} = \frac{V_0^2}{2g}.$$

5. Suppose that a projectile is sent into space from a platform at height $z > 0$ above the surface of the earth. Find the escape velocity.

Solution. In this case the conservation of energy is expressed as

$$\frac{v^2}{2} - \frac{gR^2}{R+y} = \frac{V_0^2}{2} - \frac{gR^2}{R+z}. \tag{S.55}$$

The escape velocity is found by requiring that $v^2 \geq 0$ for all $y \geq 0$. Taking the limit $y \longrightarrow \infty$ in (S.55), we find that

$$\frac{V_0^2}{2} - \frac{gR^2}{R+z} \geq 0 \tag{S.56}$$

The smallest value of V_0 which satisfies (S.56) is obtained from

$$\frac{V_0^2}{2} = \frac{gR^2}{R+z} \tag{S.57}$$

or

$$V_0 = \sqrt{\frac{2gR^2}{R+z}}. \tag{S.58}$$

6. **a.** Suppose that a projectile is sent into space from Earth at an initial velocity greater than the escape velocity: $V_0 > V_{Earth}$. Find the limiting velocity when $t \longrightarrow \infty$.

   b. Illustrate with $R = 4000$ miles and $V_0 = 10$ miles/second.

Solution. **a.** The conservation of energy is expressed as

$$\frac{v^2}{2} - \frac{gR^2}{R+y} = \frac{V_0^2}{2} - gR. \tag{S.59}$$

To find the limiting velocity v_∞ we let $y \longrightarrow \infty$ to obtain

$$\frac{v_\infty^2}{2} = \frac{V_0^2}{2} - gR. \tag{S.60}$$

Equation (S.60) is solved to obtain

$$v_\infty = \sqrt{V_0^2 - 2gR}.$$

b. If $R = 4000$ miles and $V_0 = 10$ miles/second, then $g = 32/5280 = 0.006$ miles/second^2 and

$$v_\infty = \sqrt{10^2 - 2(0.006)(4000)} = \sqrt{100 - 48} = 7.211 \text{ miles/second.}$$

Section 7.3, page 224

Solve the following problems involving LR circuits, assuming that L, R, E_0 and ω are constants.

1. $\begin{cases} L\dfrac{dI}{dt} + RI = E_0, \\ I(0) = 0 \end{cases}$

Solution. We use the solution (7.38) with $I_0 = 0$ to obtain $I(t) = \dfrac{E_0(1 - e^{-(R/L)t})}{R}$.

2. $\begin{cases} L\dfrac{dI}{dt} + RI = E_0 \cos(\omega t), \\ I(0) = 0 \end{cases}$

Solution. We apply the integral formula (7.37) with $E(t) = E_0 \cos(\omega t)$ and $I_0 = 0$. Direct computation of the integral directly yields

$$I(t) = \frac{E_0}{R^2 + \omega^2 L^2}\left(R\cos(\omega t) + L\omega\sin(\omega t) - Re^{-(R/L)t}\right).$$

3. $\begin{cases} L\dfrac{dI}{dt} + RI = E_0 \cos(\omega t), \\ I'(0) = 0 \end{cases}$

Solution. When we evaluate (7.34) at $t = 0$ we get

$$E_0 = E(0) = LI'(0) + RI_0 = RI_0.$$

Hence we must have $I_0 = E_0/R$ and the solution

$$I(t) = \frac{E_0\left(R^2\cos(\omega t) + LR\omega\sin(\omega t) + L^2\omega^2 e^{-(R/L)t}\right)}{R(R^2 + \omega^2 L^2)}.$$

4. $\begin{cases} L\dfrac{dI}{dt} + RI = E_0\left(\cos(\omega t)\right)^2, \\ I(0) = 0 \end{cases}$

Solution. We first write $\left(\cos(\omega t)\right)^2 = (1 + \cos(2\omega t))/2$. Then we apply the integral formula (7.37) separately to each of the two terms to obtain

$$I(t) = \frac{Ee^{-(R/L)t}}{2R^3 + 8\omega^2 L^2 R}\left(-2R^2 + e^{(R/L)t}R^2 - 4\omega^2 L^2 + \right.$$

$$\left. 4e^{(R/L)t}\omega^2 L^2 + e^{(R/L)t}R^2\cos(2\omega t) + 2e^{(R/L)t}\omega LR\sin(2\omega t)\right).$$

5. Suppose that an LR circuit contains an inductor with $L = 0.2$ henry and a resistor with $R = 300$ ohms in series with an impressed voltage $E_0 = 5000$ volts.

    a. Find the formula for the current if initially $I(0) = 0$.

    b. Find the limit of $I(t)$ as $t \longrightarrow \infty$.

    c. Find the time t_1 for which $|I(t_1) - I(\infty)| = I(\infty)/2$.

    d. Plot $I(t)$ for $0 \leq t \leq 0.0005$.

Solution.

    a. $I(t) = \dfrac{50(1 - e^{-15000t})}{3}$

    b. $\displaystyle\lim_{t\to\infty} I(t) = \dfrac{50}{3}$

    c. $t_1 = 0.000046$ seconds

6. Let $I(t)$ be the solution of the differential equation of Exercise 2

    a. Show that the limit of $I(t)$ as $t \longrightarrow \infty$ does not exist.

    b. Compute $\displaystyle\lim_{T\to\infty} \dfrac{1}{T} \int_0^T I(t)\, dt$.

    c. Compute $\displaystyle\lim_{T\to\infty} \dfrac{1}{T} \int_0^T I(t)^2 dt$.

Solution.

    a. Since

$$I(t) = \frac{E_0\left(R\,\cos(\omega t) + \omega L\,\sin(\omega t) - R\,e^{-(R/L)t}\right)}{R^2 + \omega^2 L^2},$$

it follows that $I(t)$ cannot approach a limit because of the trigonometric functions in the numerator.

    b. We compute

$$\frac{1}{T}\int_0^T I(t)\, dt = \int_0^T \frac{E_0\left(R\,\cos(\omega t) + \omega L\,\sin(\omega t) - R\,e^{-(R/L)t}\right)}{T(R^2 + \omega^2 L^2)}\, dt$$

$$= \left. \frac{E_0\left((R/\omega)\sin(\omega t) - L\,\cos(\omega t) + L\,e^{-(R/L)t}\right)}{T(R^2 + \omega^2 L^2)} \right|_0^T$$

$$= \frac{E_0\left((R/\omega)\sin(\omega T) - L\,\cos(\omega T) + L\,e^{-(R/L)T}\right)}{T(R^2 + \omega^2 L^2)}$$

Hence $\displaystyle\lim_{T\to\infty} \dfrac{1}{T} \int_0^T I(t)\, dt = 0$.

c. $\displaystyle \lim_{T\to\infty} \frac{1}{T}\int_0^T I(t)^2\, dt = \frac{E_0^2}{2(R^2+\omega^2 L^2)}.$

7. A resistor of 150 ohms and an inductor of 3 henrys are connected in an LR circuit with a 50 cycle voltage source having amplitude 220 volts. Assume the initial current is 1 ampere. Find the formula for the current and plot it from 0 to $\pi/10$.

Solution. We use

```
i[t_] = ODE[{3ii'+ 150 ii == 220 Sin[100 Pi t],
         ii[0] == 1},ii,t,
       Method->FirstOrderLinear,Form->Explicit,
       PostSolution->{Simplify,Cancel}]
```

to get the output

```
                  2          50 t
(15 + 44 Pi + 60 Pi  - 44 E      Pi Cos[100 Pi t] +

    50 t                      50 t        50 t   2
 22 E     Sin[100 Pi t]) / (15 (E     + 4 E     Pi ))
```

Thus

$$I(t) = \frac{(15+44\pi+60\pi^2)e^{-50t} + 22\sin(100\pi\, t) - 44\pi\cos(100\pi\, t)}{15+60\pi^2}.$$

The plot is obtained using

```
Plot[Evaluate[i[t]],{t,0,Pi/10},PlotPoints->500,
PlotRange->All];
```

8. Suppose that an LR circuit containing an inductor with $L = 0.5$henry and a resistor with $R = 500$ ohms is subjected to an external source with $E(t) = 15$ volts. Use **ODE** to plot simultaneously the currents when the initial current assumes the values $I_0 = 0, 0.1, 0.2, 0.3, 0.4, 0.5$.

Solution. We use

```
i[L_,R_,E0_,I0][t_] = ODE[{L ii' + R ii == E0,
                       ii[0] == I0},ii,t,Form->Explicit,
                       Method->FirstOrderLinear]
```

to define the current in terms of the parameters **L, R, E0** and **I0**. Then the simultaneous plot is obtained using

```
Plot[Evaluate[
Table[Evaluate[i[1/2,500,15,I0][t]],{I0,0,0.05,0.01}]],
{t,0,0.005},PlotRange->All]
```

Section 7.4, page 230

1. Suppose that a tank initially contains 5 grams of pure salt in a solution of 1000 cubic centimeters. A brine solution with a salt concentration of 0.03 grams per cubic centimeter is added at the rate of 20 cubic centimeters per second to the tank. The resulting brine solution is removed at the rate of 25 cubic centimeters per second. Find the differential equation for the mass of salt in the tank at time t, solve it and determine the limiting behavior.

Solution. In this case we have $r_{in} = 20$, $r_{out} = 25$, $\rho = 0.03$ and $V_0 = 1000$; thus, $V(t) = 1000 + (20 - 25)t = 1000 - 5t$, and we obtain the initial value problem

$$\begin{cases} x'(t) = 20(0.3) - \dfrac{25x(t)}{1000 - 5t} = 0.6 - \dfrac{25x(t)}{1000 - 5t}, \\ x(0) = 5, \end{cases} \qquad \text{(S.61)}$$

which falls into Case II. First, we rewrite the differential equation of (S.61) as

$$x' - \frac{5x}{t - 200} = \frac{3}{5}. \qquad \text{(S.62)}$$

This first-order differential equation has the integrating factor $(t - 200)^5$; thus, the general solution to (S.62) is

$$x(t) = -\frac{3}{20}(t - 200) + C(t - 200)^5. \qquad \text{(S.63)}$$

Since $x(0) = 5$, the constant of integration in (S.63) is given by $C = 25(200)^{-4}$. Hence the solution to (S.61) is

$$x(t) = 30 + \frac{3t}{20} + \frac{4 \times 10^{10}}{(200 + t)^4}.$$

We also have $b = r_{in} - r_{out} = -5$, so that $V_0/|b| = 200$. Because (S.61) falls into case 2, the solution $x(t)$ does not exist for $t > 200$. Moreover,

$$\lim_{t \to 200} x(t) = x(200) = 0.$$

The *Mathematica* commands to check these computations are

```
x[t_] = ODE[{xx'[t] == 3/5 - 25 xx[t]/(1000 - 5t),
        xx[0] == 5}, xx, t,
        Method->FirstOrderLinear, Form->Explicit]
```

yielding

$$5 + \frac{19\,t}{40} - \frac{t^2}{160} + \frac{t^3}{32000} - \frac{t^4}{12800000} + \frac{t^5}{12800000000}$$

Furthermore

x[200]

yields 0 as expected. The solution can be plotted with

Plot[Evaluate[x[t]],{t,0,200}]

yielding

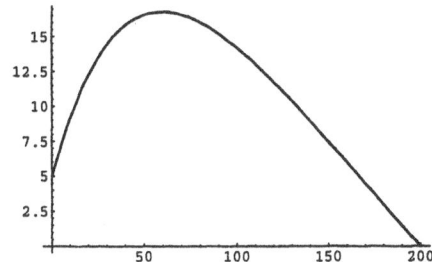

2. Suppose that a tank initially contains 5 grams of pure salt in a solution of 1000 cubic centimeters. A brine solution with a salt concentration of 0.03 grams per cubic centimeter is added at the rate of 25 cubic centimeters per second to the tank. The resulting brine solution is removed at the same rate. Find the differential equation for the mass of salt in the tank at time t, solve it and determine the limiting behavior.

Solution. In this case we have $r_{in} = r_{out} = 25$, $\rho = 0.03$ and $V_0 = 1000$; thus, $V(t) = 1000$ for all t, and we obtain the initial value problem

$$\begin{cases} x'(t) = (20)(0.3) - \dfrac{25x(t)}{1000} = 0.75 - \dfrac{x(t)}{40}, \\ x(0) = 5, \end{cases} \qquad \text{(S.64)}$$

which falls into Case III. First, we rewrite the differential equation of (S.64) as

$$x' + \frac{x}{40} = \frac{3}{4}. \qquad \text{(S.65)}$$

This first-order differential equation has the integrating factor $e^{t/400}$; thus, the general solution to (S.64) is

$$x(t) = 30 + C\,e^{-t/40}. \qquad \text{(S.66)}$$

Since $x(0) = 5$, we find that $C = 25$. Hence the solution to (S.61) is

$$x(t) = 30 + 25e^{-t/40}$$

The limiting behavior is $\lim_{t \to \infty} x(t) = 30$.

The *Mathematica* commands to check these computations are

```
x[t_] = ODE[{xx'[t] == 0.75 - 25 xx[t]/1000,xx[0] == 5},
           xx,t,Method->FirstOrderLinear,Form->Explicit]
```

yielding

```
         25.
30. - ---------
       0.025 t
      E
```

```
Limit[x[t],t->Infinity]
```

3. The pollution of a lake of constant volume is governed by the differential equation

$$\frac{dP}{dt} = \frac{P_I - P}{\tau}, \tag{S.67}$$

where $P(t)$ is the density of pollution, $P_I(t)$ is the input pollution density and the constant τ is the so-called *average retention time of water*, related to the time necessary to drain the entire lake.

    a. Solve (S.67) in case the input pollution density is constant, that is $P_I(t) = K$.

    b. Graph the solutions obtained in part **a** in case of Lake Superior ($\tau = 189$ years), Lake Michigan ($\tau = 30.8$ years), Lake Ontario ($\tau = 7.8$ years) and Lake Erie ($\tau = 2.6$ years). Consider separately the values $K/P(0) = 0, 0.25, 0.50, 0.75$.

    c. Solve (S.67) in case the input pollution density is given by $P_I(t) = K_0 e^{-\alpha t}$ with $0 < \alpha \neq \tau$.

Solution. **a.**

$$\frac{P(t)}{K} = \left(\frac{P(0)}{K_0} - 1\right) e^{-t/\tau} + 1$$

    c.

$$\frac{P(t)}{K_0} = \left(\frac{P(0)}{K_0} - \frac{1}{1 - \alpha\tau}\right) e^{-t/\tau} + \frac{e^{-\alpha t}}{1 - \alpha\tau}$$

4. A tank contains 300 gallons of water. By mistake 400 pounds of salt are poured into the tank instead of 200 pounds. To correct the mistake, the stopper is removed from the bottom of the tank allowing 4 gallons of the brine to flow out each minute. Simultaneously 4 gallons of fresh water per minute are pumped into the tank. The mixture is kept uniform by constant stirring. Write down the initial value problem for the mass of salt, solve it and determine how long it will take for the brine to contain 200 pounds of salt.

Solution. Let $x(t)$ denote the number of pounds of salt in the solution at time t. We have $r_{in} = r_{out} = 4$, $\rho = 0$ and $V_0 = 200$; thus, $V(t) = 200$ for all t, and we obtain the initial value problem

$$\begin{cases} x'(t) = -\dfrac{4x(t)}{300}, \\ x(0) = 400, \end{cases}$$

which falls into Case III. The solution of this initial value problem is $x(t) = 400e^{-t/75}$ When $x(t) = 200$, we have $t = 75 \log(2) \approx 51.986$ minutes.

The corresponding *Mathematica* commands are

```
x[t_] = ODE[{xx'[t] == -4 xx[t]/300,xx[0] == 400},
            xx,t,Method->FirstOrderLinear,Form->Explicit]

N[Solve[x[t] == 200,t]]
```

Section 7.5, page 234

1. Suppose that a truck starts at a point $(a, 0)$ on a desert and moves north. Let a police car start at the point $(0, 0)$ with speed 0 and pursue the truck across the desert. Assume that the police car travels twice as fast as the truck. Show that the police car captures the truck at the point $(a, 2a/3)$.

Solution. Taking $k = 2$ in (7.64) we find that the capture point is $(a, 2a/3)$.

2. Show that the equation for the pursuit curve of Example 7.15 in the case that the speed ratio is 1 is given by

$$y = \frac{a}{4}\left(\left(\frac{a-t}{a}\right)^2 - 1 - 2\log\left(\frac{a-t}{a}\right)\right), \qquad \text{(S.68)}$$

and that the pursuer never catches the pursuer.

Solution. The same steps as those of Example 7.15 can be used, taking $k = 1$. Everything works out the same way until (7.62); this equation becomes

$$y' = \frac{1}{2}\left(\left(\frac{a-t}{a}\right)^{-1} - \left(\frac{a-t}{a}\right)\right) = \frac{1}{2}\left(\frac{a}{a-t} - \frac{a-t}{a}\right). \qquad \text{(S.69)}$$

Integration of (S.69) yields

$$y = \frac{1}{2}\left(-a\log(a-t) + \frac{(a-t)^2}{2a}\right) + C = \frac{a}{4}\left(\left(\frac{a-t}{a}\right)^2 - 2\log(a-t)\right) + C,$$

where C is a constant. Since $y(0) = 0$, we get (S.68). The following *Mathematica* commands also work:

```
xx1 = ODE[{p' == Sqrt[1 + p^2]/(u0 - t),p[0] == 0},p,t,
          Method->Separable,Form->Explicit,CheckSolution->No]

Simplify[Integrate[xx1,{t,0,t}]]
```

The result is

```
-t    t 2     u0 Log[t - u0]     u0 Log[-u0]
-- + ----- - -------------- + -------------
 2   4 u0          2                2
```

3. Equation (7.63) can be used to define a function in *Mathematica* by

```
linearpursuit[a_,k_][t_]:= (a k)/(k^2 - 1)
    + (k(a - t)(a - t)^k^(-1))/(2a^(1/k)(1 + k))
    - (a^(1/k)k(a - t)^(1 - 1/k))/(2(k - 1))
```

Solution. The command

```
Plot[Evaluate[linearpursuit[1,1.2][t]],{t,0,1}]
```

yields the plot

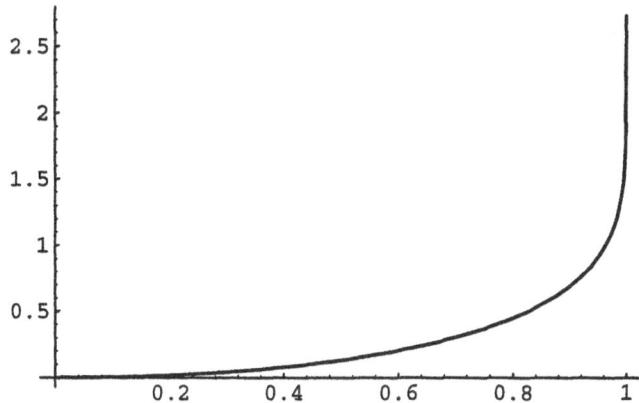

4. Generalize Example 7.15 to the case when the initial velocity of the curve α is nonzero.

Solution. The initial value problem can be solved using

```
yyy = ODE[{k p'==Sqrt[1+p^2]/(a-t),p[0]==p1},p,t,
    Method->DSolve,Form->Explicit]
```

$$\text{Sinh}\left[\frac{k \ \text{ArcSinh}[p1] + \text{Log}[-a] - \text{Log}[-a + t]}{k}\right]$$

Then

```
Integrate[yyy,{t,0,t}]//Simplify
```

yields

$$\frac{a \ k \ (k \ p1 + \text{Sqrt}[1 + p1^2])}{-1 + k^2} + (k \ (a - t)$$

$$\left(-\text{Cosh}\left[\frac{k \ \text{ArcSinh}[p1] + \text{Log}[-a] - \text{Log}[-a + t]}{k}\right] - \right.$$

$$\left. k \ \text{Sinh}\left[\frac{k \ \text{ArcSinh}[p1] + \text{Log}[-a] - \text{Log}[-a + t]}{k}\right]\right)) / (-1 + k^2)$$

5. Solve the differential equation (7.60) using **ODE**. Compare the results using

Method->Separable,

Method->IntegratingFactor,

Method->DSolve.

Solution.

```
ODE[{k p' == Sqrt[1+p^2]/(a-t),p[0] == 0},p,t,
Method->Separable]
```

yields

$$\{\{p \to \text{Sinh}[\frac{\text{Log}[-a] - \text{Log}[-a + t]}{k}]\}\}$$

```
ODE[{k p' == Sqrt[1+p^2]/(a-t),p[0] == 0},p,t,
Method->IntegratingFactor]
```

yields

$$\{\{p \to \text{Sinh}[\frac{\text{Log}[\frac{a}{a - t}]}{k}]\}\}$$

and

```
ODE[{k p' == Sqrt[1+p^2]/(a-t),p[0] == 0},p,t,
Method->DSolve]
```

yields

$$\{\{p \to \text{Sinh}[\frac{\text{Log}[-a] - \text{Log}[-a + t]}{k}]\}\}$$

Chapter 8

Section 8.1, page 239

Reduce the following second-order linear differential equations to the leading-coefficient-unity-form (8.2).

1. $\sin(t)y'' + t\,y' + 3y = \sin(t)$.

   Solution. Dividing by $\sin(t)$ and using the identity $\csc(t) = 1/\sin(t)$ yields $y'' + t\,\csc(t)y' + 3y\csc(t) = 1$

2. $t\,y'' + 4y' + t\,y = 0$

   Solution. Here we divide by t to obtain $y'' + (4/t)y' + y = 0$.

3. $\dfrac{y''}{t} + y' + t\,y = t^3$.

   Solution. In this case we multiply by t and get $y'' + t\,y' + t^2 y = t^4$.

4. $y'' + 4y' + 3y = 0$.

   Solution. This equation is already in leading-coefficient-unity-form; therefore the answer is $y'' + 4y' + 3y = 0$

Use the method of Example 8.1 to solve the following differential equations by elementary calculus.

5. $y'' = 3t^2 + 2t + \sin(t)$.

   Solution. The differential equation can be solved by integrating twice or by using

   ```
   ODE[y'' == 3t^2 + 2t + Sin[t],y,t]
   ```

   resulting in the output

   ```
            3    4
            t    t
   {{y ->   -  + -  + C[1] + t C[2] - Sin[t]}}
            3    4
   ```

   Thus, the solution is

   $$y(t) = \frac{t^4}{4} + \frac{t^3}{3} - \sin(t) + C_1 t + C_2.$$

6. $y'' = t e^t$

Solution. Integration twice by parts yields $y(t) = (t - 2)e^t + C_1 t + C_2$.
   Alternatively:

```
ODE[y'' == t E^t,y,t]
```

$$\{\{y \rightarrow -2 \text{ E}^t + \text{E}^t \text{ t} + \text{C[1]} + \text{t C[2]}\}\}$$

Use the method of Example 8.2 to solve the following differential equations, which involve only y'', y' and t.

7. $y'' + y'/t = 0$

Solution. Here the integrating factor is $\mu(t) = t$, which leads to $t\,y' = C_1$, or equivalently $y' = C_1/t$ yielding $y(t) = C_1 \log(t) + C_2$. Alternatively:

```
ODE[y'' + y'/t == 0,y,t]
```

$$\{\{y \rightarrow \text{C[1]} + \text{C[2] Log[t]}\}\}$$

8. $y'' + 4y' = 17\sin(2t)$

Solution. In this case, the integrating factor is $\mu(t) = \exp(4t)$. This leads to the equation

$$(e^{4t}y')' = 17e^{4t}\sin(2t),$$

which after integrating once, solving for y' and integrating again, yields

$$y(t) = -\frac{17}{8} + e^{-4t}C_1 + C_2 - \frac{34\cos(4t) - 17\sin(4t)}{20}.$$

Using **ODE** we get

```
ODE[y'' + 4y' == 17 Sin[2t],y,t,Method->NthOrderLinear]
```

$$\{\{y \rightarrow -(\frac{17}{8}) + \frac{\text{C[1]}}{4\,\text{t}} + \text{C[2]} - \frac{17\,\text{Cos[2 t]}}{10} - \frac{17\,\text{Sin[2 t]}}{20}\}\}$$

Use the method of Example 8.3 to solve the following differential equations, which involve only y'' and y.

9. $y'' + 9y = 0$

Solution. Here $k = 3$ which leads immediately to $y(t) = C_1 \sin(3t) + C_2 \cos(3t)$.
   Alternatively:

```
ODE[y'' + 9y == 0,y,t]

{{y -> C[2] Cos[3 t] - C[1] Sin[3 t]}}
```

10. $5y'' + y = 0$

Solution. For this problem, $k = 1/\sqrt{5}$ leading directly to

$$y(t) = C_1 \sin(t/\sqrt{5}) + C_2 \cos(t/\sqrt{5}).$$

Alternatively:

```
ODE[5y'' + y == 0,y,t,
Method->SecondOrderLinear]

                  t                        t
{{y -> C[2] Cos[-------] + C[1] Sin[-------]}}
               Sqrt[5]                  Sqrt[5]
```

Section 8.2, page 244

In which intervals can we expect to find a solution to the following initial value problems?

1. $\begin{cases} y'' + y' + y = 0, \\ y(t_0) = Y_0, \ y(t_0) = Y_1 \end{cases}$

Solution. The solution exists on any interval and since this problem is linear, this means that a solution exists everywhere.

2. $\begin{cases} y'' + ty' + y = \sin(t), \\ y(t_0) = Y_0, \ y(t_0) = Y_1 \end{cases}$

Solution. A solution exists on any interval and hence, this means that a solution exists everywhere.

3. $\begin{cases} t\,y'' + t\,y' + y = \sin(t), \\ y(t_0) = Y_0, \ y(t_0) = Y_1 \end{cases}$

Solution. After writing the differential equation in the leading-coefficient-unity-form $y'' + y' + y/t = \sin(t)/t$, we observe that the coefficient of y is not continuous at $t = 0$. Since $\sin(t)/t$ is continuous everywhere, a solution exists on any interval which does not include 0.

4.
$$\begin{cases} t\,y'' + t^2 y = \cos(t), \\ y(t_0) = Y_0, \quad y(t_0) = Y_1 \end{cases}$$

Solution. A solution exists on any interval, since t and $\cos(t)/t$ are continuous everywhere.

5.
$$\begin{cases} \sin(t)y'' + 3y = 0, \\ y(t_0) = Y_0, \quad y(t_0) = Y_1 \end{cases}$$

Solution. A solution exists on any interval which does not include $0, \pi, -\pi, 2\pi, -2\pi, \ldots$, which are the zeros of $\sin(t)$.

6.
$$\begin{cases} \cot(t)y'' + y\,t = 0, \\ y(t_0) = Y_0, \quad y(t_0) = Y_1 \end{cases}$$

Solution. A solution exists on any interval which does not include $\pi/2, -\pi/2, 3\pi/2, -3\pi/2, \ldots$, which are the zeros of $\cot(t)$.

7. Verify that (8.18) is a solution of (8.17).

Solution. Given

$$y(t) = \frac{t\,Y_0}{2t_0} + \frac{t_0 Y_0}{2t} + \frac{t\,Y_1}{2} - \frac{t_0^2 Y_1}{2t},$$

we compute

$$t^2 y'(t) = \frac{t\,Y_0}{t} - \frac{t_0^2 Y_1}{t},$$

$$t\,y'(t) = \frac{t\,Y_0}{2t_0} - \frac{t_0 Y_0}{2t} + \frac{t\,Y_1}{2} + \frac{t_0^2 Y_1}{2t}.$$

Hence, $t^2 y'' + t\,y' - y = 0$. Also, $y(t_0) = Y_0$ and $y'(t_0) = Y_1$.

Section 8.3, page 246

Find a fundamental set of solutions for the following second-order differential equations.

1. $y'' - 3y' = 0$

Solution. Using the method of Example 8.2, we find that $\mu(t) = e^{-3t}$, hence $e^{-3t} y' = \tilde{C}_1$ or $y = C_1 e^{3t} + C_2$. Therefore, $y_1(t) = 1, y_2(t) = e^{3t}$.

2. $y'' + 5y' = 0$

Solution. Using the method of Example 8.2, we find that $\mu(t) = e^{5t}$, hence $e^{5t} y' = \tilde{C}_1$ or $y = C_1 e^{-5t} + C_2$. Therefore, $y_1(t) = 1, y_2(t) = e^{-5t}$.

3. $y'' + 16y = 0$

Solution. If we use Example 8.3 and set $k = 4$, we find that $y = C_1 \sin(4t) + C_2 \cos(4t)$. Therefore, $y_1(t) = \sin(4t)$, $y_2(t) = \cos(4t)$.

4. $y'' - 9y = 0$

Solution. Since we know that $y'' - y = 0$ has the fundamental set $\{e^t, e^{-t}\}$, we may easily check that $y_1(t) = e^{3t}$, $y_2(t) = e^{-3t}$ is a fundamental set for $y'' - 9y = 0$.

Find the general solution to each of the following second-order equations.

5. $y'' - 3y' = 3$

Solution. Using the method of Example 8.2, we find that $\mu(t) = e^{-3t}$, hence $e^{-3t}y' = \tilde{C}_2$ or $y_H = C_1 + C_2 e^{3t}$. We may easily guess a particular solution $y_P = -t$. Therefore, the general solution is $y(t) = -t + C_1 + C_2 e^{3t}$.

6. $y'' + 16y = 32$

Solution. Using the method of Example 8.3, if we set $k = 4$, then $y_H = C_1 \cos(4t) + C_2 \sin(4t)$. We may easily guess a particular solution $y_P = 2$. Therefore, the general solution is $y(t) = 2 + C_1 \cos(4t) + C_2 \sin(4t)$.

7. $y'' - 9y = 18$

Solution. Using the solution to Problem 4, we have $y_H = C_1 e^{3t} + C_2 e^{-3t}$. We may easily guess a particular solution $y_P = -2$. Therefore, the general solution is $y_1(t) = -2 + C_1 e^{3t} + C_2 e^{-3t}$.

8. $4y'' + y = 0$

Solution. Using the method of Example 8.3, if we set $k = 1/2$, then $y_H = C_1 \cos(t/2) + C_2 \sin(t/2)$. Since this is a homogeneous equation, $y_P = 0$. Therefore, the general solution is $y(t) = C_1 \cos(t/2) + C_2 \sin(t/2)$.

Section 8.4, page 253

Compute the Wronskian of the following pairs of functions. First, do the computations by hand; then use *Mathematica*.

1. $y_1 = t e^t$ and $y_2 = e^t$

Solution. We compute $W(y_1, y_2)(t) = y_1(t)y_2'(t) - y_1'(t)y_2(t) = t e^{2t} - (1+t)e^{2t} = -e^{2t}$. Alternatively:

```
y1[t_]:= t Exp[t]
y2[t_]:= Exp[t]
Wronskian[y1,y2][t]
```
$$-E^{2\,t}$$

2. $y_1 = \cos(2t)$ and $y_2 = \sin(2t)$

Solution. We compute $W(y_1, y_2)(t) = y_1(t)y_2'(t) - y_1'(t)y_2(t) = \left(\cos(t)\right)^2 + \left(\sin(t)\right)^2 = 1.$ Alternatively:

```
y1[t_]:= Cos[t]
y2[t_]:= Sin[t]
Wronskian[y1,y2][t]
```
```
1
```

3. $y_1 = e^t \cos(2t)$ and $y_2 = e^t \sin(2t)$

Solution. We compute $W(y_1, y_2)(t) = y_1(t)y_2'(t) - y_1'(t)y_2(t) = e^t \cos(2t)(2\cos(2t) + \sin(2t)) - e^t \sin(2t)(cos(2t) - 2\sin(2t)) = 2e^{2t}.$ Alternatively:

```
y1[t_]:= Exp[t]Cos[2t]
y2[t_]:= Exp[t]Sin[2t]
Wronskian[y1,y2][t]
```
$$2\,E^{2\,t}$$

4. $y_1 = e^{2t}$ and $y_2 = e^{3t}$

Solution. We compute $W(y_1, y_2)(t) = y_1(t)y_2'(t) - y_1'(t)y_2(t) = 3e^{5t} - 2e^{5t} = e^{5t}.$ Alternatively:

```
y1[t_]:= Exp[2t]
y2[t_]:= Exp[3t]
Wronskian[y1,y2][t]
```
$$E^{5\,t}$$

5. Use *Mathematica*'s command **Solve** to solve (8.34).

Solution. Using

```
Solve[{2x + 3y == 1,3x - 5y == -3},{x,y}]
```

we obtain

$$\{\{x \to -(\tfrac{4}{19}),\ y \to \tfrac{9}{19}\}\}$$

Section 8.5, page 258

In each of the following problems, determine whether the pair of functions $\{y_1, y_2\}$ is linearly independent or dependent.

1. $y_1 = t^3 - 3t$ and $y_2 = t^3 + 3t$

Solution. Note that $\dfrac{t^3 - 3t}{t^3 + 3t} = \dfrac{t^2 - 3}{t^2 + 3}$ is different from a constant on any interval. Therefore, the pair is linearly independent.

2. $y_1 = e^{at} \cos(bt)$ and $y_2 = e^{at} \sin(bt)$

Solution. We see that $\dfrac{e^{at} \sin(bt)}{e^{at} \cos(bt)} = \tan(bt)$ is different from a constant on any interval. Therefore, the pair is linearly independent.

3. $y_1 = \log(t)$ and $y_2 = \log(2t)$

Solution. Note that $\dfrac{\log(2t)}{\log(t)} = \log(2)/\log(t) + 1$ is different from a constant on any interval of positive numbers. Therefore, the pair is linearly independent.

4. $y_1 = e^{at}$ and $y_2 = e^{at+b}$

Solution. Observe that $\dfrac{e^{at+b}}{e^{at}} = e^b$ is a constant. Therefore, the pair is linearly dependent.

Section 8.6, page 260

Use the method of reduction of order to find a second linearly independent solution to each of the following second-order equations when we are given the indicated first solution.

1. $\begin{cases} y'' - 3y' + 2y = 0, \\ y_1(t) = e^{2t} \end{cases}$

Solution. We first compute the Wronskian

$$W(t) = \exp\left(-\int -3\,dt\right) = Ce^{3t}.$$

Then a second linearly independent solution is obtained from

$$y_2(t) = Ce^{2t} \int \frac{e^{3t}}{(e^{2t})^2}\,dt = Ce^{2t} \int e^{-t}\,dt = -Ce^t.$$

Therefore, if we set $C = -1$, we obtain $y_2(t) = e^t$.

2. $\begin{cases} y'' + 4y' + 4y = 0, \\ y_1(t) = e^{-2t} \end{cases}$

Solution. We first compute the Wronskian

$$W(t) = \exp\left(-\int 4\,dt\right) = Ce^{-4t}.$$

Then a second linearly independent solution is obtained from

$$y_2(t) = Ce^{-2t} \int \frac{e^{-4t}}{(e^{-2t})^2}\,dt = Ce^{-2t} \int dt = Ct\,e^{-2t}.$$

Therefore, if we set $C = 1$, we obtain $y_2(t) = t\,e^{-2t}$.

3. $\begin{cases} t^2 y'' + 4t\,y' = 0, \\ y_1(t) = 6 \end{cases}$

Solution. We first compute the Wronskian

$$W(t) = \exp\left(-\int \frac{4}{t}\,dt\right) = \frac{C}{t^4}.$$

Then a second linearly independent solution is obtained from

$$y_2(t) = C6 \int \frac{1}{36t^4}\,dt = \frac{C}{6} \int \frac{1}{t^4}\,dt = -\frac{C}{3}t^{-3}.$$

Therefore, if we set $C = -3$, we obtain $y_2(t) = t^{-3}$.

4. $\begin{cases} t^2 y'' + 5t\,y' - 5y = 0, \\ y_1(t) = t \end{cases}$

Solution. We first compute the Wronskian

$$W(t) = \exp\left(-\int \frac{5}{t}\,dt\right) = \frac{C}{t^5}.$$

Then a second linearly independent solution is obtained from

$$y_2(t) = Ct \int \frac{1}{t^2 t^5}\,dt = Ct \int \frac{1}{t^7}\,dt = -\frac{Ct}{6}t^{-6}.$$

Therefore, if we set $C = -6$, we obtain $y_2(t) = t^{-5}$.

5. $\begin{cases} t^2 y'' + t\, y' + (t^2 - 1/4)y = 0, \\ y_1(t) = \sin(t)/\sqrt{t} \end{cases}$

Solution. We first compute the Wronskian

$$W(t) = \exp\left(-\int \frac{1}{t}\,dt\right) = \frac{C}{t}.$$

Then a second linearly independent solution is obtained from

$$y_2(t) = C\frac{\sin(t)}{\sqrt{t}} \int \frac{t}{t\left(\sin(t)\right)^2}\,dt$$

$$= C\frac{\sin(t)}{\sqrt{t}} \int \frac{1}{\left(\sin(t)\right)^2}\,dt = -C\frac{\cos(t)}{\sqrt{t}}.$$

Therefore, if we set $C = -1$, we obtain $y_2(t) = \cos(t)/\sqrt{t}$.

6. $\begin{cases} t\, y'' - (t+1)y' + y = 0, \\ y_1(t) = t+1 \end{cases}$

Solution. We first compute the Wronskian

$$W(t) = \exp\left(-\int \frac{-(t+1)}{t}\,dt\right) = C t\, e^t.$$

Then a second linearly independent solution is obtained from

$$y_2(t) = C(t+1) \int \frac{t e^t}{(t+1)^2}\,dt = C e^t.$$

Therefore, if we set $C = 1$, we obtain $y_2(t) = e^t$.

7. $\begin{cases} t\, y'' - (t+2)y' + 2y = 0, \\ y_1(t) = e^t \end{cases}$

Solution. We first compute the Wronskian

$$W(t) = \exp\left(-\int \frac{-(t+2)}{t}\,dt\right) = C t^2 e^t.$$

Then a second linearly independent solution is obtained from

$$y_2(t) = C e^t \int \frac{t^2 e^t}{e^{2t}}\,dt = C e^t \int t^2 e^{-t}\,dt = C(-2 - 2t - t^2).$$

Therefore, if we set $C = -1$, we obtain $y_2(t) = 2 + 2t + t^2$.

8. $\begin{cases} t\,y'' - y' + 4t^3 y = 0, \\ y_1(t) = \sin(t^2) \end{cases}$

Solution. We first compute the Wronskian

$$W(t) = \exp\left(-\int \frac{-1}{t}\,dt\right) = Ct.$$

Then a second linearly independent solution is obtained from

$$y_2(t) = C\sin(t^2)\int \frac{t}{\left(\sin(t^2)\right)^2}\,dt = -\frac{C}{2}\cos(t^2).$$

Therefore, if we set $C = -2$, we obtain $y_2(t) = \cos(t^2)$.

Section 8.7, page 262

The following pairs of functions form a fundamental set of solutions of a linear second-order differential equation $y'' + p(t)y' + q(t)y = 0$. Find the functions $p(t)$ and $q(t)$.

1. $t\,e^t$ and $t^2 e^t$

Solution. We first compute the Wronskian $W(t) = y_1(t)y_2'(t) - y_1'(t)y_2(t) = t^2 e^{2t}(2+t) - t^2 e^{2t}(1+t) = t^2 e^{2t}$. Then we compute

$$p(t) = \frac{y_1'' y_2 - y_1 y_2''}{W(t)} = \frac{-2t\,e^{2t}(1+t)}{t^2 e^{2t}} = -\frac{2(1+t)}{t}$$

and

$$q(t) = \frac{-y_1'' y_2' + y_1' y_2''}{W(t)} = \frac{e^{2t}(2+2t+t^2)}{t^2 e^{2t}} = \frac{(2+2t+t^2)}{t^2}$$

yielding

$$y'' - \frac{2(t+1)y'}{t} + \frac{(t^2+2t+2)y}{t^2} = 0$$

2. $\cos(2t)$ and $\sin(2t)$

Solution. We first compute the Wronskian $W(t) = y_1(t)y_2'(t) - y_1'(t)y_2(t) = 2(\cos(2t))^2 + 2(\sin(2t))^2 = 2$. Then we compute

$$p(t) = \frac{y_1'' y_2 - y_1 y_2''}{W(t)} = \frac{0}{2} = 0$$

and

$$q(t) = \frac{-y_1'' y_2' + y_1' y_2''}{W(t)} = \frac{8}{2} = 4$$

yielding

$$y'' + 4y = 0$$

3. $e^t \cos(2t)$ and $e^t \sin(2t)$

Solution. We first compute the Wronskian $W(t) = y_1(t)y_2'(t) - y_1'(t)y_2(t) = e^{2t} \cos(2t)(2\cos(2t) + \sin(2t)) - e^{2t} \sin(2t)(\cos(2t) - 2\sin(2t)) = 2e^{2t}$. Then we compute

$$p(t) = \frac{y_1'' y_2 - y_1 y_2''}{W(t)} = \frac{-4e^{2t}}{2e^{2t}} = -2$$

and

$$q(t) = \frac{-y_1'' y_2' + y_1' y_2''}{W(t)} = \frac{10e^{2t}}{2e^{2t}} = 5$$

yielding $y'' - 2t' + 5y = 0$

4. t and t^4

Solution. We first compute the Wronskian $W(t) = y_1(t)y_2'(t) - y_1'(t)y_2(t) = 4t^4 - t^4 = 3t^4$. Then we compute

$$p(t) = \frac{y_1'' y_2 - y_1 y_2''}{W(t)} = \frac{-12t^3}{3t^4} = -\frac{4}{t}$$

and

$$q(t) = \frac{-y_1'' y_2' + y_1' y_2''}{W(t)} = \frac{12t^2}{3t^4} = \frac{4}{t^2}$$

yielding $y'' - \dfrac{4y'}{t} + \dfrac{4y}{t^2} = 0$

5. 2 and $\log(t)$

Solution. We first compute the Wronskian $W(t) = y_1(t)y_2'(t) - y_1'(t)y_2(t) = 2/t - 0 = 2/t$. Then we compute

$$p(t) = \frac{y_1'' y_2 - y_1 y_2''}{W(t)} = \frac{2t}{2t^2} = \frac{1}{t}$$

and

$$q(t) = \frac{-y_1'' y_2' + y_1' y_2''}{W(t)} = \frac{0}{2t^2} = 0$$

yielding $y'' + \dfrac{y'}{t} = 0$

6. t^3 and t^{-3}

Solution. We first compute the Wronskian $W(t) = y_1(t)y_2'(t) - y_1'(t)y_2(t) = -3/t - 3/t = -6/t$. Then we compute

$$p(t) = \frac{y_1'' y_2 - y_1 y_2''}{W(t)} = \frac{-6t}{-6t^2} = \frac{1}{t}$$

and

$$q(t) = \frac{-y_1'' y_2' + y_1' y_2''}{W(t)} = \frac{54t}{-6t^3} = -\frac{9}{t^2}$$

yielding $y'' + \dfrac{y'}{t} - \dfrac{9y}{t^2} = 0$

Chapter 9

Section 9.1, page 275

Find a fundamental set of solutions and the general solution for the following constant-coefficient equations:

1. $y'' + 4y' + 3y = 0$

Solution. The roots of the characteristic equation $r^2 + 4r + 3 = (r+3)(r+1) = 0$ are $r_1 = -3$ and $r_2 = -1$. Therefore, a fundamental set is given by $\{e^{-3t}, e^{-t}\}$ and the general solution is $y = C_1 e^{-3t} + C_2 e^{-t}$.

2. $2y'' + 4y' - 6y = 0$

Solution. The roots of the characteristic equation $2r^2 + 4r - 6 = (r+3)(r-1) = 0$ are $r_1 = -3$ and $r_2 = 1$. Therefore, a fundamental set is given by $\{e^{-3t}, e^t\}$ and the general solution is $y = C_1 e^{-3t} + C_2 e^t$.

3. $y'' - 6y' + 9y = 0$

Solution. The roots of the characteristic equation $r^2 - 6r + 9 = (r-3)(r-3) = 0$ are $r_1 = 3$ and $r_2 = 3$. Therefore, since the roots are equal, a fundamental set is given by $\{e^{3t}, t e^{3t}\}$ and the general solution is $y = C_1 e^{3t} + C_2 t e^{3t}$.

4. $5y'' + 50y' + 250y = 0$

Solution. The roots of the characteristic equation $5r^2 + 50r + 250 = 0$ are $r_1 = -5 + 5i$ and $r_2 = -5 - 5i$. Therefore, a real fundamental set is given by $\{e^{-5t}\cos(5t), e^{-5t}\sin(5t)\}$ and the general solution is $y = C_1 e^{-5t}\cos(5t) + C_2 e^{-5t}\sin(5t)$.

5. $y'' + 4y' + 5y = 0$

Solution. The roots of the characteristic equation $r^2 + 4r + 5 = 0$ are $r_1 = -2 + i$ and $r_2 = -2 - i$. Therefore, a real fundamental set is given by $\{e^{-2t}\cos(t), e^{-2t}\sin(t)\}$ and the general solution is $y = C_1 e^{-2t}\cos(t) + C_2 e^{-2t}\sin(t)$.

6. $5y'' - 20y' + 30y = 0$

Solution. The roots of the characteristic equation $5r^2 - 20r + 30 = 0$ are $r_1 = 2 + \sqrt{2}i$ and $r_2 = 2 - \sqrt{2}i$. Therefore, a real fundamental set is given by $\{e^{2t}\cos(\sqrt{2}t), e^{2t}\sin(\sqrt{2}t)\}$ and the general solution is $y = C_1 e^{2t}\cos(\sqrt{2}t) + C_2 e^{2t}\sin(\sqrt{2}t)$.

7. $y'' + ky' = 0$

Solution. The roots of the characteristic equation $r^2 + kr = r(r + k) = 0$ are $r_1 = 0$ and $r_2 = -k$. Therefore, a fundamental set is given by $\{1, e^{-kt}\}$ and the general solution is $y = C_1 + C_2 e^{-kt}$.

8. $y'' + k^2 y = 0$

Solution. The roots of the characteristic equation $r^2 + k^2 = 0$ are $r_1 = -ki$ and $r_2 = ki$. Therefore, a real fundamental set is given by $\{\cos(kt), \sin(kt)\}$ and the general solution is $y = C_1 \cos(kt) + C_2 \sin(kt)$.

9. $y'' - k^2 y = 0$

Solution. The roots of the characteristic equation $r^2 - k^2 = 0$ are $r_1 = -k$ and $r_2 = k$. Therefore, a fundamental set is given by $\{e^{-kt}, e^{kt}\}$ and the general solution is $y = C_1 e^{-kt} + C_2 e^{kt}$.

10. $y'' - 2ky' + k^2 y = 0$

Solution. The roots of the characteristic equation $r^2 - 2kr + k^2 = (r - k)(r - k) = 0$ are $r_1 = k$ and $r_2 = k$. Therefore, since the roots are equal, a fundamental set is given by $\{e^{kt}, t\,e^{kt}\}$ and the general solution is $y = C_1 e^{kt} + C_2 t\,e^{kt}$.

Solve the following initial value problems:

11.
$$\begin{cases} y'' - 4y' + 3y = 0, \\ y(0) = 7, \quad y'(0) = 16 \end{cases}$$

Solution. The roots of the characteristic equation $r^2 - 4r + 3 = (r-1)(r-3) = 0$ are $r_1 = 1$ and $r_2 = 3$. Consequently, the general solution is $y = C_1 e^t + C_2 e^{3t}$. Using the initial conditions, we obtain the equations $7 = C_1 + C_2$ and $16 = C_1 + 3C_2$, which lead to $C_1 = 5/2$ and $C_2 = 9/2$. Therefore $y(t) = (5/2)e^t + (9/2)e^{3t}$.

12.
$$\begin{cases} 2y'' + 4y' - 6y = 0, \\ y(0) = 4, \quad y'(0) = 0 \end{cases}$$

Solution. The roots of the characteristic equation $2r^2 + 4r - 6 = (r+3)(r-1) = 0$ are $r_1 = -3$ and $r_2 = 1$. Consequently, the general solution is $y = C_1 e^{-3t} + C_2 e^t$. Using the initial conditions, we obtain the equations $4 = C_1 + C_2$ and $0 = -3C_1 + C_2$, which lead to $C_1 = 1$ and $C_2 = 3$. Therefore $y(t) = e^{-3t} + 3e^t$.

13.
$$\begin{cases} y'' - 6y' + 9y = 0, \\ y(0) = 4, \quad y'(0) = 17 \end{cases}$$

Solution. The roots of the characteristic equation $r^2 - 6r + 9 = (r-3)(r-3) = 0$ are $r_1 = 3$ and $r_2 = 3$. Consequently, the general solution is $y = C_1 e^{3t} + C_2 t\, e^t$. Using the initial conditions, we obtain the equations $4 = C_1$ and $17 = 3C_1 + C_2$, which lead to $C_1 = 4$ and $C_2 = 5$. Therefore $y(t) = 4e^{3t} + 5t\, e^{3t}$.

14.
$$\begin{cases} 5y'' + 50y' + 250y = 0, \\ y(0) = 0, \quad y'(0) = -5 \end{cases}$$

Solution. The roots of the characteristic equation $5r^2 + 50r + 250 = 0$ are $r_1 = -5 + 5i$ and $r_2 = -5 - 5i$. Consequently, the general solution is $y = C_1 e^{-5t} \cos(5t) + C_2 e^{-5t} \sin(5t)$. Using the initial conditions, we obtain the equations $0 = C_1$ and $-5 = -5C_1 + 5C_2$, which lead to $C_1 = 0$ and $C_2 = -1$. Therefore $y(t) = -e^{-5t} \sin(5t)$.

15.
$$\begin{cases} y'' + 4y' + 5y = 0, \\ y(0) = 3, \quad y'(0) = -2 \end{cases}$$

Solution. The roots of the characteristic equation $r^2 + 4r + 5 = 0$ are $r_1 = -2 + i$ and $r_2 = -2 - i$. Consequently, the general solution is $y = C_1 e^{-2t} \cos(t) + C_2 e^{-2t} \sin(t)$. Using the initial conditions, we obtain the equations $3 = C_1$ and $-2 = -2C_1 + C_2$, which lead to $C_1 = 3$ and $C_2 = 4$. Therefore $y(t) = 3e^{-2t} \cos(t) + 4e^{-2t} \sin(t)$.

16. $\begin{cases} 5y'' - 20y' + 30y = 0, \\ y(0) = 2, \ y'(0) = 4 \end{cases}$

Solution. The roots of the characteristic equation $5r^2 - 20r + 30 = 0$ are $r_1 = 2 + \sqrt{2}i$ and $r_2 = 2 - \sqrt{2}i$. Consequently, the general solution is $y = C_1 e^{2t} \cos(\sqrt{2}t) + C_2 e^{2t} \sin(\sqrt{2}t)$. Using the initial conditions, we obtain the equations $2 = C_1$ and $4 = 2C_1 + \sqrt{2}C_2$, which lead to $C_1 = 2$ and $C_2 = 0$. Therefore $y(t) = 2e^{2t} \cos(\sqrt{2}t)$.

17. $\begin{cases} y'' + k\,y' = 0, \\ y(0) = a, \ y'(0) = b \end{cases}$

Solution. The roots of the characteristic equation $r^2 + kr = r(r + k) = 0$ are $r_1 = 0$ and $r_2 = -k$. Consequently, the general solution is $y = C_1 + C_2 e^{-kt}$. Using the initial conditions, we obtain the equations $a = C_1 + C_2$ and $b = -k\,C_2$, which lead to $C_1 = a + b/k$ and $C_2 = -b/k$. Therefore $y(t) = -(b/k)e^{-kt} + a + b/k$.

18. $\begin{cases} y'' + k^2 y = 0, \\ y(0) = a, \ y'(0) = b \end{cases}$

Solution. The roots of the characteristic equation $r^2 + k^2 = 0$ are $r_1 = ki$ and $r_2 = -ki$. Consequently, the general solution is $y = C_1 \cos(kt) + C_2 \sin(kt)$. Using the initial conditions, we obtain the equations $a = C_1$ and $b = k\,C_2$, which lead to $C_1 = a$ and $C_2 = b/k$. Therefore $y(t) = a\cos(kt) + (b/k)\sin(kt)$.

19. $\begin{cases} y'' - k^2 y = 0, \\ y(0) = a, \ y'(0) = b \end{cases}$

Solution. The roots of the characteristic equation $r^2 - k^2 = 0$ are $r_1 = k$ and $r_2 = -k$. Consequently, the general solution is $y = C_1 e^{kt} + C_2 e^{-kt}$. Using the initial conditions, we obtain the equations $a = C_1 + C_2$ and $b = k(C_1 - C_2)$, which lead to $C_1 = \left(\dfrac{a}{2} + \dfrac{b}{2k}\right)$ and $C_2 = \left(\dfrac{a}{2} - \dfrac{b}{2k}\right)$. Therefore $y(t) = \left(\dfrac{a}{2} + \dfrac{b}{2k}\right)e^{kt} + \left(\dfrac{a}{2} - \dfrac{b}{2k}\right)e^{-kt}$.

20. $\begin{cases} y'' - 2k\,y' + k^2 y = 0, \\ y(0) = a, \ y'(0) = b \end{cases}$

Solution. The roots of the characteristic equation $r^2 - 2k + k^2 = (r - k)(r - k) = 0$ are $r_1 = k$ and $r_2 = k$. Consequently, the general solution is $y = C_1 e^{kt} + C_2 t\, e^{kt}$. Using the initial conditions, we obtain the equations $a = C_1$ and $b = k\,C_1 + C_2$, which lead to $C_1 = a$ and $C_2 = (b - ka)$. Therefore $y(t) = a\,e^{kt} + (b - ka)t\,e^{kt}$

Section 9.2, page 282

In Problems 1–6 find the general solution of the stated differential equation:

1. $y'' - 3y' + 3y = 0$

   Solution. The roots of the characteristic equation $r^2 - 3r + 3 = 0$ are $r_1 = 3/2 + \sqrt{3}/2i$ and $r_2 = 3/2 - \sqrt{3}/2i$. Consequently, the general complex solution is $y = C_1 e^{(3/2+\sqrt{3}/2i)t} + C_2 e^{(3/2-\sqrt{3}/2i)t}$. Using Euler's identity, we finally obtain $y(t) = e^{(3t/2)}\left(A\cos(\sqrt{3}\,t/2) + B\sin(\sqrt{3}\,t/2)\right)$.

2. $y'' + 4y = 0$

   Solution. The roots of the characteristic equation $r^2 + 4 = 0$ are $r_1 = 2i$ and $r_2 = -2i$. Consequently, the general complex solution is $y = C_1 e^{2it} + C_2 e^{-2it}$. Using Euler's identity, we finally obtain $y(t) = A\sin(2t) + B\cos(2t)$.

3. $y'' - 4y' + 8y = 0$

   Solution. The roots of the characteristic equation $r^2 - 4r + 8 = 0$ are $r_1 = 2 + 2i$ and $r_2 = 2 - 2i$. Consequently, the general complex solution is $y = C_1 e^{(2+2i)t} + C_2 e^{(2-2i)t}$. Using Euler's identity, we finally obtain $y(t) = e^{2t}\left(A\cos(2t) + B\cos(2t)\right)$.

4. $y'' - (2+4i)y' - (3-4i)y = 0$

   Solution. The roots of the characteristic equation $r^2 - (2+4i)r - (3-4i) = 0$ are $r_1 = 1 + 2i$ and $r_2 = 1 + 2i$. Therefore, the general complex solution is $y(t) = C_1 e^{(1+2i)t} + C_2 t\, e^{(1+2i)t}$.

5. $y'' - (3+2i)y' + 6i\, y = 0$

   Solution. The roots of the characteristic equation $r^2 - (3+2i)r + 6i = 0$ are $r_1 = 2i$ and $r_2 = 3$. Therefore, the general complex solution is $y(t) = C_1 e^{2it} + C_2 e^{3t}$.

6. $y'' - (1-4i)y' - (3+3i)y = 0$

   Solution. The roots of the characteristic equation $r^2 - (1-4i)r - (3+3i) = 0$ are $r_1 = -3i$ and $r_2 = 1 - i$. Therefore, the general complex solution is $y(t) = C_1 e^{-3it} + C_2 e^{(1-i)t}$.

7. Use equation (9.42) to establish equations (9.36) and (9.38).

   Solution. From
   $$e^z = e^{x+iy} = e^x\left(\cos(y) + i\,\sin(y)\right),$$
   we note that $e^x \neq 0$ for all real numbers x and that the zeros of the sine function are distinct from the zeros of the cosine function. Therefore, $e^z \neq 0$ for all complex numbers z and equation (9.36) is established.

To establish equation (9.38), we first note that

$$f'(z) = \lim_{\substack{h \to 0 \\ k \to 0}} \frac{f(z + (h + ik)) - f(z)}{h + ik}$$

$$= \lim_{h \to 0} \frac{f(z + h) - f(z)}{h} = \frac{\partial f}{\partial x}$$

$$= \lim_{k \to 0} \frac{f(z + ik) - f(z)}{ik} = -i \frac{\partial f}{\partial y}$$

Therefore, from

$$e^{az} = e^{a(x + iy)} = e^{ax}(\cos(a y) + i \sin(a y)),$$

we see that

$$\frac{de^{az}}{dz} = \frac{\partial e^{a(x + iy)}}{\partial x} = a e^{ax}(\cos(a y) + i \sin(a y)) = a e^{az}$$

$$= -i \frac{\partial e^{a(x + iy)}}{\partial y} = -i a e^{ax}(-\sin(a y) + i \cos(a y)) = a e^{az}.$$

8. The five most important numbers in mathematics are perhaps 0, 1, i, e, and π, and the three most important operations are probably addition, multiplication, and exponentiation. Show that are they are all related by the formula

$$e^{i\pi} + 1 = 0.$$

Solution. From

$$e^{x + iy} = e^x(\cos(y) + i \sin(y)),$$

we note that if $x = 0$ and $y = \pi$, then $e^{i\pi} = 1(-1 + i\,0) = -1$.

9. Suppose that a, b, and c are real numbers and that z is a complex solution of $a\,y'' + b\,y' + c\,y = 0$. Show that the real and imaginary parts of z are also solutions of $a\,y'' + b\,y' + c\,y = 0$.

Solution. If we write a root of the characteristic equation as $r = (-b/2a) + i\left(\sqrt{4ac - b^2}/2a\right) = \alpha + i\,\beta$, then from equation (9.42) we may write

$$z(t) = e^{(\alpha + i\beta)t} = \left(e^{\alpha t} \cos(\beta t)\right) + i\left(e^{\alpha t} \sin(\beta t)\right).$$

If we set $y_1(t) = \Re(z(t)) = e^{\alpha t} \cos(\beta t)$ and $y_2(t) = \Im(z(t)) = e^{\alpha t} \sin(\beta t)$, then each is easily seen to be a solution of the equation $a\,y'' + b\,y' + c\,y = 0$. For example,

$$y_1'(t) = e^{\alpha t}(\alpha \cos(\beta t) - \beta \sin(\beta t)),$$

$$y_1''(t) = e^{\alpha t}(\alpha^2 \cos(\beta t) - \beta^2 \cos(\beta t) - \alpha \beta \sin(\beta t)).$$

Section 9.3, page 292

In Exercises 1–10 use the method of undetermined coefficients to find a particular solution of the indicated equation.

1. $y'' + 3y' + 2y = e^{-5t}$

   Solution. Since the roots of the characteristic equation are $r_1 = -2$ and $r_2 = -1$, we assume that $y_P(t) = Ae^{-5t}$. Substituting this form into the equation, we obtain

   $12Ae^{-5t} = e^{-5t}$. Hence, $A = 1/12$ and $y_P(t) = \dfrac{e^{-5t}}{12}$.

2. $y'' + 3y' - 10y = e^{-5t}$

   Solution. Since the roots of the characteristic equation are $r_1 = -5$ and $r_2 = 2$, we detect first-order resonance and now assume that $y_P(t) = t(Ae^{-5t})$. Substituting this form into the equation, we obtain $-7Ae^{-5t} = e^{-5t}$. Hence, $A = -1/7$ and

   $y_P(t) = \dfrac{-t\, e^{-5t}}{7}$.

3. $y'' + 4y' + 4y = e^{-2t}$

   Solution. Since the roots of the characteristic equation are $r_1 = -2$ and $r_2 = -2$, we detect second-order resonance and now assume that $y_P(t) = t^2(Ae^{-2t})$. Substituting this form into the equation, we obtain $2Ae^{-2t} = e^{-2t}$. Hence, $A = 1/2$ and

   $y_P(t) = \dfrac{t^2 e^{-2t}}{2}$.

4. $y'' + 4y' + 4y = \cos(2t)$

   Solution. Since the roots of the characteristic equation are $r_1 = -2$ and $r_2 = -2$, we assume that $y_P(t) = A\cos(2t) + B\sin(2t)$. Substituting this form into the equation, we obtain $8\big(B\cos(2t) - A\sin(2t)\big) = \cos(2t)$. Hence, $A = 0$, $B = 1/8$ and

   $y_P(t) = \dfrac{\sin(2t)}{8}$.

5. $y'' + 4y = \cos(2t)$

   Solution. Since the roots of the characteristic equation are $r_1 = 2i$ and $r_2 = -2i$, we detect first-order resonance and now assume that $y_P(t) = t\big(A\cos(2t) + B\sin(2t)\big)$. Substituting this form into the equation, we obtain $4\big(B\cos(2t) - A\sin(2t)\big) = \cos(2t)$. Hence, $A = 0$, $B = 1/4$ and $y_P(t) = \dfrac{t\sin(2t)}{4}$.

6. $y'' + 3y' + 2y = t^2$

Solution. Since the roots of the characteristic equation are $r_1 = -2$ and $r_2 = -1$, we assume that $y_P(t) = A + Bt + Ct^2$. Substituting this form into the equation, we obtain $2A + 3B + 2C = 0$, $2B + 6C = 0$ and $2C = 1$. Hence, $A = 7/4$, $B = -3/2$, $C = 1/2$ and $y_P(t) = t^2/2 - 3t/2 + 7/4$.

7. $y'' + 3y' + 2y = t^2 e^t$

Solution. Since the roots of the characteristic equation are $r_1 = -2$ and $r_2 = -1$, we assume that $y_P(t) = (A + Bt + Ct^2)e^t$. Substituting this form into the equation, we obtain $6A + 5B + 2C = 0$, $6B + 10C = 0$ and $6C = 1$. Hence, $A = 19/108$, $B = -5/18$, $C = 1/6$ and $y_P(t) = \left(\dfrac{19}{108} - \dfrac{5t}{18} + \dfrac{t^2}{6}\right)e^t$.

8. $y'' + 3y' + 2y = t^2 e^{-t}$

Solution. Since the roots of the characteristic equation are $r_1 = -2$ and $r_2 = -1$, we detect first-order resonance and now assume that $y_P(t) = t(A + Bt + Ct^2)e^{-t}$. Substituting this form into the equation, we obtain $A + 2B = 0$, $2B + 6C = 0$ and $3C = 1$. Hence, $A = 2$, $B = -1$, $C = 1/3$ and $y_P(t) = \left(2t - t^2 + \dfrac{t^3}{3}\right)e^{-t}$.

9. $y'' + 4y' + 2y = \left(\sin(t)\right)^2$

Solution. The roots of the characteristic equation are $r_1 = -2+\sqrt{2}$ and $r_2 = -2-\sqrt{2}$. Using the identity $\sin(t)^2 = 1/2 - \cos(2t)/2$, we may consider this problem as two simpler problems. First, we assume that $y_{P_1}(t) = A$. Substituting this form into the equation, we obtain $2A = 1/2$ or $A = 1/4$ and $y_{P_1}(t) = 1/4$. Next, we assume that $y_{P_2}(t) = A\cos(2t) + B\sin(2t)$. Substituting this form into the equation, we obtain $-2A + 8B = -1/2$ and $-8A - 2B = 0$. Hence $A = 1/68$, $B = -1/17$ and $y_{P_2}(t) = \cos(2t)/68 - \sin(2t)/17$. Combining these two particular solutions, we obtain $y_P(t) = y_{P_1}(t) + y_{P_2}(t) = \dfrac{1}{4} - \dfrac{\cos(2t)}{68} + \dfrac{\sin(2t)}{17}$.

10. $y'' + 6y' + 3y = t e^t \cos(2t)$

Solution. Since the roots of the characteristic equation are $r_1 = -3 + \sqrt{6}i$ and $r_2 = -3 - \sqrt{6}i$, we assume that $y_P(t) = (A + Bt)e^t \cos(2t) + (C + Dt)e^t \sin(2t)$. Substituting this form into the equation, we obtain $6A + 8B + 16C + 4D = 0$, $6B + 16D = 1$, $-16A - 4B + 6C + 8D = 0$ and $-16B + 6D = 0$. Hence, $A = 62/5329$, $B = 3/146$, $C = -151/5329$, $D = 4/73$ and $y_P(t) = \left(\dfrac{62}{5329} + \dfrac{3t}{146}\right)e^t \cos(2t) + \left(-\dfrac{151}{5329} + \dfrac{4t}{73}\right)e^t \sin(2t)$.

In Exercises 11–20 solve the stated initial value problem

11.
$$\begin{cases} y'' + 3y' + 2y = e^{-5t}, \\ y(0) = 0, \ y'(0) = 0 \end{cases}$$

Solution. Since the roots of the characteristic equation are $r_1 = -2$ and $r_2 = -1$, we assume that $y_p(t) = A e^{-5t}$. Substituting this form into the equation, we obtain $12A = 1$. Hence, $A = 1/12$ and the general solution becomes $y(t) = e^{-5t}/12 + C_1 e^{-2t} + C_2 e^{-t}$. The initial values generate the equations $1/12 + C_1 + C_2 = 0$ and $-5/12 - 2C_1 - C_2 = 0$ which have $C_1 = -1/3$ and $C_2 = 1/4$ as the solution. Hence, $y(t) = \dfrac{e^{-5t}}{12} + \dfrac{e^{-t}}{4} - \dfrac{e^{-2t}}{3}$.

12.
$$\begin{cases} y'' + 3y' - 10y = e^{-5t}, \\ y(0) = 1, \ y'(0) = 2 \end{cases}$$

Solution. Since the roots of the characteristic equation are $r_1 = -5$ and $r_2 = 2$, we detect first-order resonance and now assume that $y_p(t) = t(A e^{-5t})$. Substituting this form into the equation, we obtain $-7A = 1$. Hence, $A = -1/7$ and the general solution becomes $y(t) = -(t e^{-5t})/7 + C_1 e^{-5t} + C_2 e^{2t}$. The initial values generate the equations $C_1 + C_2 = 1$ and $-1/7 - 5C_1 + 2C_2 = 2$ which have $C_1 = -1/49$ and $C_2 = 50/49$ as the solution. Hence, $y(t) = -\left(\dfrac{t}{7} + \dfrac{1}{49}\right)e^{-5t} + \dfrac{50e^{2t}}{49}$.

13.
$$\begin{cases} y'' + 4y' + 4y = e^{-2t}, \\ y(0) = 2, \ y'(0) = 0 \end{cases}$$

Solution. Since the roots of the characteristic equation are $r_1 = -2$ and $r_2 = -2$, we detect second-order resonance and now assume that $y_p(t) = t^2 A e^{-2t}$. Substituting this form into the equation, we obtain $2A = 1$. Hence, $A = 1/2$ and the general solution becomes $y(t) = (t^2 e^{-2t})/2 + C_1 e^{-2t} + C_2 t e^{-2t}$. The initial values generate the equations $C_1 = 2$ and $-2C_1 + C_2 = 0$ which have $C_1 = 2$ and $C_2 = 4$ as the solution. Hence, $y(t) = \left(\dfrac{t^2}{2} + 4t + 2\right)e^{-2t}$.

14.
$$\begin{cases} y'' + 4y' + 4y = \cos(2t), \\ y(0) = 3, \ y'(0) = 0 \end{cases}$$

Solution. Since the roots of the characteristic equation are $r_1 = -2$ and $r_2 = -2$, we assume that $y_p(t) = A \cos(2t) + B \sin(2t)$. Substituting this form into the equation, we obtain $-A = 0$ and $8B = 1$. Hence, $A = 0$ and $B = 1/8$ and the general solution becomes $y(t) = \sin(2t)/8 + C_1 e^{-2t} + C_2 t e^{-2t}$. The initial values generate the equations $C_1 = 3$ and $1/4 - 2C_1 + C_2 = 0$ which have $C_1 = 3$ and $C_2 = 23/4$

as the solution. Hence, $y(t) = \dfrac{\sin(2t)}{8} + \left(3 + \dfrac{23t}{4}\right)e^{-2t}$.

15. $\begin{cases} y'' + 4y = \cos(2t), \\ y(0) = 2,\ y'(0) = 5 \end{cases}$

Solution. Since the roots of the characteristic equation are $r_1 = 2i$ and $r_2 = -2i$, we detect first-order resonance and now assume that $y_p(t) = t(A\cos(2t) + B\sin(2t))$. Substituting this form into the equation, we obtain $-4A = 0$ and $4B = 1$. Hence, $A = 0$ and $B = 1/4$ and the general solution becomes $y(t) = (t\sin(2t))/4 + C_1\sin(2t) + C_2\cos(2t)$. The initial values generate the equations $2C_1 = 5$ and $C_2 = 2$ which have $C_1 = 5/2$ and $C_2 = 2$ as the solution. Hence,
$y(t) = \left(\dfrac{t}{4} + \dfrac{5}{2}\right)\sin(2t) + 2\cos(2t)$.

16. $\begin{cases} y'' + 3y' + 2y = t^2, \\ y(0) = 4,\ y'(0) = 2 \end{cases}$

Solution. Since the roots of the characteristic equation are $r_1 = -2$ and $r_2 = -1$, we assume that $y_p(t) = A + Bt + Ct^2$. Substituting this form into the equation, we obtain $2A + 3B + 2C = 0$, $2B + 6C = 0$ and $2C = 1$. Hence, $A = 7/4$, $B = -3/2$ and $C = 1/2$ and the general solution becomes $y(t) = 7/4 - 3t/2 + t^2/2 + C_1e^{-2t} + C_2e^{-t}$. The initial values generate the equations $7/4 + C_1 + C_2 = 4$ and $-3/2 - 2C_1 - C_2 = 2$ which have $C_1 = -23/4$ and $C_2 = 8$ as the solution. Hence, $y(t) = \dfrac{t^2}{2} - \dfrac{3t}{2} + \dfrac{7}{4} + 8e^{-t} - \dfrac{23e^{-2t}}{4}$.

17. $\begin{cases} y'' + 3y' + 2y = t^2e^t, \\ y(0) = 8,\ y'(0) = 2 \end{cases}$

Solution. Since the roots of the characteristic equation are $r_1 = -2$ and $r_2 = -1$, we assume that $y_p(t) = (A + Bt + Ct^2)e^t$. Substituting this form into the equation, we obtain $6A + 5B + 2C = 0$, $6B + 10C = 0$ and $6C = 1$. Hence, $A = 19/108$, $B = -5/18$ and $C = 1/6$ and the general solution becomes $y(t) = (19/108 - 5t/18 + t^2/6)e^t + C_1e^{-2t} + C_2e^{-t}$. The initial values generate the equations $19/108 + C_1 + C_2 = 8$ and $-11/108 - 2C_1 - C_2 = 2$ which have $C_1 = -268/27$ and $C_2 = 71/4$ as the solution. Hence, $y(t) = \dfrac{71e^{-t}}{4} - \dfrac{268e^{-2t}}{27} + \left(\dfrac{19}{108} - \dfrac{5t}{18} + \dfrac{t^2}{6}\right)e^t$.

18.
$$\begin{cases} y'' + 3y' + 2y = t^2 e^{-t}, \\ y(0) = 5, \ y'(0) = 2 \end{cases}$$

Solution. Since the roots of the characteristic equation are $r_1 = -2$ and $r_2 = -1$, we detect first-order resonance and now assume that $y_p(t) = t(A + Bt + Ct^2)e^{-t}$. Substituting this form into the equation, we obtain $A + 2B = 0$, $2B + 6C = 0$ and $3C = 1$. Hence, $A = 2$, $B = -1$ and $C = 1/3$ and the general solution becomes $y(t) = (2t - t^2 + t^3/3)e^{-t} + C_1 e^{-2t} + C_2 e^{-t}$. The initial values generate the equations $C_1 = -5$ and $C_2 = 10$. Hence, $y(t) = \left(10 + 2t - t^2 + \dfrac{t^3}{3}\right)e^{-t} - 5e^{-2t}$

19.
$$\begin{cases} y'' + 4y' + 2y = \big(\sin(t)\big)^2, \\ y(0) = 3, \ y'(0) = 1 \end{cases}$$

Solution. Since the roots of the characteristic equation are $r_1 = -2 - \sqrt{2}$ and $r_2 = -2 + \sqrt{2}$, we may use the results of Problem 9 and obtain $y(t) = 1/4 + \cos(2t)/68 - \sin(2t)/17 + C_1 e^{(-2-\sqrt{2})t} + C_2 e^{(-2+\sqrt{2})t}$. The initial values generate the equations $9/34 + C_1 + C_2 = 3$ and $-2/17 + (-2 - \sqrt{2})C_1 + (-2 + \sqrt{2})C_2 = 1$ which have $C_1 = (93 - 112\sqrt{2})/68$ and $C_2 = (93 + 112\sqrt{2})/68$ as the solution. Hence,

$$y(t) = \frac{1}{4} + \frac{\cos(2t)}{68} - \frac{\sin(2t)}{17} + \left(\frac{93}{68} - \frac{28\sqrt{2}}{17}\right)e^{(-2-\sqrt{2})t}$$

$$+ \left(\frac{93}{68} + \frac{28\sqrt{2}}{17}\right)e^{(-2+\sqrt{2})t}$$

$$= \frac{1}{4} + \frac{\cos(2t) - 4\sin(2t)}{68} + \frac{e^{-2t}\big(93\cosh(\sqrt{2}t) + 112\sqrt{2}\cosh(\sqrt{2}t)\big)}{68}$$

20.
$$\begin{cases} y'' + 6y' + 3y = t\,e^t \cos(2t), \\ y(0) = 2, \ y'(0) = 1 \end{cases}$$

Solution. Since the roots of the characteristic equation are $r_1 = -3 - \sqrt{6}$ and $r_2 = -3 + \sqrt{6}$, we may use the results of Problem 10 to obtain $y(t) = \left(\dfrac{62}{5329} + \dfrac{3t}{146}\right)e^t \cos(2t)$ $+ \left(-\dfrac{151}{5329} + \dfrac{4t}{73}\right)e^t \sin(2t) + C_1 \exp\big((-3 - \sqrt{6})t\big) + C_2 \exp\big((-3 + \sqrt{6})t\big)$. The initial values generate the equations $62/5329 + C_1 + C_2 = 1$ and $-261/10658 + (-3 - \sqrt{6})C_1 + (-3 + \sqrt{6})C_2 = 2$ which have $C_1 = 5298/5329 - 74495/(21316\sqrt{6})$

and $C_2 = 5298/5329 + 74495/(21316\sqrt{6})$ as the solution. Hence,

$$y(t) = \left(\frac{62}{5329} + \frac{3t}{146}\right)e^t \cos(2t) + \left(-\frac{151}{5329} + \frac{4t}{73}\right)e^t \sin(2t)$$

$$+ \left(\frac{5298}{5329} - \frac{74495}{21316\sqrt{6}}\right)e^{(-3-\sqrt{6})t} + \left(\frac{5298}{5329} + \frac{74495}{21316\sqrt{6}}\right)e^{(-3+\sqrt{6})t}.$$

In Exercises 21–30 determine a suitable form for a particular solution of the indicated equation. Do not evaluate the constants.

21. $y'' + 4y' = 3t^2 + t^4 e^{-4t} + \sin(2t)$

Solution. The roots of the characteristic equation are $r_1 = -4$ and $r_2 = 0$ which lead to the solution $y_H(t) = c_1 e^{-4t} + c_2$. The root $r_2 = 0$ indicates that we have first-order resonance corresponding to the term $3t^2$ and the root $r = -4$ indicates that we have first-order resonance corresponding to the term $t^4 e^{-4t}$. No resonance is detected for the last term. Hence, we assume that $y_P(t) = t(A_0 + A_1 t + A_2 t^2) + t(B_0 + B_1 t + B_2 t^2 + B_3 t^3 + B_4 t^4)e^{-4t} + C \cos(2t) + D \sin(2t)$.

22. $y'' + 2y = 2t + 3t \sin(4t)$

Solution. The roots of the characteristic equation are $r_1 = \sqrt{2}i$ and $r_2 = -\sqrt{2}i$ which lead to the solution $y_H(t) = c_1 \cos(\sqrt{2}t) + c_2 \sin(\sqrt{2}t)$. There is no resonance. Hence, we assume that $y_P(t) = A_0 + A_1 t + (B_0 + B_1 t) \cos(4t) + (C_0 + C_1 t) \sin(4t)$.

23. $y'' + 5y' + 6y = e^{-t} \cos(2t) + 6t^2 e^{2t} \sin(t)$

Solution. The roots of the characteristic equation are $r_1 = -3$ and $r_2 = -2$ which lead to the solution $y_H(t) = c_1 e^{-3t} + c_2 e^{-2t}$. There is no resonance. Hence, we assume that $y_P(t) = A e^{-t} \cos(2t) + B e^{-t} \sin(2t) + (C_0 + C_1 t + C_2 t^2)e^{2t} \sin(t) + (D_0 + D_1 t + D_2 t^2)e^{2t} \cos(t)$.

24. $y'' + 2y' + 2y = 3e^{-t} + 2e^{-t} \cos(t) + 4e^{-t} t^2 \sin(t)$

Solution. The roots of the characteristic equation are $r_1 = -1 + i$ and $r_2 = -1 - i$ which lead to the solution $y_H(t) = c_1 e^{-t} \cos(t) + c_2 e^{-t} \sin(t)$. These roots indicate that we have no resonance for the term e^{-t}, first-order resonance corresponding to the term $2e^{-t} \cos(t)$ and no resonance is detected for the last term. Hence, we assume that $y_P(t) = A e^{-t} + t(D_0 + D_1 t + D_2 t^2)e^{-t} \cos(t) + t(E_0 + E_1 t + E_2 t^2)e^{-t} \sin(t)$.

25. $y'' + 2y' + 5y = 4t e^{-t} \cos(2t) + 2t e^{-2t} \cos(t)$

Solution. The roots of the characteristic equation are $r_1 = -1 + 2i$ and $r_2 = -1 - 2i$ which lead to the solution $y_H(t) = c_1 e^{-t} \cos(2t) + c_2 e^{-t} \sin(2t)$. These roots indicate that we have first-order resonance corresponding to the term $4t\,e^{-t}\cos(2t)$ and no resonance is detected for the last term. Hence, we assume that $y_P(t) = t(A_0 + A_1 t)e^{-t}\cos(2t) + t(B_0 + B_1 t)e^{-t}\sin(2t)$

$+ (C_0 + C_1 t)e^{-2t}\cos(t) + (D_0 + D_1 t)e^{-2t}\sin(t)$

26. $y'' + 9y = t^2\sin(3t) + (3t + 4)\cos(3t)$

Solution. The roots of the characteristic equation are $r_1 = 3i$ and $r_2 = -3i$ which lead to the solution $y_H(t) = c_1\cos(3t) + c_2\sin(3t)$. These roots indicate that we have first-order resonance corresponding to both terms $t^2\sin(3t)$ and $(3t + 4)\cos(3t)$. Hence, we assume that $y_P(t) = t(A_0 + A_1 t + A_2 t^2)\cos(3t) + t(B_0 + B_1 t + B_2 t^2)\sin(3t)$.

27. $y'' + 6y' + 5y = (t + 1)e^t\sin(2t) + 3t^2 e^{5t} + 2t^3 e^{-5t}$

Solution. The roots of the characteristic equation are $r_1 = -5$ and $r_2 = -1$ which lead to the solution $y_H(t) = c_1 e^{-5t} + c_2 e^{-t}$. These roots indicate that we have first-order resonance corresponding to the term $2t^3 e^{-5t}$ and no resonance is detected for the other terms. Hence, we assume that $y_P(t) = (A_0 + A_1 t)e^t\cos(2t) + (B_0 + B_1 t)e^t\sin(2t) + (C_0 + C_1 t + C_2 t^2)e^{5t}$

$+ t(D_0 + D_1 t + D_2 t^2 + D_3 t^3)e^{-5t}$.

28. $y'' + 4y' + 5y = 3t^2 e^{-2t}\cos(t) + 4t\,e^{2t}\sin(t)$

Solution. The roots of the characteristic equation are $r_1 = -2 + i$ and $r_2 = -2 - i$ which lead to the solution $y_H(t) = c_1 e^{-2t}\cos(t) + c_2 e^{-2t}\sin t$. These roots indicate that we have first-order resonance corresponding to the term $3t^2 e^{-2t}\cos(t)$ and no resonance is detected for the last term. Hence, we assume that $y_P(t) = t(A_0 + A_1 t + A_2 t^2)e^{-2t}\cos(t) + t(B_0 + B_1 t + B_2 t^2)e^{-2t}\sin(t)$

$+ (C_0 + C_1 t)e^{2t}\cos(t) + (D_0 + D_1 t)e^{2t}\sin(t)$.

29. $y'' + 6y' + 8y = 4e^{-2t} + 3t\,e^{-3t} + 2t^2 e^{-4t} + t^3 e^{-5t}$

Solution. The roots of the characteristic equation are $r_1 = -4$ and $r_2 = -2$ which lead to the solution $y_H(t) = c_1 e^{-4t} + c_2 e^{-2t}$. These roots indicate that we have first-order resonance corresponding to the terms $4e^{-2t}$ and $2t^2 e^{-4t}$ and no resonance is detected for the other terms. Hence, we assume that $y_P(t) = A\,t\,e^{-2t} + (B_0 + B_1 t)e^{-3t} + t(C_0 + C_1 t + C_2 t^2)e^{-4t}$

$+ (D_0 + D_1 t + D_2 t^2 + D_3 t^3)e^{-5t}$.

30. $y'' + 4y' + 4y = t^2 e^{-2t} + 4 + t^2 e^{4t}$

Solution. The roots of the characteristic equation are $r_1 = -2$ and $r_2 = -2$ which lead to the solution $y_H(t) = c_1 e^{-2t} + c_2 t\,e^{-2t}$. These roots indicate that we have second-order resonance corresponding to the term $t^2 e^{-2t}$ and no resonance is detected for the other terms. Hence, we assume that $y_P(t) = t^2(A_0 + A_1 t + A_2 t^2)e^{-2t} + B + (C_0 + C_1 t + C_2 t^2)e^{4t}$.

31. Suppose that $c \neq 0$. Show that there is a solution of the differential equation

$$L[y] = a y'' + b y' + c y = t^n \qquad (S.70)$$

of the form

$$y_P(t) = P_n(t) = A_0 + A_1 t + A_2 t^2 + \cdots + A_n t^n, \qquad (S.71)$$

Solution. Inserting (S.71) into (S.70), we have

$$L[y_P] = a\left(2A_2 + 6A_3 t + \cdots + n(n-1)A_n t^{n-2}\right) + b\left(A_1 + 2A_2 t + \cdots + n A_n t^{n-1}\right)$$
$$+ c\left(A_0 + A_1 t + \cdots + A_n t^n\right)$$
$$= (2a\, A_2 + b\, A_1 + c\, A_0) + \cdots + (b\, n\, A_n + c\, A_{n-1})t^{n-1} + c\, A_n t^n.$$

Since $c \neq 0$ we can take $A_n = 1/c$ and determine the remaining coefficients by direct substitution, where we equate all of the lower terms to zero.

32. Suppose that $b \neq 0$. Show that there is a solution of the differential equation

$$L[y] = a y'' + b y' = t^n \qquad (S.72)$$

of the form

$$y_P(t) = t\, P_n(t) = t\left(A_0 + A_1 t + A_2 t^2 + \cdots + A_n t^n\right), \qquad (S.73)$$

Solution. Inserting (S.73) into (S.72) we have

$$L[y_P] = (2a\, A_2 + b\, A_1) + t(6a\, A_3 + 2b\, A_2) + \cdots + (n+1)b\, A_{n+1} t^n.$$

Since $b \neq 0$ then we solve for A_{n+1}:

$$A_{n+1} = \frac{1}{b(n+1)}.$$

We can determine the remaining coefficients by setting all of the lower terms to zero and solving for $A_n, \ldots, A_1$.

33. Suppose that $a \neq 0$. Show that there is a solution of the differential equation $a\, y'' = t^n$ of the form
$$y_P = A_{n+2} t^{n+2}.$$

Solution. Take
$$y_P = \frac{t^{n+2}}{a(n+1)(n+2)}$$

34. Suppose that $r = \alpha + i\beta$ is a complex number for which $a\, r^2 + br + c \neq 0$. Show that there is a solution of the differential equation
$$a\, y'' + b\, y' + c\, y = t^n\, e^{\alpha t}\, \cos(\beta t) \qquad\qquad \text{(S.74)}$$

of the form
$$y(t) = e^{\alpha t}\big(P_n(t)\cos(\beta t) + Q_n(t)\sin(\beta t)\big),$$

where $P_n(t)$ and $Q_n(t)$ are polynomials of degree n.

Solution. The proof is most effectively done by first solving the equation with a complex-valued right side $t^n e^{rt}$ where $r = \alpha + i\beta$. The real-valued solutions are then obtained by taking the real and imaginary parts of the complex-valued solution.

We look for a particular solution of the form
$$y_P = P_n(t)e^{rt},$$

where P_n is a polynomial of degree n. Applying the requisite derivatives yields
$$y_P'(t) = P_n' e^{rt} + r\, P_n e^{rt} \qquad \text{and} \qquad y_P''(t) = P_n'' e^{rt} + 2r\, P_n' e^{rt} + r^2 P_n e^{rt}.$$

Hence
$$L[y_P] = \big(a(P_n'' + 2r P_n' + r^2 P_n) + b(P_n' + r P_n) + c\, P_n\big)e^{rt}$$
$$= \big(a\, P_n'' + (2a\, r + b)P_n' + (a\, r^2 + br + c)P_n\big)e^{rt}.$$

Therefore, we can cancel the common factor of e^{rt} from the differential equation $L[y_P] = t^n e^{rt}$ and obtain the following equation for the polynomial function $P_n(t)$:
$$a\, P_n'' + (2a\, r + b)P_n' + (a\, r^2 + br + c)P_n = t^n.$$

By hypothesis the constant term $a\, r^2 + br + c \neq 0$, so that we can apply the results found for the purely polynomial right side in Exercise 31 and conclude that the polynomial $P_n(t)$ can be found by equating the coefficients of the powers of t.

35. Suppose that $r = \alpha + i\beta$ is a complex number for which $ar^2 + br + c = 0$ and $2ar + b \neq 0$. Show that there is a solution of (S.74) of the form

$$y(t) = t\, e^{\alpha t}\left(P_n(t)\cos(\beta t) + Q_n(t)\sin(\beta t)\right),$$

where $P_n(t)$ and $Q_n(t)$ are polynomials of degree n.

Solution. In this case we obtain a particular solution in the form

$$y_P = t\left(P_n(t)\cos(\beta t) + Q_n(t)\sin(\beta t)\right)e^{\alpha t}$$

$$= \left(P_{n+1}(t)\cos(\beta t) + Q_{n+1}(t)\sin(\beta t)\right)e^{\alpha t}.$$

Computing as above, we have

$$L[P_{n+1}e^{rt}] = \left(a(P_{n+1}'' + 2r\, P_{n+1}' + r^2 P_{n+1}) + b(P_{n+1}' + r\, P_{n+1}) + c\, P_{n+1}\right)e^{rt}$$

$$= \left(a\, P_{n+1}'' + (2ar + b)P_{n+1}' + (ar^2 + br + c)P_{n+1}\right)e^{rt}.$$

Therefore, we can cancel the common factor of e^{rt} from the differential equation $L[y] = t^n e^{rt}$ and obtain the following equation for the polynomial function $P_{n+1}(t)$:

$$a\, P_{n+1}'' + (2ar + b)P_{n+1}' + (ar^2 + br + c)P_{n+1} = t^n.$$

By hypothesis the constant term $ar^2 + br + c = 0$, but $2ar + b \neq 0$. Again we can apply the results found for the purely polynomial right side in Exercise 32 and conclude that the polynomial $P_{n+1}(t)$ can be found by equating the coefficients of the powers of t.

36. Suppose that $r = \alpha + i\beta$ is a complex number for which $ar^2 + br + c = 0$ and $2ar + b = 0$. Show that there is a solution of (S.74) of the form

$$y(t) = t^2\, P_n(t)e^{\alpha t},$$

where $P_n(t)$ is a polynomial of degree n.

Solution. The condition $2ar + b = 0$ shows that $\beta = 0$ so that in this case $r = -b/2a$ is real and we look for the particular solution in the form $y_P = t^2 P_n(t)e^{rt}$, where $P_n(t)$ is a polynomial of degree n. Repeating the above steps from the solution of Exercise 34 and using the hypotheses yields

$$L[t^2 P_n(t)e^{rt}] = L[P_{n+2}(t)e^{rt}] = a\, P_{n+2}''(t)e^{rt}$$

so that we must solve the equation $a\, P_{n+2}'' = t^n$, which immediately yields the desired result.

Section 9.4, page 300

Use the method of variation of parameters to find a particular solution of the indicated equation.

1. $y'' - 4y' + 4y = (t+1)e^{2t}$

   Solution. The fundamental set for this equation is $\{e^{2t}, t\,e^{2t}\}$ with a Wronskian of e^{4t}. Therefore,

   $$y_P(t) = \int \frac{y_1(s)y_2(t) - y_1(t)y_2(s)}{W(y_1, y_2)(s)} g(s)\,ds \Big|_{s \to t} = \int (1+s)(t-s)e^{2t}\,ds \Big|_{s \to t}.$$

   After integrating, we obtain $y_P(t) = \dfrac{t^2 e^{2t}}{2} + \dfrac{t^3 e^{2t}}{6}$.

2. $4y'' + 36y = \csc(3t)$

   Solution. The fundamental set for this equation is $\{\cos(3t), \sin(3t)\}$ with a Wronskian of 3. Therefore,

   $$y_P(t) = \int \frac{y_1(s)y_2(t) - y_1(t)y_2(s)}{W(y_1, y_2)(s)} g(s)\,ds \Big|_{s \to t}$$

   $$= -\int \csc(3s) \sin\big(3(s-t)\big)/3\,ds \Big|_{s \to t}.$$

   After integrating, we obtain $y_P(t) = \dfrac{\log\big(\sin(3t)\big)\sin(3t)}{9} - \dfrac{t\cos(3t)}{3}$.

3. $y'' + y = e^{-t}$

   Solution. The fundamental set for this equation is $\{\cos(t), \sin(t)\}$ with a Wronskian of 1. Therefore,

   $$y_P(t) = \int \frac{y_1(s)y_2(t) - y_1(t)y_2(s)}{W(y_1, y_2)(s)} g(s)\,ds \Big|_{s \to t} = -\int e^{-s} \sin(s-t)\,ds \Big|_{s \to t}.$$

   After integrating, we obtain $y_P(t) = \dfrac{1}{2e^t}$.

4. $y'' + 9y = 9(\sec(3t))^2$

Solution. The fundamental set for this equation is $\{\cos(3t), \sin(3t)\}$ with a Wronskian of 3. Therefore,

$$y_P(t) = \int \frac{y_1(s)y_2(t) - y_1(t)y_2(s)}{W(y_1, y_2)(s)} g(s) \, ds \Big|_{s \to t}$$

$$= -3 \int \left(\sec(3s)\right)^2 \sin\left(3(s - t)\right) ds \Big|_{s \to t}.$$

After integrating, we obtain $y_P(t) = 3t \sin(3t) + \cos(3t) \log\left(\cos(3t)\right)$.

5. $y'' + y = e^t \sin(t)$

Solution. The fundamental set for this equation is $\{\cos(t), \sin(t)\}$ with a Wronskian of 1. Therefore,

$$y_P(t) = \int \frac{y_1(s)y_2(t) - y_1(t)y_2(s)}{W(y_1, y_2)(s)} g(s) \, ds \Big|_{s \to t}$$

$$= -\int e^s \sin(s) \sin\left(3(s - t)\right) ds \Big|_{s \to t}.$$

After integrating, we obtain $y_P(t) = \dfrac{e^t \sin(t)}{5} - \dfrac{2e^t \cos(t)}{5}$.

6. $y'' + y = \sec(t)\tan(t)$

Solution. The fundamental set for this equation is $\{\cos(t), \sin(t)\}$ with a Wronskian of 1. Therefore,

$$y_P(t) = \int \frac{y_1(s)y_2(t) - y_1(t)y_2(s)}{W(y_1, y_2)(s)} g(s) \, ds \Big|_{s \to t}$$

$$= -\int \sec(s)\tan(s)\sin(s - t) \, ds \Big|_{s \to t}.$$

After integrating, we obtain $y_P(t) = t \cos(t) - \sin(t)\left(1 + \log\left(\cos(t)\right)\right)$.

7. $y'' - 2y' + y = \dfrac{e^t}{t}$

Solution. The fundamental set for this equation is $\{e^t, t\,e^t\}$ with a Wronskian of e^{2t}. Therefore,

$$y_P(t) = \int \frac{y_1(s)y_2(t) - y_1(t)y_2(s)}{W(y_1, y_2)(s)} g(s) \, ds \Big|_{s \to t} = e^t \int \frac{(t - s)}{s} \, ds \Big|_{s \to t}.$$

After integrating, we obtain $y_P(t) = t\,e^t(\log(t) - 1)$.

8. $y'' - 2y' + y = \dfrac{e^t}{t^2}$

Solution. The fundamental set for this equation is $\{e^t, t\,e^t\}$ with a Wronskian of e^{2t}. Therefore,

$$y_P(t) = \int \frac{y_1(s)y_2(t) - y_1(t)y_2(s)}{W(y_1, y_2)(s)} g(s)\, ds \bigg|_{s \to t} = e^t \int \frac{(t-s)}{s^2}\, ds \bigg|_{s \to t}.$$

After integrating, we obtain $y_P(t) = -e^t(\log(t) + 1)$.

9. $y'' + y = \tan(t)$

Solution. The fundamental set for this equation is $\{\cos(t), \sin(t)\}$ with a Wronskian of 1. Therefore,

$$y_P(t) = \int \frac{y_1(s)y_2(t) - y_1(t)y_2(s)}{W(y_1, y_2)(s)} g(s)\, ds \bigg|_{s \to t} = -\int \tan(s)\sin(s - t)\, ds \bigg|_{s \to t}.$$

After integrating, we obtain $y_P(t) = \cos(t)\log\big(\cos(t)\big) - \cos(t)\log\big(1 + \cos(t)\big)$.

10. $y'' - 3y' + 2y = \dfrac{1}{1 + e^{-t}}$

Solution. The fundamental set for this equation is $\{e^t, e^{2t}\}$ with a Wronskian of e^{3t}. Therefore,

$$y_P(t) = \int \frac{y_1(s)y_2(t) - y_1(t)y_2(s)}{W(y_1, y_2)(s)} g(s)\, ds \bigg|_{s \to t} = \int \frac{(e^t - e^s)e^{t-s}}{1 + e^s}\, ds \bigg|_{s \to t}.$$

After integrating, we obtain $y_P(t) = -e^t\big(e^t \log(1 + e^{-t}) + \log(1 + e^{-t}) - 1\big)$.

11. $y'' + 2y' + y = e^{-t}\log(t)$

Solution. The fundamental set for this equation is $\{e^{-t}, t\,e^{-t}\}$ with a Wronskian of e^{-2t}. Therefore,

$$y_P(t) = \int \frac{y_1(s)y_2(t) - y_1(t)y_2(s)}{W(y_1, y_2)(s)} g(s)\, ds \bigg|_{s \to t} = e^{-t} \int \log(s)(t - s)\, ds \bigg|_{s \to t}.$$

After integrating, we obtain $y_P(t) = \dfrac{t^2}{e^t}\left(\dfrac{\log(t)}{2} - \dfrac{3}{4}\right)$.

12. $t^2 y'' - 2t\, y' + 2y = t^3 \sin(t)$. [Hint: $y_1 = t$, $y_2 = t^2$.]

Solution. The fundamental set for this equation is given by the hint as $\{t, t^2\}$ with a Wronskian of t^2. Therefore,

$$y_P(t) = \int \frac{y_1(s)y_2(t) - y_1(t)y_2(s)}{W(y_1, y_2)(s)} g(s)\, ds \Big|_{s \to t} = t \int s^2(t - s)\sin(s)\, ds \Big|_{s \to t}.$$

After integrating, we obtain $y_P(t) = -t \sin(t)$.

13. $y'' - 2y' + y = t^n e^t$, where $n \neq -1, -2$. (Notice that n need not be an integer.)

Solution. The fundamental set for this equation is $\{e^t, t\, e^t\}$ with a Wronskian of e^{2t}. Therefore,

$$y_P(t) = \int \frac{y_1(s)y_2(t) - y_1(t)y_2(s)}{W(y_1, y_2)(s)} g(s)\, ds \Big|_{s \to t} = e^t \int s^n(t - s)\, ds \Big|_{s \to t}.$$

After integrating, we obtain $y_P(t) = \dfrac{t^{n+2} e^t}{n^2 + 3n + 2}$.

14. Find a particular solution of $y'' + 2y' + y = 4t^2 - 3 + \dfrac{e^{-t}}{t}$

Solution. The fundamental set for this equation is $\{e^{-t}, t\, e^{-t}\}$ with a Wronskian of e^{-2t}. Therefore,

$$y_P(t) = \int \frac{y_1(s)y_2(t) - y_1(t)y_2(s)}{W(y_1, y_2)(s)} g(s)\, ds \Big|_{s \to t}$$

$$= e^{-t} \int \frac{(4s^3 e^s - 3s\, e^s + 1)(t - s)}{s}\, ds \Big|_{s \to t}.$$

After integrating, we obtain $y_P(t) = 21 - 16t - \dfrac{t}{e^t} + 4t^2 + \dfrac{t \log(t)}{e^t}$.

Use the method of variation of parameters to solve each of the following initial value problems.

15. $\begin{cases} y'' - 4y' + 3y = 9t^2 + 4 \\ y(0) = 6,\ y'(0) = 8 \end{cases}$

Solution. The fundamental set for this equation is $\{e^t, t\, e^{3t}\}$ with a Wronskian of $2e^{4t}$. Therefore,

$$y_p(t) = \int \frac{y_1(s)y_2(t) - y_1(t)y_2(s)}{W(y_1, y_2)(s)} g(s)\, ds \bigg|_{s \to t}$$

$$= \frac{e^t}{2} \int e^{-3s}(e^{2t} - e^{2s})(4 + 9s^2)\, ds \bigg|_{s \to t}.$$

After integrating, we obtain $y_p(t) = 10 + 8t + 3t^2$. Therefore, $y(t) = C_1 e^t + C_2 e^{3t} + 10 + 8t + 3t^2$. The initial values generate the equations $10 + C_1 + C_2 = 6$ and $8 + C_1 + 3C_2 = 8$ which have $C_1 = -6$ and $C_2 = 2$ as the solution. Hence, $y(t) = 10 - 6e^t + 2e^{3t} + 8t + 3t^2$.

16.
$$\begin{cases} y'' + 5y' + 4y = t + e^t \\ y(0) = 0,\ y'(0) = 3 \end{cases}$$

Solution. The fundamental set for this equation is $\{e^{-4t}, t\, e^{-t}\}$ with a Wronskian of $3e^{-5t}$. Therefore,

$$y_p(t) = \int \frac{y_1(s)y_2(t) - y_1(t)y_2(s)}{W(y_1, y_2)(s)} g(s)\, ds \bigg|_{s \to t}$$

$$= \frac{e^{-4t}}{3} \int e^s(e^{3t} - e^{3s})(s + e^s)\, ds \bigg|_{s \to t}.$$

After integrating, we obtain $y_p(t) = -5/16 + e^t/10 + t/4$. Therefore, $y(t) = C_1 e^{-4t} + C_2 e^{-t} - 5/16 + e^t/10 + t/4$. The initial values generate the equations $-17/80 + C_1 + C_2 = 0$ and $7/20 - 4C_1 - C_2 = 3$ which have $C_1 = -229/240$ and $C_2 = 7/6$ as the solution. Hence, $y(t) = -\dfrac{229e^{-4t}}{240} + \dfrac{7e^{-t}}{6} - \dfrac{5}{16} + \dfrac{e^t}{10} + \dfrac{t}{4}$.

17.
$$\begin{cases} y'' - 8y' + 15y = 9t\, e^{2t} \\ y(0) = 5,\ y'(0) = 10 \end{cases}$$

Solution. The fundamental set for this equation is $\{e^{3t}, t\, e^{5t}\}$ with a Wronskian of $2e^{8t}$. Therefore,

$$y_p(t) = \int \frac{y_1(s)y_2(t) - y_1(t)y_2(s)}{W(y_1, y_2)(s)} g(s)\, ds \bigg|_{s \to t} = \frac{9e^{3t}}{2} \int s\, e^{-3s}(e^{2t} - e^{2s})\, ds \bigg|_{s \to t}.$$

After integrating, we obtain $y_p(t) = (4 + 3t)e^{2t}$. Therefore, $y(t) = C_1 e^{3t} + C_2 e^{5t} + (4 + 3t)e^{2t}$. The initial values generate the equations $4 + C_1 + C_2 = 5$ and $11 + 3C_1 + 5C_2 = 10$ which have $C_1 = 3$ and $C_2 = -2$ as the solution. Hence, $y(t) = 3e^{3t} - 2e^{5t} + (4 + 3t)e^{2t}$.

18. $\begin{cases} y'' + 7y' + 10y = 4t\,e^{-3t}\sin(t) \\ y(0) = 0,\ y'(0) = -1 \end{cases}$

Solution. The fundamental set for this equation is $\{e^{-5t}, t\,e^{-2t}\}$ with a Wronskian of $3e^{-7t}$. Therefore,

$$y_P(t) = \int \frac{y_1(s)y_2(t) - y_1(t)y_2(s)}{W(y_1, y_2)(s)} g(s)\,ds \Big|_{s\to t}$$

$$= \frac{4e^{-5t}}{3} \int s\,e^{-s}(e^{3t} - e^{3s})\sin(s)\,ds \Big|_{s\to t}.$$

After integrating, we obtain $y_P(t) = -(2e^{-3t}/25)\big((11+5t)\cos(t) - (2-15t)\sin(t)\big)$. Therefore, $y(t) = C_1 e^{-5t} + C_2 e^{-2t} - (2e^{-3t}/25)\big((11+5t)\cos(t) - (2-15t)\sin(t)\big)$. The initial values generate the equations $-22/25 + C_1 + C_2 = 0$ and $12/5 - 5C_1 - 2C_2 = -1$ which have $C_1 = 41/75$ and $C_2 = 1/3$ as the solution. Hence, $y(t) = \dfrac{41}{75e^{5t}} + \dfrac{1}{3e^{2t}} - \dfrac{\cos(t)}{25e^{3t}}(22 + 10t) + \dfrac{\sin(t)}{25e^{3t}}(4 - 30t)$.

19. $\begin{cases} 4y'' - y = t\,e^{t/2} \\ y(0) = 1,\ y'(0) = 0 \end{cases}$

Solution. The fundamental set for this equation is $\{e^{-t/2}, e^{t/2}\}$ with a Wronskian of 1. Therefore,

$$y_P(t) = \int \frac{y_1(s)y_2(t) - y_1(t)y_2(s)}{W(y_1, y_2)(s)} g(s)\,ds \Big|_{s\to t} = \frac{e^{-t/2}}{4} \int s(e^t - e^s)\,ds \Big|_{s\to t}.$$

After integrating, we obtain $y_P(t) = e^{t/2}(2 - 2t + t^2)/8$. Therefore, $y(t) = C_1 e^{-t/2} + C_2 e^{t/2} + e^{t/2}(2 - 2t + t^2)/8$. The initial values generate the equations $1/4 + C_1 + C_2 = 1$ and $-1/4 - C_1 + C_2 = 0$ which have $C_1 = 1/4$ and $C_2 = 1/2$ as the solution. Hence, $y(t) = \dfrac{1}{4e^{t/2}} + \dfrac{3e^{t/2}}{4} - \dfrac{t\,e^{t/2}}{4} + \dfrac{t^2 e^{t/2}}{8}$.

20. $\begin{cases} y'' + 2y' - 8y = 2e^{-2t} - e^{-t} \\ y(0) = 1,\ y'(0) = 0 \end{cases}$

Solution. The fundamental set for this equation is $\{e^{-4t}, e^{2t}\}$ with a Wronskian of $6e^{-2t}$. Therefore,

$$y_P(t) = \int \frac{y_1(s)y_2(t) - y_1(t)y_2(s)}{W(y_1, y_2)(s)} g(s)\,ds \Big|_{s\to t}$$

$$= \frac{e^{4t}}{6} \int (2 - e^s)(e^{6t} - e^{6s})e^{4s}\,ds \Big|_{s\to t}.$$

After integrating, we obtain $y_p(t) = -1/(4e^{2t}) + 1/(9e^t)$. Therefore, $y(t) = C_1 e^{-4t} + C_2 e^{2t} - 1/(4e^{2t}) + 1/(9e^t)$. The initial values generate the equations $-5/36 + C_1 + C_2 = 1$ and $7/18 - 4C_1 + 2C_2 = 0$ which have $C_1 = 4/9$ and $C_2 = 25/36$ as the solution. Hence, $y(t) = \dfrac{4}{9e^{4t}} - \dfrac{1}{4e^{2t}} + \dfrac{1}{9e^t} + \dfrac{25e^{2t}}{36}$.

21.
$$\begin{cases} y'' - 4y' + 4y = (12t^2 - 6t)e^{2t} \\ y(0) = 1, \; y'(0) = 0 \end{cases}$$

Solution. The fundamental set for this equation is $\{e^{2t}, t\,e^{2t}\}$ with a Wronskian of e^{4t}. Therefore,

$$y_p(t) = \int \frac{y_1(s)y_2(t) - y_1(t)y_2(s)}{W(y_1, y_2)(s)} g(s)\, ds \bigg|_{s \to t} = 6e^{2t} \int s(1 - 2s)(s - t)\, ds \bigg|_{s \to t}$$

After integrating, we obtain $y_p(t) = -t^3 e^{2t} + t^4 e^{2t}$. Therefore, $y(t) = C_1 e^{2t} + C_2 t\, e^{2t} - t^3 e^{2t} + t^4 e^{2t}$. The initial values generate the equations $C_1 = 1$ and $2C_1 + C_2 = 0$ which have $C_1 = -2$ and $C_2 = 1$ as the solution. Hence, $y(t) = e^{2t} - 2t\, e^{2t} - t^3 e^{2t} + t^4 e^{2t}$.

22.
$$\begin{cases} y'' - 2y' + y = e^t \arctan(t) \\ y(0) = 0, \; y'(0) = 1 \end{cases}$$

Solution. The fundamental set for this equation is $\{e^t, t\,e^t\}$ with a Wronskian of e^{2t}. Therefore,

$$y_p(t) = \int \frac{y_1(s)y_2(t) - y_1(t)y_2(s)}{W(y_1, y_2)(s)} g(s)\, ds \bigg|_{s \to t} = e^t \int (t - s) \arctan(s)\, ds \bigg|_{s \to t}.$$

After integrating, we obtain $y_p(t) = (e^t/2)\big(t - \arctan(t) + t^2 \arctan(t) - t\,\log(1 + t^2)\big)$. Therefore, $y(t) = C_1 e^t + C_2 t\, e^t + (e^t/2)\big(t - \arctan(t) + t^2 \arctan(t) - t\,\log(1 + t^2)\big)$. The initial values generate the equations $C_1 = 0$ and $C_1 + C_2 = 1$ which have $C_1 = 0$ and $C_2 = 1$ as the solution. Hence, $y(t) = \dfrac{3t\,e^t}{2} - \dfrac{e^t \arctan(t)}{2} + \dfrac{t^2 e^t \arctan(t)}{2} - \dfrac{t\,e^t \log(1 + t^2)}{2}$.

23.
$$\begin{cases} y'' + 3y' + 2y = \sin(e^t) \\ y(0) = 1, \; y'(0) = 1 \end{cases}$$

Solution. The fundamental set for this equation is $\{e^{-2t}, e^{-t}\}$ with a Wronskian of e^{-3t}. Therefore,

$$y_P(t) = \int \frac{y_1(s)y_2(t) - y_1(t)y_2(s)}{W(y_1, y_2)(s)} g(s)\, ds \Big|_{s \to t}$$

$$= e^{-2t} \int e^s(e^t - e^s)\sin(e^s)\, ds \Big|_{s \to t}.$$

After integrating, we obtain $y_P(t) = -e^{-2t}\sin(e^t)$. Therefore, $y(t) = C_1 e^{-2t} + C_2 e^{-t} - e^{-2t}\sin(e^t)$. The initial values generate the equations $C_1 + C_2 - \sin(1) = 1$ and $-2C_1 - C_2 - \cos(1) + 2\sin(1) = 1$ which have $C_1 = -2 - \cos(1) + \sin(1)$ and $C_2 = 3 + \cos(1)$ as the solution. Hence, $y(t) = \dfrac{3}{e^t} - \dfrac{2}{e^{2t}} + \dfrac{\cos(1)}{e^t}$

$-\dfrac{\cos(1)}{e^{2t}} + \dfrac{\sin(1)}{e^{2t}} - \dfrac{\sin(e^t)}{e^{2t}}.$

24. $\begin{cases} y'' + 3y' + 2y = (1 - e^t)^{3/2} \\ y(0) = 1,\ y'(0) = 0 \end{cases}$

Solution. The fundamental set for this equation is $\{e^{-2t}, e^{-t}\}$ with a Wronskian of e^{-3t}. Therefore,

$$y_P(t) = -\int (e^{2s-2t} - e^{3s-2t} - e^{s-t} + e^{2s-t})\sqrt{1 - e^s}\, ds \Big|_{s \to t}.$$

After integrating, we obtain $y_P(t) = 4e^{-2t}(1 - e^t)^{7/2}/35$. Therefore, $y(t) = C_1 e^{-2t} + C_2 e^{-t} + 4e^{-2t}(1 - e^t)^{7/2}/35$. The initial values generate the equations $C_1 + C_2 = 1$ and $-2C_1 - C_2 = 0$ which have $C_1 = -1$ and $C_2 = 2$ as the solution. Hence, $y(t) = -e^{-2t} + 2e^{-t} + 4e^{-2t}(1 - e^t)^{7/2}/35$.

The method of variation of parameters can be used to find a particular solution to $y'' + p(t)y' + q(t)y = r(t)$ whether or not $p(t)$ and $q(t)$ are constant, provided that a fundamental set of solutions $\{y_1, y_2\}$ is known. Methods for finding such a fundamental set for a wide variety of equations will be given in Chapters 20 and 21. In the meantime, find the general solutions to the following differential equations for which the fundamental set is provided.

25. $\begin{cases} t^2 y'' - 2t\, y' + 2y = t \\ y_1(t) = t,\ y_2(t) = t^2 \end{cases}$

Solution. Using the given fundamental set, the Wronskian is t^2. Therefore,

$$y_P(t) = \int \frac{y_1(s)y_2(t) - y_1(t)y_2(s)}{W(y_1, y_2)(s)} g(s)\, ds \Big|_{s \to t} = \int \frac{t^2}{s^2} - \frac{t}{s}\, ds \Big|_{s \to t}.$$

After integrating, we obtain $y_p(t) = -t - t\,\log(t)$. Hence,
$$y(t) = C_1 t + C_2 t^2 - t - t\,\log(t).$$

26.
$$\begin{cases} t^2 y'' - 4t\,y' + 6y = \log(t) \\ y_1(t) = t^2,\ y_2(t) = t^3 \end{cases}$$

Solution. Using the given fundamental set, the Wronskian is t^4. Therefore,
$$y_p(t) = \int \frac{y_1(s)y_2(t) - y_1(t)y_2(s)}{W(y_1, y_2)(s)} g(s)\,ds \Big|_{s \to t} = \int \frac{t^3 \log(s)}{s^4} - \frac{t^2 \log(s)}{s^3}\,ds \Big|_{s \to t}.$$

After integrating, we obtain $y_p(t) = 5/36 + \log(t)/6$. Hence,
$$y(t) = C_1 t^2 + C_2 t^3 + \frac{5}{36} + \frac{\log(t)}{6}.$$

27.
$$\begin{cases} t^2 y'' - 2t\,y' + 2y = \big(\log(t)\big)^2 \\ y_1(t) = t,\ y_2(t) = t^2 \end{cases}$$

Solution. Using the given fundamental set, the Wronskian is t^2. Therefore,
$$y_p(t) = \int \frac{y_1(s)y_2(t) - y_1(t)y_2(s)}{W(y_1, y_2)(s)} g(s)\,ds \Big|_{s \to t}$$
$$= -\int \frac{t\big(\log(s)\big)^2}{s^2} - \frac{t^2\big(\log(s)\big)^2}{s^3}\,ds \Big|_{s \to t}.$$

After integrating, we obtain $y_p(t) = 7/4 + 3\log(t)/2 + \big(\log(t)\big)^2/2$. Hence,
$$y(t) = C_1 t + C_2 t^2 + \frac{7}{4} + \frac{3}{2}\log(t) + \frac{1}{2}\big(\log(t)\big)^2.$$

28.
$$\begin{cases} (t-1)y'' - t\,y' + y = 1 \\ y_1(t) = t,\ y_2(t) = e^t \end{cases}$$

Solution. Using the given fundamental set, the Wronskian is $e^t(t-1)$. Therefore,
$$y_p(t) = \int \frac{y_1(s)y_2(t) - y_1(t)y_2(s)}{W(y_1, y_2)(s)} g(s)\,ds \Big|_{s \to t} = \int \frac{s\,e^{t-s} - t}{(s-1)^2}\,ds \Big|_{s \to t}.$$

After integrating, we obtain $y_p(t) = 1$. Hence,
$$y(t) = C_1 t + C_2 e^t + 1.$$

29.
$$\begin{cases} (1+t)y'' - (2+t)y' = e^t \\ y_1(t) = 1,\ y_2(t) = t\,e^t \end{cases}$$

Solution. Using the given fundamental set, the Wronskian is $e^t(1+t)$. Therefore,

$$y_P(t) = \int \frac{y_1(s)y_2(t) - y_1(t)y_2(s)}{W(y_1, y_2)(s)} g(s)\, ds \Big|_{s\to t} = \int \frac{t\, e^t - s\, e^s}{(1+s)^2}\, ds \Big|_{s\to t}.$$

After integrating, we obtain $y_P(t) = -e^t$. Hence,
$$y(t) = C_1 + C_2\, t\, e^t - e^t.$$

30. $$\begin{cases} \big(\cos(t) - \sin(t)\big)y'' + 2\sin(t)y' - \big(\cos(t) + \sin(t)\big)y = 1 \\ y_1(t) = \sin(t),\ y_2(t) = e^t \end{cases}$$

Solution. Using the given fundamental set, the Wronskian is $e^t\big(\sin(t) - \cos(t)\big)$. Therefore,

$$y_P(t) = \int \frac{y_1(s)y_2(t) - y_1(t)y_2(s)}{W(y_1, y_2)(s)} g(s)\, ds \Big|_{s\to t} = \int \frac{\sin(t) - e^{t-s}\sin(s)}{1 - \sin(2s)}\, ds \Big|_{s\to t}.$$

After integrating, we obtain $y_P(t) = -\sin(t)/2 - \cos(t)/2$. Hence,
$$y(t) = C_1\,\sin(t) + C_2\, e^t - \frac{\sin(t)}{2} - \frac{\cos(t)}{2}.$$

Solve the following initial value problems.

31. $$\begin{cases} t^2 y'' - 2t\, y' + 2y = t^2 \\ y(1) = 0,\ y'(1) = 0 \\ y_1(t) = t,\ y_2(t) = t^2 \end{cases}$$

Solution. Using the given fundamental set, the Wronskian is t^2. Therefore,

$$y_P(t) = \int \frac{y_1(s)y_2(t) - y_1(t)y_2(s)}{W(y_1, y_2)(s)} g(s)\, ds \Big|_{s\to t} = -\int t - \frac{t^2}{s}\, ds \Big|_{s\to t}.$$

After integrating, we obtain $y_P(t) = -t^2 + t^2 \log(t)$. Therefore, $y(t) = C_1 t + C_2 t^2 - t^2 + t^2 \log(t)$. The initial values generate the equations $-1 + C_1 + C_2 = 0$ and $-1 + C_1 + 2C_2 = 0$ which have $C_1 = 1$ and $C_2 = 0$ as the solution. Hence, $y(t) = t - t^2 + t^2 \log(t)$

32. $$\begin{cases} t^2 y'' - 3t\, y' + 3y = \log(t) \\ y(1) = 0,\ y'(1) = 0 \\ y_1(t) = t,\ y_2(t) = t^3 \end{cases}$$

Solution. Using the given fundamental set, the Wronskian is $2t^3$. Therefore,

$$y_p(t) = \int \frac{y_1(s)y_2(t) - y_1(t)y_2(s)}{W(y_1, y_2)(s)} g(s)\, ds \bigg|_{s \to t} = \frac{1}{2} \int \frac{t^3 \log(s) - t\, \log(s)}{s^2}\, ds \bigg|_{s \to t}.$$

After integrating, we obtain $y_p(t) = 4/9 + \log(t)/3$. Therefore, $y(t) = C_1 t + C_2 t^3 + 4/9 + \log(t)/3$. The initial values generate the equations $4/9 + C_1 + C_2 = 0$ and $1/3 + C_1 + 3C_2 = 0$ which have $C_1 = -1/2$ and $C_2 = 1/18$ as the solution. Hence,

$$y(t) = \frac{4}{9} - \frac{t}{2} + \frac{t^3}{18} + \frac{\log(t)}{3}.$$

33. $\begin{cases} t^2 y'' - 2y = \left(\log(t)\right)^2 \\ y(1) = 0,\ y'(1) = 0 \\ y_1(t) = 1/t,\ y_2(t) = t^2 \end{cases}$

Solution. Using the given fundamental set, the Wronskian is 3. Therefore,

$$y_p(t) = \int \frac{y_1(s)y_2(t) - y_1(t)y_2(s)}{W(y_1, y_2)(s)} g(s)\, ds \bigg|_{s \to t}$$

$$= \frac{1}{3} \int \left(\frac{t^2}{3s^2} - \frac{1}{3t} \right) \left(\log(s)\right)^2 ds \bigg|_{s \to t}.$$

After integrating, we obtain $y_p(t) = -3/4 + \log(t)/2 - \left(\log(t)\right)^2/2$. Therefore, $y(t) = C_1/t + C_2 t^2 - 3/4 + \log(t)/2 - \left(\log(t)\right)^2/2$. The initial values generate the equations $-3/4 + C_1 + C_2 = 0$ and $1/2 - C_1 + 2C_2 = 0$ which have $C_1 = 2/3$ and $C_2 = 1/12$ as the solution. Hence, $y(t) = -\dfrac{3}{4} + \dfrac{2}{3t} + \dfrac{t^2}{12} + \dfrac{\log(t)}{2} - \dfrac{\left(\log(t)\right)^2}{2}$.

34. $\begin{cases} (t-1)y'' - t\, y' + y = 1 \\ y(0) = 0,\ y'(0) = 0 \\ y_1(t) = t,\ y_2(t) = e^t \end{cases}$

Solution. Using the given fundamental set, the Wronskian is $e^t(t-1)$. Therefore,

$$y_p(t) = \int \frac{y_1(s)y_2(t) - y_1(t)y_2(s)}{W(y_1, y_2)(s)} g(s)\, ds \bigg|_{s \to t} = \int \frac{s\, e^{t-s} - t}{(s-1)^2}\, ds \bigg|_{s \to t}.$$

After integrating, we obtain $y_p(t) = 1$. Therefore, $y(t) = C_1 t + C_2 e^t + 1$. The initial values generate the equations $1 + C_2 = 0$ and $C_1 + C_2 = 0$ which have $C_1 = 1$ and $C_2 = -1$ as the solution. Hence, $y(t) = 1 + t - e^t$.

$$
35. \quad
\begin{cases}
(1+t)y'' - (2+t)y' = e^t \\
y(0) = 0, \ y'(0) = 0 \\
y_1(t) = 1, \ y_2(t) = t\,e^t
\end{cases}
$$

Solution. Using the given fundamental set, the Wronskian is $e^t(1+t)$. Therefore,

$$
y_P(t) = \int \frac{y_1(s)y_2(t) - y_1(t)y_2(s)}{W(y_1, y_2)(s)} g(s)\, ds \bigg|_{s \to t} = \int \frac{t\,e^t - s\,e^s}{(1+s)^2}\, ds \bigg|_{s \to t}.
$$

After integrating, we obtain $y_P(t) = -e^t$. Therefore, $y(t) = C_1 + C_2 t\, e^t - e^t$. The initial values generate the equations $-1 + C_1 = 0$ and $-1 + C_2 = 0$ which have $C_1 = 1$ and $C_2 = 1$ as the solution. Hence, $y(t) = 1 - e^t + t\, e^t$.

$$
36. \quad
\begin{cases}
\big(\cos(t) - \sin(t)\big)y'' + 2\sin(t)y' - \big(\cos(t) + \sin(t)\big)y = 1 \\
y(0) = 0, \ y'(0) = 0 \\
y_1(t) = \sin(t), \ y_2(t) = e^t
\end{cases}
$$

Solution. Using the given fundamental set, the Wronskian is $e^t\big(\sin(t) - \cos(t)\big)$. Therefore,

$$
y_P(t) = \int \frac{y_1(s)y_2(t) - y_1(t)y_2(s)}{W(y_1, y_2)(s)} g(s)\, ds \bigg|_{s \to t} = \int \frac{\sin(t) - e^{t-s}\sin(s)}{1 - \sin(2s)}\, ds \bigg|_{s \to t}.
$$

After integrating, we obtain $y_P(t) = -\cos(t)/2 - \sin(t)/2$. Therefore, $y(t) = C_1 \sin(t) + C_2 e^t - \cos(t)/2 - \sin(t)/2$. The initial values generate the equations $-1/2 + C_2 = 0$ and $-1/2 + C_1 + C_2 = 0$ which have $C_1 = 0$ and $C_2 = 1/2$ as the solution. Hence, $y(t) = \dfrac{e^t}{2} - \dfrac{\cos(t)}{2} - \dfrac{\sin(t)}{2}$.

Chapter 10

Section 10.1, page 314

Use **ODE** to find the general solution of each of the following differential equations.

1. $3y'' + 4y' - 2y = 0$

Solution. The roots of the characteristic equation $3r^2 + 4r - 2 = 0$ are $r_1 = -2/3 + \sqrt{10}/3$ and $r_2 = -2/3 - \sqrt{10}/3$. Consequently, the general solution is $y = C_1 e^{(-2/3 - \sqrt{10}/3)t} + C_2 e^{(-2/3 + \sqrt{10}/3)t}$. Using **ODE**, we enter

```
ODE[3y'' + 4y' - 2y == 0,y,t,Method->SecondOrderLinear]
```

2. $y'' + 24y' - 2y = 0$

    Solution. The roots of the characteristic equation $r^2 + 24r - 2 = 0$ are $r_1 = -12 - \sqrt{146}$ and $r_2 = -12 + \sqrt{146}$. Consequently, the general solution is $y = C_1 e^{(-12-\sqrt{146})t} + C_2 e^{(-12+\sqrt{146})t}$. Using **ODE**, we enter

    ```
    ODE[y'' + 24y' - 2y == 0,y,t,Method->SecondOrderLinear]
    ```

3. $32y'' + 25y' - 4y = 0$

    Solution. The roots of the characteristic equation $32r^2 + 25r - 4 = 0$ are $r_1 = -25/64 - \sqrt{1137}/64$ and $r_2 = -25/64 + \sqrt{1137}/64$. Consequently, the general solution is $y = C_1 e^{(-25/64-\sqrt{1137}/64)t} + C_2 e^{(-25/64+\sqrt{1137}/64)t}$. Using **ODE**, we enter

    ```
    ODE[32y'' + 25y' - 4y == 0,y,t,
    Method->SecondOrderLinear]
    ```

4. $2y'' - y' + 5y = 0$

    Solution. The roots of the characteristic equation $2r^2 - r + 5 = 0$ are $r_1 = 1/4 - \sqrt{39}/4\,i$ and $r_2 = 1/4 + \sqrt{39}/4\,i$. Consequently, the general solution is $y = C_1 e^{t/4} \sin\left((\sqrt{39}/4)t\right) + C_2 e^{t/4} \cos\left((\sqrt{39}/4)t\right)$. Using **ODE**, we enter

    ```
    ODE[2y'' - y' + 5y == 0,y,t,Method->SecondOrderLinear]
    ```

5. $y'' + 16y = e^t - 9t^2$

    Solution. The roots of the characteristic equation $r^2 + 16 = 0$ are $r_1 = -4i$ and $r_2 = 4i$. Consequently, the general solution to the homogeneous equation is $y_H(t) = C_1 \sin(4t) + C_2 \cos(4t)$. The particular solution can be computed from the form $y_P(t) = Ae^t + (B_0 + B_1 t + B_2 t^2)$ using undetermined coefficients. With **ODE**, we simply enter

    ```
    ODE[y'' + 16y == E^t - 9t^2,y,t,
    Method->SecondOrderLinear]
    ```

6. $y'' + 2i\,y' - 3y = 0$

    Solution. The roots of the characteristic equation $r^2 + 2ir - 3 = 0$ are $r_1 = -\sqrt{2} - i$ and $r_2 = \sqrt{2} - i$. Consequently, the general solution is $y = C_1 e^{(-\sqrt{2}-i)t} + C_2 e^{(\sqrt{2}-i)t}$. Using **ODE**, we enter

    ```
    ODE[y'' + 2I y' - 3y == 0,y,t,Method->SecondOrderLinear]
    ```

7. $i\,y'' + y' - 3y = \sin(i\,t)$

    Solution. The roots of the characteristic equation $i\,r^2 + r - 3 = 0$ are $r_1 = i/2 - \sqrt{-1 - 12i}/2$ and $r_2 = i/2 + \sqrt{-1 - 12i}/2$. Consequently, the general solution to the homogeneous equation is $y_H(t) = C_1 e^{(i/2 - \sqrt{-1-12i}/2)t} + C_2 e^{(i/2 + \sqrt{-1-12i}/2)t}$. The particular solution can be computed from the form $y_P(t) = A \sin(it) + B \cos(it)$ using undetermined coefficients. With **ODE**, we simply enter

```
ODE[I y'' + y' - 3y == Sin[I t],y,t,
Method->SecondOrderLinear]
```

8. $y'' + 6y' + 9 = t^2 e^{-3t}$

Solution. The roots of the characteristic equation $r^2 + 6r + 9 = 0$ are $r_1 = -3$ and $r_2 = -3$. Consequently, the general solution to the homogeneous equation is $y_H(t) = C_1 e^{-3t} + C_2 t\, e^{-3t}$. Because of second-order resonance, the particular solution is computed from the form $y_P(t) = t^2 (A_0 + A_1 t + A_2 t^2) e^{-3t}$ using undetermined coefficients. With **ODE**, we simply enter

```
ODE[y''+6y'+9 == t^2 E^(-3t),y,t,
Method->SecondOrderLinear]
```

9. $y'' - 4y' + 12y = 15e^t - 4\cos(t)$

Solution. The roots of the characteristic equation $r^2 - 4r + 12 = 0$ are $r_1 = 2 - 2\sqrt{2}i$ and $r_2 = 2 + 2\sqrt{2}i$. Consequently, the general solution to the homogeneous equation is $y_H(t) = C_1 e^{2t} \sin(2\sqrt{2}t) + C_2 e^{2t} \cos(2\sqrt{2}t)$. The particular solution can be computed from the form $y_P(t) = Ae^t + B\cos(t) + C\sin(t)$ using undetermined coefficients. With **ODE**, we simply enter

```
ODE[y'' - 4y' + 12y == 15E^t -4 Cos[t],y,t,
Method->SecondOrderLinear]
```

10. $y'' + 12y = \sin(12t)$

Solution. The roots of the characteristic equation $r^2 + 12 = 0$ are $r_1 = -2\sqrt{3}i$ and $r_2 = 2\sqrt{3}i$. Consequently, the general solution to the homogeneous equation is $y_H(t) = C_1 \sin(2\sqrt{3}t) + C_2 \cos(2\sqrt{3}t)$. The particular solution can be computed from the form $y_P(t) = A\cos(12t) + B\sin(12t)$ using undetermined coefficients. With **ODE**, we simply enter

```
ODE[y'' + 12y == Sin[12t],y,t,Method->SecondOrderLinear]
```

Use **ODE** to solve and plot each of the following initial value problems. This requires some experimentation to find a good interval to plot the solution over. In each problem begin by using the interval $-1 < t < 1$, and then try to determine a larger or smaller interval for which the graph of the solution is more interesting.

11. $\begin{cases} 2y'' - 10y' - 3y = 0, \\ y(0) = 5, \quad y'(0) = -3 \end{cases}$

Solution. The roots of the characteristic equation $2r^2 - 10r - 3 = 0$ are $r_1 = 5/2 - \sqrt{31}/2$ and $r_2 = 5/2 + \sqrt{31}/2$. Consequently, the general solution is $y = C_1 e^{((5-\sqrt{31})/2)t} + C_2 e^{((5+\sqrt{31})/2)t}$. Using the initial conditions, we obtain the equations $C_1 + C_2 = 5$ and $((5 - \sqrt{31})/2)C_1 + ((5 + \sqrt{31})/2)C_2 = 3$, which lead to $C_1 = 5/2 + 19/(2\sqrt{31})$ and $C_2 = 5/2 - 19/(2\sqrt{31})$. Hence, $y(t) = (5/2 + 19/(2\sqrt{31}))e^{((5-\sqrt{31})/2)t} + (5/2 - 19/(2\sqrt{31}))e^{((5+\sqrt{31})/2)t}$. With **ODE**, we choose the interval $[-1, 1]$ and enter

```
ODE[{2y'' - 10y' - 3y == 0,y[0] == 5,y'[0] == -3},y,t,
    Method->SecondOrderLinear,
    PlotSolution->{{t,-1,1}}]
```

12. $\begin{cases} y'' + 10y = 0, \\ y(0) = 1, \ y'(0) = -1 \end{cases}$

Solution. The roots of the characteristic equation $r^2 + 10 = 0$ are $r_1 = -\sqrt{10}i$ and $r_2 = \sqrt{10}i$. Consequently, the general solution is $y = C_1 \sin(\sqrt{10}t) + C_2 \cos(\sqrt{10}t)$. Using the initial conditions, we obtain the equations $\sqrt{10}C_1 = -1$ and $C_2 = 1$, which lead to $C_1 = -1/\sqrt{10}$ and $C_2 = 1$. Hence, $y(t) = -\sin(\sqrt{10}t)/\sqrt{10} + \cos(\sqrt{10}t)$. With **ODE**, we choose the interval $[-1, 1]$ and enter

```
ODE[{y'' + 10y == 0,y[0] == 1,y'[0] == -1},y,t,
    Method->SecondOrderLinear,
    PlotSolution->{{t,-1,1}}]
```

13. $\begin{cases} y'' - 3y' + 10y = 0, \\ y(0) = 1, \ y'(0) = -1 \end{cases}$

Solution. The roots of the characteristic equation $r^2 - 3r + 10 = 0$ are $r_1 = 3/2 - \sqrt{31}i$ and $r_2 = 3/2 + \sqrt{31}i$. Consequently, the general solution is $y = C_1 e^{3t/2} \sin(\sqrt{31}t) + C_2 e^{3t/2} \cos(\sqrt{31}t)$. Using the initial conditions, we obtain the equations $\sqrt{31}C_1 + 3C_2 = -2$ and $C_2 = 1$, which lead to $C_1 = -5/\sqrt{31}$ and $C_2 = 1$. Hence, $y(t) = -(5/\sqrt{31})e^{3t/2} \sin(\sqrt{31}t) + e^{3t/2} \cos(\sqrt{31}t)$. With **ODE**, we choose the interval $[-3, 3]$ and enter

```
ODE[{y'' - 3y' + 10y == 0,y[0] == 1,y'[0] == -1},y,t,
    Method->SecondOrderLinear,
    PlotSolution->{{t,-3,3}}]
```

14. $\begin{cases} 8y'' + 3y' - 5y = 0, \\ y(0) = 1, \ y'(0) = 2 \end{cases}$

Solution. The roots of the characteristic equation $8r^2 + 3r - 5 = 0$ are $r_1 = -1$ and $r_2 = 5/8$. Consequently, the general solution is $y = C_1 e^{-t} + C_2 e^{5t/8}$. Using the initial conditions, we obtain the equations $C_1 + C_2 = 1$ and $-C_1 + (5/8)C_2 = 2$, which lead to $C_1 = -11/13$ and $C_2 = 24/13$. Hence, $y(t) = -(11/13)e^{-t} + (24/13)e^{5t/8}$. With **ODE**, we choose the interval $[-3, 3]$ and enter

```
ODE[{8y'' + 3y' - 5y == 0,y[0] == 1,y'[0] == 2},y,t,
    Method->SecondOrderLinear,
    PlotSolution->{{t,-3,3}}]
```

15. $\begin{cases} 8y'' + 5y = t^2, \\ y(0) = 1, \quad y'(0) = \pi/\sqrt{5/2} \end{cases}$

Solution. The roots of the characteristic equation $8r^2 + 5 = 0$ are $r_1 = -(\sqrt{5/2}/2)i$ and $r_2 = (\sqrt{5/2}/2)i$. Consequently, the general solution to the homogeneous equation is $y_H(t) = C_1 \sin\left((\sqrt{5/2}/2)t\right) + C_2 \cos\left((\sqrt{5/2}/2)t\right)$. The particular solution is computed from the form $y_P(t) = A_0 + A_1 t + A_2 t^2$ using undetermined coefficients. Solving for the coefficients, we obtain $A_0 = -16/25$, $A_1 = 0$ and $A_2 = 1/5$. Using the initial conditions, we obtain the equations $\sqrt{5/2}C_1 = 2\sqrt{2/5}\pi$ and $-(16/25) + C_2 = 1$, which lead to $C_1 = 4\pi/5$ and $C_2 = 41/25$. Hence, $y(t) = -(16/25) + (5/25)t^2 + (41/25) \cos\left((\sqrt{5/2}/2)t\right) + (4\pi/5) \sin\left((\sqrt{5/2}/2)t\right)$. With **ODE**, we choose the interval $[-10, 10]$ and enter

```
ODE[{8y'' + 5y == t^2,y[0] == 1,y'[0] == Pi/Sqrt[5/2]},
    y,t,Method->SecondOrderLinear,
    PlotSolution->{{t,-10,10}}]
```

16. $\begin{cases} y'' + 5y = t \sin(t), \\ y(0) = 1, \quad y'(0) = \pi \end{cases}$

Solution. The roots of the characteristic equation $r^2 + 5 = 0$ are $r_1 = -\sqrt{5}i$ and $r_2 = \sqrt{5}i$. Consequently, the general solution to the homogeneous equation is $y_H(t) = C_1 \sin(\sqrt{5}t) + C_2 \cos(\sqrt{5}t)$. The particular solution is computed from the form $y_P(t) = (A_0 + A_1 t) \cos(t) + (B_0 + B_1 t) \sin(t)$ using undetermined coefficients. Solving for the coefficients, we obtain $A_0 = -1/8$, $A_1 = 0$, $B_0 = 0$ and $B_1 = 1/4$. Using the initial conditions, we obtain the equations $\sqrt{5}C_1 = \pi$ and $-1/8 + C_2 = 1$, which lead to $C_1 = \pi/\sqrt{5}$ and $C_2 = 9/8$. Hence, $y(t) = -\cos(t)/8 + (t/4) \sin(t) + (9/8) \cos(\sqrt{5}t) + (\pi/\sqrt{5}) \sin(\sqrt{5}t)$. With **ODE**, we choose the interval $[-10, 10]$ and enter

```
ODE[{y'' + 5y == t Sin[t],y[0] == 1,y'[0] == Pi},y,t,
    Method->SecondOrderLinear,
    PlotSolution->{{t,-10,10}}]
```

17. $\begin{cases} y'' - 4y = (1+t)\cos(10t), \\ y(0) = 0, \ y'(0) = 0 \end{cases}$

Solution. The roots of the characteristic equation $r^2 - 4 = 0$ are $r_1 = -2$ and $r_2 = 2$. Consequently, the general solution to the homogeneous equation is $y_H(t) = C_1 e^{-2t} + C_2 e^{2t}$. The particular solution is computed from the form $y_P(t) = (A_0 + A_1 t)\cos(10t) + (B_0 + B_1 t)\sin(10t)$ using undetermined coefficients. Solving for the coefficients, we obtain $A_0 = -1/104$, $A_1 = -1/104$, $B_0 = 5/2704$ and $B_1 = 0$. Using the initial conditions, we obtain the equations $-1/104 + C_1 + C_2 = 0$ and $3/338 - 2C_1 + 2C_2 = 0$, which lead to $C_1 = 19/2704$ and $C_2 = 7/2704$. Hence, $y(t) = -((1+t)/104)\cos(10t) + (5/2704)\sin(10t) + (19/2704)e^{-2t} + (7/2704)e^{2t}$. With **ODE**, we choose the interval $[-1, 1]$ and enter

```
ODE[{y'' - 4y == (1 + t)Cos[10t],y[0] == 0,y'[0] == 0},
    y,t,Method->SecondOrderLinear,
    PlotSolution->{{t,-1,1}}]
```

18. $\begin{cases} y'' - 8y = (1+t-t^2)\,e^t, \\ y(0) = 0, \ y'(0) = 1 \end{cases}$

Solution. The roots of the characteristic equation $r^2 - 8 = 0$ are $r_1 = -2\sqrt{2}$ and $r_2 = 2\sqrt{2}$. Consequently, the general solution to the homogeneous equation is $y_H(t) = C_1 e^{-2\sqrt{2}t} + C_2 e^{2\sqrt{2}t}$. The particular solution is computed from the form $y_P(t) = (A_0 + A_1 t + A_2 t^2)e^t$ using undetermined coefficients. Solving for the coefficients, we obtain $A_0 = -41/343$, $A_1 = -3/49$ and $A_2 = 1/7$. Using the initial conditions, we obtain the equations $-(41/343) + C_1 + C_2 = 0$ and $-(62/343) - 2\sqrt{2}C_1 + 2\sqrt{2}C_2 = 1$, which lead to $C_1 = (41/686) - 405/(1372\sqrt{2})$ and $C_2 = (41/686) + 405/(1372\sqrt{2})$. Hence, $y(t) = \left(-41/343 - 3t/49 + t^2/7\right)e^t + \left(41/686 - 405/(1372\sqrt{2})\right)e^{-2\sqrt{2}t} + \left(41/686 + 405/(1372\sqrt{2})\right)e^{2\sqrt{2}t}$. With **ODE**, we choose the interval $[-1, 1]$ and enter

```
ODE[{y'' - 8y == (1 + t - t^2) Exp[t],
    y[0] == 0,y'[0] == 1},y,t,Method->SecondOrderLinear,
    PlotSolution->{{t,-1,1}}]
```

19. $\begin{cases} y'' - 2y = \sin(t) + e^t, \\ y(0) = 1, \ y'(0) = 0 \end{cases}$

Solution. The roots of the characteristic equation $r^2 - 2 = 0$ are $r_1 = -\sqrt{2}$ and $r_2 = \sqrt{2}$. Consequently, the general solution to the homogeneous equation is

$y_H(t) = C_1 e^{-\sqrt{2}t} + C_2 e^{\sqrt{2}t}$. The particular solution is computed from the form $y_P(t) = A\sin(t) + B\cos(t) + Ce^t$ using undetermined coefficients. Solving for the coefficients, we obtain $A = -1/3$, $B = 0$ and $C = -1$. Using the initial conditions, we obtain the equations $-1 + C_1 + C_2 = 1$ and $-4/3 - \sqrt{2}C_1 + \sqrt{2}C_2 = 0$, which lead to $C_1 = 1 - \sqrt{2}/3$ and $C_2 = 1 + \sqrt{2}/3$. Hence, $y(t) = -\sin(t)/3 - e^t + (1 - \sqrt{2}/3)e^{-\sqrt{2}t} + (1 + \sqrt{2}/3)e^{\sqrt{2}t}$. With **ODE**, we choose the interval $[-1, 1]$ and enter

```
ODE[{y'' - 2y == Sin[t] + Exp[t],y[0] == 1,y'[0] == 0},
    y,t,Method->SecondOrderLinear,
    PlotSolution->{{t,-1,1}}]
```

20. $\begin{cases} y'' - 2y' + y = t\sin(4t) + e^t\cos(t), \\ y(0) = 1, \quad y'(0) = 0 \end{cases}$

Solution. The roots of the characteristic equation $r^2 - 2r + 1 = 0$ are $r_1 = 1$ and $r_2 = 1$. Consequently, the general solution to the homogeneous equation is $y_H(t) = C_1 e^t + C_2 t\, e^t$. The particular solution is computed from the form $y_P(t) = (A_0 + A_1 t)\sin(4t) + (B_0 + B_1 t)\cos(4t) + Ce^t\cos(t)$ using undetermined coefficients. Solving for the coefficients, we obtain $A_0 = -94/4913$, $A_1 = -15/289$, $B_0 = -104/4913$, $B_1 = 8/289$, and $C = -1$. Using the initial conditions, we obtain the equations $-(5017/4913) + C_1 = 1$ and $-(5153/4913) + C_1 + C_2 = 0$, which lead to $C_1 = 9930/4913$ and $C_2 = -281/289$. Hence, $y(t) = -(94/4913 + 15/289t)\sin(4t) - (104/4913 - 8/289t)\cos(4t) - e^t\cos(t) + 9930/4913e^t - 281/289t\, e^t$. With **ODE**, we choose the interval $[-3, 3]$ and enter

```
ODE[{y'' - 2y' + y == t Sin[4t] + Exp[t] Cos[t],
    y[0] == 1,y'[0] == 0},y,t,Method->SecondOrderLinear,
    PlotSolution->{{t,-3,3}}]
```

Use **ODE** to solve and stack plot families of solutions to each of the following initial value problems. This requires some experimentation to find a good interval to plot the solution over. Also, it may be necessary to adjust the view point using the option **ViewPoint**. In each problem begin by using the interval $-1 < t < 1$, and then try to determine a larger or smaller interval for which the graph of the solution is more interesting.

21. $\begin{cases} y'' + y = \sin(2t), \\ y(0) = 0, \quad 0 \le y'(0) \le 4 \end{cases}$

Solution. The roots of the characteristic equation $r^2 + 1 = 0$ are $r_1 = -i$ and $r_2 = i$. Consequently, the general solution to the homogeneous equation is $y_H(t) = C_1\sin(t) + C_2\cos(t)$. The particular solution is computed from the form $y_P(t) = A\sin(2t) + B\cos(2t)$ using undetermined coefficients. Solving for the coefficients, we obtain $A = -1/3$ and $B = 0$. Using the initial conditions, we obtain the equations $-2/3 + C_1 = y'(0)$ and $C_2 = 0$, which lead to $C_1 = 2/3 + y'(0)$ and $C_2 = 0$. Hence, $y(t) = -\sin(2t)/3 + (2/3 + y'(0))\sin(t)$. With **ODE**, we choose the interval $-2 \le t \le 2$ and enter

```
ODE[{y'' + y == Sin[2t],y[0] == 0,y'[0] == a},y,t,
    Method->SecondOrderLinear,Parameters->{{a,0,4,0.1}},
    StackPlotSolution->{{t,-2,2},PlotPoints->30},
    ViewPoint->{-1.582, 2.225, 2.000}]
```

22. $\begin{cases} y'' - 2y' + y = t, \\ -2 \le y(0) \le 2, \ y'(0) = 0 \end{cases}$

Solution. The roots of the characteristic equation $r^2 - 2r + 1 = 0$ are $r_1 = 1$ and $r_2 = 1$. Consequently, the general solution to the homogeneous equation is $y_H(t) = C_1 e^t + C_2 t\, e^t$. The particular solution is computed from the form $y_p(t) = A + Bt$ using undetermined coefficients. Solving for the coefficients, we obtain $A = 2$ and $B = 1$. Using the initial conditions, we obtain the equations $2 + C_1 = y(0)$ and $1 + C_1 + C_2 = 0$, which lead to $C_1 = -2 + y(0)$ and $C_2 = 1 - y(0)$. Hence, $y(t) = 2 + t - (2 - y(0))e^t + (1 - y(0))t\, e^t$. With **ODE**, we choose the interval $-2 \le t \le 2$ and enter

```
ODE[{y'' - 2y' + y == t,y[0] == a,y'[0] == 0},
    y,t,Method->SecondOrderLinear,
    Parameters->{{a,-2,2,0.1}},
    StackPlotSolution->{{t,-2,2},PlotPoints->50},
    ViewPoint->{-1.582, 2.225, 2.000}]
```

23. $\begin{cases} 3y'' - 2y' + 15y = 0, \\ 0 \le y(0) \le 4, \ y'(0) = -1 \end{cases}$

Solution. The roots of the characteristic equation $3r^2 - 2r + 15 = 0$ are $r_1 = 1/3 - (2\sqrt{11}/3)i$ and $r_2 = 1/3 + (2\sqrt{11}/3)i$. Consequently, the general solution is $y(t) = C_1 e^{t/3}\sin((2\sqrt{11}/3)t) + C_2 e^{t/3}\cos((2\sqrt{11}/3)t)$. Using the initial conditions, we obtain the equations $2\sqrt{11}C_1 + C[2] = -3$ and $C_2 = y(0)$, which lead to $C_1 = -(3 + y(0))/(2\sqrt{11})$ and $C_2 = y(0)$. Hence, $y(t) = -(3 + y(0))/(2\sqrt{11})e^{t/3}\sin((2\sqrt{11}/3)t) + y(0)e^{t/3}\cos((2\sqrt{11}/3)t)$. With **ODE**, we choose the interval $-2\pi \le t \le 2\pi$ and enter

```
ODE[{3y'' - 2y' + 15y == 0,y[0] == a,y'[0] == -1},
    y,t,Method->SecondOrderLinear,
    Parameters->{{a,0,4,0.2}},
```

```
StackPlotSolution->{{t,-2Pi,2Pi},PlotPoints->50},
ViewPoint->{0.109, -2.980, 1.600}]
```

24. $\begin{cases} y'' + 5y = t\sin(t), \\ 0 \le y(0) \le 4, \quad y'(0) = \pi \end{cases}$

Solution. The roots of the characteristic equation $r^2 + 5 = 0$ are $r_1 = -\sqrt{5}i$ and $r_2 = \sqrt{5}i$. Consequently, the general solution to the homogeneous equation is $y_H(t) = C_1 \sin(\sqrt{5}t) + C_2 \cos(\sqrt{5}t)$. The particular solution is computed from the form $y_P(t) = (A_0 + A_1 t)\sin(t) + (B_0 + B_1 t)\cos(t)$ using undetermined coefficients. Solving for the coefficients, we obtain $A_0 = 0$, $A_1 = 1/4$, $B_0 = -1/8$ and $B_1 = 0$. Using the initial conditions, we obtain the equations $\sqrt{5}C_1 = \pi$ and $-1/8 + C_2 = y(0)$, which lead to $C_1 = \pi/\sqrt{5}$ and $C_2 = 1/8 + y(0)$. Hence, $y(t) = (\pi/\sqrt{5})\sin(\sqrt{5}t) + (1/8 + y(0))\cos(\sqrt{5}t)$. With ODE, we choose the interval $-2\pi \le t \le 2\pi$ and enter

```
ODE[{y'' + 5y == t Sin[t],y[0] == a,y'[0] == Pi},y,t,
   Method->SecondOrderLinear,Parameters->{{a,0,4,0.2}},
   StackPlotSolution->{{t,-2Pi,2Pi},PlotPoints->50},
   ViewPoint->{0.232, -3.035, 1.479}]
```

25. $\begin{cases} y'' + y = \cos(kt), 5 \le k \le 10 \\ y(0) = 1, \quad y'(0) = 0 \end{cases}$

Solution. The roots of the characteristic equation $r^2 + 1 = 0$ are $r_1 = -i$ and $r_2 = i$. Consequently, the general solution to the homogeneous equation is $y_H(t) = C_1 \sin(t) + C_2 \cos(t)$. The particular solution is computed from the form $y_P(t) = A\cos(t) + B\sin(t)$ using undetermined coefficients. Solving for the coefficients, we obtain $A = 1/(1 - k^2)$ and $B = 0$. Using the initial conditions, we obtain the equations $C_1 = 0$ and $C_2 = k^2/(k^2 - 1)$. Hence, $y(t) = (k^2/(k^2 - 1))\cos(t) - \cos(kt)/(k^2 - 1)$. With ODE, we choose the interval $-2\pi \le t \le 2\pi$ and enter

```
ODE[{y'' + y == Cos[a t],y[0] == 1,y'[0] == 0},y,t,
   Method->SecondOrderLinear,Parameters->{{a,5,10,.2}},
   StackPlotSolution->{{t,-2Pi,2Pi},PlotPoints->100},
   ViewPoint->{0.052, -2.647, 2.108}]
```

26. $\begin{cases} y'' + ky = \exp(t), 1 \le k \le 5 \\ y(0) = 0, \quad y'(0) = 0 \end{cases}$

Solution. The roots of the characteristic equation $r^2 + k = 0$ are $r_1 = -\sqrt{k}i$ and $r_2 = \sqrt{k}i$. Consequently, the general solution to the homogeneous equation is $y_H(t) = C_1 \sin(\sqrt{k}t) + C_2 \cos(\sqrt{k}t)$. The particular solution is computed from the form $y_P(t) = Ae^t$ using undetermined coefficients. Solving for the co-efficients, we obtain $A = 1/(1+k)$. Using the initial conditions, we obtain the equations $C_1 = -1/(\sqrt{k}(1+k))$ and $C_2 = -1/(1+k)$. Hence, $y(t) = e^t/(1+k) - 1/(\sqrt{k}(1+k)) \sin(\sqrt{k}t) - 1/(1+k) \cos(\sqrt{k}t)$. With ODE, we choose the interval $-2\pi \le t \le 2\pi$ and declare the parameter a to be real and positive and enter

```
a /: Im[a] = 0;
a /: Positive[a] = True;
ODE[{y'' + a y == Exp[t],y[0] == 0,y'[0] == 0},y,t,
    Method->SecondOrderLinear,Parameters->{{a,1,5,0.1}},
    StackPlotSolution->{{t,-2Pi,2Pi},PlotPoints->100},
    ViewPoint->{-1.028, -2.728, 1.718}]
```

27. $\begin{cases} y'' + y = e^{kt}, 0 \le k \le 5 \\ y(0) = 0, \ y'(0) = 0 \end{cases}$

Solution. The roots of the characteristic equation $r^2 + 1 = 0$ are $r_1 = -i$ and $r_2 = i$. Consequently, the general solution to the homogeneous equation is $y_H(t) = C_1 \sin(t) + C_2 \cos(t)$. The particular solution is computed from the form $y_P(t) = Ae^{kt}$ using undetermined coefficients. Solving for the coefficients, we obtain $A = 1/(1+k^2)$. Using the initial conditions, we obtain the equations $C_1 = -k/(1+k^2)$ and $C_2 = -1/(1+k^2)$. Hence, $y(t) = e^{kt}/(1+k^2) - 1/(1+k^2) \cos(t) - k/(1+k^2) \sin(t)$. With ODE, we choose the interval $-2\pi \le t \le 2\pi$ and enter

```
ODE[{y'' + y == Exp[a t],y[0] == 0,y'[0] == 0},
    y,t,Method->SecondOrderLinear,
    Parameters->{{a,0,5,0.2}},
    StackPlotSolution->{{t,-1,1},PlotPoints->100},
    ViewPoint->{0.479, -2.975, 1.540}]
```

28. $\begin{cases} k y'' + y' + k y = 0, -1 \le k \le 1 \\ y(0) = 0, \ y'(0) = 1 \end{cases}$

Solution. The roots of the characteristic equation $kr^2 + r + k = 0$ are $r_1 = (-1 - \sqrt{1-4k^2})/2k$ and $r_2 = (-1+\sqrt{1-4k^2})/2k$. Consequently, the general solution is $y(t) = C_1 e^{((-1-\sqrt{1-4k^2})/2k)t} + C_2 e^{((-1+\sqrt{1-4k^2})/2k)t}$. Using the initial conditions, we obtain the equations $C_1 = -k/\sqrt{1-4k^2}$ and $C_2 = k/\sqrt{1-4k^2}$. Hence, $y(t) = -k/\sqrt{1-4k^2} e^{((-1-\sqrt{1-4k^2})/2k)t} + k/\sqrt{1-4k^2} e^{((-1+\sqrt{1-4k^2})/2k)t}$. With ODE, we choose the interval $-2\pi \le t \le 2\pi$ and declare the parameter a to be real and positive and enter

```
a /: Im[a] = 0;
a /: Positive[a] = True;
ODE[{a y'' + y' + a y == 0,y[0] == 0,y'[0] == 1},
    y,t,Method->SecondOrderLinear,
    Parameters->{{a,-1.05,1.05,0.1}},
    StackPlotSolution->{{t,-1,1},PlotPoints->100},
    ViewPoint->{0.805, -2.994, 1.356}]
```

Find and plot the solutions to the following initial value problems, whose solutions involve nonelementary integrals.

29. $\begin{cases} y'' - 4y' + 4y = \sqrt{t}, \\ y(0) = y'(0) = 0 \end{cases}$

Solution. The roots of the characteristic equation $r^2 - 4r + 4 = 0$ are $r_1 = 2$ and $r_2 = 2$. Consequently, the general solution to the homogeneous equation is $y_H(t) = C_1 e^{2t} + C_2 t e^{2t}$. Using the fundamental set, the Wronskian is e^{4t} and variation of parameters leads to

$$y_P(t) = -e^{2t} \int s^{3/2} e^{-2s} \, ds + t e^{2t} \int \sqrt{s} e^{-2s} \, ds \Big|_{s \to t}.$$

Therefore, $y(t) = C_1 e^{2t} + C_2 t e^{2t} + y_P(t)$. The initial values lead to $C_1 = 0$ and $C_2 = 0$. Hence, $y(t) = y_P(t)$. *Mathematica* will express this integral in terms of the error function

$$\text{erf}(t) = \frac{2}{\sqrt{\pi}} \int_{-\infty}^{t} e^{-s^2} \, ds.$$

With **ODE**, we choose the interval $0 \le t \le 2$ and enter

```
ODE[{y'' - 4y' +4y == Sqrt[t],y[0] == 0,y'[0] == 0},y,t,
Method->SecondOrderLinear,PlotSolution->{{t,0,2}}]
```

30. $\begin{cases} y'' - y = e^{-t^2}, \\ y(0) = y'(0) = 0 \end{cases}$

Solution. The roots of the characteristic equation $r^2 - 1 = 0$ are $r_1 = -1$ and $r_2 = 1$. Consequently, the general solution to the homogeneous equation is $y_H(t) = C_1 e^{-t} + C_2 e^{t}$. Using the fundamental set, the Wronskian is 2 and variation of parameters leads to

$$y_P(t) = -\frac{e^{-t}}{2} \int e^{s-s^2} \, ds + \frac{e^{t}}{2} \int e^{-s-s^2} \, ds \Big|_{s \to t}.$$

Therefore, $y(t) = C_1 e^{-t} + C_2 e^t + y_P(t)$. The initial values lead to $C_1 = 0$ and $C_2 = 0$. Hence, $y(t) = y_P(t)$. *Mathematica* will express this integral in terms of the error function

$$\mathbf{erf}(t) = (2/\sqrt{\pi}) \int_{-\infty}^{t} e^{-s^2} \, ds.$$

With **ODE**, we choose the interval $0 \le t \le 3$ and enter

```
ODE[{y'' - y == E^(-t^2),y[0] == 0,y'[0] == 0},y,t,
Method->SecondOrderLinear,PlotSolution->{{t,0,3}}]
```

31.
$$\begin{cases} y'' - y = e^{t^2}, \\ y(0) = y'(0) = 0 \end{cases}$$

Solution. The roots of the characteristic equation $r^2 - 1 = 0$ are $r_1 = -1$ and $r_2 = 1$. Consequently, the general solution to the homogeneous equation is $y_H(t) = C_1 e^{-t} + C_2 e^t$. Using the fundamental set, the Wronskian is 2 and variation of parameters leads to

$$y_P(t) = -\frac{e^{-t}}{2} \int e^{s+s^2} \, ds + \frac{e^t}{2} \int e^{-s+s^2} \, ds \Big|_{s \to t}.$$

Therefore, $y(t) = C_1 e^{-t} + C_2 e^t + y_P(t)$. The initial values lead to $C_1 = 0$ and $C_2 = 0$. Hence, $y(t) = y_P(t)$. *Mathematica* will express this integral in terms of the error function

$$\mathbf{erf}(t) = (2/\sqrt{\pi}) \int_{-\infty}^{t} e^{-s^2} \, ds.$$

With **ODE**, we choose the interval $0 \le t \le 3$ and enter

```
ODE[{y'' - y == E^(t^2),y[0] == 0,y'[0] == 0},y,t,
Method->SecondOrderLinear,PlotSolution->{{t,0,3}}]
```

32.
$$\begin{cases} y'' + y' - 2y = \log t, \\ y(1) = y'(1) = 0 \end{cases}$$

Solution. The roots of the characteristic equation $r^2 + r - 2 = 0$ are $r_1 = -2$ and $r_2 = 1$. Consequently, the general solution to the homogeneous equation is $y_H(t) = C_1 e^{-2t} + C_2 e^t$. Using the fundamental set, the Wronskian is $3e^{-t}$ and variation of parameters leads to

$$y_P(t) = -\frac{e^{-2t}}{3} \int e^{-s} \log(s) \, ds + \frac{e^t}{3} \int e^{-s} \log(s) \, ds \Big|_{s \to t}.$$

Therefore, $y(t) = C_1 e^{-t} + C_2 e^t + y_p(t)$. *Mathematica* will express this integral in terms of the second exponential integral function

$$\text{Ei}(t) = -\int_{-t}^{\infty} \frac{e^{-s}}{s} \, ds.$$

With **ODE**, we choose the interval $0 \le t \le 3$ and enter

```
ODE[{y'' + y'  - 2y == Log[t],y[1]  == 0,y'[1] == 0},
y,t,Method->SecondOrderLinear,
PlotSolution->{{t,0,3},PlotRange->All}]
```

Section 10.2, page 320

In each of the following problems use **HomogeneousSecondOrderLinear** to find the general solution to the following homogeneous second-order linear differential equations.

1. $y'' + y = 0$
Solution.

```
HomogeneousSecondOrderLinear[1,0,1,C][t]
```

2. $y'' - y = 0$
Solution.

```
HomogeneousSecondOrderLinear[1,0,-1,C][t]
```

3. $y'' + 3y' + 2y = 0$
Solution.

```
HomogeneousSecondOrderLinear[1,3,2,C][t]
```

4. $y'' + y' = 0$
Solution.

```
HomogeneousSecondOrderLinear[1,1,0,C][t]
```

5. $y'' - 3y' + 2y = 0$
Solution.

```
HomogeneousSecondOrderLinear[1,-3,2,C][t]
```

6. $y'' + k^2 y = 0$
Solution.

```
HomogeneousSecondOrderLinear[1,0,k^2,C][t]
```

7. $y'' + i y = 0$

Solution.

```
HomogeneousSecondOrderLinear[1,0,I,C][t]
```

8. $y'' + 3i y' - 2y = 0$

Solution.

```
HomogeneousSecondOrderLinear[1,3I,-2,C][t]
```

In each of the following problems use the command **VariationOfParameters** to find a particular solution to the following second-order linear differential equations which have the functions y_1 and y_2 as solutions of the corresponding homogeneous equation.

9. $\begin{cases} y'' + y = \sec(t) \\ y_1 = \sin(t), \quad y_2 = \cos(t) \end{cases}$

Solution.

```
VariationOfParameters[{Sin,Cos},Sec][t]
```

10. $\begin{cases} y'' - 4y' + 3y = \dfrac{e^t}{1 + e^t} \\ y_1 = e^t, \quad y_2 = e^{3t} \end{cases}$

Solution.

```
VariationOfParameters[{Exp,Exp[3#]&},
Exp[#]/(1+Exp[#])&][t]
```

11. $\begin{cases} y'' - y = \cosh(t) \\ y_1 = e^t, \quad y_2 = e^{-t} \end{cases}$

Solution.

```
VariationOfParameters[{Exp[-#]&,Exp[#]&},Cosh[#]&][t]
```

12. $\begin{cases} y'' + 3y' + 2y = \sin(e^t) \\ y_1 = e^{-t}, \quad y_2 = e^{-2t} \end{cases}$

Solution.

```
VariationOfParameters[{Exp[-#]&,Exp[-2#]&},
Sin[Exp[#]]&][t]
```

13. $\begin{cases} y'' - y = 2^t \\ y_1 = e^t, \quad y_2 = e^{-t} \end{cases}$

Solution.

```
VariationOfParameters[{Exp[#]&,Exp[-#]&},2^#&][t]
```

14. $\begin{cases} y'' + y = \sec(t)\tan(t) \\ y_1 = \sin(t), \quad y_2 = \cos(t) \end{cases}$

Solution.

```
VariationOfParameters[{Sin[#]&,Cos[#]&},
Sec[#]Tan[#]&][t]
```

15. $\begin{cases} y'' - 2y' + 2y = e^t \sec(t) \\ y_1 = e^t \sin(t), \quad y_2 = e^t \cos(t) \end{cases}$

Solution.

```
VariationOfParameters[{Exp[#]Sin[#]&,Exp[#]Cos[#]&},
Exp[#]Sec[#]&][t]
```

16. $\begin{cases} y'' - 2y' + y = 4 e^t \log(t) \\ y_1 = e^t, \quad y_2 = t e^t \end{cases}$

Solution.

```
VariationOfParameters[{Exp[#]&,#Exp[#]&},
4Exp[#]Log[#]&][t]
```

Section 10.3, page 322

Use the command **ReductionOfOrder** to find a second linearly independent solution to each of the following second-order equations when given the indicated first solution.

1. $\begin{cases} t^2 y'' - 6t\, y' + 12y = 0, \\ y_1(t) = t^3 \end{cases}$

Solution.

```
ODE[12y/t^2 - 6y'/t + y'' == 0,y,t,
    Method->ReductionOfOrder,
    KnownSolution->t^3]
```

2. $\begin{cases} 2t^2y'' + 4t\,y' - 2y = 0, \\ y_1(t) = t^{(-1+\sqrt{5})/2} \end{cases}$

Solution.

```
ODE[2t^2 y'' + 4t y' - 2y == 0,y,t,
    Method->ReductionOfOrder,
    KnownSolution->t^((-1 +Sqrt[5])/2)]
```

3. $\begin{cases} y'' + \left(-\dfrac{1}{t} + \dfrac{3}{16t^2}\right)y = 0, \\ y_1(t) = t^{1/4}\exp\left(2t^{1/2}\right) \end{cases}$

Solution.

```
ODE[y'' - (1/t - 3/(16t^2))y == 0,y,t,
    Method->ReductionOfOrder,
    KnownSolution->t^(1/4)Exp[2Sqrt[t]]]
```

4. $\begin{cases} t\,y'' - (2 - t\tan(t))y' \\ \quad + \left(\dfrac{2}{t} - \tan(t)\right)y = 0, \\ y_1(t) = t \end{cases}$

Solution.

```
ODE[t y'' - (2 - t Tan[t])y' +
    (2/t - Tan[t])y == 0,y,t,
    Method->ReductionOfOrder,
    KnownSolution->t]
```

5. $\begin{cases} t^2y'' - 2y = 0, \\ y_1(t) = t^2 \end{cases}$

Solution.

```
ODE[t^2 y'' - 2y == 0,y,t,
    Method->ReductionOfOrder,
    KnownSolution->t^2]
```

6. $\begin{cases} y'' - t\,y' - y = 0, \\ y_1(t) = \exp(t^2/2) \end{cases}$

Solution.

```
ODE[y'' - t y' - y == 0,
    y,t,Method->ReductionOfOrder,
    KnownSolution->Exp[t^2/2]]
```

7. $\begin{cases} (t^2 - t)y'' + (3t - 1)y' + y = 0, \\ y_1(t) = 1/(1 - t) \end{cases}$

Solution.

```
ODE[(t^2 - t) y'' + (3t - 1)y' + y == 0,
    y,t,Method->ReductionOfOrder,
    KnownSolution->1/(1 - t)]
```

8. $\begin{cases} t\,y'' + (1 - t)y' + 2y = 0, \\ y_1(t) = t^2 - 4t + 2 \end{cases}$

Solution.

```
ODE[t y'' + (1 - t)y' + 2y == 0,
    y,t,Method->ReductionOfOrder,
    KnownSolution->t^2 - 4t + 2]
```

Section 10.4, page 324

In each of the following problems use **EquationFromSolutions** to find a second-order linear differential equation that has the functions y_1 and y_2 as solutions.

1. $\begin{cases} y_1 = \sin(2t) \\ y_2 = \cos(2t) \end{cases}$

Solution.

```
EquationFromSolutions[Sin[2#]&,Cos[2#]&][t,y]
```

2. $\begin{cases} y_1 = t\sin(t) \\ y_2 = \cos(t) \end{cases}$

Solution.

```
EquationFromSolutions[#Sin[#]&,Cos][t,y]
```

3.
$$\begin{cases} y_1 = t^2 \\ y_2 = \log(t) \end{cases}$$

Solution.

```
EquationFromSolutions[#^2&,Log][t,y]
```

4.
$$\begin{cases} y_1 = e^t \\ y_2 = \log(t)^2 \end{cases}$$

Solution.

```
EquationFromSolutions[E^#&,(Log[#])^2&][t,y]
```

5.
$$\begin{cases} y_1 = t\,e^t \\ y_2 = \log(t) \end{cases}$$

Solution.

```
EquationFromSolutions[#E^#&,Log[#]&][t,y]
```

6.
$$\begin{cases} y_1 = t\,e^t \\ y_2 = \sin(t) \end{cases}$$

Solution.

```
EquationFromSolutions[#E^#&,Sin[#]&][t,y]
```

7.
$$\begin{cases} y_1 = t^2 + t \\ y_2 = \cos(t) \end{cases}$$

Solution.

```
EquationFromSolutions[#^2+#&,Cos[#]&][t,y]
```

8.
$$\begin{cases} y_1 = \sin(t) \\ y_2 = \exp(t)\cos(t) \end{cases}$$

Solution

```
EquationFromSolutions[Sin[#]&,Exp[#]Cos[#]&][t,y]
```

Chapter 11

Section 11.1, page 346

Problems 1–2 are concerned with Hooke's constant.

1. A weight of 3 pounds stretches a spring by the amount 4 inches. Find Hooke's constant of the spring.

   Solution. We have $w = m\,g = 3$ pounds and $\Delta l = 4$ inches $= (1/3)$foot. Hence using (11.1) we get

$$k = \frac{m\,g}{\Delta l} = \frac{3}{1/3} = 9 \text{ pounds/foot.} \quad \blacksquare$$

2. Determine Hooke's constant for a spring of natural length 15 inches that is stretched to a distance of 18 inches by an object weighing 5 pounds.

   Solution. We have $w = m\,g = 5$ pounds and $\Delta l = (18 - 15)$ inches $= (1/4)$foot. Hence using (11.1) we get

$$k = \frac{m\,g}{\Delta l} = \frac{5}{1/4} = 20 \text{ pounds/foot.} \quad \blacksquare$$

In Exercises 3–8 put $u(t)$ into the amplitude-phase form $u(t) = R \cos(\omega_0 t - \alpha)$.

3. $u(t) = 3 \cos(2t) - 7 \sin(2t)$

   Solution. We observe that $\omega_0 = 2$. After computing $R = \sqrt{3^2 + 7^2} = 7.616$ and $\alpha = \tan^{-1}(-7/3) = -1.166$, we obtain $u(t) = 7.616 \cos(2t + 1.166)$.

4. $u(t) = \sqrt{3} \cos(5t) + \sin(5t)$

   Solution. We observe that $\omega_0 = 5$. After computing $R = \sqrt{\sqrt{3}^2 + 1} = 2$ and $\alpha = \tan^{-1}(1/\sqrt{3}) = .524$, we obtain $u(t) = 2 \cos(5t - 0.524)$.

5. $u(t) = 1.2 \cos(2t)$

   Solution. We observe that $\omega_0 = 2$. After computing $R = \sqrt{1.2^2} = 1.2$ and $\alpha = \tan^{-1}(0/1.2) = 0$, we obtain $u(t) = 1.2 \cos(2t)$.

6. $u(t) = -6 \cos(7t) + 6 \sin(7t)$

   Solution. We observe that $\omega_0 = 7$. After computing $R = \sqrt{(-6)^2 + 6^2} = 8.485$ and $\alpha = \tan^{-1}\left(6/(-6)\right) = -\pi/4 = -.7854$, we note that we are in the second quadrant. Thus, $\alpha = -.7854 + \pi = 2.356$ and we obtain $u(t) = 8.485 \cos(7t - 2.356)$.

7. $u(t) = -4\cos(t) - 4\sqrt{3}\sin(t)$

Solution. We observe that $\omega_0 = 1$. After computing $R = \sqrt{(-4)^2 + (-4\sqrt{3})^2} = 8$ and $\alpha = \tan^{-1}\left(-4\sqrt{3}/(-4)\right) = 1.0472$, we note that we are in the forth quadrant. Thus, $\alpha = 1.0472 - \pi = -2.0944$ and we obtain $u(t) = 8\cos(t + 2.0944)$.

8. $u(t) = 7\sin(11t)$

Solution. We observe that $\omega_0 = 11$. After computing $R = \sqrt{7^2} = 7$ and $\alpha = \tan^{-1}(7/0) = \pi/2 = 1.571$, we obtain $u(t) = 7\cos(11t - 1.571)$.

Problems 9–19 deal with undamped mass-spring systems ($c = 0$).

9. A mass of 50 grams is attached to a spring of natural length 100 centimeters. Suppose that the spring is stretched an additional 10 centimeters by the addition of this mass. If the mass is started in motion with an initial velocity of 20 centimeters per second in the vertical direction, find the formula for the position of the mass.

Solution. Since $m = 50$ grams and $\Delta l = 10$ centimeters, we have

$$k = \frac{m\,g}{\Delta l} = \frac{50 \times 980}{10} = 4900 \text{ grams/centimeter}$$

and

$$\omega_0 = \sqrt{\frac{k}{m}} = \sqrt{\frac{4900}{50}} = 2\sqrt{7} \approx 9.90 \text{ radians/second.}$$

Also, $U_0 = 0$ and $U_1 = 20$ centimeters/second. Substitution of these values into the general formula (11.15) yields

$$u(t) = \frac{10\sqrt{2}}{7}\sin(7\sqrt{2}t) \approx 2.02\sin(9.9t).$$

The **ODE**-generated exact solution is found with

```
ODE[{50u'' + (50 980/10)u == 0,u[0] == 0,u'[0] == 20},
u,t,Method->SecondOrderLinear]
```

yielding

```
        10 Sqrt[2] Sin[7 Sqrt[2] t]
{{u ->  ---------------------------}}
                    7
```

Similarly, the **ODE**-generated approximate solution is found with

```
ODE[{50u'' + (50 980/10)u == 0,u[0] == 0,u'[0] == 20},
u,t,Method->ApproximateNthOrderLinear]
```

yielding

```
{{u -> 2.02 Sin[9.9 t]}}
```

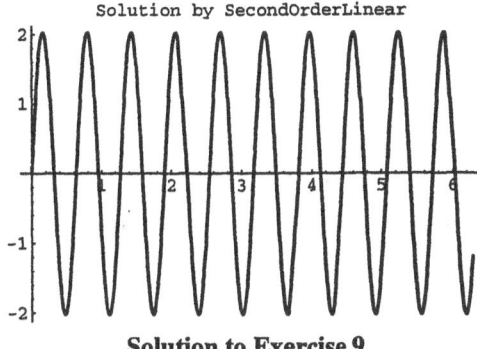

Solution to Exercise 9

10. Suppose that the weight on the spring of Exercise 9 is released from a height of 10 centimeters *below* the equilibrium position with an initial velocity of 2 centimeters per second in the downward direction. Find the formula for the position of the weight.

Solution. We can use the data from Exercise 9, but changing the initial conditions to $U_0 = 10$ centimeters and $U_1 = 2$ centimeters/second. The result is

$$u(t) = 10\cos(7\sqrt{2}t) + \frac{\sqrt{2}\sin(7\sqrt{2}t)}{7}$$

$$\approx 10.0\cos(9.9t) + 0.202\sin(9.9t) \approx 10.002\cos(9.9t - 0.0202).$$

The **ODE**-generated exact solution is found with

```
ODE[{50u'' + (50 980/10)u == 0,u[0] == 10,u'[0] == 2},
u,t,Method->SecondOrderLinear]
```

yielding

```
                                    Sqrt[2] Sin[7 Sqrt[2] t]
{{u -> 10 Cos[7 Sqrt[2] t] + -------------------------}}
                                             7
```

Similarly, the **ODE**-generated approximate solution using the amplitude-phase representation is found with

```
ODE[{50u'' + (50 980/10)u == 0,u[0] == 10,u'[0] == 2},
u,t,Method->ApproximateNthOrderLinear,
PostSolution->{AmplitudePhaseAngle}]
```

yielding

```
{{u -> 10. Cos[0.0202 - 9.9 t]}}
```

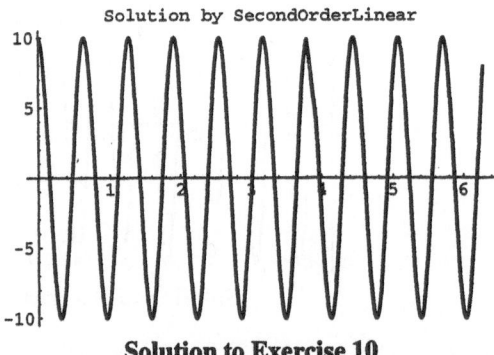

Solution by SecondOrderLinear

Solution to Exercise 10

11. Suppose that the weight on the spring of Exercise 9 is released from a height of 10 centimeters *above* the equilibrium position with an initial velocity of 2 centimeters per second in the downward direction. Find the formula for the position of the weight.

Solution. We can use the data from Exercise 9, but changing the initial conditions to $U_0 = -10$ centimeters and $U_1 = 2$ centimeters/second. The result is

$$u(t) = -10\cos(7\sqrt{2}\,t) + \frac{\sqrt{2}\sin(7\sqrt{2}\,t)}{7}$$

$$\approx -10.0\cos(9.9t) + 0.202\sin(9.9t) \approx 10.002\cos(9.9t - 3.12).$$

The **ODE**-generated exact solution is found with

```
ODE[{50u'' + (50 980/10)u == 0,u[0] == -10,u'[0] == 2},
u,t,Method->SecondOrderLinear]
```

yielding

```
                          Sqrt[2] Sin[7 Sqrt[2] t]
{{u -> -10 Cos[7 Sqrt[2] t] + ───────────────────────}}
                                       7
```

Similarly, the **ODE**-generated approximate solution using the amplitude-phase representation is found with

```
ODE[{50u'' + (50 980/10)u == 0,u[0] == -10,u'[0] == 2},
u,t,Method->ApproximateNthOrderLinear,
PostSolution->{AmplitudePhaseAngle}]
```

yielding

```
{{u -> 10. Cos[3.12 - 9.9 t]}}
```

12. Solve the undamped mass-spring equation $m\,u'' + k\,u = 0$ for the initial conditions $u(0) = U_0$ and $u'(0) = 0$ and determine the amplitude and phase lag. Graph the solution in case $m = 20$, $k = 5$ and $U_1 = 0$, where U_0 takes on the values ± 6, ± 4, ± 2 and 0.

Solution. We use the general formula (11.15) with $U_1 = 0$ to get

$$u(t) = U_0 \cos(\omega_0 t).$$

Hence the amplitude is $R = U_0$, and the phase lag is $\alpha = 0$. The **ODE**-generated exact solution is found and plotted with

```
ODE[{20u'' + 5u == 0,u[0] == U0,u'[0] == 0},u,t,
Method->SecondOrderLinear,Parameters->{{U0,-6,6,2}},
PlotSolution->{{t,0,8Pi}}]
```

```
               t
{{u -> U0 Cos[-]}}
               2
```

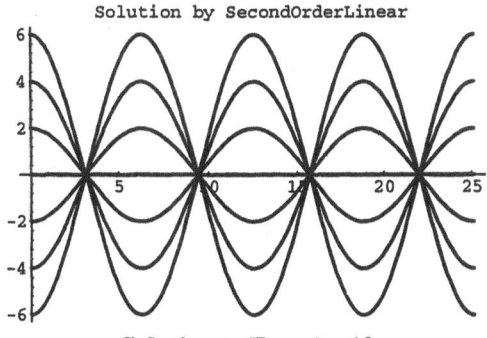

Solution by SecondOrderLinear

Solution to Exercise 12

13. Solve the undamped mass-spring equation $m\,u'' + k\,u = 0$ for the initial conditions $u(0) = 0$ and $u'(0) = U_1$, and determine the amplitude and phase lag. Graph the solution in case $m = 18$, $k = 2$ and $U_0 = 0$, where U_1 takes on the values $\pm 3, \pm 2, \pm 1$ and 0.

Solution. We use the general formula (11.15) with $U_0 = 0$ and $\omega_0 = \sqrt{k/m}$ to get

$$u(t) = \frac{U_1 \sin(\omega_0 t)}{\omega_0}.$$

With the data of the exercise, this equation reduces to

$$u(t) = 3U_1 \sin\left(\frac{t}{3}\right) = 3U_1 \cos\left(\frac{t}{3} - \frac{\pi}{2}\right).$$

Hence the amplitude is $R = 3U_1$, and the phase lag is $\alpha = \pi/2$. The **ODE**-generated exact solution is found and plotted with

```
ODE[{18u'' + 2u == 0,u[0] == 0,u'[0] == U1},u,t,
Method->SecondOrderLinear,Parameters->{{U1,-3,3,1}},
PlotSolution->{{t,0,9Pi}}]

                      t
{{u -> 3 U1 Sin[-]}}
                      3
```

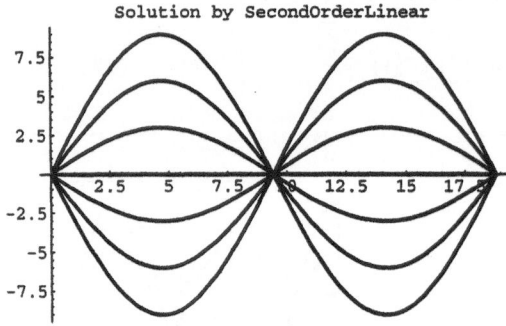

Solution by SecondOrderLinear

Solution to Exercise 13

14. Solve the undamped mass-spring equation $m\,u'' + k\,u = 0$ for the initial conditions $u(0) = U_0$ and $u'(0) = U_1$, and determine the amplitude and phase lag. Graph the solution in case $m = 25$ and $k = 1$, where U_0 takes on the values ± 50 and 0, and U_1 takes on the values ± 25 and 0.

Solution. We use the general formula (11.15) to get

$$u(t) = U_0 \cos(\omega_0 t) + \frac{U_1 \sin(\omega_0 t)}{\omega_0}.$$

With the data of the exercise, this equation reduces to

$$u(t) = U_0 \cos\left(\frac{t}{5}\right) + 5U_1 \sin\left(\frac{t}{5}\right).$$

Hence the amplitude is $R = \sqrt{U_0^2 + 5U_1^2}$, and the phase lag is $\alpha = \arctan(5U_1/U_0)$. The **ODE**-generated exact solution is found and plotted with

```
ODE[{25u'' + u == 0,u[0] == U0,u'[0] == U1},u,t,
Method->SecondOrderLinear,
Parameters->{{U0,-50,50,50},{U1,-25,25,25}},
PlotSolution->{{t,0,25Pi}}]

                t                t
{{u -> U0 Cos[-] + 5 U1 Sin[-]}}
                5                5
```

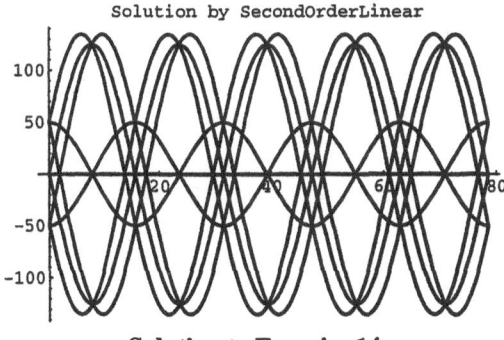

Solution by SecondOrderLinear

Solution to Exercise 14

15. Write the solution of Exercise 14 in the form $u(t) = R \cos(\omega_0 t - \alpha)$, where $R \geq 0$, $-\pi < \alpha \leq \pi$ and $\omega_0 > 0$.

Solution. From Exercise 14 we have

$$u(t) = \sqrt{\frac{k U_0^2 + m U_1^2}{k}} \cos\left(\omega_0 t - \arctan\left(\frac{U_1 \omega_0}{U_0}\right)\right)$$

$$= \sqrt{U_0^2 + 25U_1^2} \cos\left(\frac{t}{5} - \arctan\left(\frac{5U_1}{U_0}\right)\right).$$

To get the amplitude-phase representation with **ODE**, we use

```
ODE[{25u'' + u == 0,u[0] == U0,u'[0] == U1},u,t,
Method->SecondOrderLinear,
PostSolution->{AmplitudePhaseAngle}]
```

obtaining

```
          2       2       t
{{u -> Sqrt[U0  + 25 U1 ] Cos[-  - ArcTan[U0, 5 U1]]}}
                              5
```

16. Use **ODE** to find the general solution to solve the initial value problem

$$\begin{cases} m\,u''(t) + k\,u(t) = 0, \\ u(0) = U_0, u'(0) = U_1 \end{cases}$$

Solution. Let $\omega_0 = \sqrt{k/m}$. We enter the following commands to make m and k positive real numbers.

```
m /: Im[m]=0;
m /: Positive[m]=True;
k /: Im[k]=0;
k /: Positive[k]=True;
```

Then

```
ODE[{m u'' + k u == 0,u[0] == U0,u'[0] == U1},u,t,
Method->SecondOrderLinear,Form->Explicit,
PostSolution->{(# /. k ->omega0^2 m)&,PowerExpand}]
```

yields

```
                     U1 Sin[omega0 t]
U0 Cos[omega0 t] +  ------------------              or
                        omega0
```

$$u(t) = U_0 \cos(\omega_0 t) + \frac{U_1 \sin(\omega_0 t)}{\omega_0}.$$

17. The **total energy** $E(t)$ of a mass-spring system is defined by

$$E(t) = \frac{1}{2} m\, u'(t)^2 + \frac{1}{2} k\, u(t)^2.$$

Show that for an undamped mass-spring system the total energy is a constant independent of t.

Solution. We compute

$$\frac{d}{dt}E(t) = m\,u'(t)u''(t) + k\,u(t)u'(t) = u'(t)\big(m\,u''(t) + k\,u(t)\big) = 0.$$

Since its derivative is zero, the function $E(t)$ must be a constant.

18. With reference to Exercise 17, show that in the amplitude-phase representation the amplitude R can be computed in terms of the total energy by the formula

$$R = \sqrt{\frac{2E}{k}}.$$

Solution. Writing $u(t) = R\cos(\omega_0 t - \alpha)$, we have $u'(t) = -R\,\omega_0\sin(\omega_0 t - \alpha)$. Thus

$$2E = m\,u'(t)^2 + k\,u(t)^2$$

$$= m\,R^2\omega_0^2\big(\sin(\omega_0 t - \alpha)\big)^2 + k\,R^2\big(\cos(\omega_0 t - \alpha)\big)^2$$

$$= k\,R^2\big(\big(\sin(\omega_0 t - \alpha)\big)^2 + \big(\cos(\omega_0 t - \alpha)\big)^2\big) = k\,R^2.$$

19. The **phase portrait** of a second-order differential equation is constructed from a Cartesian coordinate system (x_1, x_2) where the horizontal x_1 axis represents the displacement $u(t)$ and the vertical x_2 axis represents the velocity $u'(t)$. Show that a mass-spring system in the absence of friction moves along an ellipse, and find the lengths of the axes of this ellipse.

Solution. Writing $x_1 = u(t) = R\cos(\omega_0 t - \alpha)$, we have

$$x_2 = u'(t) = -R\,\omega_0\sin(\omega_0 t - \alpha).$$

Thus

$$\frac{x_1^2}{R^2} + \frac{x_2^2}{R^2\omega_0^2} = \big(\cos(\omega_0 t - \alpha)\big)^2 + \big(\sin(\omega_0 t - \alpha)\big)^2 = 1, \qquad \text{(S.75)}$$

which is the equation of an ellipse. From (S.75) it follows that the lengths of the axes of the ellipse are

$$2R = 2\sqrt{U_0^2 + \frac{U_1^2}{\omega_0^2}} \qquad \text{and} \quad 2R\,\omega_0 = 2\sqrt{\omega_0^2 U_0^2 + U_1^2}.$$

20. Suppose that the particle of a mass-spring system of mass m and spring constant k is set into motion from a displacement of U_0 with initial velocity U_1.

   a. Find the maximum displacement which the particle attains and the velocity at that point.

   b. Find the maximum velocity the particle attains and the position at which this occurs.

Solution. We use the notation of Exercise 19. From (S.75) it follows that if the displacement $x_1 = u$ is a maximum, then $x_2 = 0$. Hence

$$u = x_1 = \pm R = \pm\sqrt{U_0^2 + \frac{U_1^2}{\omega_0^2}} = \pm\sqrt{\frac{k\,U_0^2 + m\,U_1^2}{k}}.$$

Similarly, (S.75) implies that if the velocity $u' = x_2$ is a maximum, then $x_1 = 0$, so that

$$u' = x_2 = \pm R\,\omega_0 = \pm\sqrt{\omega_0^2 U_0^2 + U_1^2} = \pm\sqrt{\frac{k\,U_0^2 + m\,U_1^2}{m}}.$$

Exercises 21–32 deal with damped mass-spring systems.

21. Find the solution of a mass-spring system for which $m = 2$ grams, $c = 3.5$ gram seconds/centimeter and $k = 4.7$ grams/centimeter. The initial conditions are $u(0) = 3$ centimeters and $u'(0) = -5$ centimeters/second. Find the quasiperiod and the relaxation time and graph the solution.

Solution. The roots of the characteristic equation are $-0.875 \pm 1.26i$, hence this is an underdamped system whose general solution is

$$u(t) = e^{-0.875t}\left(C_1 \cos(1.26t) + C_2 \sin(1.26t)\right).$$

If we use **ODE**, we obtain

```
ODE[{2u'' + 3.5u' + 4.7u == 0,u[0] == 3, u'[0] == -5},
u,t,Method->SecondOrderLinear,
PlotSolution->{{t,0,6},PlotRange->All}]
```

to get

$$\{\{u \to \frac{3.\,\mathrm{Cos}[1.26\ t]}{\mathrm{E}^{0.875\ t}} - \frac{1.89\,\mathrm{Sin}[1.26\ t]}{\mathrm{E}^{0.875\ t}}\}\}$$

The solution is $u(t) \approx e^{-0.875t}(3\cos(1.26t) - 1.89\sin(1.26t))$.

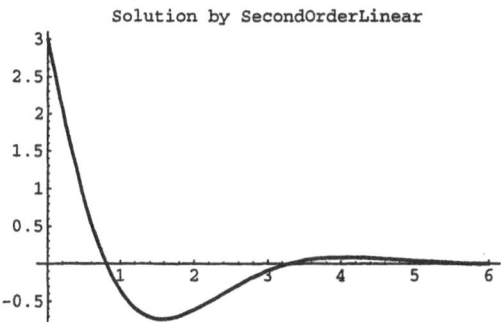

Solution by SecondOrderLinear

Solution to Exercise 21

The quasiperiod is given by

$$T_c = \frac{4\pi\, m}{\sqrt{4m\,k - c^2}} = \frac{4\pi \times 2}{\sqrt{4 \times 2 \times 4.7 - 3.5^2}} = 4.991 \text{ seconds.}$$

and the relaxation time is

$$\tau = \frac{1}{0.875} = 1.143 \text{ seconds.}$$

22. Find the solution of a mass-spring system for which $m = 1$ gram, $c = 4$ gram seconds/centimeter and $k = 4$ grams/centimeter. The initial conditions are $u(0) = -1$ centimeter and $u'(0) = 6$ centimeters/second. Find the relaxation time and graph the solution.

Solution. The roots of the characteristic equation are -2 and -2, hence this is a critically damped system whose general solution is

$$u(t) = C_1 e^{-2t} + C_2 t\, e^{-2t}.$$

If we use **ODE**, we obtain

```
ODE[{u'' + 4u' + 4u == 0,u[0] == -1, u'[0] == 6},
u,t,Method->SecondOrderLinear,
PlotSolution->{{t,0,4},PlotRange->All}]
```

to get

```
          -2 t     4 t
{{u -> -E      +  ----}}
                  2 t
                 E
```

The solution is $u(t) = e^{-2t}(-1 + 4t)$ and the relaxation time is $\tau = 2m/c = 0.5$ second.

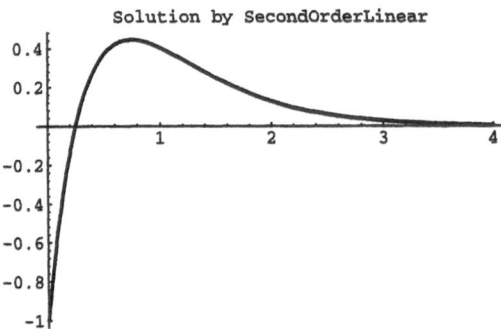

Solution by SecondOrderLinear

Solution to Exercise 22

23. Find the solution of a mass-spring system for which $m = 3$ grams, $c = 12$ gram seconds/centimeter and $k = 4$ grams/centimeter. The initial conditions are $u(0) = 1$ centimeter and $u'(0) = -2$ centimeters/second. Find the relaxation time and graph the solution.

Solution. The roots of the characteristic equation are $-2 \pm 2\sqrt{2/3}$, hence this is an overdamped system whose general solution is

$$u(t) = C_1 e^{-2-2\sqrt{2/3}t} + C_2 e^{-2+2\sqrt{2/3}t}$$

If we use **ODE**, we obtain

```
ODE[{3u'' + 12u' + 4u == 0 ,u[0] == 1, u'[0] == -2},
u,t,Method->SecondOrderLinear,
PlotSolution->{{t,0,4},PlotRange->All}]
```

to get

```
            -2 t - 2 Sqrt[2/3] t      -2 t + 2 Sqrt[2/3] t
           E                         E
  {{u ->  ------------------------  +  ------------------------}}
                    2                           2
```

The solution is

$$u(t) = e^{-2t} \cosh\left(2\sqrt{\frac{2}{3}}t\right),$$

and the relaxation time is

$$\tau = -\frac{1}{r_1} = \frac{1}{2 - 2\sqrt{2/3}} \text{ seconds} \approx 2.725 \text{ seconds.}$$

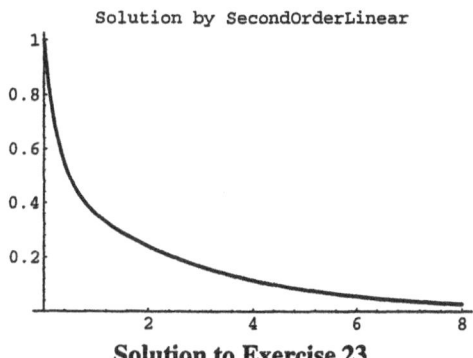

Solution by SecondOrderLinear

Solution to Exercise 23

24. Assume that a mass-spring system is governed by

$$\begin{cases} u'' + c\,u' + 36u = 0, \\ u(0) = 1,\ u'(0) = 0, \end{cases}$$

where $c = 3, 13, 23$. Find the equation of motion and graph the three solutions simultaneously.

Solution. We use

```
u[3][t_] = ODE[{uu'' + 3 uu' + 36uu == 0,
                uu[0] == 1, uu'[0] == 0},uu,t,
           Method->SecondOrderLinear,Form->Explicit]
```

obtaining

$$\frac{Cos\left[\dfrac{3\ Sqrt[15]\ t}{2}\right]}{E^{(3\ t)/2}} + \frac{Sin\left[\dfrac{3\ Sqrt[15]\ t}{2}\right]}{Sqrt[15]\ E^{(3\ t)/2}}$$

```
u[12][t_] = ODE[{uu'' + 12 uu' + 36uu == 0,
                 uu[0] == 1, uu'[0] == 0},uu,t,
            Method->SecondOrderLinear,Form->Explicit]
```

obtaining

$$E^{-6\ t} + \frac{6\ t}{E^{6\ t}}$$

and

```
u[21][t_] = ODE[{uu'' + 21 uu' + 36uu == 0,
                uu[0] == 1, uu'[0] == 0},uu,t,
                Method->SecondOrderLinear,Form->Explicit]
```

obtaining

```
 (-21 t)/2 - (3 Sqrt[33] t)/2
E
─────────────────────────────── -
                2

    (-21 t)/2 - (3 Sqrt[33] t)/2
  7 E
  ─────────────────────────────── +
          2 Sqrt[33]

   (-21 t)/2 + (3 Sqrt[33] t)/2
 E
 ─────────────────────────────── +
              2

    (-21 t)/2 + (3 Sqrt[33] t)/2
  7 E
  ───────────────────────────────
          2 Sqrt[33]
```

Then the simultaneous plot is found using

```
Plot[Evaluate[{u[3][t],u[12][t],u[21][t]}],
{t,0,3},PlotRange->All]
```

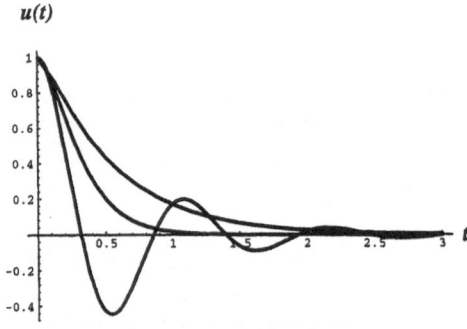

Solution to Exercise 24

25. Suppose that a mass-spring system is overdamped: $c^2 > 4m\,k$. Show that the point $(u(t), u'(t))$ moves along a curve defined implicitly by

$$(x_2 - r_1 x_1)^{r_1} = C(x_2 - r_2 x_1)^{r_2},$$

where C is a constant and r_1 and r_2 are the roots of the associated characteristic equation $m r^2 + c r + k = 0$. (See Exercise 18.) Compute the constant C in terms of the initial conditions.

Solution. Writing $x_1 = u(t) = C_1 e^{r_1 t} + C_2 e^{r_2 t}$, we have

$$x_2 = u'(t) = r_1 C_1 e^{r_1 t} + r_2 C_2 e^{r_2 t}.$$

Hence

$$|x_2 - r_1 x_1| = |(r_2 - r_1) C_2| e^{r_2 t} \qquad \text{and} \qquad |x_2 - r_2 x_1| = |(r_1 - r_2) C_1| e^{r_1 t},$$

so that

$$|x_2 - r_1 x_1|^{r_1} = |(r_2 - r_1) C_2|^{r_1} e^{r_1 r_2 t} \qquad \text{and} \qquad |x_2 - r_2 x_1|^{r_2} = |(r_1 - r_2) C_1|^{r_2} e^{r_1 r_2 t},$$

which proves the assertion. The constant C is written in terms of the initial conditions as

$$C = \frac{|u'(0) - r_1 u(0)|^{r_1}}{|u'(0) - r_2 u(0)|^{r_2}}.$$

26. Suppose that a mass-spring system is critically damped: $c^2 = 4m k$. Find the implicit equation of the curve along which the point $\big(u(t), u'(t)\big)$ moves.

Solution. In the underdamped case we have

$$
\begin{aligned}
u(t) &= C_1 e^{rt} + C_2 t e^{rt} \\
u'(t) &= r C_1 e^{rt} + (1 + rt) C_2 e^{rt} = r e^{rt}(C_1 + C_2 t) + C_2 e^{rt}
\end{aligned}
$$

so that

$$u' - ru = C_2 e^{rt}, \qquad u'/u = r + \frac{C_2}{C_1 + C_2 t}$$

Therefore, we can solve for t and substitute above, as follows:

$$t = -\frac{u}{u' - ru} - \frac{C_1}{C_2}$$

$$\log|u' - ru| = \log|C_2| + r\left(\frac{u}{u' - ru} - \frac{C_1}{C_2}\right)$$

27. Suppose that a mass-spring system is underdamped: $c^2 < 4m k$. Show that the quasiperiod T_c and the relaxation time τ are related by the equation

$$\frac{1}{4\pi^2 \tau^2} + \frac{1}{T_c^2} = \frac{1}{T_0^2},$$

where $T_0 = 2\pi\sqrt{m/k}$ is the period of the mass-spring system in the absence of friction.

Solution. The quasiperiod satisfies $\dfrac{1}{T_c^2} = \dfrac{4mk - c^2}{4\pi^2 \, 4m^2}$, while the relaxation time satisfies

$\dfrac{1}{4\pi^2\tau^2} = \dfrac{c^2}{4\pi^2 4m^2}$. Adding the two gives $\dfrac{4mk}{4\pi^2 4m^2} = \dfrac{1}{T_0^2}$.

28. Suppose that a mass of 1 grams has a quasiperiod of 24 seconds and a relaxation time of 56 seconds when attached to a spring under the influence of a frictional force. Use the result of Exercise 27 to find the spring constant.

Solution. From Exercise 27 the spring constant is given by

$$k = \frac{4\pi^2 m^2}{T_0^2} = 4\pi^2 m^2 \left(\frac{1}{T_c^2} + \frac{1}{4\pi^2\tau^2} \right)$$

Substitution of the given values and simplification yields

$$k = \frac{144 + (56\pi)^2}{451584} = 0.0689$$

29. Compute the total energy $E(t)$ (see Exercise 17) of a mass-spring system with damping, and show that it is a decreasing function of time.

Solution. We have

$$E'(t) = m\,u(t)u''(t) + k\,u(t)u'(t) = u'(t)\big(9m\,u''(t) + k\,u(t)\big)$$

$$= u'(t)\big(-c\,u'(t)\big) = -c\,u'(t)^2 \le 0.$$

Since the derivative is less than or equal to zero, the function $E(t)$ must be a decreasing function of time.

30. The **Lyapunov function** of an underdamped mass-spring system is defined as $L(t) = m\,u'(t)^2 + c\,u(t)u'(t) + k\,u(t)^2$. Show that $L(t) = L(0)e^{-ct/m}$.

Solution. We use the product rule and the differential equation $m\,u''(t) + c\,u'(t) + k\,u(t) = 0$ to write

$$L'(t) = 2m\,u'u'' + c\,u\,u'' + c\,u'^2 + 2k\,u\,u'$$

$$= (2m\,u' + c\,u)\left(-\frac{c\,u' + k\,u}{m}\right) + c\,u'^2 + 2k\,u\,u'$$

$$= -c\,u'^2 - \frac{c^2}{m}u\,u' - \frac{ck}{m}u^2 = -\frac{c}{m}\left(m\,u'^2 + c\,u\,u' + k\,u^2\right)$$

$$= -\frac{c}{m}L(t).$$

Therefore, $L(t)$ satisfies the differential equation $L'(t) = -(c/m)L(t)$, whose unique solution is $L(t) = L(0)e^{-ct/m}$

31. Suppose that we have an underdamped mass-spring system, with

$$0 < c^2 < 4m\,k, \qquad \mu = \frac{\sqrt{4m\,k - c^2}}{2c} \qquad \text{and} \qquad \lambda = \frac{c}{2m}.$$

Define a system of polar coordinates (r, θ) in the phase plane (see Exercise 18) by the formulas

$$\begin{cases} \mu\,u = r\cos(\theta), \\ u' + \lambda\,u = r\sin(\theta). \end{cases}$$

Show that in the phase plane the motion takes place along a logarithmic spiral curve whose equation is of the form

$$r = C\,e^{\lambda\theta/\mu},$$

where C is a constant. Find the constant C in terms of the initial conditions and in terms of the amplitude-phase representation of the solution.

Solution. If $u(t) = R\,e^{-\lambda t}\cos(\mu\,t - \alpha)$, then $u' + \lambda\,u = -\mu\,R\,e^{-\lambda t}\sin(\mu\,t - \alpha)$. Therefore

$$r^2 = (\mu\,u)^2 + (u' + \lambda\,u)^2 = \mu^2 R^2 e^{-2\lambda t} \qquad \text{and} \qquad \log(r) = \log(\mu R) - \lambda\,t.$$

But

$$\tan(\theta) = \frac{u' + \lambda\,u}{\mu\,u} = -\tan(\mu\,t - \alpha),$$

so that $-\theta = \mu\,t - \alpha - m\pi$ for some $m = 0, \pm 1, \pm 2, \ldots$. This proves that $r = \mu R e^{\lambda\theta/\mu + D}$ for a constant D, from which the polar equation follows. C can be computed by taking logarithms: $\log C = \log r(0) - (\lambda/\mu)\theta(0)$ The initial values are computed from $r(0) = \sqrt{(\mu\,u)^2 + (u' + \lambda\,u)^2}$, $\theta(0) = \arctan\frac{u' + \lambda u}{\mu u}$.

32. Consider the initial value problem

$$\begin{cases} u'' + 5u' + 4u = 0, \\ u(0) = 1, \, u'(0) = a. \end{cases}$$

Find the values of a for which $u(t) \neq 0$ for all $t \geq 0$ and find the values of a for which $u(t_0) = 0$ for some $t_0 \geq 0$.

Solution. The general solution of the differential equation is $u(t) = A e^{-4t} + B e^{-t}$. To satisfy the initial conditions we must have $1 = A + B$ and $a = -4A - B$, which is solved to yield $B = (4 + a)/3$ and $A = -(1 + a)/3$. Hence

$$u(t) = -\frac{1+a}{3} e^{-t} + \frac{4+a}{3} e^{-4t}.$$

In order that $u(t_0) = 0$ for some $t_0 > 0$, we must have

$$\frac{4+a}{1+a} = e^{-3t_0}$$

Therefore, the quantity $(4 + a)/(1 + a)$ must be positive and less than one. If we graph this function using

```
Plot [(4 + a)/(1 + a),{a,0,5}]
```

```
Plot [(4 + a)/(1 + a),{a,-10,-2}]
```

we see that the graph lies in the vertical interval $0 < y < 1$ only when a satisfies $a < -4$. This can also be seen algebraically by noting that the equation $(4+a)/(1+a) = 1$ has no solution and that for large a it is nearly equal to 1, so that the graph must lie above the vertical interval $0 \leq y \leq 1$ for all $a > -1$. Below that point the graph has a vertical asymptote at $a = -1$, crosses the y-axis at $a = -4$, and tends to 1 when $a \longrightarrow -\infty$.

Section 11.2, page 364

1. A mass weighing 12 pounds is attached to a spring with spring constant equal to 2 pounds/inch and is set into motion at $t = 0$ from a height of 1 inch with no velocity by an external force of $8 \cos(9t)$ pounds. Assume there is no friction. Find and plot the formula for the position of the mass.

Solution. The roots of the characteristic equation $(12/32)r^2 + 2 = 0$ are $r_1 = -4i/\sqrt{3}$ and $r_2 = 4i/\sqrt{3}$. Consequently, the general solution to the homogeneous equation is $y_H(t) = C_1 \cos(4t\sqrt{3}) + C_2 \sin(4t\sqrt{3})$. The particular solution is computed from the form $y_p(t) = A\cos(9t) + B\sin(9t)$ using undetermined coefficients. Solving for the coefficients, we obtain $A = -64/227$, and $B = 0$. Using the initial conditions, we obtain the equations $-64/227 + C_2 = -1$ and $(4/\sqrt{3})C_1 = 0$, which lead to $C_1 = 0$ and $C_2 = -163/227$. With ODE, we choose the interval $[0, 2\pi]$ and enter

```
ODE[{(12/32)u'' + 2u == 8Cos[9t],u[0] == -1,u'[0] == 0},
u,t,Method->SecondOrderLinear,Form->Explicit,
PlotSolution->{{t,0,2Pi}}]
```

to get

$$\frac{-64\ \text{Cos}[9\ t]}{227} - \frac{163\ \text{Cos}[\dfrac{4\ t}{\text{Sqrt}[3]}]}{227} \qquad \text{and the plot}$$

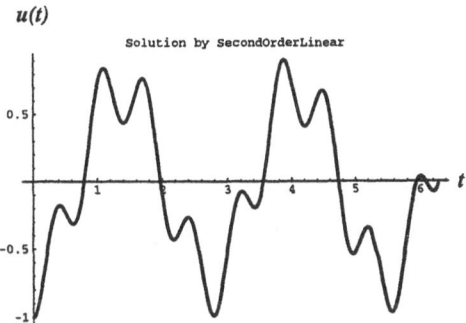

$u(t)$

Solution by SecondOrderLinear

Solution to Exercise 1

Thus $u(t) = -\dfrac{163}{227}\cos\left(\dfrac{4t}{\sqrt{3}}\right) - \dfrac{163\cos(9t)}{227}$.

2. A spring is stretched 3 inches by a mass that weighs 4 pounds. The resulting system experiences a frictional force with damping constant equal to 0.10 pound seconds/foot. Further, an external force of $2\cos(2t)$ pounds is applied. Find the transient and steady-state solutions of the resulting initial value problem.

Solution. The roots of the characteristic equation $(4/32)r^2 + r/10 + 16 = 0$ are $r_1 = -2/5 - (2\sqrt{799}/5)i$ and $r_2 = -2/5 + (2\sqrt{799}/5)i$. Consequently, the general solution to the homogeneous equation is $y_H(t) = e^{-2t/5}\left(C_1\cos(2\sqrt{799}t/5) + C_2\sin(2\sqrt{799}t/5)\right)$. The particular solution is computed from the form $y_p(t) = A\cos(2t) + B\sin(2t)$ using undetermined coefficients. Solving for the coefficients, we obtain $A = -3100/24029$, and $B = 40/24029$. Using the initial conditions, we obtain the equations $3100/24029 + C_2 = 0$ and $80/24029 + (2\sqrt{799}/5)C_1 - (2/5)C_2 = 10$, which lead to $C_1 = 597425/(24095\sqrt{799})$ and $C_2 = -3100/24029$. With **ODE**, we choose the interval $[0, 2\pi]$ and enter

```
ODE[{(4/32)u'' + u'/10 + (4/(1/4))u == 2Cos[2t],
u[0] == 0,u'[0] == 10},u,t,
Method->SecondOrderLinear,Form->Explicit,
PlotSolution->{{t,0,2Pi}}]
```

to get

```
                      2 Sqrt[799] t
            3100 Cos[-------------]
3100 Cos[2 t]               5                40 Sin[2 t]
------------- - ---------------------   +   ------------  +
   24029              (2 t)/5                  24029
                24029 E


          2 Sqrt[799] t
597425 Sin[-------------]
                5
-------------------------
                   (2 t)/5
24029 Sqrt[799] E
```

and the plot

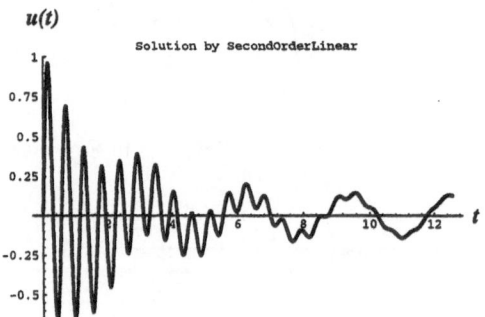

u(t)

Solution by SecondOrderLinear

Solution to Exercise 2

Thus, the transient solution is

$$u_H(t) = e^{-2t/5}\left(-\frac{3100}{24029}\cos\left(\frac{2\sqrt{799}\,t}{5}\right) + \frac{597425}{24029\sqrt{799}}\sin\left(\frac{2\sqrt{799}\,t}{5}\right)\right)$$

and the steady-state solution is

$$u_P(t) = \frac{3100\cos(2t)}{24029} + \frac{40\sin(2t)}{24029}.$$

3. Write each of the following functions in the form $A \cos(\omega t - \alpha) \cos(\nu t - \beta)$ for suitable values of ω, ν, α, β:

    a. $\cos(8t) - \cos(6t)$ **c.** $\sin(9t) - \sin(7t)$

    b. $\cos(8t) + \cos(6t)$ **d.** $\sin(9t) + \sin(7t)$

Solution. Writing $\cos(8t) = \cos(7t + t) = \cos(7t)\cos(t) - \sin(7t)\sin(t)$, and $\cos(6t) = \cos(7t - t) = \cos(7t)\cos(t) + \sin(7t)\sin(t)$ we have $\cos(8t) - \cos(6t) = -2\sin(7t)\sin(t)$ and $\cos(8t) + \cos(6t) = 2\cos(7t)\cos(t)$. Similarly, $\sin(9t) = \sin(8t + t) = \sin(8t)\cos(t) + \cos(8t)\sin(t)$, $\sin(7t) = \sin(8t - t) = \sin(8t)\cos(t) - \cos(8t)\sin(t)$, so that $\sin(9t) - \sin(7t) = 2\cos(8t)\sin(t)$ and $\sin(9t) + \sin(7t) = 2\sin(8t)\cos(t)$. The results are re-written in the form

    a. $-2\cos(7t - \pi/2)\,\cos(t - \pi/2)$ **c.** $2\cos(8t)\,\cos(t - \pi/2)$

    b. $2\cos(7t)\,\cos(t)$ **d.** $2\cos(8t - \pi/2)\,\cos(t)$

4. Show that a function of the form

$$u(t) = C_1 \cos(ct) + C_2 \sin(ct) + D_1 \cos(ft) + D_2 \sin(ft)$$

with $c < f$ can be written in the form

$$u(t) = A \cos(\omega t - \alpha) \cos(\nu t - \beta),$$

where $\omega = (c + f)/2$, $\nu = (f - c)/2$ provided that the constants C_1, C_2, D_1, D_2 satisfy the relation $C_1^2 + C_2^2 = D_1^2 + D_2^2$. [Hint: Apply the amplitude-phase representation twice and then use the trigonometric identities for $\sin(x + y)$ and $\cos(x + y)$.]

Solution. From the amplitude-phase representation we see that

$$u(t) = R\cos(ct - \alpha_1) + R\cos(ft - \beta_1),$$

where $R^2 = C_1^2 + C_2^2 = D_1^2 + D_2^2$, $\alpha_1 = \arctan(C_2/C_1)$ and $\beta_1 = \arctan(D_2/D_1)$. Let $c = \omega - \nu$ and $f = \omega + \nu$. Defining $\alpha = (\alpha_1 + \beta_1)/2$ and $\beta = (\beta_1 - \alpha_1)/2$, we can apply the trigonometric addition formulas for $\cos\big((\omega t - \alpha) \pm (\nu t - \beta)\big)$ to obtain

$$
\begin{aligned}
u(t) &= R\big(\cos\big((\omega - \nu)t - (\alpha - \beta)\big) + \cos\big((\omega + \nu)t - (\alpha + \beta)\big)\big) \\
&= R\big(\cos\big((\omega t - \alpha) - (\nu t - \beta)\big) + \cos\big((\omega t - \alpha) + (\nu t - \beta)\big)\big) \\
&= 2R\cos(\omega t - \alpha)\cos(\nu t - \beta).
\end{aligned}
$$

5. (Converse to Exercise 4)

    a. Show that any function of the form

$$u(t) = A \cos(\omega t - \alpha) \cos(v t - \beta)$$

can be written in the form

$$u(t) = C_1 \cos(c t) + C_2 \sin(c t) + D_1 \cos(f t) + D_2 \sin(f t)$$

for a suitable choice of the constants C_1, C_2, D_1, D_2, c, f, where $C_1^2 + C_2^2 = D_1^2 + D_2^2$.
[Hint: Use the trigonometric identities

$$2 \cos(x) \cos(y) = \cos(x - y) + \cos(x + y),$$

$$2 \sin(x) \sin(y) = \cos(x - y) - \cos(x + y),$$

$$2 \sin(x) \cos(y) = \sin(x + y) + \sin(x - y).]$$

    b. Conclude that the set of functions of the form

$$u(t) = A \cos(\omega t - \alpha) \cos(v t - \beta)$$

is identical to the set of functions of the form

$$u(t) = C_1 \cos(c t) + C_2 \sin(c t) + D_1 \cos(f t) + D_2 \sin(f t)$$

where $C_1^2 + C_2^2 = D_1^2 + D_2^2$.

Solution. For part **a** we write

$$
\begin{aligned}
u(t) &= A \cos(\omega t - \alpha) \cos(v t - \beta) \\
&= A\big(\cos((\omega - v)t - (\alpha - \beta)) + \cos((\omega + v)t - (\alpha + \beta)) \big) \\
&= A \cos(\alpha - \beta) \cos\big((\omega - v)t\big) + A \sin(\alpha - \beta) \sin\big((\omega - v)t\big) \\
&\quad + A \cos(\alpha + \beta) \cos\big((\omega + v)t\big) + A \sin(\alpha + \beta)) \sin\big((\omega + v)t\big),
\end{aligned}
$$

which is of the required form. To do part **b**, we combine the result of part **a** with Exercise 4.

6. Use **ODE** to solve the initial value problem

$$
\begin{cases}
m u''(t) + k u(t) = F_0 \cos(\omega t), \\
u(0) = U_0, \, u'(0) = U_1
\end{cases}
$$

where $\omega \neq \sqrt{k/m}$.

Solution. Let $\omega_0 = \sqrt{k/m}$. We enter the following commands to make m, k and c positive real numbers.

```
Clear[m,k,c];
m /: Im[m]=0;
m /: Positive[m]=True;
k /: Im[k]=0;
k /: Positive[k]=True;
c /: Im[c]=0;
c /: Positive[c]=True;
```

Then

```
ODE[{m u'' + k u == F0 Cos[omega t],
u[0] == U0,u'[0] == U1},u,t,
Method->SecondOrderLinear,Form->Explicit,
PostSolution->{(# /. k ->omega0^2 m)&,
PowerExpand,Expand,Collect[#,{U0,U1,F0}]&,Cancel}]
```

yields

```
                     F0 (-Cos[omega t] + Cos[omega0 t])
U0 Cos[omega0 t] +   ───────────────────────────────────  +
                               2            2
                         m omega   - m omega0

   U1 Sin[omega0 t]
   ────────────────
       omega0
```

Thus

$$u(t) = \left(U_0 - \frac{F_0}{m(\omega_0^2 - \omega^2)}\right)\cos(\omega_0 t) + \frac{U_1 \sin(\omega_0 t)}{\omega_0} + \frac{F_0 \cos(\omega t)}{m(\omega_0^2 - \omega^2)}.$$

7. Use **ODE** to solve the initial value problem

$$\begin{cases} m\,u''(t) + k\,u(t) = F_1 \sin(\omega t), \\ u(0) = U_0,\ u'(0) = U_1 \end{cases}$$

where $\omega \neq \sqrt{k/m}$.

Solution. We modify the solution method of Exercise 6; specifically

```
ODE[{m u'' + k u == F1 Sin[omega t],
u[0] == U0,u'[0] == U1},u,t,
Method->SecondOrderLinear,Form->Explicit,
PostSolution->{(# /. k ->omega0^2 m)&,
PowerExpand,Expand,Collect[#,{U0,U1,F1}]&,Cancel}]
```

yields

```
                      U1 Sin[omega0 t]
UO Cos[omega0 t] +   ------------------    +
                           omega0

   F1 (-(omega0 Sin[omega t]) + omega Sin[omega0 t])
  --------------------------------------------------
                   2              2
        omega0 (m omega   - m omega0 )
```

Thus

$$u(t) = U_0 \cos(\omega_0 t) + \left(U_1 - \frac{F_1 \omega}{m(\omega_0^2 - \omega^2)} \right) \frac{\sin(\omega_0 t)}{\omega_0} + \frac{F_1 \sin(\omega t)}{m(\omega_0^2 - \omega^2)}.$$

8. Use **ODE** to solve the initial value problem

$$\begin{cases} m\, u''(t) + k\, u(t) = F_0 + F_1 t + F_2 t^2, \\ u(0) = U_0, \, u'(0) = U_1 \end{cases}$$

where F_0, F_1, F_2, k and m are constants.

Solution. We modify the solution method of Exercise 6; specifically

```
ODE[{m u'' + k u == F0 + F1 t + F2 t^2,
u[0] == U0,u'[0] == U1},u,t,
Method->SecondOrderLinear,Form->Explicit,
PostSolution->{(# /. k ->omega0^2 m)&,PowerExpand,
Collect[#,{U0,U1,F0,F1,F2}]&,Cancel}]
```

yields

```
FO (1 - Cos[omega0 t])
----------------------  + UO Cos[omega0 t] +
          2
    m omega0

             2  2
    F2 (-2 + omega0  t  + 2 Cos[omega0 t])
   ---------------------------------------  +
                     4
               m omega0

   F1 (omega0 t - Sin[omega0 t])     U1 Sin[omega0 t]
  -------------------------------  + ------------------
                  3                        omega0
            m omega0
```

Thus

$$u(t) = U_0 \cos(\omega_0 t) + \frac{U_1 \sin(\omega_0 t)}{\omega_0} + \frac{F_0(1 - \cos(\omega_0 t))}{m\,\omega_0^2}$$

$$+ \frac{F_1(\omega_0 t - \sin(\omega_0 t))}{m\,\omega_0^3} + \frac{F_2(-2 + \omega_0^2 t^2 + 2\cos(\omega_0 t))}{m\,\omega_0^4}.$$

9. A mass-spring system with constants m and $k > 0$ is acted upon by the forcing function $F(t) = F_0\big(\cos(\omega t)\big)^2$, where $\omega \neq \sqrt{k/m}$. Find the formula for the motion, given the initial conditions $u(0) = 4$ and $u'(0) = 0$.

Solution. The roots of the characteristic equation $m r^2 + k = 0$ are $r_1 = -\sqrt{k/m}\, i$ and $r_2 = \sqrt{k/m}\, i$. Consequently, the general solution to the homogeneous equation is $y_H(t) = C_1 \cos(\sqrt{k/m}\, t) + C_2 \sin(\sqrt{k/m}\, t)$. Using the identity $\big(\cos(\omega t)\big)^2 = 1/2 + \cos(2\omega t)/2$, the particular solution is computed from the form $y_P(t) = A + B\cos(2\omega t) + C\sin(2\omega t)$ using undetermined coefficients. Solving for the coefficients, we obtain $A = F_0/(2k)$, $B = F_0/\big(2(k - 4m\omega^2)\big)$ and $C = 0$. Using the initial conditions, we obtain $C_1 = 0$ and $C_2 = (-k + 4k^2 + 2m\omega^2 - 16km\omega^2)/(k^2 - 4km\omega^2)$. After some simplifying, we obtain

$$u(t) = \frac{F_0(1 - \cos(t\sqrt{k/m}))}{2k} + \frac{F_0\big(\cos(2\omega t) - \cos(t\sqrt{k/m})\big)}{4m\,\omega^2 - k} + 4\cos(t\sqrt{k/m})$$

10. Let $c^2 < 4m\,k$. Find the solution of the initial value problem

$$\begin{cases} m\,u''(t) + c\,u'(t) + k\,u(t) = F_0 \cos(\omega t) \\ u(0) = U_0,\ u'(0) = U_1 \end{cases}$$

Solution.

$$u(t) = e^{\lambda t}\big(C_1 \cos(\mu t) + C_2 \sin(\mu t)\big) + \frac{F_0 \cos(\omega t - \delta)}{\sqrt{D}},$$

where

$$\lambda = -\frac{c}{2m}, \qquad \mu = \frac{\sqrt{4m\,k - c^2}}{2m},$$

$$\omega_0 = \sqrt{\frac{k}{m}}, \qquad D = m^2(\omega_0^2 - \omega^2)^2 + c^2\omega^2,$$

and

$$C_1 = U_0 - \frac{F_0 \cos(\delta)}{\sqrt{D}}, \qquad C_2 = \frac{U_1 - \lambda U_0}{\lambda} + \frac{F_0(\lambda \cos(\delta) - \omega \sin(\delta))}{\mu\sqrt{D}}.$$

11. Consider a mass-spring system with sinusoidal forcing, described by the differential equation

$$m\,y'' + c\,y' + k\,y = F_0 \cos(\omega\,t) \tag{S.76}$$

with steady-state solution

$$y(t) = \frac{F_0 \cos(\omega t - \delta)}{\sqrt{D}}. \tag{S.77}$$

The **amplification factor** is defined to be

$$\rho = \frac{k}{\sqrt{D}} = \frac{k}{\sqrt{(k - m\,\omega^2)^2 + c^2\omega^2}} = \frac{1}{\sqrt{\left(1 - \dfrac{\omega^2}{\omega_0^2}\right)^2 + \dfrac{c^2\omega^2}{\omega_0^2}}};$$

it is the factor F_0/k must be multiplied by to get the amplitude of the steady-state solution of (S.76). Consider ρ to be a function of ω/ω_0 for given values of c.

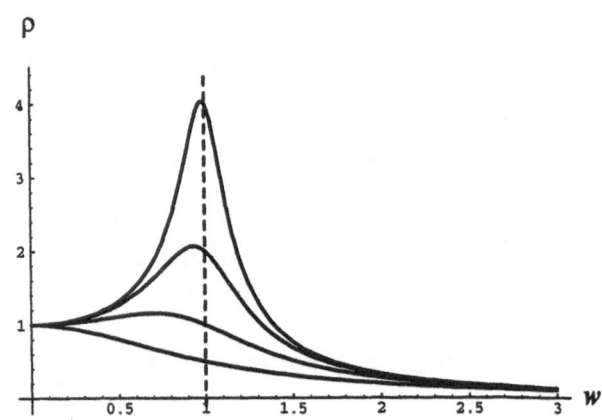

The amplification factor as a function of $w = \omega/\omega_0$

a. Show that in the limiting case $c = 0$ the amplification factor is infinite when $\omega = \omega_0 = \sqrt{k/m}$.

b. Show that if $0 < c^2 < 2m\,k$, then the amplification factor is maximum when $\omega^2 = \omega_0^2 - c^2/(2m^2)$.

c. Show that if $c^2 \geq 2m\,k$, then the amplification factor is maximum when $\omega = 0$.

Solution.

a. If $c = 0$ the amplification factor is $\dfrac{k}{\sqrt{m\,\omega^2 - k}}$, which is infinite when $\omega = \sqrt{k/m}$.

b. If $0 < c^2 < 2mk$, then the amplification factor is $k/\sqrt{D}$, where $D = m^2(\omega^2 - \omega_0^2)^2 + c^2\omega^2$. The maximum is obtained by finding the minimum value of D. This is a quadratic polynomial in ω^2 with an interior minimum that occurs when $2m^2\omega^2 + (c^2 - 2m^2\omega_0^2) = 0$.

c. If $c^2 \geq 2mk$, then $m^2(\omega^2 - \omega_0^2)^2 + c^2\omega^2$ has no interior minimum. At the end point $\omega = 0$ we have $D = m^2\omega_0^4$, whereas $D \longrightarrow \infty$ for $\omega \longrightarrow \infty$. Hence the absolute minimum of D occurs when $\omega = 0$.

12. Consider a mass-spring system with sinusoidal forcing, described by the differential equation (S.76) with the steady-state solution (S.77). The **steady-state phase lag** δ is defined for $c > 0$ by

$$\tan(\delta) = \frac{c\,\omega}{k - m\,\omega^2} = \frac{(\omega/\omega_0)(c/\sqrt{m\,k})}{1 - (\omega/\omega_0)^2},$$

with the requirement that $0 \leq \delta < \pi$. This is considered as a function of ω for given values of c, m, k and F_0.

a. Show that in the limiting case $c = 0$, the steady-state phase lag has the value $\delta = 0$ for $\omega < \omega_0$ and has the value $\delta = \pi$ for $\omega > \omega_0$.

b. Show that for $c > 0$ the steady-state phase lag is a strictly increasing function of ω for $\omega > 0$ with $\delta(0) = 0$, $\delta(\omega_0) = \pi/2$ and $\delta(\omega) \longrightarrow \pi$ as $\omega \longrightarrow \infty$.

Solution.

a. The tangent function is positive in the interval $(0, \pi/2)$ and negative in the interval $(\pi/2, \pi)$. If $0 < \omega < \omega_0$, then $c\,\omega/(m(\omega_0^2 - \omega^2)) > 0$, and tends to zero when c tends to zero. Hence we must have $0 < \delta < \pi/2$ with $\delta \longrightarrow 0$. If $\omega > \omega_0$, then $c\,\omega/(m(\omega_0^2 - \omega^2)) < 0$, and tends to zero when c tends to zero. Hence we must have $\pi/2 < \delta < \pi$ with $\delta \longrightarrow \pi$.

b. The function $\omega \longrightarrow c\,\omega/(m(\omega_0^2 - \omega^2))$ is strictly increasing for $0 < \omega < \omega_0$ and tends to $+\infty$ when $\omega \longrightarrow \omega_0$. The inverse tangent function is also strictly increasing and tends to $\pi/2$. Therefore, δ is strictly increasing for $0 < \omega < \omega_0$ with $\delta(\omega_0) = \pi/2$. The function $\omega \longrightarrow c\,\omega/(m(\omega_0^2 - \omega^2))$ is also strictly increasing for $\omega > \omega_0$. Since the inverse tangent function is strictly increasing for negative argument, we conclude that $\delta(\omega)$ is also strictly increasing for $\omega > \omega_0$

13. An undamped mass-spring system with impulsive forcing is described by the differential equation

$$m\,y'' + k\,y = F(t),$$

where $F(t) = F_0$ for $0 < t < \varepsilon$ and $F(t) = 0$ for $t > \varepsilon$. Assume the initial conditions $y(0) = 0$ and $y'(0) = 0$.

    a. Solve the initial value problem for all t, taking care to maintain the continuity of $y(t)$ and $y'(t)$ at the point $t = \varepsilon$.

    b. Suppose that $\varepsilon \longrightarrow 0$ and $F_0 \longrightarrow \infty$, so that $\varepsilon F_0 \longrightarrow I_0 > 0$. Find the limiting value of the solution $y(t)$ for any t. [Hint: First, find the limiting values of $y(\varepsilon)$ and $y'(\varepsilon)$.]

Solution.

    a. For $0 < t \le \varepsilon$ we have $y(t) = \dfrac{F_0}{k}\left(1 - \cos(\omega_0 t)\right)$, whereas for $t > \varepsilon$ we

have $y(t) = y(\varepsilon)\cos(\omega_0(t - \varepsilon)) + \dfrac{y'(\varepsilon)}{\omega_0} \sin(\omega_0(t - \varepsilon))$, where

$$y(\varepsilon) = \frac{F_0\left(1 - \cos(\omega_0\varepsilon)\right)}{k} \quad \text{and} \quad y'(\varepsilon) = \frac{F_0\omega_0 \sin(\omega_0\varepsilon)}{k}.$$

    b. When $\varepsilon \longrightarrow 0$, $F_0 \longrightarrow \infty$ with $F_0\varepsilon \longrightarrow I_0$, we have

$$y(\varepsilon) \longrightarrow 0 \quad \text{and} \quad y'(\varepsilon) \longrightarrow \frac{I_0\omega_0^2}{k} = \frac{I_0}{m},$$

by l'Hôspital's rule, for example. Therefore, $y(t)$ tends to $(I_0/m\,\omega_0) \sin(\omega_0 t)$.

14. An underdamped $(c^2 < 4m\,k)$ mass-spring system with impulsive forcing is described by the differential equation

$$m\,y'' + c\,y' + k\,y = F(t),$$

where $F(t) = F_0$ for $0 < t < \varepsilon$ and $F(t) = 0$ for $t > \varepsilon$. Assume the initial conditions $y(0) = 0$ and $y'(0) = 0$.

    a. Solve the initial value problem for all t, taking care to maintain the continuity of $y(t)$ and $y'(t)$ at the point $t = \varepsilon$.

    b. Suppose that $\varepsilon \longrightarrow 0$ and $F_0 \longrightarrow \infty$, so that $\varepsilon F_0 \longrightarrow I_0 > 0$. Find the limiting value of the solution $y(t)$ for any t. [Hint: First, find the limiting values of $y(\varepsilon)$ and $y'(\varepsilon)$.]

Solution.

    a. For $0 < t \le \varepsilon$ we have $y(t) = \dfrac{F_0}{k}\left(1 - e^{-\lambda t}\cos(\mu\,t) - \dfrac{\lambda}{\mu}e^{-\lambda t}\sin(\mu\,t)\right)$,

whereas for $t > \varepsilon$ we have

$$y(t) = y(\varepsilon)e^{-\lambda(t-\varepsilon)}\cos(\mu(t - \varepsilon)) + \left(y'(\varepsilon) + \lambda\,y(\varepsilon)\right)e^{-\lambda(t-\varepsilon)}\frac{\sin(\mu(t - \varepsilon))}{\mu}$$

b. When $\varepsilon \longrightarrow 0$ we have $y(\varepsilon) \longrightarrow 0$ and $y'(\varepsilon) \longrightarrow \frac{I_0}{k}(\mu^2 + \lambda^2) = \frac{I_0}{m}$.

Therefore $y(t) \longrightarrow \frac{I_0}{m\,\mu} e^{-\lambda t} \sin(\mu\, t)$.

Section 11.3, page 374

1. Use **ODE** to solve the initial value problem

$$\begin{cases} L\, Q''(t) + Q(t)/C = 0, \\ Q(0) = Q_0, \; Q'(0) = Q_1. \end{cases}$$

Solution. We enter the following commands to make L and Cp positive real numbers.

```
Clear[Cp,L]
L /: Im[L] = 0;
Cp /: Im[Cp] = 0;
L /: Positive[L] = True;
Cp /: Positive[Cp] = True;
```

Then

```
ODE[{L Q'' + Q/Cp == 0,Q[0] == Q0,Q'[0] == Q1},Q,t,
Method->SecondOrderLinear,Form->Explicit,
PostSolution->{(# /. Cp ->1/(omega0^2 L)&),PowerExpand}]
```

yields

```
                    Q1 Sin[omega0 t]
Q0 Cos[omega0 t] +  ────────────────          or
                        omega0
```

$$Q(t) = Q_0 \cos(\omega_0 t) + \frac{Q_1 \sin(\omega_0 t)}{\omega_0}.$$

2. Use **ODE** to solve the initial value problem

$$\begin{cases} L\, Q''(t) + Q(t)/C = E_0 \cos(\omega t), \\ Q(0) = Q_0, \; Q'(0) = Q_1. \end{cases}$$

where $\omega \neq 1/\sqrt{LC}$.

Solution. We use

```
ODE[{L Q'' + Q/Cp == E0 Cos[omega t],Q[0] == Q0,
Q'[0] == Q1},Q,t,
Method->SecondOrderLinear,Form->Explicit,
PostSolution->{(# /. Cp ->1/(omega0^2 L)&),
PowerExpand,Expand,Collect[#,{Q0,Q1,E0}]&,Cancel}]
```

to get

```
     E0 Cos[omega t]            E0 Cos[omega0 t]
 - (-------------------) +   -------------------------  +
      2          2              2           2
    L omega  - L omega0       L omega   - L omega0               or

                         Q1 Sin[omega0 t]
    Q0 Cos[omega0 t] +   ----------------
                             omega0
```

$$Q(t) = \left(Q_0 - \frac{E_0}{L(\omega_0^2 - \omega^2)} \right) \cos(\omega_0 t) + \frac{Q_1 \sin(\omega_0 t)}{\omega_0} + \frac{E_0 \cos(\omega t)}{L(\omega_0^2 - \omega^2)}.$$

3. Use **ODE** to solve the initial value problem

$$\begin{cases} L\,Q''(t) + Q(t)/C = E_1 \sin(\omega t), \\ Q(0) = Q_0, \; Q'(0) = Q_1. \end{cases}$$

where $\omega \neq 1/\sqrt{LC}$.

Solution. We use

```
ODE[{L Q'' + Q/Cp == E1 Sin[omega t],Q[0] == Q0,
Q'[0] == Q1},Q,t,
Method->SecondOrderLinear,Form->Explicit,
PostSolution->{(# /. Cp ->1/(omega0^2 L)&),
PowerExpand,Expand,Collect[#,{Q0,Q1,E1}]&,Cancel}]
```

to get

```
                         Q1 Sin[omega0 t]
    Q0 Cos[omega0 t] +   ----------------   +
                             omega0

    E1 (-(omega0 Sin[omega t]) + omega Sin[omega0 t])              or
    ------------------------------------------------
                       2             2
         omega0 (L omega   - L omega0 )
```

$$Q(t) = Q_0 \cos(\omega_0 t) + \left(Q_1 - \frac{E_1 \omega}{L(\omega_0^2 - \omega^2)} \right) \frac{\sin(\omega_0 t)}{\omega_0} + \frac{E_1 \sin(\omega t)}{L(\omega_0^2 - \omega^2)}.$$

4. Find the form of the solution of the differential equation

$$L I''(t) + R I'(t) + \frac{I(t)}{C} = 0$$

in the following three cases:

    a. $R > \sqrt{4L/C}$ (overdamped case)

    b. $R = \sqrt{4L/C}$ (critically damped case)

    c. $0 < R < \sqrt{4L/C}$ (underdamped case)

Solution. The roots of the characteristic equation $L r^2 + R r + 1/C = 0$ are $r_1 = -R/(2L) - \sqrt{C^2 R^2 - 4CL}/(2CL)$ and $r_2 = -R/(2L) + \sqrt{C^2 R^2 - 4CL}/(2CL)$. Consequently, the general solutions can be written

    a. $I(t) = C_1 e^{r_1 t} + C_2 e^{r_2 t}$, where $r_1, r_2 = \dfrac{-R \pm \sqrt{R^2 - 4L/C}}{2L}$

    b. $I(t) = C_1 e^{rt} + C_2 t\, e^{rt}$, where $r = -R/2L$

    c. $I(t) = C_1 e^{\lambda t} \cos(\mu t) + C_2 e^{\lambda t} \sin(\mu t)$, where $\lambda = -R/2L$ and

$$\mu = \sqrt{\frac{R^2}{4L^2} - \frac{1}{LC}}$$

5. Suppose that a circuit contains an inductance, a capacitor and a resistor in series, with no impressed voltage. Find the formula for the charge if the constants have the following values:

    a. $L = 1/5$ henrys, $R = 300$ ohms, $C = 1/10,000$ farad

    b. $L = 1/5$ henry, $R = 100$ ohms, $C = 1/200$ farad

    c. $L = 10$ henrys, $R = 100$ ohms, $C = 1/500$ farad

Solution. Using the results from Exercise 4,

    a. $r_1 = -750 - 50\sqrt{205}, r_2 = -750 + 50\sqrt{205}$

    b. $r_1 = -250 - 10\sqrt{615}, r_2 = -250 + 10\sqrt{615}$

    c. $r_1 = -5 - 5i, r_2 = -5 + 5i$

Consequently, the solutions can be written as

    a. $Q(t) = C_1 e^{-(750 - 50\sqrt{205}) t} + C_2 e^{-(750 + 50\sqrt{205}) t}$

b. $Q(t) = C_1 e^{-(250-10\sqrt{615})t} + C_2 e^{-(250+10\sqrt{615})t}$

c. $Q(t) = e^{-5t}(C_1 \sin(5t) + C_2 e^{-5t} \cos(5t))$

6. Suppose that a circuit contains an inductance, a capacitor and a resistor in series, with the given impressed voltage. Find the formula for the charge if the constants have the following values:

    a. $L = 10$ henrys, $R = 30$ ohms, $C = 2/100$ farad,
$E(t) = 50 \sin(2t)$ volts

    b. $L = 3$ henrys, $R = 13$ ohms, $C = 1/4$ farad,
$E(t) = 10 \cos(2t)$ volts

    c. $L = 1$ henrys, $R = 4$ ohms, $C = 1/5$ farad,
$E(t) = 7 \cos(t) + 8 \sin(t)$ volts

Solution.

    a. The roots of the characteristic equation $L r^2 + Rr + 1/C = 0$ are $r_1 = -3/2 - (\sqrt{11}/2) i$ and $r_2 = -3/2 + (\sqrt{11}/2) i$. Consequently, the general solution to the homogeneous equation is $y_H(t) = e^{-3t/2}\left(C_1 \sin \left((\sqrt{11}/2)t\right)\right.$

$\left. +C_2 \cos \left((\sqrt{11}/2)t\right)\right)$. The particular solution can be computed from the form $y_P(t) = A \cos(2t) + B \sin(2t)$ using undetermined coefficients. Solving for the coefficients, we obtain $A = -30/37$ and $B = 5/37$. Hence

$$Q(t) = e^{-3t/2}\left(C_1 \sin\left(\frac{\sqrt{11}\,t}{2}\right) + C_2 \cos\left(\frac{\sqrt{11}\,t}{2}\right)\right) + \frac{5 \sin(2t) - 30\cos(2t)}{37}.$$

    b. The roots of the characteristic equation $L r^2 + Rr + 1/C = 0$ are $r_1 = -4$ and $r_2 = -1/3$. Consequently, the general solution to the homogeneous equation is $y_H(t) = C_1 e^{-4t} + C_2 e^{-t/3}$. The particular solution can be computed from the form $y_P(t) = A \cos(2t) + B \sin(2t)$ using undetermined coefficients. Solving for the coefficients, we obtain $A = -4/37$ and $B = 15/37$. Hence

$$Q(t) = C_1 e^{-4t} + C_2 e^{-t/3} + \frac{13 \sin(2t) - 4 \cos(2t)}{37}.$$

    c. The roots of the characteristic equation $L r^2 + Rr + 1/C = 0$ are $r_1 = -2-i$ and $r_2 = -2+i$. Consequently, the general solution to the homogeneous equation is $y_H(t) = e^{-2t} (C_1 \sin(t) + C_2 \cos(t))$. The particular solution can be computed from the form $y_P(t) = A \cos(t) + B \sin(t)$ using undetermined coefficients. Solving for the coefficients, we obtain $A = -1/37$ and $B = 15/37$. Hence

$$Q(t) = C_1 e^{-2t} \cos(t) + C_2 e^{-2t} \sin(t) + \frac{15 \sin(2t) - \cos(2t)}{37}.$$

7. Suppose that a circuit contains an inductance, a capacitor and a resistor in series, with the given impressed voltage. Find the current $I(t)$ given the initial current (in amperes) and charge on the capacitor (in coulombs).

    a. $L = 2$ henrys, $R = 16$ ohms, $C = 2/100$ farad,
$E(t) = 100$ volts, $I(0) = 0$, $Q(0) = 5$.

    b. $L = 2$ henrys, $R = 60$ ohms, $C = 25/10000$ farad,
$E(t) = 100 \exp(-t)$ volts, $I(0) = Q(0) = 0$.

    c. $L = 2$ henrys, $R = 60$ ohms, $C = 25/10000$ farad,
$E(t) = 200 \exp(-10t)$ volts, $I(0) = 0$, $Q(0) = 1$.

Solution.

    a. The roots of the characteristic equation $L\,r^2 + Rr + 1/C = 0$ are $r_1 = -4 - 3i$ and $r_2 = -4 + 3i$. Consequently, the general solution to the homogeneous equation is $y_H(t) = e^{-4t}(C_1 \sin(3t) + C_2 \cos(3t))$. The particular solution can be computed from the form $y_P(t) = A$ using undetermined coefficients. Solving for the coefficient, we obtain $A = 2$. Hence, the solution for the charge in the circuit becomes

$$Q(t) = e^{-4t}(C_1 \sin(3t) + C_2 \cos(3t)) + 2.$$

Due to the relationship $dQ/dt = I$, the initial conditions can be written $Q(0) = 5$ and $Q'(0) = 0$. Using these conditions, we find that $C_1 = 4$ and $C_2 = 3$. Thus,

$$I(t) = \frac{dQ}{dt} = -25\,e^{-4t} \sin(3t).$$

    b. The roots of the characteristic equation $L\,r^2 + Rr + 1/C = 0$ are $r_1 = -20$ and $r_2 = -10$. Consequently, the general solution to the homogeneous equation is $y_H(t) = C_1 e^{-20t} + C_2 e^{-10t}$. The particular solution can be computed from the form $y_P(t) = Ae^{-t}$ using undetermined coefficients. Solving for the coefficient, we obtain $A = 50/171$. Hence, the solution for the charge in the circuit becomes

$$Q(t) = C_1 e^{-20t} + C_2 e^{-10t} + \frac{50}{171} e^{-t}.$$

Due to the relationship $dQ/dt = I$, the initial conditions can be written $Q(0) = 0$ and $Q'(0) = 0$. Using these conditions, we find that $C_1 = 5/19$ and $C_2 = -5/9$. Thus,

$$I(t) = \frac{dQ}{dt} = -\frac{100e^{-20t}}{19} + \frac{50e^{-10t}}{9} - \frac{50e^{-t}}{171}.$$

    c. The roots of the characteristic equation $L\,r^2 + Rr + 1/C = 0$ are $r_1 = -20$ and $r_2 = -10$. Consequently, the general solution to the homogeneous equation is $y_H(t) = C_1 e^{-20t} + C_2 e^{-10t}$. Because we have first-order resonance, the particular solution is computed from the form $y_P(t) = A t e^{-10t}$ using undetermined coefficients. Solving for the coefficient, we obtain $A = 10$. Hence, the solution for the charge in the circuit becomes

$$Q(t) = C_1 e^{-20t} + C_2 e^{-10t} + 10te^{-10t}.$$

Due to the relationship $dQ/dt = I$, the initial conditions can be written $Q(0) = 1$ and $Q'(0) = 0$. Using these conditions, we find that $C_1 = 0$ and $C_2 = 1$. Thus,

$$I(t) = \frac{dQ}{dt} = -100 t e^{-10t}.$$

8. If $L = 1/5$ henrys and $C = 8/100,000$ farad, determine the resistance for which the circuit is critically damped.

Solution. $R = \sqrt{4L/C} = 100$ ohms

9. Determine the steady-state current in a series circuit if $L = 1$ henry, $R = 5000$ ohms, $C = 2.5/10,000,000$ farad, and the impressed voltage is $E(t) = 110\cos(120\pi t)$ volts.

Solution.

$$I(t) = A\cos(120\pi t) + B\sin(120\pi t) \text{ with}$$

$$A = \frac{(120\pi)^2 - 4 \times 10^6)}{(120\pi)^2 - 4 \times 10^6 + 25 \times 10^6 (120\pi)^2}$$

and

$$B = \frac{5000 \times 110}{(120\pi)^2 - 4 \times 10^6 + 25 \times 10^6 (120\pi)^2}.$$

10. Suppose that we have a series circuit with given values of L, R, C, and an impressed voltage of the form $E_0 \cos(\omega t)$. For what value of ω will the amplitude of the steady-state current be a maximum?

Solution. This is obtained by minimizing the denominator $D = L^2(\omega^2 - 1/L\,C) + R^2\omega^2$ exactly as in Exercise 11 of Section 11.2.

11. A series LRC circuit has $L = 0.5$ henry, $R = 20$ ohms, $C = 10^{-2}$ farad and an impressed voltage $E = 12$volts. Assume no initial current and no initial charge. Find and plot the subsequent charge and current.

Solution. We define the charge with

```
Q[t_] = ODE[{(1/2)QQ'' + 20QQ' + QQ/10^-2 == 15,
          QQ[0] == 0,QQ'[0] == 0},QQ,t,
      Method->SecondOrderLinear,Form->Explicit]
```

Then evaluation of `Q[t]` and `Q'[t]//Expand` yields

```
 3     3 E                              3 E
 -- -  ------------------------------ + ------------------------------ -
 20              40                              20 Sqrt[2]
```

$$\frac{3}{20} - \frac{3\,E^{-20\ t\ -\ 10\ \text{Sqrt}[2]\ t}}{40} + \frac{3\,E^{-20\ t\ -\ 10\ \text{Sqrt}[2]\ t}}{20\ \text{Sqrt}[2]} -$$

$$\frac{3\,E^{-20\ t\ +\ 10\ \text{Sqrt}[2]\ t}}{40} - \frac{3\,E^{-20\ t\ +\ 10\ \text{Sqrt}[2]\ t}}{20\ \text{Sqrt}[2]}$$

and

$$\frac{-3\,E^{-20\ t\ -\ 10\ \text{Sqrt}[2]\ t}}{2\ \text{Sqrt}[2]} + \frac{3\,E^{-20\ t\ +\ 10\ \text{Sqrt}[2]\ t}}{2\ \text{Sqrt}[2]}$$

Thus

$$Q(t) = \frac{3}{20} - \frac{3e^{-20t}\left(\cosh(10\sqrt{2}\,t) + \sqrt{2}\sinh(10\sqrt{2}\,t)\right)}{40}$$

and

$$I(t) = \frac{3\sqrt{2}\,e^{-20t}\sinh(10\sqrt{2}\,t)}{4}.$$

The charge and the current can be plotted with

`Plot[Q[t]//Evaluate,{t,0,1}];`

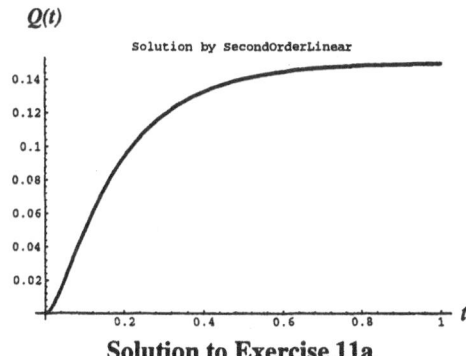

Solution to Exercise 11a

and

`Plot[Q'[t]//Evaluate,{t,0,1}];`

Q'(t)

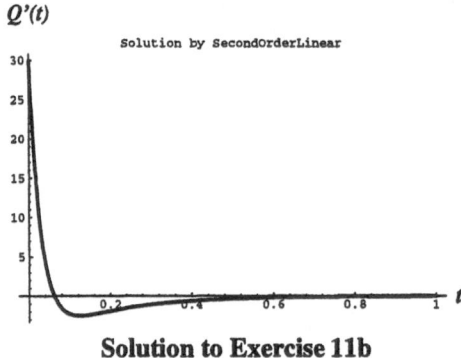

Solution to Exercise 11b

12. A series LRC circuit has $L = 1$ henry, $R = 20$ ohms, $C = 5/1000$ farad and an impressed voltage $E = 6$volts. Assume no initial current and no initial charge. Find and plot the subsequent charge and current.

Solution. We define the charge with

```
Q[t_] = ODE[{QQ'' + 20QQ' + QQ/(5/1000) == 6,
         QQ[0] == 0,QQ'[0] ==0},QQ,t,
         Method->SecondOrderLinear,Form->Explicit];
```

Then evaluation of `Q[t]` and `Q'[t]` yields

```
 3     3 Cos[10 t]    3 Sin[10 t]
---  - ----------- - -----------
100       10 t          10 t
       100 E          100 E
```

and

```
3 Sin[10 t]
-----------
   10 t
  5 E
```

Thus

$$Q(t) = \frac{3}{100} - \frac{3e^{-10t}\left(\cos(10t) + \sin(10t)\right)}{100} \qquad \text{and} \qquad I(t) = \frac{3e^{-10t}\sin(10t)}{5}.$$

The charge and the current can be plotted with

```
Plot[Q[t]//Evaluate,{t,0,1}];
```

$Q(t)$

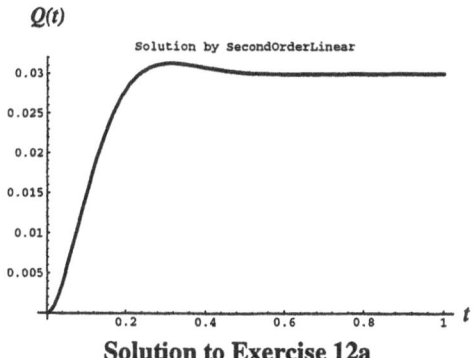

Solution to Exercise 12a

and

```
Plot[Q'[t]//Evaluate,{t,0,1}];
```

$Q'(t)$

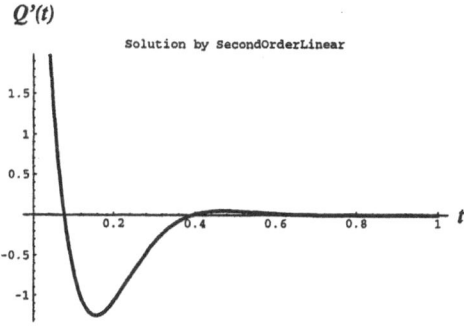

Solution to Exercise 12b

13. A series LRC circuit has $L = 1/2$ henry, $R = 30$ ohms, $C = 4/1000$ farad and an impressed voltage $E = 6$volts. Assume an initial charge of $1/10$ coulomb and an initial current of 2 amperes Find and plot the subsequent charge and current.

Solution. We define the charge with

```
Q[t_] = ODE[{(1/2)QQ'' + 30QQ' + QQ/(4 10^-3) == 0,
        QQ[0] == 2 10^-2,QQ'[0] == 2},QQ,t,
        Method->SecondOrderLinear,Form->Explicit];
```

Then evaluation of **Q[t]** and **Q'[t]//Expand** yields

$$\frac{-11}{200\ E^{50\ t}} + \frac{3}{40\ E^{10\ t}}$$

and

$$\frac{11}{4\ E^{50\ t}} - \frac{3}{4\ E^{10\ t}}$$

Thus

$$Q(t) = \frac{15e^{-10t} - 11e^{-50t}}{200} \qquad \text{and} \qquad I(t) = \frac{-3e^{-10t} + 11e^{-50t}}{4}.$$

The charge and the current can be plotted with

```
Plot[Q[t]//Evaluate,{t,0,0.5}];
```

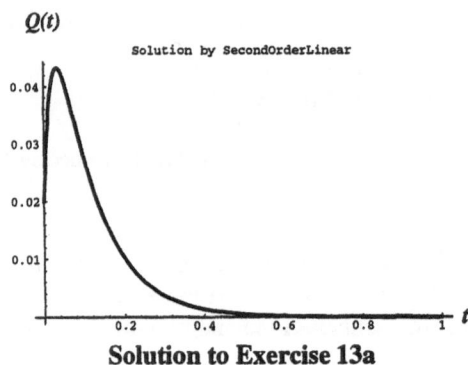

Solution to Exercise 13a

and

```
Plot[Q'[t]//Evaluate,{t,0,0.5}];
```

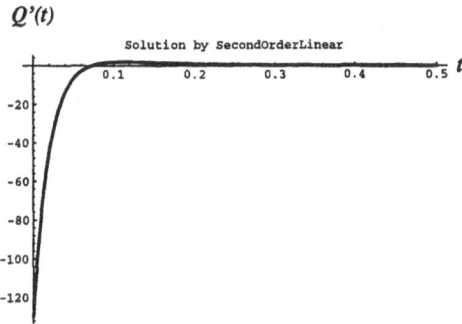

Solution to Exercise 13b

14. A series LRC circuit has $L = 5/100$ henry, $R = 10$ ohms, $C = 4/1000$ farad and an impressed voltage $E = 110$volts. Assume no initial current and no initial charge. Find and plot the subsequent charge and current.

Solution. We define the charge with

```
Q[t_] = ODE[{(5/100)QQ'' + 10QQ' + QQ/(4 10^-4) == 110,
         QQ[0] == 0,QQ'[0] == 0},QQ,t,
         Method->SecondOrderLinear,Form->Explicit];
```

Then evaluation of `Q[t]` and `Q'[t]//Expand` yields

```
11     11 Cos[200 t]     11 Sin[200 t]
---  - -------------  -  -------------
250       100 t            100 t
       250 E              500 E
```

and

```
11 Sin[200 t]
-------------
   100 t
  E
```

Thus

$$Q(t) = \frac{11}{250} - \frac{22e^{-100t}\left(\cos(200t) + \sin(200t)\right)}{500}$$

and

$$I(t) = 11e^{-100t}\sin(200t).$$

The charge and the current can be plotted with

```
Plot[Q[t]//Evaluate,{t,0,0.7}];
```

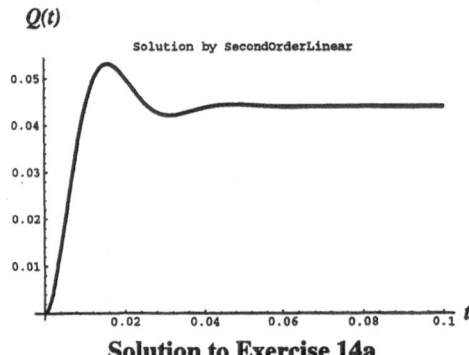

Solution to Exercise 14a

and

```
Plot[Q'[t]//Evaluate,{t,0,0.7}];
```

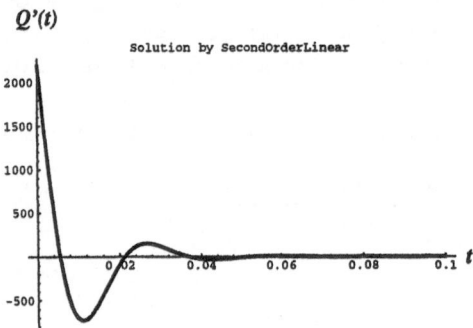

Solution to Exercise 14b

15. A series LRC circuit has $L = 5/100$ henry, $R = 10$ ohms, $C = 4/1000$ farad and an impressed voltage $E = 110\cos(100t)$ volts. Assume no initial current and no initial charge. Find and plot the subsequent charge and current.

Solution. We define the charge with

```
Q[t_] = ODE[{(5/100)QQ'' + 10 QQ' + QQ/(4 10^-4) ==
        110 Cos[100 t],QQ[0] == 0,QQ'[0] == 0},QQ,t,
        Method->SecondOrderLinear,Form->Explicit]
```

Then evaluation of `Q[t]` and `Q'[t]//Expand` yields

$$\frac{11\ \text{Cos}[100\ t]}{250} - \frac{11\ \text{Cos}[200\ t]}{250\ \text{E}^{100\ t}} + \frac{11\ \text{Sin}[100\ t]}{500} - \frac{33\ \text{Sin}[200\ t]}{1000\ \text{E}^{100\ t}}$$

and

$$\frac{11\ \text{Cos}[100\ t]}{5} - \frac{11\ \text{Cos}[200\ t]}{5\ \text{E}^{100\ t}} - \frac{22\ \text{Sin}[100\ t]}{5} + \frac{121\ \text{Sin}[200\ t]}{10\ \text{E}^{100\ t}}$$

Thus

$$Q(t) = \frac{22\cos(100t) + 11\sin(100t)}{500} - \frac{e^{-100t}\left(44\cos(200t) + 33\sin(200t)\right)}{1000}$$

and

$$I(t) = \frac{11\cos(100t) - 22\sin(100t)}{5} + \frac{e^{-100t}\left(-22\cos(200t) + 121\sin(200t)\right)}{10}.$$

The charge and the current can be plotted with

`Plot[Q[t]//Evaluate,{t,0,0.2}];`

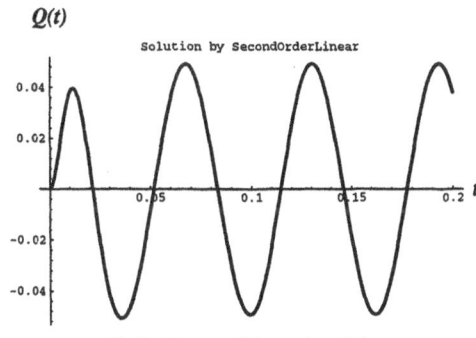

Solution to Exercise 15a

and

`Plot[Q'[t]//Evaluate,{t,0,0.2}];`

$Q'(t)$

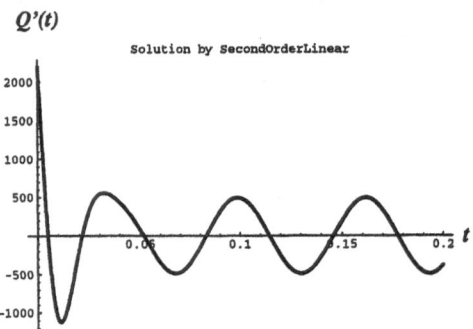

Solution to Exercise 15b

16. A series LRC circuit has $L = 20$ henry, $R = 180$ ohms, $C = 1/280$ farad and an impressed voltage $E = 2\sin(t)$ volts. Assume an initial current of one ampere and no initial charge. Find and plot the subsequent charge and current.

Solution. We define the charge with

```
Q[t_] = ODE[{20 QQ'' + 180 QQ' + 280QQ == 2 Sin[2t],
         QQ[0] == 0,QQ'[0] == 1},QQ,t,
         Method->SecondOrderLinear,Form->Explicit]
```

Then evaluation of **Q[t]** and **Q'[t]//Expand** yields

$$\frac{-266}{1325\,E^{7\,t}} + \frac{41}{200\,E^{2\,t}} - \frac{9\,\mathrm{Cos}[2\,t]}{2120} + \frac{\mathrm{Sin}[2\,t]}{424}$$

and

$$\frac{1862}{1325\,E^{7\,t}} - \frac{41}{100\,E^{2\,t}} + \frac{\mathrm{Cos}[2\,t]}{212} + \frac{9\,\mathrm{Sin}[2\,t]}{1060}$$

Thus

$$Q(t) = \frac{41e^{-2t}}{200} - \frac{266e^{-7t}}{1325} - \frac{9\cos(2t)}{2120} + \frac{\sin(2t)}{424}$$

and

$$I(t) = -\frac{41e^{-2t}}{100} + \frac{1862e^{-7t}}{1325} + \frac{\cos(2t)}{212} + \frac{9\sin(2t)}{1060}.$$

The charge and the current can be plotted with

```
Plot[Q[t]//Evaluate,{t,0,2Pi}];
```

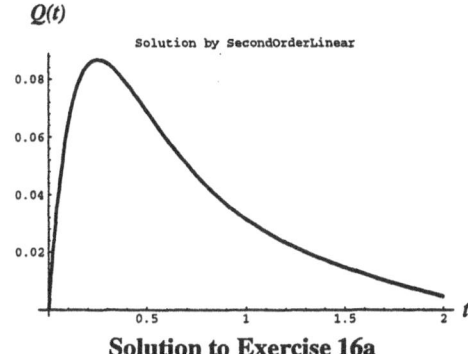

Q(*t*)

Solution by SecondOrderLinear

Solution to Exercise 16a

and

```
Plot[Q'[t]//Evaluate,{t,0,2Pi}];
```

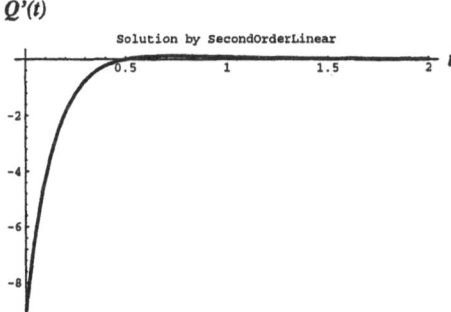

Q'(*t*)

Solution by SecondOrderLinear

Solution to Exercise 16b

Chapter 12

Section 12.1, page 385

Reduce each of the following higher-order equations to the leading-coefficient-unity-form (12.2).

12.1. $\log(t)y^{(4)} + e^t\, y''' + 3y'' + y = 0$

   Solution. $y^{(4)} + \left(e^t/\log(t)\right)y''' + \left(3/\log(t)\right)y'' + y/\log(t) = 0$

12.1. $t\, y^{(5)} + 4y^{(4)} + t\, y''' = t\, \sin(t)$

   Solution. $y^{(5)} + (4/t)y^{(4)} + y''' = \sin(t)$

12.1. $\dfrac{y'''}{t} + y'' + t\, y' + y = 1$

   Solution. $y''' + t\, y'' + t^2\, y' + t\, y = t$

12.1. $t^8 y^{(8)} + 4t^4 y^{(4)} + 4y = 0$

   Solution. $y^{(8)} + (4/t^4)y^{(4)} + (4/t^8)y = 0$

For each of the following differential equations, determine intervals on which a solution is certain to exist.

5. $y^{(8)} + 2y^{(4)} + y = t^2 e^{-t}$

   Solution. Any interval

6. $t^3 y''' + t^2 y'' + t\, y' + y = 0$

   Solution. Any interval not containing 0.

7. $t^5 y^{(5)} = 1$

   Solution. Any interval not containing 0

8. $t(t-1)y''' + y = t^2$

   Solution. Any interval not containing 0 or 1.

Use *Mathematica* to compute the Wronskian of each of the following sets of functions.

9. $\{e^t, t\, e^t, e^{3t}\}$.

   Solution.

```
Wronskian[{E^#&,#E^#&,E^(3#)&}][t]
```

$4e^{5t}$

10. $\{\sin(t), t^3 \sin(t), t^5 \sin(t), t^7 \sin(t)\}$

   Solution.

   `Wronskian[{Sin[#]&,#^3Sin[#]&,#^5Sin[#]&,#^7Sin[#]&}][t]`

   $1680t^9 \sin(t)^4$

11. Show that the operator L defined by

$$L[y] = y^{(n)} + p_{n-1}(t)y^{(n-1)} + \cdots + p_0(t)y$$

   is linear in the following sense:

$$L[C_1 y_1 + C_2 y_2] = C_1 L[y_1] + C_2 L[y_2],$$

   where C_1 and C_2 are constants and y_1 and y_2 are differentiable functions.

   Solution. The operation of differentiation is linear in the sense that for $k = 0, 1, 2, \ldots, n$, we have

$$(C_1 y_1 + C_2 y_2)^{(k)} = C_1 y_1^{(k)} + C_2 y_2^{(k)} \tag{S.78}$$

   We multiply (S.78) by $p_k(t)$, sum for $k = 0, 1, \ldots, n-1$ and add to (S.78) for $k = n$ to obtain the required result.

Section 12.2, page 391

Find the general solution to each of the following differential equations:

1. $y''' + y' = 0$

   Solution. The roots of the characteristic equation $r^3 + r = r(r^2 + 1) = 0$ are $r_1 = 0$, $r_2 = -i$ and $r_3 = i$. Consequently, the general solution to this equation is $y(t) = C_1 + C_2 \cos(t) + C_3 \sin(t)$.

2. $y''' - y'' + y' - y = 0$

   Solution. The roots of the characteristic equation $r^3 - r^2 + r - 1 = (r-1)(r^2+1) = 0$ are $r_1 = 1$, $r_2 = -i$ and $r_3 = i$. Consequently, the general solution to this equation is $y(t) = C_1 e^t + C_2 \cos(t) + C_3 \sin(t)$.

3. $y^{(4)} - 25y = 0$

   Solution. The roots of the characteristic equation $r^4 - 25 = (r^2 - 5)(r^2 + 5) = 0$ are $r_1 = -\sqrt{5}$, $r_2 = \sqrt{5}$, $r_3 = -\sqrt{5}i$ and $r_4 = \sqrt{5}i$. Consequently, the general solution to this equation is $y(t) = C_1 e^{-\sqrt{5}t} + C_2 e^{\sqrt{5}t} + C_3 \cos(\sqrt{5}t) + C_4 \sin(\sqrt{5}t)$.

4. $y^{(7)} + 2y^{(5)} - y''' - 2y' = 0$

Solution. The roots of the characteristic equation $r^7 + 2r^5 - r^3 - 2r = r(r + 1)(r - 1)(r^2 + 1)(r^2 + 2) = 0$ are $r_1 = 0$, $r_2 = -1$, $r_3 = 1$, $r_4 = -i$, $r_5 = i$, $r_6 = -\sqrt{2}i$ and $r_7 = \sqrt{2}i$. Consequently, the general solution to this equation is $y(t) = C_1 + C_2 e^{-t} + C_3 e^t + C_4 \cos(t) + C_5 \sin(t) + C_6 \cos(\sqrt{2}t) + C_7 \sin(\sqrt{2}t)$.

5. $y^{(6)} - 6y^{(5)} + 15y^{(4)} - 20y''' + 15y'' - 6y' + y = 0$

Solution. The roots of the characteristic equation $r^6 - 6r^5 + 15r^4 - 20r^3 + 15r^2 - 6r + 1 = (r - 1)^6 = 0$ are $r = 1$. Consequently, the general solution to this equation is $y(t) = C_1 e^t + C_2 t e^t + C_3 t^2 e^t + C_4 t^3 e^t + C_5 t^4 e^t + C_6 t^5 e^t$.

6. $y^{(4)} - 10y''' + 35y'' - 50y' + 24y = 0$

Solution. The roots of the characteristic equation $r^4 - 10r^3 + 35r^2 - 50r + 24 = (r-1)(r-2)(r-3)(r-4) = 0$ are $r_1 = 1$, $r_2 = 2$, $r_3 = 3$ and $r_4 = 4$. Consequently, the general solution to this equation is $y(t) = C_1 e^t + C_2 e^{2t} + C_3 e^{3t} + C_4 e^{4t}$.

7. $y^{(8)} - 2y^{(6)} - 3y^{(4)} + 4y'' + 4y = 0$

Solution. The roots of the characteristic equation $r^8 - 2r^6 - 3r^4 + 4r^2 + 4 = (r^2 + 1)^2(r^2 - 2)^2 = 0$ are $r_1 = -\sqrt{2}$, $r_2 = -\sqrt{2}$, $r_3 = \sqrt{2}$, $r_4 = \sqrt{2}$, $r_5 = -i$, $r_6 = -i$, $r_7 = i$ and $r_8 = i$. Consequently, the general solution to this equation is $y(t) = C_1 e^{-\sqrt{2}t} + C_2 t e^{-\sqrt{2}t} + C_3 e^{\sqrt{2}t} + C_4 t e^{\sqrt{2}t}$

$+ C_5 \cos(t) + C_6 \sin(t) + C_7 t \cos(t) + C_8 t \sin(t)$.

8. $y^{(8)} - 3y^{(7)} + 2y^{(6)} - 6y^{(5)} + y^{(4)} - 3y''' = 0$

Solution. The roots of the characteristic equation $r^8 - 3r^7 + 2r^6 - 6r^5 + r^4 - 3r^3 = r^3(r - 3)(r^2 + 1)^2 = 0$ are $r_1 = 0$, $r_2 = 0$, $r_3 = 0$, $r_4 = 3$, $r_5 = -i$, $r_6 = -i$, $r_7 = i$ and $r_8 = i$. Consequently, the general solution to this equation is $y(t) = C_1 + C_2 t + C_3 t^2 + C_4 e^{3t} + C_5 \cos(t) + C_6 \sin(t) + C_7 t \cos(t) + C_8 t \sin(t)$.

9. $y^{(6)} - 12y^{(5)} + 55y^{(4)} - 120y''' + 135y'' - 108y' + 81y = 0$

Solution. The roots of the characteristic equation $r^6 - 12r^5 + 55r^4 - 120r^3 + 135r^2 - 108r + 81 = (r - 3)^4(r^2 + 1) = 0$ are $r_1 = 3$, $r_2 = 3$, $r_3 = 3$, $r_4 = 3$, $r_5 = -i$ and $r_6 = i$. Consequently, the general solution to this equation is $y(t) = C_1 e^{3t} + C_2 t e^{3t} + C_3 t^2 e^{3t} + C_4 t^3 e^{3t} + C_5 \cos(t) + C_6 \sin(t)$.

10. $y^{(8)} - 2y^{(7)} + 7y^{(6)} - 12y^{(5)} + 17y^{(4)} - 22y''' + 17y'' - 12y' + 6y = 0$

Solution. The roots of the characteristic equation $r^8 - 2r^7 + 7r^6 - 12r^5 + 17r^4 - 22r^3 + 17r^2 - 12r + 6 = (r - 1)^2(r^2 + 1)(r^2 + 2)(r^2 + 3) = 0$ are $r_1 = 1$, $r_2 = 1$, $r_3 = -i$, $r_4 = i$, $r_5 = -\sqrt{2}i$, $r_6 = \sqrt{2}i$, $r_7 = -\sqrt{3}i$ and $r_8 = \sqrt{3}i$. Consequently, the general solution to this equation is $y(t) = C_1 e^t + C_2 t e^t + C_3 \cos(t) + C_4 \sin(t) + C_5 \cos(\sqrt{2}\,t) + C_6 \sin(\sqrt{2}\,t) + C_7 \cos(\sqrt{3}\,t) + C_8 \sin(\sqrt{3}\,t)$.

Solve each of the following initial value problems and plot the solution.

11. $\begin{cases} y''' + y'' + y' + y = 0, \\ y(0) = 0,\ y'(0) = 1,\ y''(0) = 0 \end{cases}$

Solution. The roots of the characteristic equation $r^3 + r^2 + r + 1 = (r+1)(r^2+1) = 0$ are $r_1 = -1$, $r_2 = -i$ and $r_3 = i$. The general solution to this equation is $y(t) = C_1 e^{-t} + C_2 \cos(t) + C_3 \sin t$. After incorporating the initial conditions, we obtain $C_1 + C_2 = 0$, $-C_1 + C_3 = 1$ and $C_1 - C_2 = 0$ with the solution $C_1 = C_2 = 0$ and $C_3 = 1$. Hence, $y(t) = \sin(t)$.

With **ODE**, we choose the interval $[0, 2\pi]$ and enter

```
ODE[{y''' + y'' + y' + y == 0,
y[0] == 0,y'[0] == 1,y''[0] == 0},y,t,
Method->NthOrderLinear,
PlotSolution->{{t,0,2Pi}}]
```

to get

```
{{y -> Sin[t]}}
```

and the plot

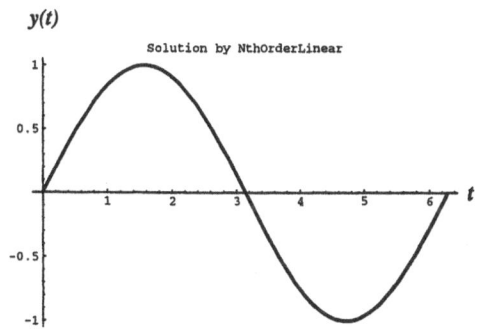

Solution to Exercise 11

12. $\begin{cases} y^{(4)} - 5y'' + 4y = 0, \\ y(0) = 1,\ y'(0) = 1,\ y''(0) = 0,\ y'''(0) = 0 \end{cases}$

Solution. The roots of the characteristic equation $r^4 - 5r^2 + 4 = (r+2)(r+1)(r-1)(r-2) = 0$ are $r_1 = -2$, $r_2 = -1$, $r_3 = 1$ and $r_4 = 2$. The general solution to this equation is $y(t) = C_1 e^{-2t} + C_2 e^{-t} + C_3 e^t + C_4 e^{2t}$. After incorporating the initial conditions, we obtain $C_1 + C_2 + C_3 + C_4 = 1$, $-2C_1 - C_2 + C_3 + 2C_4 = 1$, $4C_1 + C_2 + C_3 + 4C_4 = 0$ and $-8C_1 - C_2 + C_3 + 8C_4 = 0$ with the solution $C_1 = -1/12$, $C_2 = 0$, $C_3 = 4/3$ and $C_4 = -1/4$. Hence, $y(t) = -\dfrac{e^{-2t}}{12} + \dfrac{4e^t}{3} - \dfrac{e^{2t}}{4}$.

With **ODE**, we choose the interval $[0, 3]$ and enter

```
ODE[{y'''' - 5y'' + 4y == 0,
y[0] == 1,y'[0] == 1,y''[0] == 0,y'''[0] == 0},y,t,
Method->NthOrderLinear,
PlotSolution->{{t,0,3}}]
```

to get

```
 -1           t      2 t
------   +  4 E     E
 2 t        ---  -  ---
12 E         3       4
```

and the plot

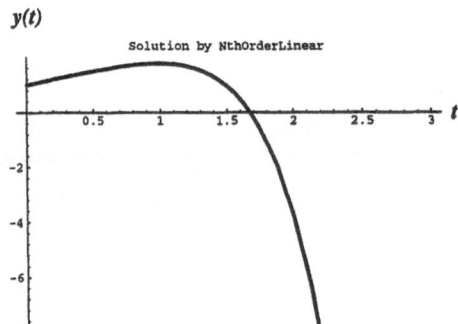

Solution to Exercise 12

13. $\begin{cases} y^{(5)} + 4y''' + 4y' = 0, \\ y(0) = 1,\, y'(0) = -1,\, y''(0) = 1,\, y'''(0) = 1,\, y^{(4)}(0) = 1 \end{cases}$

Solution. The roots of the characteristic equation $r^5 + 4r^3 + 4r = r(r^2 + 2)^2 = 0$ are $r_1 = 0$, $r_2 = -\sqrt{2}i$, $r_3 = -\sqrt{2}i$, $r_4 = \sqrt{2}i$ and $r_5 = \sqrt{2}i$. The general solution to this equation is $y(t) = C_1 + (C_2 + C_4 t)\cos(\sqrt{2}t) + (C_3 + C_5 t)\sin(\sqrt{2}t)$. After incorporating the initial conditions, we obtain $C_1 + C_2 = 1$, $\sqrt{2}C_3 + C_4 = -1$, $-2C_2 + 2\sqrt{2}C_5 = 1$, $-2\sqrt{2}C_3 - 6C_4 = 1$ and $-4C_2 - 8\sqrt{2}C_5 = 1$ with the solution $C_1 = 9/4$, $C_2 = -5/4$, $C_3 = -5/(4\sqrt{2})$, $C_4 = 1/4$ and $C_5 = -3/(4\sqrt{2})$. Hence,

$$y(t) = \frac{9 - (5-t)\cos(\sqrt{2}t)}{4} - \frac{(5-3t)\sin(\sqrt{2}t)}{4\sqrt{2}}.$$

With **ODE**, we choose the interval [0, 10] and enter

```
ODE[{y''''' + 4y''' + 4y' == 0,
y[0 ]== 1,y'[0] == -1,y''[0] == 1,y'''[0 ]== 1,
y''''[0] == 1},y,t,
Method->NthOrderLinear,
PlotSolution->{{t,0,10}}]
```

to get

```
9    5 Cos[Sqrt[2] t]     t Cos[Sqrt[2] t]
- - ----------------- + ----------------- -
4           4                   4

   5 Sin[Sqrt[2] t]     3 t Sin[Sqrt[2] t]
   ----------------- - ------------------
      4 Sqrt[2]            4 Sqrt[2]
```

and the plot

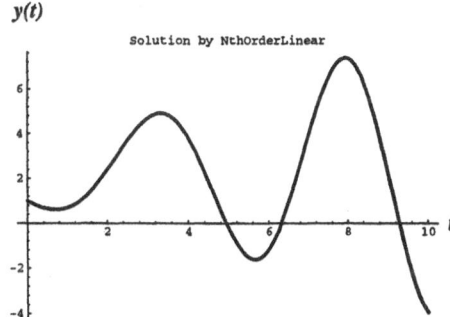

Solution to Exercise 13

14. $$\begin{cases} y^{(7)} + 6y^{(5)} + 9y''' = 0, \\ y(0) = 1,\ y'(0) = -1,\ y''(0) = 0,\ y'''(0) = 1,\ y^{(4)}(0) = 1, \\ y^{(5)}(0) = 1,\ y^{(6)}(0) = 1 \end{cases}$$

Solution. The roots of the characteristic equation $r^7 + 6r^5 + 9r^3 = r^3(r^2 + 3)^2 = 0$ are $r_1 = 0$, $r_2 = 0$, $r_3 = 0$, $r_4 = -\sqrt{3}i$, $r_5 = -\sqrt{3}i$, $r_6 = \sqrt{3}i$ and $r_7 = \sqrt{3}i$. The general solution to this equation is $y(t) = C_1 + C_2 t + C_3 t^2 + (C_4 + C_6 t)\cos(\sqrt{3}t) + (C_5 + C_7 t)\cos(\sqrt{3}t)$. After incorporating the initial conditions, we obtain $C_1 + C_4 = 1$, $C_2 + \sqrt{3}C_5 + C_6 = -1$, $2C_3 - 3C_4 + 2\sqrt{3}C_7 = 0$, $-3\sqrt{3}C_5 - 9C_6 = 1$, $9C_4 - 12\sqrt{3}C_7 = 1$, $9\sqrt{3}C_5 + 45C_6 = 1$ and $-27C_4 + 54\sqrt{3}C_7 = 1$ with the solution $C_1 = 16/27$, $C_2 = -2/9$, $C_3 = 7/18$, $C_4 = 11/27$, $C_5 = -1/\sqrt{3}$, $C_6 = 2/9$ and $C_7 = 2/(9\sqrt{3})$. Hence,

$$y(t) = \frac{21t^2 - 12t + 32}{54} + \frac{(6t + 11)\cos(\sqrt{3}t)}{27} + \frac{(2t - 9)\sin(\sqrt{3}t)}{9\sqrt{3}}.$$

With **ODE**, we choose the interval [0, 10] and enter

```
ODE[{y'''''' + 6y''''' + 9y''' == 0,
y[0] == 1,y'[0] == -1,y''[0] == 0,y'''[0] == 1,
y''''[0] == 1,y'''''[0] == 1,y''''''[0 ]== 1},y,t,
Method->NthOrderLinear,
PlotSolution->{{t,0,10}}]
```

to get

$$\frac{16}{27} - \frac{2\,t}{9} + \frac{7\,t^2}{18} + \frac{11\,\text{Cos[Sqrt[3] t]}}{27} +$$

$$\frac{2\,t\,\text{Cos[Sqrt[3] t]}}{9} - \frac{\text{Sin[Sqrt[3] t]}}{\text{Sqrt[3]}} + \frac{2\,t\,\text{Sin[Sqrt[3] t]}}{9\,\text{Sqrt[3]}}$$

and the plot

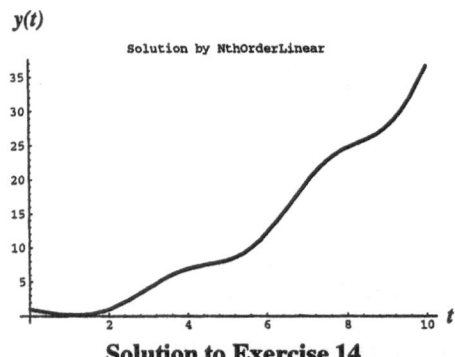

Solution to Exercise 14

15.
$$\begin{cases} y^{(8)} - 2y^{(7)} + 10y^{(6)} - 18y^{(5)} + 36y^{(4)} - 54y''' + 54y'' - 54y' + 27y = 0, \\ y(0) = 0,\ y'(0) = -10,\ y''(0) = 10,\ y'''(0) = 1,\ y^{(4)}(0) = 0, \\ y^{(5)}(0) = 0,\ y^{(6)}(0) = 0,\ y^{(7)}(0) = 0 \end{cases}$$

Solution. The roots of the characteristic equation $r^8 - 2r^7 + 10r^6 - 18r^5 + 36r^4 - 54r^3 + 54r^2 - 54r + 27 = (r-1)^2(r^2+3)^3 = 0$ are $r_1 = 1,\ r_2 = 1,\ r_3 = -\sqrt{3}i,$ $r_4 = -\sqrt{3}i,\ r_5 = -\sqrt{3}i,\ r_6 = \sqrt{3}i,\ r_7 = \sqrt{3}i$ and $r_8 = \sqrt{3}i$. The general solution to this equation is $y(t) = C_1 e^t + C_2 t e^t + (C_3 + C_5 t + C_6 t^2)\cos(\sqrt{3}t) + (C_4 + C_6 t + C_8 t^2)\sin(\sqrt{3}t)$. After incorporating the initial conditions, we obtain $C_1 + C_3 = 0,\ C_1 + C_2 + \sqrt{3}C_4 + C_5 = -10,\ C_1 + 2C_2 - 3C_3 + 2\sqrt{3}C_6 + 2C_7 = 10,$ $C_1 + 3C_2 - 3\sqrt{3}C_4 - 9C_5 + 6\sqrt{3}C_8 = 1,\ C_1 + 4C_2 + 9C_3 - 12\sqrt{3}C_6 - 36C_7 = 0,$ $C_1 + 5C_2 + 9\sqrt{3}C_4 + 45C_5 - 60\sqrt{3}C_8 = 0,\ C_1 + 6C_2 - 27C_3 + 54\sqrt{3}C_6 + 270C_7 = 0$ and $C_1 + 7C_2 - 27\sqrt{3}C_4 - 189C_5 + 378\sqrt{3}C_8 = 0$ with the solution $C_1 = 2079/128,\ C_2 = -513/64,\ C_3 = -2079/128,\ C_4 = -11333/(384\sqrt{3}),$ $C_5 = 2167/192, C_6 = -479\sqrt{3}/64,\ C_7 = 191/64$ and $C_8 = 233/(64\sqrt{3})$. Hence,
$$y(t) = \frac{(2079 - 1026t)e^t}{128} + \frac{(1146t^2 + 4334t - 6237)\cos(\sqrt{3}\,t)}{384}$$
$$+ \frac{(1389t^2 - 8622t - 11333)\sin(\sqrt{3}\,t)}{384\sqrt{3}}.$$

With **ODE**, we choose the interval [0, 2] and enter

```
ODE[{y'''''''' - 2y''''''' + 10y'''''' - 18y''''' +
36y'''' - 54y''' + 54y'' - 54y' + 27y == 0,
y[0] == 0,y'[0] == -10,y''[0] ==10,y'''[0] == 1,
y''''[0] == 0,y'''''[0] == 0,y''''''[0] == 0,
y'''''''[0] == 0},y,t,
Method->NthOrderLinear,
PlotSolution->{{t,0,2}}]
```

to get

```
       t          t
2079 E     513 E  t    2079 Cos[Sqrt[3] t]
--------- - -------- - ------------------- +
   128        64             128

       2167 t Cos[Sqrt[3] t]   191 t  Cos[Sqrt[3] t]
       --------------------- + ------------------- -
               192                     64

       11333 Sin[Sqrt[3] t]   479 Sqrt[3] t Sin[Sqrt[3] t]
       -------------------- - ---------------------------- +
           384 Sqrt[3]                    64

          2
```

```
233 t  Sin[Sqrt[3] t]
```
————————————————
```
   64 Sqrt[3]
```

and the plot

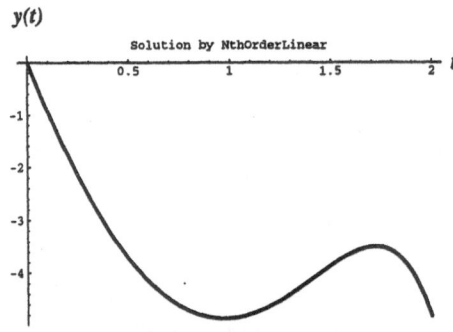

Solution by NthOrderLinear

Solution to Exercise 15

16.
$$\begin{cases} y^{(6)} - 2y^{(5)} - y^{(4)} + 4y''' - y'' - 2y' + y = 0, \\ y(0) = 10,\, y'(0) = 10,\, y''(0) = -10,\, y'''(0) = 0, \\ y^{(4)}(0) = 0,\, y^{(5)}(0) = 0 \end{cases}$$

Solution. The roots of the characteristic equation $r^6 - 2r^5 - r^4 + 4r^3 - r^2 - 2r + 1 = (r+1)^2(r-1)^4 = 0$ are $r_1 = -1$, $r_2 = -1$, $r_3 = -1$, $r_4 = 1$, $r_5 = 1$ and $r_6 = 1$. The general solution to this equation is $y(t) = (C_1 + C_2 t)e^{-t} + (C_3 + C_4 t + C_5 t^2 + C_6 t^3)e^t$. After incorporating the initial conditions, we obtain $C_1 + C_3 = 10$, $-C_1 + C_2 + C_3 + C_4 = 10$, $C_1 - 2C_2 + C_3 + 2C_4 + 2C_5 = -10$, $-C_1 + 3C_2 + C_3 - +3C_4 + 6C_5 + 6C_6 = 0$, $C_1 - 4C_2 + C_3 + 4C_4 + 12C_5 + 24C_6 = 0$, and $-C_1 + 5C_2 + C_3 + 5C_4 + 20C_5 + 60C_6 = 0$ with the solution $C_1 = -85/8$, $C_2 = -5/2$, $C_3 = 165/8$, $C_4 = -75/4$, $C_5 = 25/4$ and $C_6 = -5/6$. Hence,
$$y(t) = -\frac{(20t+85)e^{-t}}{8} + \frac{(495 - 450t + 150t^2 - 20t^3)e^t}{24}.$$
With **ODE**, we choose the interval $[0, 3]$ and enter

```
ODE[{y'''''' - 2y''''' - y'''' + 4y''' -
y'' - 2y' + y == 0,y[0] == 10,y'[0] == 10,
y''[0] == -10,y'''[0] == 0,y''''[0] == 0,
y'''''[0] == 0},y,t,
Method->NthOrderLinear,
PlotSolution->{{t,0,3}}]
```

to get

$$-\frac{85}{8\,E^t} + \frac{165\,E^t}{8} - \frac{5\,t}{2\,E^t} - \frac{75\,E^t\,t}{4} + \frac{25\,E^t\,t^2}{4} - \frac{5\,E^t\,t^3}{6}$$

and the plot

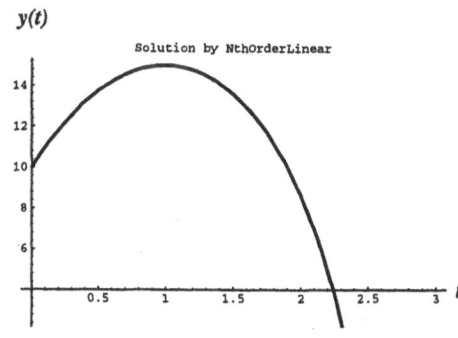

y(t)

Solution by NthOrderLinear

Solution to Exercise 16

Section 12.3, page 397

First, use the method of variation of parameters to find a particular solution of each of the following differential equations. Then find the general solution.

1. $y''' - 2y'' - y' + 2y = e^t$

Solution. The roots of the characteristic equation $r^3 - 2r^2 - r + 2 = (r+1)(r-1)(r-2) = 0$ are $r_1 = -1$, $r_2 = 1$ and $r_3 = 2$. Consequently, a fundamental set of solutions is $\{e^{-t}, e^t, e^{2t}\}$ and the general solution to the homogeneous equation is

$$y_H = \tilde{C}_1 e^{-t} + \tilde{C}_2 e^t + \tilde{C}_3 e^{2t}.$$

Using Cramer's rule we obtain

$$u_1(t) = \int \frac{\begin{vmatrix} 0 & e^t & e^{2t} \\ 0 & e^t & 2e^{2t} \\ e^t & e^t & 4e^{2t} \end{vmatrix}}{\begin{vmatrix} e^{-t} & e^t & e^{2t} \\ -e^{-t} & e^t & 2e^{2t} \\ e^{-t} & e^t & 4e^{2t} \end{vmatrix}} dt = \int \frac{e^{4t}}{6e^{2t}} dt = \frac{e^{2t}}{12},$$

$$u_2(t) = \int \frac{\begin{vmatrix} e^{-t} & 0 & e^{2t} \\ -e^{-t} & 0 & 2e^{2t} \\ e^{-t} & e^t & 4e^{2t} \end{vmatrix}}{\begin{vmatrix} e^{-t} & e^t & e^{2t} \\ -e^{-t} & e^t & 2e^{2t} \\ e^{-t} & e^t & 4e^{2t} \end{vmatrix}} dt = \int \frac{-3e^{2t}}{6e^{2t}} dt = -\frac{t}{2},$$

and

$$u_3(t) = \int \frac{\begin{vmatrix} e^{-t} & e^t & 0 \\ -e^{-t} & e^t & 0 \\ e^{-t} & e^t & e^t \end{vmatrix}}{\begin{vmatrix} e^{-t} & e^t & e^{2t} \\ -e^{-t} & e^t & 2e^{2t} \\ e^{-t} & e^t & 4e^{2t} \end{vmatrix}} dt = \int \frac{2e^t}{6e^{2t}} dt = -\frac{1}{3e^t}.$$

Hence,

$$y_p(t) = u_1 y_1 + u_2 y_2 + u_3 y_3 = -\left(\frac{1}{4} + \frac{t}{2}\right) e^t.$$

We may combine $-e^t/4$ with $y_H(t)$ and write

$$y(t) = C_1 e^{-t} + C_2 e^t + C_3 e^{2t} - \frac{t\, e^t}{2}.$$

2. $y''' - y'' + y' - y = \sin(2t)$

Solution. The roots of the characteristic equation $r^3 - r^2 + r - 1 = (r-1)(r^2+1) = 0$ are $r_1 = 1$, $r_2 = -i$ and $r_3 = i$. Consequently, a fundamental set of solutions is $\{e^t, \sin(t), \cos(t)\}$ and the general solution to the homogeneous equation is

$$y_H = C_1 e^t + C_2 \sin(t) + C_3 \cos(t).$$

Using Cramer's rule we obtain

$$u_1(t) = \int \frac{\begin{vmatrix} 0 & \sin(t) & \cos(t) \\ 0 & \cos(t) & -\sin(t) \\ \sin(2t) & -\sin(t) & -\cos(t) \end{vmatrix}}{\begin{vmatrix} e^t & \sin(t) & \cos(t) \\ e^t & \cos(t) & -\sin(t) \\ e^t & -\sin(t) & -\cos(t) \end{vmatrix}} \, dt = \int \frac{\sin(2t)}{6e^t} \, dt$$

$$= -\frac{e^{-t}}{10} \left(2\cos(2t) + \sin(2t) \right),$$

$$u_2(t) = \int \frac{\begin{vmatrix} e^t & 0 & \cos(t) \\ e^t & 0 & -\sin(t) \\ e^t & \sin(2t) & -\cos(t) \end{vmatrix}}{\begin{vmatrix} e^t & \sin(t) & \cos(t) \\ e^t & \cos(t) & -\sin(t) \\ e^t & -\sin(t) & -\cos(t) \end{vmatrix}} \, dt$$

$$= \frac{1}{4} \int \cos(3t) - \cos(t) - \sin(3t) - \sin(t) \, dt$$

$$= \frac{1}{12} \left(\cos(3t) + 3\cos(t) + \sin(3t) - 3\sin(t) \right),$$

and

$$u_3(t) = \int \frac{\begin{vmatrix} e^t & \sin(t) & 0 \\ e^t & \cos(t) & 0 \\ e^t & -\sin(t) & \sin(2t) \end{vmatrix}}{\begin{vmatrix} e^t & \sin(t) & \cos(t) \\ e^t & \cos(t) & -\sin(t) \\ e^t & -\sin(t) & -\cos(t) \end{vmatrix}} \, dt$$

$$= \frac{1}{4} \int \cos(t) - \cos(3t) - \sin(t) - \sin(3t) \, dt$$

$$= \frac{1}{12} \left(\cos(3t) + 3\cos(t) - \sin(3t) + 3\sin(t) \right).$$

Hence,

$$y_p(t) = u_1 y_1 + u_2 y_2 + u_3 y_3 = \frac{2\cos(2t) + \sin(2t)}{15}.$$

Finally, we obtain

$$y(t) = C_1 e^t + C_2 \sin(t) + C_3 \cos(t) + \frac{2\cos(2t) + \sin(2t)}{15}.$$

3. $y^{(4)} + 5y'' + 4y = \sec(t)$

Solution. The roots of the characteristic equation $r^4 + 5r^2 + 4 = (r^2 + 1)(r^2 + 4) = 0$ are $r_1 = -i$, $r_2 = i$, $r_3 = -2i$ and $r_4 = 2i$. Consequently, a fundamental set of solutions is $\{\sin(t), \cos(t), \sin(2t), \cos(2t)\}$ and the general solution to the homogeneous equation is

$$y_H = C_1 \sin(t) + C_2 \cos(t) + C_3 \sin(2t) + C_4 \cos(2t).$$

Using Cramer's rule we obtain

$$u_1(t) = \int \frac{\begin{vmatrix} 0 & \cos(t) & \sin(2t) & \cos(2t) \\ 0 & -\sin(t) & 2\cos(2t) & -2\sin(2t) \\ 0 & -\cos(t) & -4\sin(2t) & -4\cos(2t) \\ \sec(t) & \sin(t) & -8\cos(2t) & 8\sin(2t) \end{vmatrix}}{\begin{vmatrix} \sin(t) & \cos(t) & \sin(2t) & \cos(2t) \\ \cos(t) & -\sin(t) & 2\cos(2t) & -2\sin(2t) \\ -\sin(t) & -\cos(t) & -4\sin(2t) & -4\cos(2t) \\ -\cos(t) & \sin(t) & -8\cos(2t) & 8\sin(2t) \end{vmatrix}} dt = \frac{1}{3}\int dt = \frac{t}{3},$$

$$u_2(t) = \int \frac{\begin{vmatrix} \sin(t) & 0 & \sin(2t) & \cos(2t) \\ \cos(t) & 0 & -2\sin(2t) & \\ -\sin(t) & -\cos(t) & 0 & -4\cos(2t) \\ -\cos(t) & \sin(t) & \sec(t) & 8\sin(2t) \end{vmatrix}}{\begin{vmatrix} \sin(t) & \cos(t) & \sin(2t) & \cos(2t) \\ \cos(t) & -\sin(t) & 2\cos(2t) & -2\sin(2t) \\ -\sin(t) & -\cos(t) & -4\sin(2t) & -4\cos(2t) \\ -\cos(t) & \sin(t) & -8\cos(2t) & 8\sin(2t) \end{vmatrix}} dt = -\frac{1}{3}\int \tan(t)\, dt$$

$$= \frac{\log\left(\cos(t)\right)}{3},$$

$$u_3(t) = \int \frac{\begin{vmatrix} \sin(t) & \cos(t) & 0 & \cos(2t) \\ \cos(t) & -\sin(t) & 0 & -2\sin(2t) \\ -\sin(t) & -\cos(t) & -4\sin(2t) & 0 \\ -\cos(t) & \sin(t) & \sec(t) & 8\sin(2t) \end{vmatrix}}{\begin{vmatrix} \sin(t) & \cos(t) & \sin(2t) & \cos(2t) \\ \cos(t) & -\sin(t) & 2\cos(2t) & -2\sin(2t) \\ -\sin(t) & -\cos(t) & -4\sin(2t) & -4\cos(2t) \\ -\cos(t) & \sin(t) & -8\cos(2t) & 8\sin(2t) \end{vmatrix}} dt$$

$$= -\frac{1}{6}\int \cos(2t)\sec(t)\,dt$$

$$= -\frac{1}{6}\left(\log\left(\frac{1+\sin(t)}{\cos(t)}\right) + 2\sin(t)\right),$$

and

$$u_4(t) = \int \frac{\begin{vmatrix} \sin(t) & \cos(t) & \sin(2t) & 0 \\ \cos(t) & -\sin(t) & 2\cos(2t) & 0 \\ -\sin(t) & -\cos(t) & -4\sin(2t) & 0 \\ -\cos(t) & \sin(t) & -8\cos(2t) & \sec(t) \end{vmatrix}}{\begin{vmatrix} \sin(t) & \cos(t) & \sin(2t) & \cos(2t) \\ \cos(t) & -\sin(t) & 2\cos(2t) & -2\sin(2t) \\ -\sin(t) & -\cos(t) & -4\sin(2t) & -4\cos(2t) \\ -\cos(t) & \sin(t) & -8\cos(2t) & 8\sin(2t) \end{vmatrix}} dt$$

$$= \frac{1}{3}\int \sin(t)\,dt = -\frac{\cos(t)}{3}.$$

Since the term

$$-\frac{1}{3}\Big(\sin(2t)\sin(t) + \cos(2t)\cos(t)\Big) = -\frac{\cos(t)}{3}$$

may be combined with $y_H(t)$, we may write

$$y_p(t) = \frac{(-1+\cos(t))\log(\cos(t))}{3} + \frac{t\sin(t)}{3} - \frac{\sin(2t)}{6}\log\left(\frac{1+\sin(t)}{\cos(t)}\right).$$

4. $y^{(4)} - y = t\sin(2t)$

Solution. The roots of the characteristic equation $r^4 - 1 = (r+1)(r-1)(r^2+1) = 0$ are $r_1 = -1, r_2 = 1, r_3 = -i$ and $r_4 = i$. Consequently, a fundamental set of solutions is $\{e^{-t}, e^t, \sin(t), \cos(t)\}$ and the general solution to the homogeneous equation is

$$y_H = C_1 e^{-t} + C_2 e^t + C_3 \sin(t) + C_4 \cos(t).$$

Using Cramer's rule we obtain

$$u_1(t) = \int \frac{\begin{vmatrix} 0 & e^t & \sin(t) & \cos(t) \\ 0 & e^t & \cos(t) & -\sin(t) \\ 0 & e^t & -\sin(t) & -\cos(t) \\ t\sin(2t) & e^t & -\cos(t) & \sin(t) \end{vmatrix}}{\begin{vmatrix} e^{-t} & e^t & \sin(t) & \cos(t) \\ -e^{-t} & e^t & \cos(t) & -\sin(t) \\ e^{-t} & e^t & -\sin(t) & -\cos(t) \\ -e^{-t} & e^t & -\cos(t) & \sin(t) \end{vmatrix}} dt = -\frac{1}{4}\int t e^t \sin(2t)\, dt,$$

$$= \frac{-4e^t \cos(2t) + 10t\, e^t \cos(2t) - 3e^t \sin(2t) - 5t\, e^t \sin(2t)}{100},$$

$$u_2(t) = \int \frac{\begin{vmatrix} e^{-t} & 0 & \sin(t) & \cos(t) \\ -e^{-t} & 0 & \cos(t) & -\sin(t) \\ e^{-t} & 0 & -\sin(t) & -\cos(t) \\ -e^{-t} & t\sin(2t) & -\cos(t) & \sin(t) \end{vmatrix}}{\begin{vmatrix} e^{-t} & e^{t} & \sin(t) & \cos(t) \\ -e^{-t} & e^{t} & \cos(t) & -\sin(t) \\ e^{-t} & e^{t} & -\sin(t) & -\cos(t) \\ -e^{-t} & e^{t} & -\cos(t) & \sin(t) \end{vmatrix}} dt = \frac{1}{4} \int te^{-t} \sin(2t)\, dt,$$

$$= \frac{-4e^{-t}\cos(2t) - 10t\,e^{-t}\cos(2t) + 3e^{-t}\sin(2t) - 5t\,e^{-t}\sin(2t)}{100},$$

$$u_3(t) = \int \frac{\begin{vmatrix} e^{-t} & e^{t} & 0 & \cos(t) \\ -e^{-t} & e^{t} & 0 & -\sin(t) \\ e^{-t} & e^{t} & 0 & -\cos(t) \\ -e^{-t} & e^{t} & t\sin(2t) & \sin(t) \end{vmatrix}}{\begin{vmatrix} e^{-t} & e^{t} & \sin(t) & \cos(t) \\ -e^{-t} & e^{t} & \cos(t) & -\sin(t) \\ e^{-t} & e^{t} & -\sin(t) & -\cos(t) \\ -e^{-t} & e^{t} & -\cos(t) & \sin(t) \end{vmatrix}} dt = -\int t\sin(t)(\cos(t))^2\, dt$$

$$= \frac{9t\cos(t) + 3t\cos(3t) - 9\sin(t) - \sin(3t)}{36},$$

and

$$u_4(t) = \int \frac{\begin{vmatrix} e^{-t} & e^t & \sin(t) & 0 \\ -e^{-t} & e^t & \cos(t) & 0 \\ e^{-t} & e^t & -\sin(t) & 0 \\ -e^{-t} & e^t & -\cos(t) & t\sin(2t) \\ e^{-t} & e^t & \sin(t) & \cos(t) \\ -e^{-t} & e^t & \cos(t) & -\sin(t) \\ e^{-t} & e^t & -\sin(t) & -\cos(t) \\ -e^{-t} & e^t & -\cos(t) & \sin(t) \end{vmatrix}}{} \, dt = \int t\cos(t)\big(\sin(t)\big)^2 \, dt$$

$$= \frac{9\cos(t) - 3\cos(3t) + 9t\sin(t) - 3t\sin(3t)}{36}.$$

Hence,

$$y_p(t) = \frac{32\cos(2t) + 15t\sin(2t)}{225}.$$

5. $y''' - 4y'' + 3y' = te^t$

Solution. The roots of the characteristic equation $r^3 - 4r^2 + 3r = r(r-1)(r-3) = 0$ are $r_1 = 0$, $r_2 = 1$ and $r_3 = 3$. Consequently, a fundamental set of solutions is $\{1, e^t, e^{3t}\}$ and the general solution to the homogeneous equation is

$$y_H = \tilde{C}_1 + \tilde{C}_2 e^t + \tilde{C}_3 e^{3t}.$$

Using Cramer's rule we obtain

$$u_1(t) = \int \frac{\begin{vmatrix} 0 & e^t & e^{3t} \\ 0 & e^t & 3e^{3t} \\ te^t & e^t & 9e^{3t} \\ 1 & e^t & e^{3t} \\ 0 & e^t & 3e^{3t} \\ 0 & e^t & 9e^{3t} \end{vmatrix}}{} \, dt = \frac{1}{3} \int t\,e^t \, dt = \frac{e^t(t-1)}{3},$$

$$u_2(t) = \int \frac{\begin{vmatrix} 1 & 0 & e^{3t} \\ 0 & 0 & 3e^{3t} \\ 0 & te^t & 9e^{3t} \end{vmatrix}}{\begin{vmatrix} 1 & e^t & e^{3t} \\ 0 & e^t & 3e^{3t} \\ 0 & e^t & 9e^{3t} \end{vmatrix}} \, dt = -\frac{1}{2} \int t \, dt = -\frac{t^2}{4},$$

and

$$u_3(t) = \int \frac{\begin{vmatrix} 1 & e^t & 0 \\ 0 & e^t & 0 \\ 0 & e^t & te^t \end{vmatrix}}{\begin{vmatrix} 1 & e^t & e^{3t} \\ 0 & e^t & 3e^{3t} \\ 0 & e^t & 9e^{3t} \end{vmatrix}} \, dt = \frac{1}{6} \int t e^{-2t} \, dt = -\frac{(2t+1)e^{-2t}}{24}.$$

Hence,

$$y_p(t) = u_1 y_1 + u_2 y_2 + u_3 y_3 = \frac{e^t(-3 + 2t - 2t^2)}{8}.$$

We may combine $-3e^t/8$ with $y_H(t)$ and write

$$y(t) = C_1 + C_2 e^t + C_3 e^{3t} + \frac{e^t(t - t^2)}{4}.$$

6. $y''' - 4y'' + 3y' = \dfrac{e^t}{1 + e^t}$

Solution. The roots of the characteristic equation $r^3 - 4r^2 + 3r = r(r-1)(r-3) = 0$ are $r_1 = 0$, $r_2 = 1$ and $r_3 = 3$. Consequently, a fundamental set of solutions is $\{1, e^t, e^{3t}\}$ and the general solution to the homogeneous equation is

$$y_H = \tilde{C}_1 + \tilde{C}_2 e^t + \tilde{C}_3 e^{3t}.$$

Using Cramer's rule we obtain

$$u_1(t) = \int \frac{\begin{vmatrix} 0 & e^t & e^{3t} \\ 0 & e^t & 3e^{3t} \\ \frac{e^t}{1+e^t} & e^t & 9e^{3t} \end{vmatrix}}{\begin{vmatrix} 1 & e^t & e^{3t} \\ 0 & e^t & 3e^{3t} \\ 0 & e^t & 9e^{3t} \end{vmatrix}} dt = \frac{1}{3} \int \frac{e^t}{1+e^t} dt = \frac{\log(1+e^t)}{3},$$

$$u_2(t) = \int \frac{\begin{vmatrix} 1 & 0 & e^{3t} \\ 0 & 0 & 3e^{3t} \\ 0 & \frac{e^t}{1+e^t} & 9e^{3t} \end{vmatrix}}{\begin{vmatrix} 1 & e^t & e^{3t} \\ 0 & e^t & 3e^{3t} \\ 0 & e^t & 9e^{3t} \end{vmatrix}} dt = -\frac{1}{2} \int \frac{1}{1+e^t} dt = \frac{\log(1+e^t) - t}{2},$$

and

$$u_3(t) = \int \frac{\begin{vmatrix} 1 & e^t & 0 \\ 0 & e^t & 0 \\ 0 & e^t & \frac{e^t}{1+e^t} \end{vmatrix}}{\begin{vmatrix} 1 & e^t & e^{3t} \\ 0 & e^t & 3e^{3t} \\ 0 & e^t & 9e^{3t} \end{vmatrix}} dt = \frac{1}{6} \int \frac{e^{-2t}}{1+e^t} dt = -\frac{e^{-2t}}{12} + \frac{e^{-t}}{6} - \frac{\log(1+e^{-t})}{6}.$$

Hence,

$$y_p(t) = -\frac{(1+6t)e^t}{12} + \frac{e^{2t}}{6} + \frac{e^{3t}\log(1+e^{-t})}{6} + \frac{(2+3e^t)\log(1+e^t)}{6}.$$

We may combine $-e^t/12$ with $y_H(t)$ and write

$$y(t) = C_1 + C_2 e^t + C_3 e^{3t} - \frac{t\,e^t}{2} + \frac{e^{2t}}{6} + \frac{e^{3t}\log(1+e^{-t})}{6} + \frac{(2+3e^t)\log(1+e^t)}{6}.$$

7. $y^{(6)} - 6y^{(5)} + 15y^{(4)} - 20y''' + 15y'' - 6y' + y = t$

Solution. The roots of the characteristic equation $r^6 - 6r^5 + 15r^4 - 20r^3 + 15r^2 - 6r + 1 = (r-1)^6 = 0$ are $r_1 \ldots r_6 = 1$. Consequently, a fundamental set of solutions is $\{e^t, t\,e^t, t^2\,e^t, t^3\,e^t, t^4\,e^t, t^5\,e^t\}$ and the general solution to the homogeneous equation is

$$y_H = C_1 e^t + C_2 t\,e^t + C_3 t^2\,e^t + C_4 t^3\,e^t + C_5 t^4\,e^t + C_6 t^5\,e^t.$$

Using Cramer's rule we obtain

$$u_1(t) = -\frac{1}{120}\int t^6 e^{-t}\,dt = e^{-t}\left(6 + 6t + 3t^2 + t^3 + \frac{t^4}{4} + \frac{t^5}{20} + \frac{t^6}{120}\right),$$

$$u_2(t) = \frac{1}{24}\int t^5 e^{-t}\,dt = -e^{-t}\left(5 + 5t + \frac{5}{2}t^2 + \frac{5}{6}t^3 + \frac{5}{24}t^4 + \frac{t^5}{24}\right),$$

$$u_3(t) = -\frac{1}{12}\int t^4 e^{-t}\,dt = e^{-t}\left(2 + 2t + t^2 + \frac{1}{3}t^3 + \frac{1}{12}t^4\right),$$

$$u_4(t) = \frac{1}{12}\int t^3 e^{-t}\,dt = -e^{-t}\left(\frac{1}{2} + \frac{t}{2} + \frac{t^2}{4} + \frac{t^3}{12}\right),$$

$$u_5(t) = -\frac{1}{24}\int t^2 e^{-t}\,dt = e^{-t}\left(\frac{1}{12} + \frac{t}{12} + \frac{t^2}{24}\right),$$

$$u_6(t) = \frac{1}{120}\int t\,e^{-t}\,dt = -e^{-t}\left(\frac{1}{120} + \frac{t}{120}\right).$$

Hence,

$$y_p(t) = 6 + t.$$

Finally,

$$y(t) = C_1 e^t + C_2 t\,e^t + C_3 t^2\,e^t + C_4 t^3\,e^t + C_5 t^4\,e^t + C_6 t^5\,e^t + 6 + t.$$

8. $y^{(4)} - 10y''' + 35y'' - 50y' + 24y = \cosh(t)$

Solution. The roots of the characteristic equation $r^4 - 10r^3 + 35r^2 - 50r + 24 = (r-1)(r-2)(r-3)(r-4) = 0$ are $r_1 = 1$, $r_2 = 2$, $r_3 = 3$ and $r_4 = 4$. Consequently, a fundamental set of solutions is $\{e^t, e^{2t}, e^{3t}, e^{4t}\}$ and the general solution to the homogeneous equation is

$$y_H = \tilde{C}_1 e^t + \tilde{C}_2 e^{2t} + \tilde{C}_3 e^{3t} + \tilde{C}_4 e^{4t}.$$

Using Cramer's rule we obtain

$$
u_1(t) = \int \frac{\begin{vmatrix} 0 & e^{2t} & e^{3t} & e^{4t} \\ 0 & 2e^{2t} & 3e^{3t} & 4e^{4t} \\ 0 & 4e^{2t} & 9e^{3t} & 16e^{4t} \\ \cosh(t) & 8e^{2t} & 27e^{3t} & 64e^{4t} \end{vmatrix}}{\begin{vmatrix} e^t & e^{2t} & e^{3t} & e^{4t} \\ e^t & 2e^{2t} & 3e^{3t} & 4e^{4t} \\ e^t & 4e^{2t} & 9e^{3t} & 16e^{4t} \\ e^t & 8e^{2t} & 27e^{3t} & 64e^{4t} \end{vmatrix}} \, dt = -\frac{1}{6} \int e^{-t} \cosh(t) \, dt,
$$

$$
= \frac{e^{-2t} - 2t}{24},
$$

$$
u_2(t) = \int \frac{\begin{vmatrix} e^t & 0 & e^{3t} & e^{4t} \\ e^t & 0 & 3e^{3t} & 4e^{4t} \\ e^t & 0 & 9e^{3t} & 16e^{4t} \\ e^t & \cosh(t) & 27e^{3t} & 64e^{4t} \end{vmatrix}}{\begin{vmatrix} e^t & e^{2t} & e^{3t} & e^{4t} \\ e^t & 2e^{2t} & 3e^{3t} & 4e^{4t} \\ e^t & 4e^{2t} & 9e^{3t} & 16e^{4t} \\ e^t & 8e^{2t} & 27e^{3t} & 64e^{4t} \end{vmatrix}} \, dt = \frac{1}{2} \int e^{-2t} \cosh(t) \, dt,
$$

$$
= -\frac{e^{-3t} + 3e^{-t}}{12},
$$

$$u_3(t) = \int \frac{\begin{vmatrix} e^t & e^{2t} & 0 & e^{4t} \\ e^t & 2e^{2t} & 0 & 4e^{4t} \\ e^t & 4e^{2t} & 0 & 16e^{4t} \\ e^t & 8e^{2t} & \cosh(t) & 64e^{4t} \end{vmatrix}}{\begin{vmatrix} e^t & e^{2t} & e^{3t} & e^{4t} \\ e^t & 2e^{2t} & 3e^{3t} & 4e^{4t} \\ e^t & 4e^{2t} & 9e^{3t} & 16e^{4t} \\ e^t & 8e^{2t} & 27e^{3t} & 64e^{4t} \end{vmatrix}}\, dt = -\frac{1}{2}\int e^{-3t}\cosh(t)\, dt$$

$$= \frac{e^{-4t} + 2e^{-2t}}{16},$$

and

$$u_4(t) = \int \frac{\begin{vmatrix} e^t & e^{2t} & ?^{3t} & 0 \\ e^t & 2e^{2t} & 3e^{3t} & 0 \\ e^t & 4e^{2t} & 9e^{3t} & 0 \\ e^t & 8e^{2t} & 27e^{3t} & \cosh(t) \end{vmatrix}}{\begin{vmatrix} e^t & e^{2t} & e^{3t} & e^{4t} \\ e^t & 2e^{2t} & 3e^{3t} & 4e^{4t} \\ e^t & 4e^{2t} & 9e^{3t} & 16e^{4t} \\ e^t & 8e^{2t} & 27e^{3t} & 64e^{4t} \end{vmatrix}}\, dt = \frac{1}{6}\int e^{-4t}\cosh(t)\, dt$$

$$= -\frac{5e^{-3t} + 3e^{-5t}}{180}.$$

Hence,

$$y_p(t) = \frac{e^{-t}}{240} - \frac{11e^t}{72} - \frac{t\,e^t}{12}.$$

We may combine $11e^t/72$ with $y_H(t)$ and write

$$y(t) = C_1 e^t + C_2 e^{2t} + C_3 e^{3t} + C_4 e^{4t} + \frac{e^{-t}}{240} - \frac{t\,e^t}{12}.$$

9. $y^{(4)} - 2y''' - 13y'' + 14y' + 24y = t\sinh(t)$

Solution. The roots of the characteristic equation $r^4 - 2r^3 - 13r^2 + 14r + 24 = (r+3)(r+1)(r-2)(r-4) = 0$ are $r_1 = -3$, $r_2 = -1$, $r_3 = 2$ and $r_4 = 4$. Consequently, a fundamental set of solutions is $\{e^{-3t}, e^{-t}, e^{2t}, e^{4t}\}$ and the general solution to the homogeneous equation is

$$y_H = \tilde{C}_1 e^{-3t} + \tilde{C}_2 e^{-t} + \tilde{C}_3 e^{2t} + \tilde{C}_4 e^{4t}.$$

Using Cramer's rule we obtain

$$u_1(t) = \int \frac{\begin{vmatrix} 0 & e^{-t} & e^{2t} & e^{4t} \\ 0 & -e^{-t} & 2e^{2t} & 4e^{4t} \\ 0 & e^{-t} & 4e^{2t} & 16e^{4t} \\ t\sinh(t) & -e^{-t} & 8e^{2t} & 64e^{4t} \end{vmatrix}}{\begin{vmatrix} e^{-3t} & e^{-t} & e^{2t} & e^{4t} \\ -3e^{-3t} & -e^{-t} & 2e^{2t} & 4e^{4t} \\ 9e^{-3t} & e^{-t} & 4e^{2t} & 16e^{4t} \\ -27e^{-3t} & -e^{-t} & 8e^{2t} & 64e^{4t} \end{vmatrix}} dt = -\frac{1}{70} \int t\, e^{3t} \sinh(t)\, dt,$$

$$= e^{4t} \left(\frac{1}{2240} - \frac{t}{560} \right) + e^{2t} \left(\frac{t}{280} - \frac{1}{560} \right),$$

$$u_2(t) = \int \frac{\begin{vmatrix} e^{-3t} & 0 & e^{2t} & e^{4t} \\ -3e^{-3t} & 0 & 2e^{2t} & 4e^{4t} \\ 9e^{-3t} & 0 & 4e^{2t} & 16e^{4t} \\ -27e^{-3t} & t\sinh(t) & 8e^{2t} & 64e^{4t} \end{vmatrix}}{\begin{vmatrix} e^{-3t} & e^{-t} & e^{2t} & e^{4t} \\ -3e^{-3t} & -e^{-t} & 2e^{2t} & 4e^{4t} \\ 9e^{-3t} & e^{-t} & 4e^{2t} & 16e^{4t} \\ -27e^{-3t} & -e^{-t} & 8e^{2t} & 64e^{4t} \end{vmatrix}} dt = \frac{1}{30} \int t\, e^t \sinh(t)\, dt,$$

$$= e^{2t}\left(\frac{t}{120} - \frac{1}{240}\right) - \frac{t^2}{120},$$

$$u_3(t) = \int \frac{\begin{vmatrix} e^{-3t} & e^{-t} & 0 & e^{4t} \\ -3e^{-3t} & -e^{-t} & 0 & 4e^{4t} \\ 9e^{-3t} & e^{-t} & 0 & 16e^{4t} \\ -27e^{-3t} & -e^{-t} & t\sinh(t) & 64e^{4t} \end{vmatrix}}{\begin{vmatrix} e^{-3t} & e^{-t} & e^{2t} & e^{4t} \\ -3e^{-3t} & -e^{-t} & 2e^{2t} & 4e^{4t} \\ 9e^{-3t} & e^{-t} & 4e^{2t} & 16e^{4t} \\ -27e^{-3t} & -e^{-t} & 8e^{2t} & 64e^{4t} \end{vmatrix}} \, dt = -\frac{1}{30}\int t\, e^{-2t}\sinh(t)\, dt$$

$$= -e^{-3t}\left(\frac{1}{540} + \frac{t}{180}\right) + e^{-t}\left(\frac{1}{60} + \frac{t}{60}\right),$$

and

$$u_4(t) = \int \frac{\begin{vmatrix} e^{-3t} & e^{-t} & e^{2t} & 0 \\ -3e^{-3t} & -e^{-t} & 2e^{2t} & 0 \\ 9e^{-3t} & e^{-t} & 4e^{2t} & 0 \\ -27e^{-3t} & -e^{-t} & 8e^{2t} & t\sinh(t) \end{vmatrix}}{\begin{vmatrix} e^{-3t} & e^{-t} & e^{2t} & e^{4t} \\ -3e^{-3t} & -e^{-t} & 2e^{2t} & 4e^{4t} \\ 9e^{-3t} & e^{-t} & 4e^{2t} & 16e^{4t} \\ -27e^{-3t} & -e^{-t} & 8e^{2t} & 64e^{4t} \end{vmatrix}} \, dt = \frac{1}{70}\int t\, e^{-4t}\sinh(t)\, dt$$

$$= -e^{-3t}\left(\frac{1}{1260} + \frac{t}{420}\right) + e^{-5t}\left(\frac{1}{3500} + \frac{t}{700}\right).$$

Hence,

$$y_p(t) = -\frac{(181 + 30t + 450t^2)e^{-t}}{54000} + \frac{(7 + 12t)e^t}{576}.$$

We may combine $-181e^{-t}/54000$ with $y_H(t)$ and write

$$y(t) = C_1 e^{-3t} + C_2 e^{-t} + C_3 e^{2t} + C_4 e^{4t} - \frac{(t + 15t^2)e^{-t}}{1800} + \frac{(7 + 12t)e^t}{576}.$$

10. $y^{(6)} - 12y^{(5)} + 55y^{(4)} - 120y''' + 135y'' - 108y' + 81y = \sin(t) + \cos(t)$

Solution. $y_p(t) = \dfrac{(483 + 17t)\cos(t) + (219 + 31t)\sin(t)}{5000}$

11. $y^{(6)} - 3y^{(5)} + 4y^{(4)} - 6y''' + 5y'' - 3y' + 2y = t^2 e^t$

Solution. $y_p(t) = \dfrac{(-6t + 3t^2 - t^3)e^t}{12}$

12. $y^{(8)} - 2y^{(7)} + 3y^{(6)} - 4y^{(5)} + 3y^{(4)} - 2y''' + y'' = t\, e^t \sin(t)$

Solution. $y_p(t) = \dfrac{e^t \big((127 - 15t)\cos(t) - (11 - 20t)\sin(t) \big)}{250}$

13. If $y_1 = t$, $y_2 = t\log(t)$ and $y_3 = t(\log(t))^2$ are solutions to the homogeneous equation associated with

$$t^3 y''' + t\, y' - y = 24t\, \log(t), \qquad t > 0,$$

find the general solution.

Solution. Since the set $\{t, t\log(t), t(\log(t))^2\}$ is given as a fundamental set of solutions, the general solution to the homogeneous equation is

$$y_H = C_1 t + C_2 t\, \log(t) + C_3 t(\log(t))^2.$$

Using Cramer's rule we obtain

$$
u_1(t) = \int \frac{\begin{vmatrix} 0 & t\log(t) & t(\log(t))^2 \\ 0 & 1+\log(t) & 2\log(t)+(\log(t))^2 \\ 24t\log(t) & \dfrac{1}{t} & \dfrac{2+2\log(t)}{t} \end{vmatrix}}{\begin{vmatrix} t & t\log(t) & t(\log(t))^2 \\ 1 & 1+\log(t) & 2\log(t)+(\log(t))^2 \\ 0 & \dfrac{1}{t} & \dfrac{2+2\log(t)}{t} \end{vmatrix}} \, dt
$$

$$
= 12 \int t^2 (\log(t))^3 \, dt,
$$

$$
= \frac{-8\,t^3}{9} + \frac{8\,t^3\,\log(t)}{3} - 4\,t^3\log(t)^2 + 4\,t^3\log(t)^3.
$$

$$
u_2(t) = \int \frac{\begin{vmatrix} t & 0 & t(\log(t))^2 \\ 1 & 0 & 2\log(t)+(\log(t))^2 \\ 0 & 24t\log(t) & \dfrac{2+2\log(t)}{t} \end{vmatrix}}{\begin{vmatrix} t & t\log(t) & t(\log(t))^2 \\ 1 & 1+\log(t) & 2\log(t)+(\log(t))^2 \\ 0 & \dfrac{1}{t} & \dfrac{2+2\log(t)}{t} \end{vmatrix}} \, dt = -24 \int t^2 (\log(t))^2 \, dt,
$$

$$
= \frac{-16\,t^3}{9} + \frac{16\,t^3\,\log(t)}{3} - 8\,t^3\log(t)^2.
$$

and

$$u_3(t) = \int \frac{\begin{vmatrix} t & t\log(t) & 0 \\ 1 & 1+\log(t) & 0 \\ 0 & \dfrac{1}{t} & 24t\log(t) \end{vmatrix}}{\begin{vmatrix} t & t\log(t) & t\left(\log(t)\right)^2 \\ 1 & 1+\log(t) & 2\log(t)+\left(\log(t)\right)^2 \\ 0 & \dfrac{1}{t} & \dfrac{2+2\log(t)}{t} \end{vmatrix}} \, dt = 12\int t^2 \log(t)\, dt,$$

$$= \frac{-4\,t^3}{3} + 4\,t^3 \log(t).$$

Hence,

$$y_p(t) = u_1 y_1 + u_2 y_2 + u_3 y_3 = \frac{8\,t^4\,(\log(t)-1)}{9}$$

and

$$y(t) = C_1\,t + C_2\,t\,\log(t) + C_3\,t\left(\log(t)\right)^2 + \frac{8\,t^4\,(\log(t)-1)}{9}.$$

14. Assume that f is a continuous function and consider

$$t^3 y''' - 3t^2 y'' + 6t\,y' - 6y = t^3 f(t), \qquad t > 0.$$

By trying $y = t^m$, find solutions to the homogeneous equation and then find the general solution in terms of f. Does this pattern continue to an nth-order equation?

Solution. After substituting $y = t^m$ into the homogeneous equation, we find that for this form to work, m must be a root of the equation $m(m-1)(m-2) - 3m(m-1) + 6m - 6 = m^3 - 6m^2 + 11m - 6 = (m-1)(m-2)(m-3) = 0$. Therefore, $m_1 = 1, m_2 = 2$ and $m_3 = 3$. Consequently, a fundamental set of solutions is $\{t, t^2, t^3\}$ and the general solution to the homogeneous equation is

$$y_H = C_1\,t + C_2\,t^2 + C_3\,t^3.$$

Using Cramer's rule we obtain

$$u_1(t) = \int \frac{\begin{vmatrix} 0 & t^2 & t^3 \\ 0 & 2t & 3t^2 \\ t^3 f(t) & 2 & 6t \end{vmatrix}}{\begin{vmatrix} t & t^2 & t^3 \\ 1 & 2t & 3t^2 \\ 0 & 2 & 6t \end{vmatrix}}\, dt = \frac{1}{2}\int t^4 f(t)\, dt,$$

$$u_2(t) = \int \frac{\begin{vmatrix} t & 0 & t^3 \\ 1 & 0 & 3t^2 \\ 0 & t^3 f(t) & 6t \end{vmatrix}}{\begin{vmatrix} t & t^2 & t^3 \\ 1 & 2t & 3t^2 \\ 0 & 2 & 6t \end{vmatrix}}\, dt = -\int t^3 f(t)\, dt,$$

and

$$u_3(t) = \int \frac{\begin{vmatrix} t & t^2 & 0 \\ 1 & 2t & 0 \\ 0 & 2 & t^3 f(t) \end{vmatrix}}{\begin{vmatrix} t & t^2 & t^3 \\ 1 & 2t & 3t^2 \\ 0 & 2 & 6t \end{vmatrix}}\, dt = \frac{1}{2}\int t^2 f(t)\, dt,$$

Hence,

$$y_p(t) = u_1 y_1 + u_2 y_2 + u_3 y_3 = \frac{t^3 \int t^2 g(t)\, dt - 2t^2 \int t^3 g(t)\, dt + t \int t^4 g(t)\, dt}{2}$$

and

$$y(t) = C_1 t + C_2 t^2 + C_3 t^3 + \frac{t^3 \int t^2 g(t)\, dt - 2t^2 \int t^3 g(t)\, dt + t \int t^4 g(t)\, dt}{2}.$$

Yes, this pattern continues. For example, the next equation would be

$$t^4 y'''' - 4t^3 y''' + 12t^2 y'' - 24t\, y' + 24y = t^4 f(t), \qquad t > 0.$$

with the solution to the homogeneous equation

$$y_H(t) = C_1\, t + C_2\, t^2 + C_3\, t^3 + C_4\, t^4$$

and with particular solution

$$y_p(t) = \frac{t^4 \int t^3 g(t)\, dt - 3t^3 \int t^4 g(t)\, dt + 3t^2 \int t^5 g(t)\, dt - t \int t^6 g(t)\, dt}{6}.$$

Section 12.5, page 405

Use **Method->NthOrderLinear** or **Method->ApproximateNthOrderLinear** to find seminumerical approximations to the general solution of each of the following differential equations:

1. $y^{(5)} + y''' + y'' - y' + y = 0$

Solution. Using **ODE**, we enter

```
ODE[y''''' + y''' + y'' - y' + y == 0,y,t,
Method->NthOrderLinear]
```

Since *Mathematica* cannot find the exact roots of the characteristic equation, the approximate roots are computed automatically using **NSolve** and we obtain

```
Approximating the roots of the characteristic equation

           1. C[1]        0.485 t
{{y ->  ----------- + 1. E          C[4] Cos[0.56 t] +
          1.14 t
         E

          0.0863 t
    1. E            C[2] Cos[1.26 t] +

          0.485 t
    1. E            C[5] Sin[0.56 t] +

          0.0863 t
    1. E            C[3] Sin[1.26 t]}}
```

Hence,

$$y(t) \approx C_1 e^{-1.14t} + e^{0.0863t}\left(C_2 \cos(1.26t) + C_3 \sin(1.26t)\right)$$
$$+ e^{0.485t}\left(C_4 \cos(0.56t) + C_5 \sin(0.56t)\right)$$

2. $y^{(6)} + y' + y = 0$

Solution. Using **ODE**, we enter

```
ODE[y'''''' + y' + y == 0,y,t,
Method->NthOrderLinear]
```

Since *Mathematica* cannot find the exact roots of the characteristic equation, the approximate roots are computed automatically using **NSolve** and we obtain

```
Approximating the roots of the characteristic equation

            1. C[1] Cos[0.301 t]
{{y ->   ─────────────────────   +
                0.791 t
               E

          0.945 t
    1. E           C[5] Cos[0.612 t] +

    1. C[3] Cos[1.04 t]      1. C[2] Sin[0.301 t]
    ───────────────────  +   ────────────────────   +
          0.155 t                  0.791 t
         E                        E

          0.945 t                 1. C[4] Sin[1.04 t]
    1. E           C[6] Sin[0.612 t] +  ──────────────────}}
                                              0.155 t
                                             E
```

Hence,

$$y(t) \approx e^{-0.791t}\left(C_1 \cos(0.301\,t) + C_2 \sin(0.3001\,t)\right)$$
$$+ e^{-0.155t}\left(C_3 \cos(1.04\,t) + C_4 \sin(1.04\,t)\right)$$
$$+ e^{0.945t}\left(C_5 \cos(0.612\,t) + C_6 \sin(0.612\,t)\right)$$

3. $y^{(6)} + 2y''' + y'' - y' + y = 0$

Solution. Using **ODE**, we enter

```
ODE[y'''''' + 2y''' + y'' - y' + y == 0,y,t,
Method->NthOrderLinear]
```

Since *Mathematica* cannot find the exact roots of the characteristic equation, the approximate roots are computed automatically using **NSolve** and we obtain

```
Approximating the roots of the characteristic equation
            1. C[1] Cos[0.468 t]
```

```
{{y ->  ————————————  +
              1.07 t
            E

       0.385 t
   1. E            C[3] Cos[0.534 t] +

       0.681 t
   1. E            C[5] Cos[1.11 t] +

   1. C[2] Sin[0.468 t]
   —————————————————————  +
          1.07 t
        E

       0.385 t
   1. E            C[4] Sin[0.534 t] +

       0.681 t
   1. E            C[6] Sin[1.11 t]}}
```

Hence,

$$y(t) \approx e^{-1.07\,t}\big(C_1 \cos(0.468\,t) + C_2 \sin(0.468\,t)\big)$$
$$+ e^{0.385\,t}\big(C_3 \cos(0.534\,t) + C_4 \sin(0.534\,t)\big)$$
$$+ e^{0.681\,t}\big(C_5 \cos(1.11\,t) + C_6 \sin(1.11\,t)\big)$$

4. $y^{(7)} + y''' + y = 0$

Solution. Using **ODE**, we enter

ODE[y'''''' + y''' + y == 0,y,t,
Method->NthOrderLinear]

Since *Mathematica* cannot find the exact roots of the characteristic equation, the approximate roots are computed automatically using **NSolve** and we obtain

```
Approximating the roots of the characteristic equation

        1. C[1]          0.872 t
{{y ->  —————————  + 1. E          C[6] Cos[0.579 t] +
        0.863 t
      E

      1. C[2] Cos[0.845 t]
      ————————————————————  +
             0.748 t
           E

          0.307 t
      1. E            C[4] Cos[0.858 t] +

          0.872 t
```

```
1. E          C[7] Sin[0.579 t] +

1. C[3] Sin[0.845 t]          0.307 t
─────────────────── + 1. E          C[5] Sin[0.858 t]}}
     0.748 t
    E
```

Hence,

$$y(t) \approx C_1 e^{-0.863\,t} + e^{-0.748\,t}(C_2 \cos(0.845\,t) + C_3 \sin(0.845\,t))$$

$$+ e^{0.307\,t}(C_4 \cos(0.858\,t) + C_5 \sin(0.858\,t))$$

$$+ e^{0.872\,t}(C_6 \cos(0.579\,t) + C_7 \sin(0.579\,t))$$

5. $y^{(7)} + 2y^{(4)} + 3y''' + 4y = 0$

Solution. Using **ODE**, we enter

ODE[y''''''' + 2y'''' + 3y''' + 4y == 0,y,t,
Method->NthOrderLinear]

Since *Mathematica* cannot find the exact roots of the characteristic equation, the approximate roots are computed automatically using **NSolve** and we obtain

```
Approximating the roots of the characteristic equation

         1. C[1]          0.613 t
{{y ->  ───────── + 1. E          C[4] Cos[0.862 t] +
         1.17 t
        E

    1. C[2] Cos[0.87 t]          0.88 t
    ─────────────────── + 1. E          C[6] Cos[1.08 t] +
         0.91 t
        E

         0.613 t
    1. E          C[5] Sin[0.862 t] +

    1. C[3] Sin[0.87 t]          0.88 t
    ─────────────────── + 1. E          C[7] Sin[1.08 t]}}
         0.91 t
        E
```

Hence,

$$y(t) \approx C_1 e^{-1.17\,t} + e^{-0.91\,t}(C_2 \cos(0.87\,t) + C_3 \sin(0.87\,t))$$

$$+ e^{0.613\,t}(C_4 \cos(0.862\,t) + C_5 \sin(0.862\,t))$$

$$+ e^{0.88\,t}(C_6 \cos(1.078\,t) + C_7 \sin(1.08\,t))$$

6. $y^{(8)} + y''' + y = 0$

Solution. Using **ODE**, we enter

```
ODE[y'''''''' + y''' + y == 0,y,t,
Method->ApproximateNthOrderLinear]
```

Since *Mathematica* cannot find the exact roots of the characteristic equation, the approximate roots are computed automatically using **NSolve** and we obtain

```
Approximating the roots of the characteristic equation

        1. C[1] Cos[0.25 t]
{{y ->  ───────────────────   +
             0.929 t
            E

        0.934 t
   1. E          C[7] Cos[0.506 t] +

        0.382 t
   1. E          C[5] Cos[0.795 t] +

   1. C[3] Cos[1.04 t]    1. C[2] Sin[0.25 t]
   ───────────────────  +  ───────────────────  +
         0.387 t                0.929 t
        E                      E

        0.934 t
   1. E          C[8] Sin[0.506 t] +

        0.382 t                      1. C[4] Sin[1.04 t]
   1. E          C[6] Sin[0.795 t] +  ───────────────────}}
                                            0.387 t
                                           E
```

Hence,

$$y(t) \approx e^{-0.929\,t}(C_1 \cos(0.25\,t) + C_2 \sin(0.25\,t))$$
$$+ e^{-0.387\,t}(C_3 \cos(1.04\,t) + C_4 \sin(1.04\,t))$$
$$+ e^{0.382\,t}(C_5 \cos(0.795\,t) + C_6 \sin(0.795\,t))$$
$$+ e^{0.934\,t}(C_7 \cos(0.506\,t) + C_8 \sin(0.506\,t))$$

7. $y^{(8)} + y^{(4)} - y''' + y = 0$

Solution. Using **ODE**, we enter

```
ODE[y'''''''' + y'''' - y''' + y == 0,y,t,
Method->NthOrderLinear]
```

Since *Mathematica* cannot find the exact roots of the characteristic equation, the approximate roots are computed automatically using **NSolve** and we obtain

```
Approximating the roots of the characteristic equation

              0.877 t
{{y -> 1. E          C[7] Cos[0.367 t] +

    1. C[1] Cos[0.647 t]     1. C[3] Cos[0.715 t]
    ────────────────────  +  ────────────────────  +
          0.902 t                  0.469 t
         E                        E

         0.494 t
    1. E          C[5] Cos[0.991 t] +

         0.877 t
    1. E          C[8] Sin[0.367 t] +

    1. C[2] Sin[0.647 t]     1. C[4] Sin[0.715 t]
    ────────────────────  +  ────────────────────  +
          0.902 t                  0.469 t
         E                        E

          0.494 t
     1. E          C[6] Sin[0.991 t]}}
```

Hence,
$$y(t) \approx e^{-0.902\,t}\left(C_1 \cos(0.647\,t) + C_2 \sin(0.647\,t)\right)$$
$$+ e^{-0.469\,t}\left(C_3 \cos(0.715\,t) + C_4 \sin(0.715\,t)\right)$$
$$+ e^{0.494\,t}\left(C_5 \cos(0.991\,t) + C_6 \sin(0.991\,t)\right)$$
$$+ e^{0.877\,t}\left(C_7 \cos(0.367\,t) + C_8 \sin(0.367\,t)\right)$$

8. $y^{(10)} + y^{(5)} - y''' + y = 0$

Solution. Using **ODE**, we enter

```
ODE[y'''''''''' + y''''' - y''' + y == 0,y,t,
Method->NthOrderLinear]
```

Since *Mathematica* cannot find the exact roots of the characteristic equation, the approximate roots are computed automatically using **NSolve** and we obtain

```
Approximating the roots of the characteristic equation

              0.887 t
{{y -> 1. E          C[9] Cos[0.296 t] +

    1. C[1] Cos[0.301 t]    1. C[3] Cos[0.775 t]
```

$$\frac{}{E^{1.01\,t}} + \frac{}{E^{0.394\,t}} +$$

$$1.\,E^{0.732\,t}\;\;C[7]\;Cos[0.816\,t] +$$

$$\frac{1.\;C[5]\;Cos[1.05\,t]}{E^{0.217\,t}} +$$

$$1.\,E^{0.887\,t}\;\;C[10]\;Sin[0.296\,t] +$$

$$\frac{1.\;C[2]\;Sin[0.301\,t]}{E^{1.01\,t}} + \frac{1.\;C[4]\;Sin[0.775\,t]}{E^{0.394\,t}} +$$

$$1.\,E^{0.732\,t}\;\;C[8]\;Sin[0.816\,t] + \frac{1.\;C[6]\;Sin[1.05\,t]}{E^{0.217\,t}}\}\}$$

Hence,

$$y(t) \approx e^{-1.01\,t}\left(C_1\cos(0.301\,t) + C_2\sin(0.301\,t)\right)$$
$$+\, e^{-0.394\,t}\left(C_3\cos(0.775\,t) + C_4\sin(0.775\,t)\right)$$
$$+\, e^{-0.217\,t}\left(C_5\cos(1.05\,t) + C_6\sin(1.05\,t)\right)$$
$$+\, e^{0.732\,t}\left(C_7\cos(0.816\,t) + C_8\sin(0.816\,t)\right)$$
$$+\, e^{0.887\,t}\left(C_9\cos(0.296\,t) + C_{10}\sin(0.296\,t)\right)$$

Use **Method->NthOrderLinear** or **Method->ApproximateNthOrderLinear** to solve and plot each of the following initial value problems:

9. $$\begin{cases} y''' + 4y' = \sec(2t) \\ y(0) = y'(0) = 1,\; y''(0) = 0 \end{cases}$$

Solution. With **ODE**, we choose the interval [0, 5] and enter

```
ODE[{y''' + 4y' == Sec[2t],
y[0] == 1,y'[0] == 1,y''[0] == 0},y,t,
Method->NthOrderLinear,PlotSolution->{{t,0,5}}]
```

to get

$$\{\{y \rightarrow 1 - \frac{t\;Cos[2\,t]}{4} - \frac{Log[Cos[t] - Sin[t]]}{8} +$$

$$\frac{\text{Log}[\text{Cos}[t] + \text{Sin}[t]]}{8} + \frac{\text{Sin}[2\ t]}{2} +$$

$$\frac{\text{Log}[\text{Cos}[2\ t]]\ \text{Sin}[2\ t]}{8}\}\}$$

Hence,

$$y(t) = 1 - \frac{t\cos(2t)}{4} + \frac{\sin(2t)}{2} + \frac{1}{8}\log\big((1+\sin(2t))(1+\tan(2t))\big)$$

and the plot is

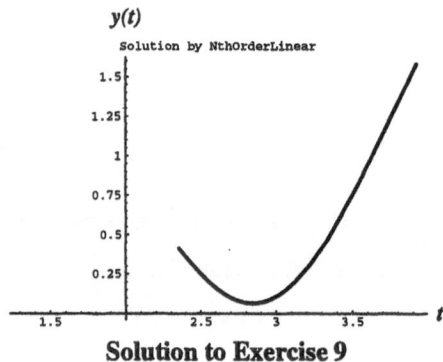

y(t)

Solution by NthOrderLinear

Solution to Exercise 9

10. $\begin{cases} 2y''' - 6y'' = t^2 \\ y(0) = y'(0) = 1,\ y''(0) = 0 \end{cases}$

Solution. With **ODE**, we choose the interval $[-3, 2]$ and enter

```
ODE[{2y''' - 6y'' == t^2,
y[0] == 1,y'[0] == 1,y''[0] == 0},y,t,
Method->NthOrderLinear,PlotSolution->{{t,-3,2}}]
```

to get

$$\{\{y \to \frac{242}{243} + \frac{E^{3\ t}}{243} + \frac{80\ t}{81} - \frac{t^2}{54} - \frac{t^3}{54} - \frac{t^4}{72}\}\}$$

Hence,

$$y(t) = \frac{242 + e^{3t}}{243} + \frac{80t}{81} - \frac{t^2 + t^3}{54} - \frac{t^4}{72}$$

and the plot is

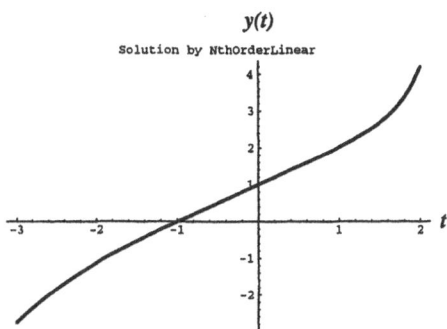

y(t)

Solution by NthOrderLinear

Solution to Exercise 10

11. $$\begin{cases} y^{(4)} + 2y'' + y = t, \\ y(0) = y'(0) = y''(0) = y'''(0) = 0 \end{cases}$$

Solution. With **ODE**, we choose the interval $[-15, 15]$ and enter

```
ODE[{y'''' + 2y'' + y == t,
y[0] == 0,y'[0] == 0,y''[0] == 0,y'''[0] == 0},y,t,
Method->NthOrderLinear,PlotSolution->{{t,-15,15}}]
```

to get

```
                t Cos[t]    3 Sin[t]
{{y -> t +      --------  - --------}}
                   2           2
```

Hence,

$$y(t) = t + \frac{t\cos(t) - 3\sin(t)}{2}$$

and the plot is

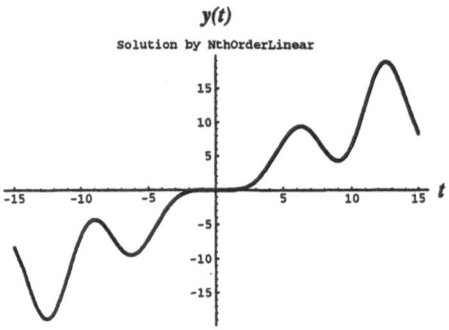

y(t)

Solution by NthOrderLinear

Solution to Exercise 11

12. $\begin{cases} y^{(5)} + y''' + y = 0, \\ y(0) = 1, \, y'(0) = y''(0) = y'''(0) = y^{(4)}(0) = 0 \end{cases}$

Solution. With **ODE**, we choose the interval [0, 5] and enter

```
ODE[{y''''' + y''' + y == 0,y[0] == 1,y'[0] == 0,
y''[0] == 0,y'''[0] == 0,y''''[0] == 0},y,t,
Method->NthOrderLinear,PlotSolution->{{t,0,5}}]
```

to get

```
Approximating the roots of the characteristic equation
         0.261              0.637 t
{{y -> -------- + 0.487 E          Cos[0.665 t] +
         0.838 t
        E

    0.251 Cos[1.17 t]           0.637 t
    ------------------ + 0.131 E          Sin[0.665 t] -
         0.218 t
        E

    0.106 Sin[1.17 t]
    -----------------}}
         0.218 t
        E
```

Hence,

$$y(t) \approx 0.261\, e^{-0.838\,t} + e^{0.637\,t}(0.487\, \cos(0.665\,t) + 0.13098\, \sin(0.665\,t))$$
$$- e^{-0.218\,t}(0.251\, \cos(1.167\,t) + 0.106\, \sin(1.167\,t))$$

and the plot is

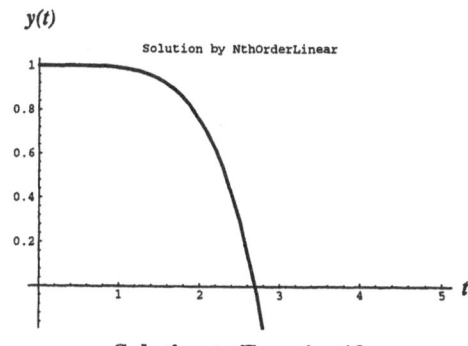

Solution to Exercise 12

13. $\begin{cases} y^{(5)} + y''' + y'' + y = 0, \\ y(0) = y'(0) = 1, \ y''(0) = y'''(0) = y^{(4)}(0) = 0 \end{cases}$

Solution. With **ODE**, we choose the interval $[0, 5]$ and enter

```
ODE[{y''''' + y''' + y'' + y == 0,
y[0] == 1,y'[0] == 1,y''[0] == 0,
y'''[0] == 0,y''''[0] == 0},
y,t,Method->NthOrderLinear,PlotSolution->{{t,0,5}}]
```

to get

$$\{\{y \to Cos[t] + \frac{2\ E^{t/2}\ Sin[\frac{Sqrt[3]\ t}{2}]}{Sqrt[3]}\}\}$$

Hence,

$$y(t) = \cos(t) + \frac{2e^{t/2}\sin(\sqrt{3}\,t/2)}{\sqrt{3}}$$

and the plot is

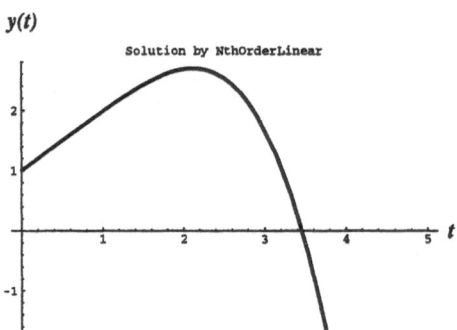

y(t)

Solution by NthOrderLinear

Solution to Exercise 13

14. $\begin{cases} y^{(4)} + 2y'' + y = \cosh(t), \\ y(0) = y'(0) = y''(0) = y'''(0) = 0 \end{cases}$

Solution. With **ODE**, we choose the interval $[-2\pi, 2\pi]$ and enter

```
ODE[{y'''' + 2y'' + y == Cosh[t],
y[0] == 0,y'[0] == 0,y''[0] == 0,y'''[0] == 0},
y,t,Method->NthOrderLinear,PlotSolution->{{t,-2Pi,2Pi}}]
```

to get

$$\{\{y -> \frac{-\text{Cos}[t]}{4} + \frac{\text{Cosh}[t]}{4} - \frac{t\ \text{Sin}[t]}{4}\}\}$$

Hence,

$$y(t) = \frac{-\cos(t) + \cosh(t) - t\sin(t)}{4}$$

and the plot is

y(t)

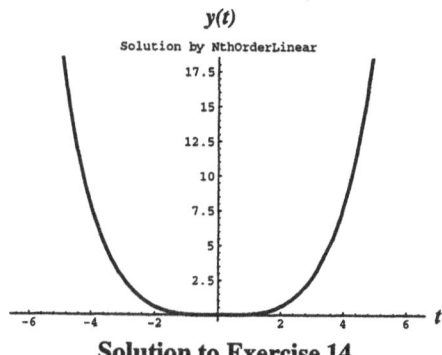

Solution to Exercise 14

15. $\begin{cases} y^{(4)} - 2y''' - 13y'' + 14y' + 24y = \sinh(t), \\ y(0) = y'(0) = 1, \ y''(0) = y'''(0) = 0 \end{cases}$

Solution. With **ODE**, we choose the interval $[-1, 1]$ and enter

```
ODE[{y'''' - 2y''' - 13y'' + 14y' + 24y == Sinh[t],
y[0] == 1,y'[0] == 1,y''[0] == 0,y'''[0] == 0},
y,t,Method->NthOrderLinear,PlotSolution->{{t,-1,1}}]
```

to get

```
            -81         103      E    37 E    82 E
{{y ->   --------  +  -------  + --  + ----- - ----- -
           3 t           t      48     45      525
        560 E         225 E

           t
        ------}}
           t
        60 E
```

Hence,

$$y(t) = -\frac{81e^{-3t}}{560} + \frac{103e^{-t}}{225} + \frac{e^t}{48} + \frac{37\,e^{2t}}{45} - \frac{82\,e^{4t}}{525} - \frac{t\,e^{-t}}{60}$$

and the plot is

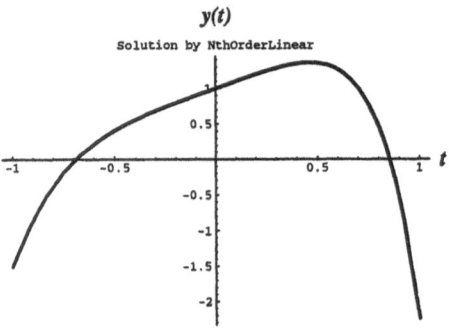

Solution to Exercise 15

16. $\begin{cases} y^{(5)} + y''' + y = 0, \\ y(0) = y'(0) = 1, \ y''(0) = y'''(0) = y^{(4)}(0) = 0 \end{cases}$

Solution. With **ODE**, we choose the interval $[-6, 6]$ and enter

```
ODE[{y''''' + y''' + y == 0,
y[0] == 1,y'[0] == 1,y''[0] == 0,
y'''[0] == 0,y''''[0] == 0},
y,t,Method->NthOrderLinear,PlotSolution->{{t,-6,6}}]
```

to get

```
Approximating the roots of the characteristic equation

       -0.0507            0.637 t
{{y -> --------- + 0.751 E        Cos[0.665 t] +
       0.838 t
      E

    0.3 Cos[1.17 t]           0.637 t
    --------------- + 0.612 E        Sin[0.665 t] +
       0.218 t
      E

    0.119 Sin[1.17 t]
    -----------------}}
       0.218 t
      E
```

Hence,

$$y(t) \approx -0.0507\, e^{-0.838\, t} + e^{0.637\, t} \left(0.751\ \cos(0.665\, t) + 0.612\ \sin(0.665\, t)\right)$$
$$+ e^{-0.2178\, t} \left(0.3\ \cos(1.17\, t) + 0.119\ \sin(1.167\, t)\right)$$

and the plot is

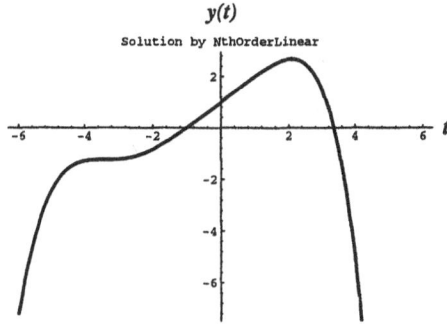

y(t)

Solution by NthOrderLinear

Solution to Exercise 16

Chapter 13

Section 13.1, page 413

In the following exercises, perform the indicated computations by hand, retaining only four significant digits at each step of the calculations.

For a first-order initial value problem defined by

$$\begin{cases} y' = f(t, y), \\ y(t_0) = y_0, \end{cases}$$

if h denotes the step size, all problems in Section 13.1 use the Euler method, defined by the following recursive relation:

$$\begin{cases} Y_0 = y_0, \\ Y_{n+1} = Y_n + hf(t_n, Y_n). \end{cases}$$

1. Use the Euler method to compute an approximate solution to the initial value problem $y' = \sqrt{y+1}$, $y(0) = 0$ with step size $h = 0.1$. Compare with the exact solution $y(t) = t(4+t)/4$.

Solution.

n	t_n	Y_n	$y_{\text{exact}}(t_n)$
0	0.0	0.0	0.0
1	0.1	0.10	0.1025
2	0.2	0.2049	0.21
3	0.3	0.3146	0.3225
4	0.4	0.4293	0.44
5	0.5	0.5489	0.5625
6	0.6	0.6733	0.69
7	0.7	0.8027	0.8225
8	0.8	0.9369	0.96
9	0.9	1.076	1.1025
10	1.0	1.22	1.25

2. Consider the following initial value problem

$$\begin{cases} y' = y - t y^2, \\ y(0) = 2. \end{cases}$$

   a. Use the Euler method to compute an approximate solution at $t = 1.0$. Use the values $n = 2, 4, 8$, which correspond to step sizes of $h = 1/2, 1/4, 1/8$.

b. Determine the actual solution and compare the value of $y(1)$ with the results from part **a**.

Solution.

$$y(t) = \frac{2e^t}{3 - 2e^t + 2t\,e^t}$$

h	t_n	Y_n	$y_{exact}(t_n)$
1/2	1.0	2.25	1.812
1/4	1.0	1.948	1.812
1/8	1.0	1.882	1.812

Repeat Exercise 2 for the following initial value problems:

3. $\begin{cases} y' + 2y = t, \\ y(0) = 1 \end{cases}$

Solution.

$$y(t) = \frac{5e^{2t}}{4} - \frac{t}{2} - \frac{1}{4}$$

h	t_n	Y_n	$y_{exact}(t_n)$
1/2	1.0	0.25	0.4192
1/4	1.0	0.3281	0.4192
1/8	1.0	0.3751	0.4192

4. $\begin{cases} y' = t + 1, \\ y(0) = 1 \end{cases}$

Solution.

$$y(t) = 1 + t + \frac{t^2}{2}$$

h	t_n	Y_n	$y_{exact}(t_n)$
1/2	1.0	2.25	2.5
1/4	1.0	2.375	2.5
1/8	1.0	2.438	2.5

5. $\begin{cases} y' = t^3 e^{-2y}, \\ y(0) = 0 \end{cases}$

Solution.

$$y(t) = \log\left(\frac{\sqrt{2 + t^4}}{\sqrt{2}}\right)$$

h	t_n	Y_n	$y_{exact}(t_n)$
1/2	1.0	0.0625	0.2027
1/4	1.0	0.1333	0.2027
1/8	1.0	0.1691	0.2027

6.
$$\begin{cases} y' = \dfrac{e^t}{y}, \\ y(0) = 1 \end{cases}$$

Solution.

$y(t) = e^{e^t - 1}$

h	t_n	Y_n	$y_{exact}(t_n)$
1/2	1.0	2.05	2.106
1/4	1.0	2.078	2.106
1/8	1.0	2.092	2.106

7. Consider the following initial value problem

$$\begin{cases} y' = y + t^2, \\ y(0) = 1. \end{cases}$$

    a. Find the solution $y(t)$ and evaluate it for $t = 0.2, 0.4, \ldots, 1.0$.

    b. Using the Euler method, with step size of $h = 0.2$, find approximate values for the solution at the t values in part a.

    c. Repeat part b using $h = 0.1$.

    d. Compare the results of part b with those of part c and the exact values. (The differences in the results for $h = 0.2$ and $h = 0.1$ tend to indicate whether a smaller step size must be used for the desired range of t values. The general rule of thumb is to use the smaller step size if the two solutions agree to the desired accuracy. If they do not agree, h should be reduced and the calculations repeated. This gives an indication but not a proof of the accuracy of the result.)

Solution.

$y(t) = -2 + 3e^t - 2t - t^2$

t_n	$Y_n(h = 0.2)$	$Y_n(h = 0.1)$	$y_{exact}(t_n)$
0.0	1.0	1.0	1.0
0.1		1.1	1.106
0.2	1.2	1.211	1.224
0.3		1.336	1.36
0.4	1.448	1.479	1.515
0.5		1.643	1.696
0.6	1.77	1.832	1.906
0.7		2.051	2.151
0.8	2.196	2.305	2.437
0.9		2.6	2.769
1.0	2.763	2.941	3.155

The initial value problems in the following exercises cannot be solved by symbolic methods. Use the Euler method with step size $h = 0.1$ to approximate the solution at $t = 1$ to three significant digits (see Exercise 7).

8. $\begin{cases} y' = \sin(y) + e^t, \\ y(0) = 0 \end{cases}$

Solution. 2.192

9. $\begin{cases} y' = y^{3/2} + t, \\ y(0) = 1 \end{cases}$

Solution. 4.256

10. $\begin{cases} y' = \sin(t) + \cos(y), \\ y(0) = 1 \end{cases}$

Solution. 1.688

11. $\begin{cases} y' = e^{t^3}, \\ y(0) = 1 \end{cases}$

Solution. 2.263

12. Consider the initial value problem

$$\begin{cases} y' = t^2 y, \\ y(0) = 1. \end{cases}$$

a. Find the exact solution at $t = 1.0$. Express this value to four decimal places.

Solution. $\begin{cases} y(t) = (e^{t^3})^{1/3}, \\ y(1.0) = 1.396. \end{cases}$

b. Use the Euler method with $h = 1/8$ to approximate the solution at $t = 1.0$. Compute the absolute error.

c. Repeat part b with $h = 1/16, h = 1/32$. Create a table and a graph showing the absolute errors corresponding to the various step sizes. A theoretical analysis for the Euler method suggests a linear relationship between the absolute error and the step size. Do the numbers agree with the theory?

d. Observe in part c that the error is roughly proportional to the step size. Use these data to estimate the constant of proportionality.

Solution.

h	t_n	Y_n	$|\text{error}(t_n)|$	$|\text{error}(t_n)|/h$
1/8	1.0	1.303	0.0922	0.7380
1/16	1.0	1.347	0.0490	0.7846
1/32	1.0	1.370	0.0253	0.8102

13. Apply the Euler method with successively smaller step sizes on the interval $0 \le t \le 1$ to verify empirically that the solution of the initial value problem

$$\begin{cases} y' = t^2 + y^2, \\ y(0) = 1. \end{cases}$$

has a vertical asymptote near $t = 0.97$.

Solution.

h	t_n	Y_n
1/10	1.0	7.19
1/20	1.0	12.32
1/30	1.0	17.8
1/40	1.0	23.93
1/50	1.0	30.92
1/60	1.0	39.03
1/70	1.0	48.59
1/80	1.0	59.99
1/90	1.0	73.8
1/100	1.0	90.76

Section 13.2, page 417

In the following exercises, perform the indicated computations by hand, retaining only four significant digits at each step of the calculations.

For a first-order initial value problem defined by

$$\begin{cases} y' = f(t, y), \\ y(t_0) = y_0, \end{cases}$$

if h denotes the step size, all problems in Section 13.2 use the Heun method, defined by the following recursive relation:

$$\begin{cases} Y_0 = y_0, \\ Y_{n+1} = Y_n + \dfrac{h}{2}\left(f(t_n, Y_n) + f\left(t_{n+1}, Y_n + h\, f(t_n, Y_n)\right) \right). \end{cases}$$

1. Use the Heun method to compute an approximate solution to the initial value problem $y' = y + 1$, $y(0) = 0$ with step size $h = 0.1$. Compare with the exact solution $y(t) = e^t - 1$ and with the Euler method approximation from

Solution.

n	t_n	Y_n	$y_{exact}(t_n)$
0	0.0	0.0	0.0
1	0.1	0.105	0.1052
2	0.2	0.221	0.2214
3	0.3	0.3492	0.3499
4	0.4	0.4909	0.4918
5	0.5	0.6474	0.6487
6	0.6	0.8204	0.8221
7	0.7	1.012	1.014
8	0.8	1.223	1.226
9	0.9	1.456	1.460
10	1.0	1.714	1.718

2. Consider the following initial value problem

$$\begin{cases} y' = y - t y^2, \\ y(0) = 2. \end{cases}$$

   a. Use the Heun algorithm to compute an approximate solution at $t = 1.0$. Use the values $n = 2, 4, 8$, which correspond to step sizes of $h = 1/2, 1/4, 1/8$.

   b. Determine the actual solution and compare the value of $y(1)$ with the results from part a.

Solution.

$$y(t) = \frac{2 e^t}{3 - 2 e^t + 2 t e^t}$$

h	t_n	Y_n	$y_{exact}(t_n)$
1/2	1.0	1.547	1.812
1/4	1.0	1.787	1.812
1/8	1.0	1.809	1.812

Repeat Exercise 2 for the following initial value problems:

3. $$\begin{cases} y' + 2y = t, \\ y(0) = 1 \end{cases}$$

Solution.

$$y(t) = \frac{5e^{2t}}{4} - \frac{t}{2} - \frac{1}{4}$$

h	t_n	Y_n	$y_{exact}(t_n)$
1/2	1.0	0.5625	0.4192
1/4	1.0	0.4407	0.4192
1/8	1.0	0.4235	0.4192

4. $\begin{cases} y' = t+1, \\ y(0) = 1 \end{cases}$

Solution.

$$y(t) = 1 + t + \frac{t^2}{2}$$

h	t_n	Y_n	$y_{exact}(t_n)$
1/2	1.0	2.5	2.5
1/4	1.0	2.5	2.5
1/8	1.0	2.5	2.5

5. $\begin{cases} y' = t^3 e^{-2y}, \\ y(0) = 0 \end{cases}$

Solution.

$$y(t) = \log\left(\frac{\sqrt{2+t^4}}{\sqrt{2}}\right)$$

h	t_n	Y_n	$y_{exact}(t_n)$
1/2	1.0	0.2694	0.2027
1/4	1.0	0.2156	0.2027
1/8	1.0	0.2055	0.2027

6. $\begin{cases} y' = \dfrac{e^t}{y}, \\ y(0) = 1 \end{cases}$

Solution.

$$y(t) = e^{e^t - 1}$$

h	t_n	Y_n	$y_{exact}(t_n)$
1/2	1.0	2.124	2.106
1/4	1.0	2.111	2.106
1/8	1.0	2.107	2.106

7. Consider the following initial value problem

$$\begin{cases} y' = y + t^2, \\ y(0) = 1. \end{cases}$$

a. Find the solution $y(t)$ and evaluate it for $t = 0.2, 0.4, \ldots, 1.0$.

b. Using the Heun method, with step size of $h = 0.2$, find approximate values for the solution at the t values in part **a**.

c. Repeat part **b** using $h = 0.1$.

d. Compare the results of part **b** with those of part **c** and the exact values. (The differences in the results for $h = 0.2$ and $h = 0.1$ tend to indicate whether a smaller step size must be used for the desired range of t values. The general rule of thumb is to use the smaller step size if the two solutions agree to the desired accuracy. If they do not agree, reduce h and repeat the calculations. This gives an indication but not a proof of the accuracy of the result.)

Solution.

t_n	$Y_n (h = 0.2)$	$Y_n (h = 0.1)$	$y_{\text{exact}}(t_n)$
0.0	1.0	1.0	1.0
0.1		1.105	1.106
0.2	1.224	1.224	1.224
0.3		1.359	1.36
0.4	1.514	1.515	1.515
0.5		1.695	1.696
0.6	1.902	1.905	1.906
0.7		2.15	2.151
0.8	2.428	2.434	2.437
0.9		2.765	2.769
1.0	3.139	3.15	3.155

$y(t) = -2 + 3e^t - 2t - t^2$

The initial value problems in the following exercises cannot be solved by exact methods. Use the Heun method with step size $h = 0.1$ to approximate the solution at $t = 1$ to three significant digits (see Exercise 7).

8.
$$\begin{cases} y' = \sin(y) + e^t, \\ y(0) = 0 \end{cases}$$

Solution. 2.332

9.
$$\begin{cases} y' = y^{3/2} + t, \\ y(0) = 1 \end{cases}$$

Solution. 5.286

10.
$$\begin{cases} y' = \sin(t) + \cos(y), \\ y(0) = 1 \end{cases}$$

Solution. 1.692

11. $\begin{cases} y' = e^{t^3}, \\ y(0) = 1 \end{cases}$

Solution. 2.349

12. Consider the initial value problem

$$\begin{cases} y' = t^2 y, \\ y(0) = 1. \end{cases}$$

    a. Find the exact solution at $t = 1.0$. Express this value to four decimal places.

Solution. $\begin{cases} y(t) = (e^{t^3})^{1/3}, \\ y(1.0) = 1.396. \end{cases}$

    b. Use the Heun method with $h = 1/8$ to approximate the solution at $t = 1.0$. Compute the absolute error.

    c. Repeat part b with $h = 1/16$, $h = 1/32$. Create a table and a graph showing the absolute errors corresponding to the various step sizes. A theoretical analysis for the Heun method suggests a linear relationship between the absolute error and the square of the step size. Do the numbers agree with the theory?

    d. Observe in part c that the error is roughly proportional to the square of the step size. Use the data to estimate the constant of proportionality.

Solution.

h	t_n	Y_n	$|\text{error}(t_n)|$	$|\text{error}(t_n)|/h^2$
1/8	1.0	1.398	0.002739	0.1753
1/16	1.0	1.396	0.0007293	0.1867
1/32	1.0	1.396	0.0001884	0.1929

13. Apply the Heun method with successively smaller step sizes on the interval $0 \le t \le 1$ to verify empirically that the solution of the initial value problem

$$\begin{cases} y' = t^2 + y^2, \\ y(0) = 1. \end{cases}$$

has a vertical asymptote near $t = 0.97$.

Solution.

h	t_n	Y_n
1/10	1.0	3.813×10^1
1/20	1.0	1.440×10^2
1/30	1.0	5.562×10^2
1/40	1.0	3.015×10^3
1/50	1.0	3.103×10^4
1/60	1.0	9.073×10^5
1/70	1.0	1.355×10^8
1/80	1.0	2.458×10^{11}
1/90	1.0	1.993×10^{16}
1/100	1.0	5.313×10^{23}

Section 13.3, page 422

In the following exercises, perform the indicated computations by hand, retaining only four significant digits at each step of the calculations.

For a first-order initial value problem defined by

$$\begin{cases} y' = f(t, y), \\ y(t_0) = y_0, \end{cases}$$

if h denotes the step size, all problems in Section 13.3 use the Runge-Kutta method, defined by the following recursive relation:

$$\begin{cases} Y_0 = y_0, \\ Y_{n+1} = Y_n + \dfrac{h}{6}(a_{1,n} + 2a_{2,n} + 2a_{3,n} + a_{4,n}), \end{cases}$$

where

$$a_{1,n} = f(t_n, Y_n),$$

$$a_{2,n} = f\left(t_n + \frac{h}{2}, Y_n + \frac{h}{2}a_{1,n}\right),$$

$$a_{3,n} = f\left(t_n + \frac{h}{2}, Y_n + \frac{h}{2}a_{2,n}\right),$$

$$a_{4,n} = f(t_n + h, Y_n + h\,a_{3,n}).$$

1. Use the Runge-Kutta method to compute an approximate solution to the initial value problem $y' = y + 1$, $y(0) = 0$ with step size $h = 0.1$. Compare with the exact solution $y(t) = e^t - 1$ and with the Euler method approximation from Example 13.1.

Solution.

n	t_n	Y_n	$y_{exact}(t_n)$
0	0.0	0.0	0.0
1	0.1	0.1052	0.1052
2	0.2	0.2214	0.2214
3	0.3	0.3499	0.3499
4	0.4	0.4918	0.4918
5	0.5	0.6487	0.6487
6	0.6	0.8221	0.8221
7	0.7	1.014	1.014
8	0.8	1.226	1.226
9	0.9	1.46	1.460
10	1.0	1.718	1.718

2. Consider the following initial value problem

$$\begin{cases} y' = y - t y^2, \\ y(0) = 2. \end{cases}$$

    a. Use the Runge-Kutta algorithm to compute an approximate solution at $t = 1.0$. Use the values $n = 2, 4, 8$, which correspond to step sizes of $h = 1/2, 1/4, 1/8$.

    b. Determine the actual solution and compare the value of $y(1)$ with the results from part **a**.

Solution.

$$y(t) = \frac{2 e^t}{3 - 2 e^t + 2 t e^t}$$

h	t_n	Y_n	$y_{exact}(t_n)$
1/2	1.0	1.793	1.812
1/4	1.0	1.812	1.812
1/8	1.0	1.812	1.812

Repeat Exercise 1 for the following initial value problems:

3. $$\begin{cases} y' + 2y = t, \\ y(0) = 1 \end{cases}$$

Solution.

$$y(t) = \frac{5 e^{2t}}{4} - \frac{t}{2} - \frac{1}{4}$$

h	t_n	Y_n	$y_{exact}(t_n)$
1/2	1.0	0.4258	0.4192
1/4	1.0	0.4194	0.4192
1/8	1.0	0.4192	0.4192

4. $\begin{cases} y' = t + 1, \\ y(0) = 1 \end{cases}$

Solution.

$y(t) = 1 + t + \dfrac{t^2}{2}$

h	t_n	Y_n	$y_{exact}(t_n)$
1/2	1.0	2.5	2.5
1/4	1.0	2.5	2.5
1/8	1.0	2.5	2.5

5. $\begin{cases} y' = t^3 e^{-2y}, \\ y(0) = 0 \end{cases}$

Solution.

$y(t) = \log\left(\dfrac{\sqrt{2 + t^4}}{\sqrt{2}}\right)$

h	t_n	Y_n	$y_{exact}(t_n)$
1/2	1.0	0.2036	0.2027
1/4	1.0	0.2028	0.2027
1/8	1.0	0.2027	0.2027

6. $\begin{cases} y' = \dfrac{e^t}{y}, \\ y(0) = 1 \end{cases}$

Solution.

$y(t) = e^{e^t - 1}$

h	t_n	Y_n	$y_{exact}(t_n)$
1/2	1.0	2.106	2.106
1/4	1.0	2.106	2.106
1/8	1.0	2.106	2.106

7. Consider the following initial value problem

$$\begin{cases} y' = y + t^2, \\ y(0) = 1. \end{cases}$$

a. Find the solution $y(t)$ and evaluate it for $t = 0.2, 0.4, \ldots, 1.0$.

b. Using the Runge-Kutta method, with step size of $h = 0.2$, find approximate values for the solution at the t values in part **a**.

c. Repeat part **b** using $h = 0.1$.

d. Compare the results of part **b** with those of part **c** and the exact values. (The differences in the results for $h = 0.2$ and $h = 0.1$ tend to indicate whether a smaller step size must be used for the desired range of t values. The general rule of thumb is to use the smaller step size if the two solutions agree to the desired accuracy. If they do not agree, reduce h and repeat the calculations. This gives an indication but not a proof of the accuracy of the result.)

Solution.

t_n	$Y_n(h = 0.2)$	$Y_n(h = 0.1)$	$y_{exact}(t_n)$
0.0	1.0	1.0	1.0
0.1		1.106	1.106
0.2	1.224	1.224	1.224
0.3		1.36	1.36
0.4	1.515	1.515	1.515
0.5		1.696	1.696
0.6	1.906	1.906	1.906
0.7		2.151	2.151
0.8	2.437	2.437	2.437
0.9		2.769	2.769
1.0	3.155	3.155	3.155

$y(t) = -2 + 3e^t - 2t - t^2$

The initial value problems in the following exercises cannot be solved by exact methods. Use the Runge-Kutta method with step size $h = 0.1$ to approximate the solution at $t = 1$ to three significant digits (see Exercise 7).

8.
$$\begin{cases} y' = \sin(y) + e^t, \\ y(0) = 0 \end{cases}$$

Solution. 2.336

9.
$$\begin{cases} y' = y^{3/2} + t, \\ y(0) = 1 \end{cases}$$

Solution. 5.372

10.
$$\begin{cases} y' = \sin(t) + \cos(y), \\ y(0) = 1 \end{cases}$$

Solution. 1.692

11.
$$\begin{cases} y' = e^{t^3}, \\ y(0) = 1 \end{cases}$$

Solution. 2.342

12. Consider the initial value problem

$$\begin{cases} y' = t^2 y, \\ y(0) = 1. \end{cases}$$

a. Find the exact solution at $t = 1.0$. Express this value to four decimal places.

Solution. $\begin{cases} y(t) = (e^{t^3})^{1/3}, \\ y(1.0) = 1.396. \end{cases}$

b. Use the Runge-Kutta method with $h = 1/8$ to approximate the solution at $t = 1.0$. Compute the absolute error.

c. Repeat part **b** with $h = 1/16$, $h = 1/32$. Create a table and a graph showing the absolute errors corresponding to the various step sizes. A theoretical analysis for the Runge-Kutta method suggests a linear relationship between the absolute error and the forth power of the step size. Do the numbers agree with the theory?

d. Observe in part **c** that the error is roughly proportional to the forth power of the step size. Use the data to estimate the constant of proportionality.

Solution.

h	t_n	Y_n	$|\text{error}(t_n)|$	$|\text{error}(t_n)|/h^4$
1/8	1.0	1.396	$5.568\ 10^{-7}$	0.002281
1/16	1.0	1.396	$2.607\ 10^{-8}$	0.001708
1/32	1.0	1.396	$1.328\ 10^{-9}$	0.001392

13. Apply the Runge-Kutta method with successively smaller step sizes on the interval $0 \le t \le 1$ to verify empirically that the solution of the initial value problem

$$\begin{cases} y' = t^2 + y^2, \\ y(0) = 1. \end{cases}$$

has a vertical asymptote near $t = 0.97$.

Solution.

h	t_n	Y_n
1/10	1.0	3.813×10^1
1/20	1.0	1.440×10^2
1/30	1.0	5.562×10^2
1/40	1.0	3.015×10^3
1/50	1.0	3.103×10^4
1/60	1.0	9.073×10^5
1/70	1.0	1.355×10^8
1/80	1.0	2.458×10^{11}
1/90	1.0	1.993×10^{16}
1/100	1.0	5.313×10^{23}

14. This exercise describes the relation between the Runge-Kutta formula (13.17) and Simpson's rule.

    a. Suppose that $f(t)$ is a continuous function for $t_0 \leq t \leq t_0 + h$. Find a quadratic polynomial $y = a t^2 + bt + c$ which agrees with $f(t)$ at the points t_0, $t_0 + h/2$ and $t_0 + h$.

    b. Find the integral of this polynomial: $A = \int_{t_0}^{t_0+h} (a t^2 + bt + c)\, dt$.

    c. Show that

$$A = \frac{h(f(t_0) + 4 f(t_0 + h/2) + f(t_0 + h))}{6},$$

consistent with (13.17).

Solution. Solving the following system of linear equations

$$\begin{cases} a t_0^2 + b t_0 + c = f(t_0), \\ a(t_0 + h/2)^2 + b(t_0 + h/2) + c = f(t_0 + h/2), \\ a(t_0 + h)^2 + b(t_0 + h) + c = f(t_0 + h), \end{cases}$$

we obtain the following values for the coefficients of a quadratic polynomial

$$\begin{cases} a = \dfrac{2\,(f(t_0) - 2 f(h/2 + t_0) + f(h + t_0))}{h^2}, \\[3mm] b = \dfrac{-f(t_0) + f(h + t_0)}{h} - \dfrac{2\,(h + 2 t_0)\,(f(t_0) - 2 f(h/2 + t_0) + f(h + t_0))}{h^2}, \\[3mm] c = \dfrac{h^2 f(t_0) + 3 h t_0 f(t_0) + 2 t_0^2 f(t_0) - 4 h t_0 f(h/2 + t_0) - 4 t_0^2 f(h/2 + t_0)}{h^2} \\[3mm] \qquad + \dfrac{h t_0 f(h + t_0) + 2 t_0^2 f(h + t_0)}{h^2}. \end{cases}$$

Integrating $a t^2 + b t + c$ over the interval $t_0 < t_0 + h$ produces the desired result.

Section 13.5, page 433

Use **ODE** with **Method->Euler** to plot the numerical solutions of the following differential equations. In each case use a step size $h = 0.01$ and do 100 steps. Try to find an exact solution for each problem using **ODE**. For those problems for which **ODE** yields an exact solution, compare the exact solution with the solution found by the Euler method.

1.
$$\begin{cases} y' = 1 - t\,y^{10}, \\ y(0) = 1 \end{cases}$$

Solution. The numerical solution is found with the command

```
ODE[{y' == 1 - t y^10,y[0] == 1},y,{t,0,1},
Method->Euler,StepSize->0.01,PlotSolution->{{t,0,1}}];
```

which produces

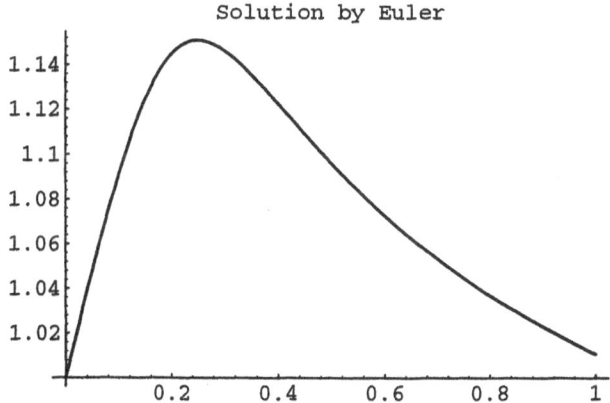
Solution by Euler

ODE cannot find an exact solution.

2.
$$\begin{cases} y' = \sin(t\,y), \\ y(0) = 1 \end{cases}$$

Solution. The numerical solution is found with the command

```
ODE[{y' == Sin[t y],y[0] == 1},y,{t,0,1},
Method->Euler,StepSize->0.01,PlotSolution->{{t,0,1}}];
```

which produces

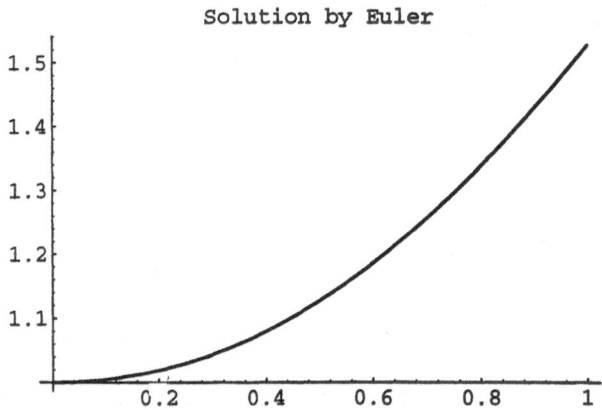

Solution by Euler

ODE cannot find an exact solution.

3. $\begin{cases} y' = t - \sin(y), \\ y(0) = 1 \end{cases}$

Solution. The numerical solution is found with the command

```
ODE[{y' == t - y,y[0] == 1},y,{t,0,1},
Method->Euler,StepSize->0.01,
GraphLabel->EX3Euler,PlotSolution->{{t,0,1},
PlotStyle->{{RGBColor[1,0,0]}}}];
```

The exact solution is found with the command

```
ODE[{y' == t - y,y[0] == 1},y,t,
Method->FirstOrderLinear,
GraphLabel->EX3Exact,PlotSolution->{{t,0,1},
PlotStyle->{{RGBColor[0,1,0]}}}];
```

$$y(t) = t - 1 + 2e^{-t}.$$

The two plots can be combined with

```
Show[Graph[EX3Exact],Graph[EX3Euler],PlotLabel->None];
```

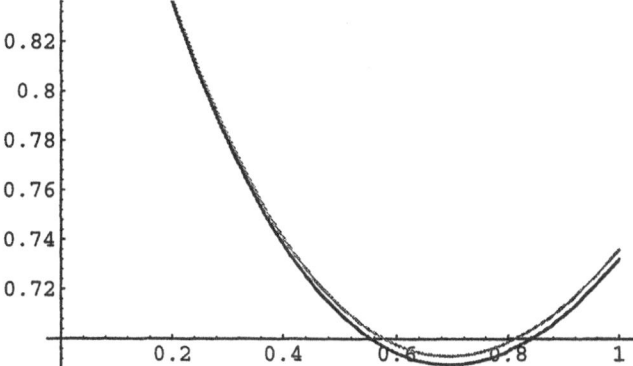

4. $\begin{cases} y' = \dfrac{t}{y^5} - y, \\ y(0) = 1 \end{cases}$

Solution. The numerical solution is found with the command

```
ODE[{y' == t/y^5 - y,y[0] == 1},y,{t,0,1},
Method->Euler,StepSize->0.01,
GraphLabel->EX4Euler,PlotSolution->{{t,0,1},
PlotStyle->{{RGBColor[1,0,0]}}}];
```

The exact solution is found with the command

```
ODE[{y' == t/y^5 - y,y[0] == 1},y,{t,0,1},
Method->Bernoulli,StepSize->0.01,
GraphLabel->EX4Exact,PlotSolution->{{t,0,1},
PlotStyle->{{RGBColor[0,1,0]}}}];
```

$$y(t) = 6^{-1/6}(6t - 1 + 7e^{-6t})^{1/6}.$$

The two plots can be combined with

```
Show[Graph[EX4Exact],Graph[EX4Euler],PlotLabel->None];
```

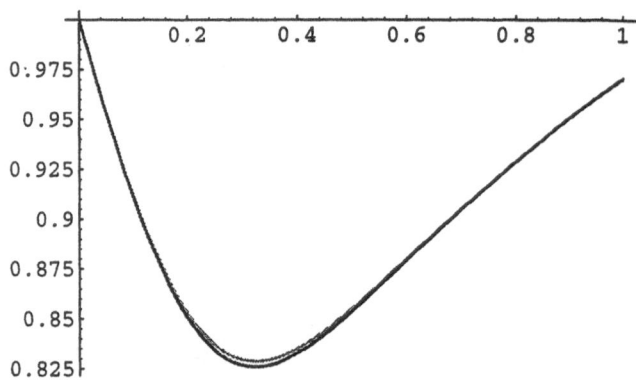

Use *Mathematica* to plot the numerical solutions of the following differential equations using the Euler, Heun and Runge-Kutta methods. In each problem use the indicated step size **h** and number of steps **steps**, and combine the 3 graphs.

5. $y' = 1 - t\,y^{10}$, $\qquad y(0) = 1 \qquad$ **h=0.1, steps = 10**.

Solution. The three commands to produce the numerical solutions are

```
ODE[{y' == 1 - t y^10,y[0] == 1},y,{t,0,1},
Method->Euler,StepSize->0.1,GraphLabel->EX5Euler,
PlotSolution->{{t,0,1},
PlotStyle->{{RGBColor[1,0,0]}}}];

ODE[{y' == 1 - t y^10,y[0] == 1},y,{t,0,1},
Method->Heun,StepSize->0.1,GraphLabel->EX5Heun,
PlotSolution->{{t,0,1},
PlotStyle->{{RGBColor[0,1,0]}}}];

ODE[{y' == 1 - t y^10,y[0] == 1},y,{t,0,1},
Method->RungeKutta4,StepSize->0.1,GraphLabel->EX5RK,
PlotSolution->{{t,0,1},
PlotStyle->{{RGBColor[0,0,1]}}}];
```

The three plots can be combined with

```
Show[{Graph[EX5Euler],Graph[EX5Heun],Graph[EX5RK]},
PlotLabel->None];
```

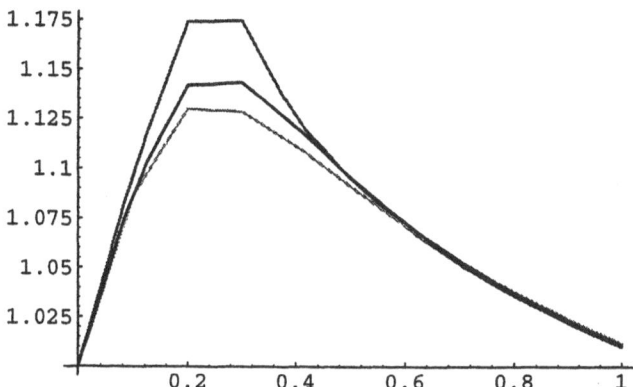

6. $y' = \sin(t\, y)$, $y(0) = 1$, **h=0.1, steps = 10**.

Solution. The three commands to produce the numerical solutions are

```
ODE[{y' == Sin[t y],y[0] == 1},y,{t,0,2},
Method->Euler,StepSize->0.2,GraphLabel->EX6Euler,
PlotSolution->{{t,0,2},
PlotStyle->{{RGBColor[1,0,0]}}}];


ODE[{y' == Sin[t y],y[0] == 1},y,{t,0,2},
Method->Heun,StepSize->0.2,GraphLabel->EX6Heun,
PlotSolution->{{t,0,2},
PlotStyle->{{RGBColor[0,1,0]}}}];


ODE[{y' == Sin[t y],y[0] == 1},y,{t,0,2},
Method->RungeKutta4,StepSize->0.2,GraphLabel->EX6RK,
PlotSolution->{{t,0,2},
PlotStyle->{{RGBColor[0,0,1]}}}];
```

The three plots can be combined with

```
Show[{Graph[EX6Euler],Graph[EX6Heun],Graph[EX6RK]},
PlotLabel->None];
```

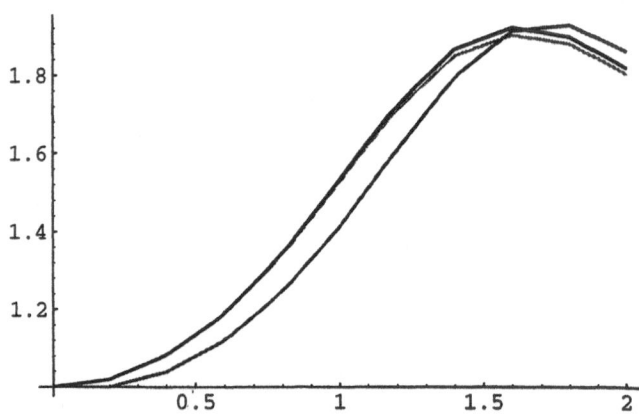

7. $y' = t - \sin(y)$, $\quad y(0) = 1$, $\quad$ **h=0.1, steps = 10**.

Solution. The three commands to produce the numerical solutions are

```
ODE[{y' == t - Sin[y],y[0] == 1},y,{t,0,2},
Method->Euler,StepSize->0.2,GraphLabel->EX7Euler,
PlotSolution->{{t,0,2},
PlotStyle->{{RGBColor[1,0,0]}}}];


ODE[{y' == t - Sin[y],y[0] == 1},y,{t,0,2},
Method->Heun,StepSize->0.2,GraphLabel->EX7Heun,
PlotSolution->{{t,0,2},
PlotStyle->{{RGBColor[0,1,0]}}}];


ODE[{y' == t - Sin[y],y[0] == 1},y,{t,0,2},
Method->RungeKutta4,StepSize->0.2,GraphLabel->EX7RK,
PlotSolution->{{t,0,2},
PlotStyle->{{RGBColor[0,0,1]}}}];
```

The three plots can be combined with

```
Show[{Graph[EX7Euler],Graph[EX7Heun],Graph[EX7RK]},
PlotLabel->None];
```

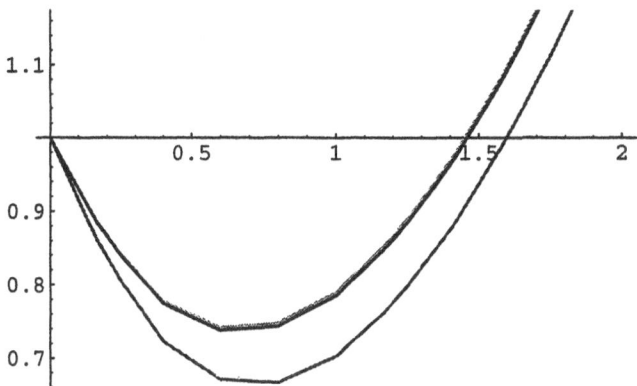

8. $y' = t/y^5 - y$, $y(0) = 1$,   h=0.1, steps = 15.

Solution. The three commands to produce the numerical solutions are

```
ODE[{y' == t/y^5 - y,y[0] == 1},y,{t,0,3},
Method->Euler,StepSize->0.2,GraphLabel->EX8Euler,
PlotSolution->{{t,0,3},
PlotStyle->{{RGBColor[1,0,0]}}}];


ODE[{y' == t - Sin[y],y[0] == 1},y,{t,0,2},
Method->Heun,StepSize->0.2,GraphLabel->EX8Heun,
PlotSolution->{{t,0,2},
PlotStyle->{{RGBColor[0,1,0]}}}];


ODE[{y' == t/y^5 - y,y[0] == 1},y,{t,0,3},
Method->RungeKutta4,StepSize->0.2,GraphLabel->EX8RK,
PlotSolution->{{t,0,3},
PlotStyle->{{RGBColor[0,0,1]}}}];
```

The three plots can be combined with

```
Show[{Graph[EX8Euler],Graph[EX8Heun],Graph[EX8RK]},
PlotLabel->None];
```

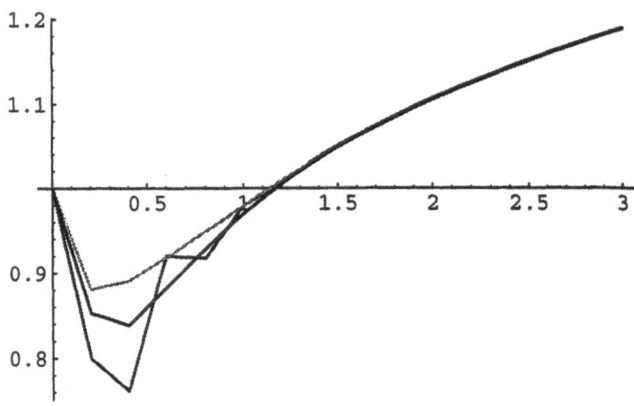

For each of the initial value Problems 9–12, use *Mathematica* with the Euler, Heun and Runge-Kutta methods to obtain a six-decimal approximation to $y(0.5)$. In each problem use the step size $h = 0.1$.

9. $\begin{cases} y' = e^{-y}, \\ y(0) = 0 \end{cases}$

Solution. We name the equation with

```
eq = {y' == Exp[-y],y[0] == 0};
```

Then

```
Last[ODE[eq,y,{t,0,0.5},Method->Euler,StepSize->0.1,
    ODEDigits->6]]
Last[ODE[eq,y,{t,0,0.5},Method->Heun,StepSize->0.1,
    ODEDigits->6]]
Last[ODE[eq,y,{t,0,0.5},Method->RungeKutta4,StepSize->0.1,
    ODEDigits->6]]
```

tells us the following approximate values for $y(0.5)$:

```
Euler      = 0.419761
Heun       = 0.405281
RungeKutta = 0.405465
```

10. $\begin{cases} y' = t + y^2, \\ y(0) = 0 \end{cases}$

Solution. We name the equation with

```
eq = {y' == t + y^2,y[0] == 0};
```

The command used in Exercise 9 tells us the following approximate values for $y(0.5)$:

```
Euler        = 0.100461
Heun         = 0.126556
RungeKutta = 0.126588
```

11. $\begin{cases} y' = t\,y + \sqrt{y}, \\ y(0) = 1 \end{cases}$

Solution. We name the equation with

```
eq = {y' == t y + Sqrt[y],y[0] == 1};
```

The command used in Exercise 9 tells us the following approximate values for $y(0.5)$:

```
Euler        = 1.69024
Heun         = 1.75569
RungeKutta = 1.75609
```

12. $\begin{cases} y' = y - y^2, \\ y(0) = 0.5 \end{cases}$

Solution. We name the equation with

```
eq = {y' == y - y^2,y[0] == 0.5};
```

The command used in Exercise 9 tells us the following approximate values for $y(0.5)$:

```
Euler        = 0.623148
Heun         = 0.622407
RungeKutta = 0.622459
```

Section 13.6, page 437

In Exercises 1–10 plot the solutions to following initial value problems using **ODE** with the option **Method->NDSolve**. [Hint: It may be necessary to use *Mathematica's* **?** to determine the name and syntax of certain functions.]

1. $\begin{cases} y' = \sin(t\,y^2)/(t - 21), \\ y(0) = -1 \end{cases}$

Solution.

```
ODE[{y' == Sin[t y^2]/(t - 21),y[0] == -1},y,{t,0,20},
Method->NDSolve,PlotSolution->{{t,0,20}}];
```

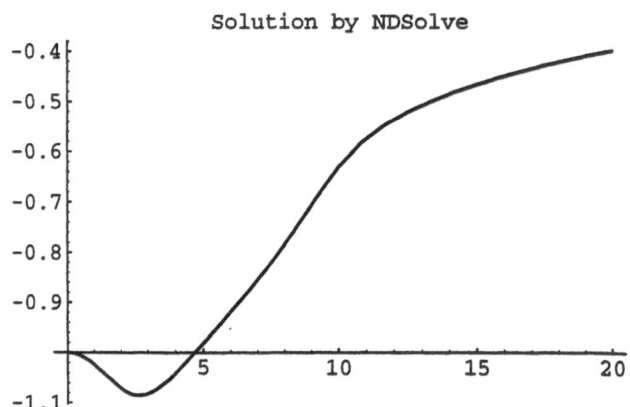

$$2. \quad \begin{cases} y' = 1/t + 1/y, \\ y(1) = 1 \end{cases}$$

Solution.

```
ODE[{y' == 1/t + 1/y,y[1] == 1},y,{t,1,10},
Method->NDSolve,PlotSolution->{{t,1,10}}];
```

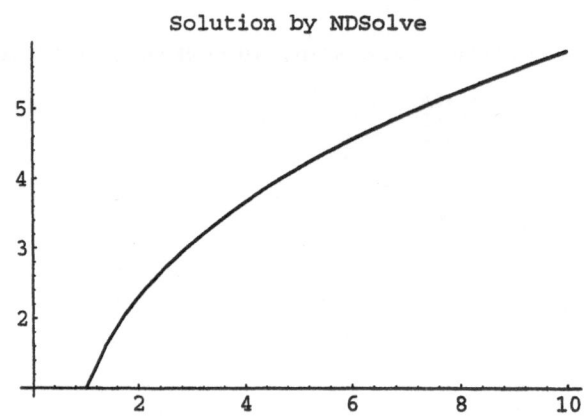

$$3. \quad \begin{cases} y' = t\,y/(t^2 - 1), \\ y(0) = 1 \end{cases}$$

Solution.

```
ODE[{y' == t y/(t^2 - 1),y[0] == 1},y,{t,0,2},
Method->NDSolve,PlotSolution->{{t,0,2}}];
```

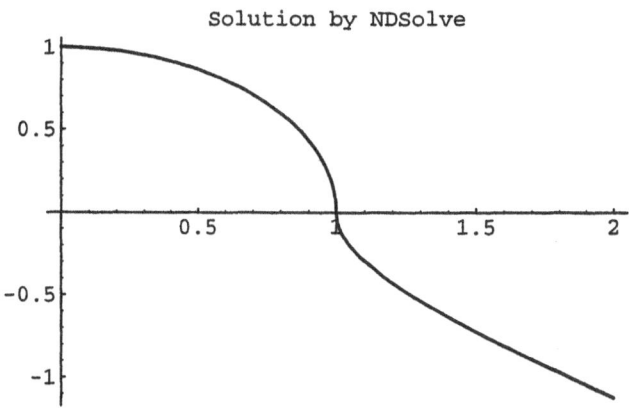

Solution by NDSolve

4. $\begin{cases} y' = (t - \sqrt{1 - y^2})/y, \\ y(1) = 1 \end{cases}$

Solution.

```
ODE[{y' == (t-Sqrt[1 - y^2])/y,y[1] == 1},y,{t,0,2},
Method->NDSolve,PlotSolution->{{t,0,2}}];
```

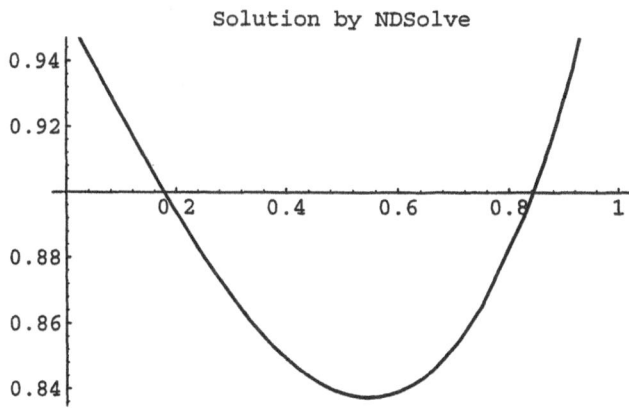

Solution by NDSolve

5. $\begin{cases} y' = \log\left(\sqrt{t^2 + y^2}\right), \\ y(0) = 1 \end{cases}$

Solution.

```
ODE[{y' == Log[Sqrt[t^2 + y^2]],y[0] == 1},y,{t,0,5},
Method->NDSolve,PlotSolution->{{t,0,5}}];
```

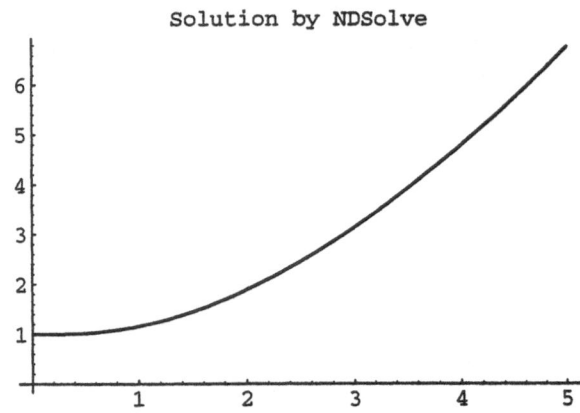

6. $\begin{cases} y' = t^2 + y^2, \\ y(0) = 1 \end{cases}$

Solution.

```
ODE[{y' == t^2 + y^2,y[0] == 1},y,{t,0,1},
Method->NDSolve,PlotSolution->{{t,0,1}}];
```

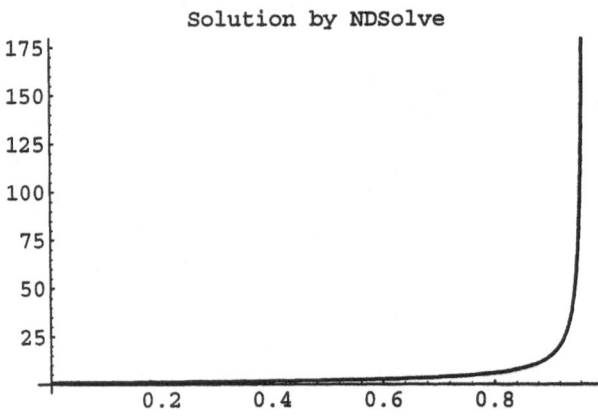

7. $\begin{cases} y' = \max(t, y), \\ y(0) = 0 \end{cases}$

Solution.

```
ODE[{y' == Max[t,y],y[0] == 0},y,{t,0,1},
Method->NDSolve,PlotSolution->{{t,0,1}}];
```

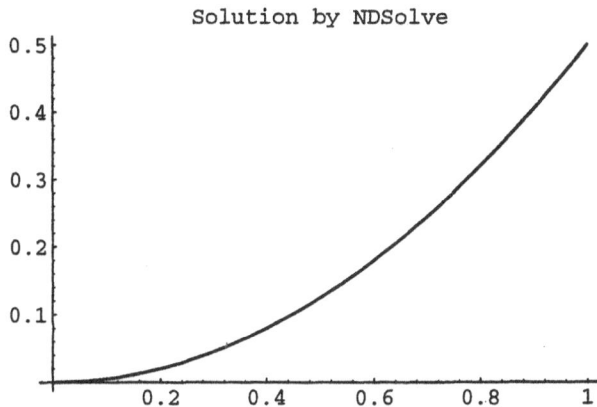

Solution by NDSolve

8. $\begin{cases} y' = (y-t)^2, \\ y(0) = 0 \end{cases}$

Solution.

```
ODE[{y' == (y - t)^2,y[0] == 0},y,{t,0,5},
Method->NDSolve,PlotSolution->{{t,0,5}}];
```

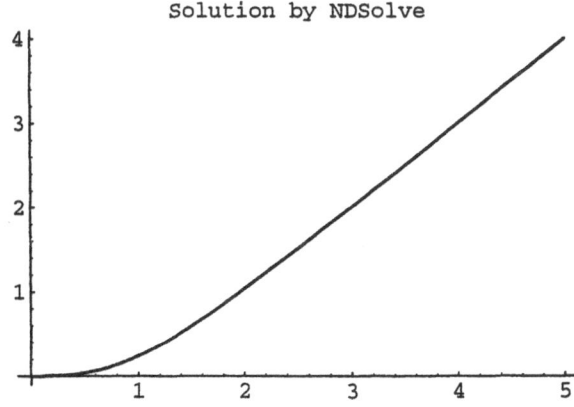

Solution by NDSolve

9. $\begin{cases} y' = t^2/|y|, \\ y(1) = 1 \end{cases}$

Solution.

```
ODE[{y' == t^2/Abs[y],y[1] == 1},y,{t,1,5},
Method->NDSolve,PlotSolution->{{t,1,5}}];
```

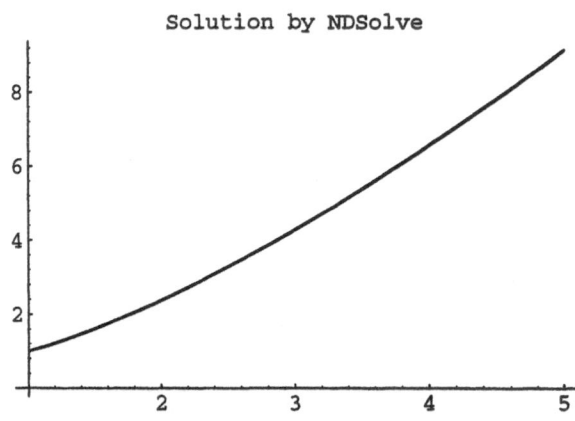

Solution by NDSolve

10. $\begin{cases} y' = \text{sign}(t\,y)\sqrt{5\log(|y|)}, \\ y(1) = 2 \end{cases}$

Solution.

```
ODE[{y' == Sign[t y]Sqrt[5Log[Abs[y]]],y[1] == 2},y,
{t,1,5},Method->NDSolve,PlotSolution->{{t,1,5}}];
```

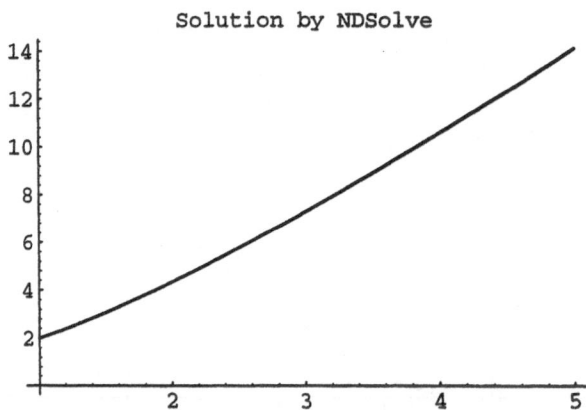

Solution by NDSolve

11. Consider the initial value problem $y' = t^2 + y^2$, $y(0) = 1$. Use **ODE** and adjust the necessary **NDSolve** options to find the first positive vertical asymptote correct to six digits.

Solution.

```
ODE[{y' == t^2 + y^2,y[0] == 1},y,{t,0,1},
Method->NDSolve,MaxSteps->1000,
PlotSolution->{{t,0,1}}];
```

12. Consider the initial value problem $y' = 1/(t^2 + y^2)$, $y(0) = 1$. Use **ODE** with the option **Method->NDSolve** to estimate the set of $t > 0$ for which $y' < 10^{-3}$, sometimes referred to as the pseudo steady-state region.

Solution.

```
f[t_] = ODE[{y' == 1/(t^2 + y^2),y[0] == 1},y,{t,0,100},
        Form->Explicit,Method->NDSolve,MaxSteps->1000];
Plot[Evaluate[f'[t]],{t,0,10}];
FindRoot[Evaluate[f'[t]] == 10^(-3),{t,10,100}]
```

NDSolve can also be used to solve higher-order differential equation numerically. Use **ODE** with the option **Method->NDSolve** to solve the following second-order initial value problems. Plot each solution on an interval containing the initial point.

13. $\begin{cases} y'' = t\,y, \\ y(0) = 1, \ y'(0) = 1 \end{cases}$

Solution.

```
ODE[{y'' == t y,y[0] == 1,y'[0] == 1},y,{t,0,1},
Method->NDSolve,PlotSolution->{{t,0,1}}];
```

14. $\begin{cases} t^2 y'' + t\,y' + (t^2 - 9)y = 0, \\ y(1) = 1, \ y'(1) = 1 \end{cases}$

Solution.

```
ODE[{t^2 y'' + t y' + (t^2 - 9)y == 0,y[1] == 0,
y'[1] == 1},y,{t,1,20},Method->NDSolve,
MaxSteps->1000,PlotSolution->{{t,1,20}}];
```

15. $\begin{cases} t^2 y'' + t\,y' - (t^2 + 4)y = 0, \\ y(1) = 0, \ y'(1) = 1 \end{cases}$

Solution.

```
ODE[{t^2 y'' + t y' - (t^2 + 4)y == 0,y[1] == 0,
y'[1] == 1},y,{t,1,2},Method->NDSolve,
PlotSolution->{{t,1,2}}];
```

16.
$$\begin{cases} y'' + \sin(y) = 0, \\ y(0) = 1,\ y'(0) = 0 \end{cases}$$

Solution.

```
ODE[{y'' + Sin[y] == 0,y[0] == 1,y'[0] == 0},y,
{t,0,3Pi},Method->NDSolve,
PlotSolution->{{t,0,3Pi}}];
```

17.
$$\begin{cases} y'' - t\,y' + 7y = 0, \\ y(0) = 1,\ y'(0) = 1 \end{cases}$$

Solution.

```
ODE[{y'' - t y' + 7y == 0,y[0] == 1,y'[0] == 0},y,
{t,0,5},Method->NDSolve,PlotSolution->{{t,0,5}}];
```

18.
$$\begin{cases} t\,y'' + \left(\dfrac{1}{2} - t\right)y' - \dfrac{3y}{4} = 0, \\ y(1) = 1,\ y'(1) = 1 \end{cases}$$

Solution.

```
ODE[{t y'' + (1/2 - t)y' - 3y/4 == 0,y[1] == 1,
y'[1] == 0},y,{t,1,3},
Method->NDSolve,PlotSolution->{{t,1,3}}];
```

19.
$$\begin{cases} (1 - t^2)y'' - 2t\,y' + \left(\dfrac{8 + 8t^2 - 25}{1 - t^2}\right)y = 0, \\ y(0) = 1,\ y'(0) = 1 \end{cases}$$

Solution.

```
ODE[{(1 - t^2) y'' - 2t y' + (6 - 25/(1 - t^2))y == 0,
y[0] == 1,y'[0] == 0},y,{t,0,0.5},
Method->NDSolve,PlotSolution->{{t,0,0.5}}];
```

20.
$$\begin{cases} (1 - t^2)y'' + \left(\dfrac{1 - 3t^2}{t}\right)y' - y = 0, \\ y(1/2) = 1,\ y'(1/2) = 1 \end{cases}$$

Solution.

```
ODE[{(1 - t^2)y'' + (1 - 3t^2)/t y' - y == 0,
y[1/2] == 1,y'[1/2] == 0},y,{t,0,1},
Method->NDSolve,PlotSolution->{{t,0,1}}];
```

21. Consider the initial value problem

$$\begin{cases} y'' + 0.1y' - y + y^5 = 0, \\ y(0) = -2.0, \ y'(0) = 0. \end{cases}$$

   a. Plot a numerical solution from $t = 0$ to $t = 50$.

   b. Experiment with **NDSolve's** option **MaxSteps** to obtain an error free result. The option **PlotPoints** may also be needed to render an accurate plot.

   c. Use the **Parameter** option to produce a plot for each of the initial values $y(0) = -2.0, -1.5, \dots, 2.0$.

   d. What can be inferred from this plot concerning the dependence of the long term behavior of the solutions on the initial values?

Solution.

```
ODE[{y'' + 0.1 y' - y + y^5 == 0,y[0] == a,y'[0] == 0},
y,{t,0,50},Method->NDSolve,MaxSteps->10000,
Parameters->{{a,-2,2,0.5}},
PlotSolution->{{t,0,50},PlotPoints->50}];
```

It appears that $\lim\limits_{t \to \infty} y(t) = \pm 1$.

22. Consider the initial value problem

$$\begin{cases} y'' + y + k\,y^3 = \cos(1.5t), \\ y(0) = 0, \ y'(0) = 0. \end{cases}$$

   a. Plot a numerical solution from $t = 0$ to $t = 100$ using $k = 0$.

   b. Experiment with **NDSolves**'s option **MaxSteps** to obtain an error free result. The option **PlotPoints** may also be needed to render an accurate plot.

   c. Use the **Table** command to produce a plot for each of the values $k = 0, -0.1, \dots, -0.5$.

   d. What can be inferred from this plot concerning the values of k for which the solution is periodic?

   e. What can be inferred from this plot concerning the dependence of the long term behavior of the solutions on the parameter k?

Solution.

```
Table[ODE[{y'' + y + k y^3 == Cos[1.5 t],y[0] == 0,
y'[0] == 0},y,{t,0,100},Method->NDSolve,
MaxSteps->10000,PlotSolution->{{t,0,100},
PlotPoints->50}],{k,0.0,-0.5,-0.1}]
```

The solution appears to change from one which is periodic to one for which $\lim_{t \to \infty} y(t) = -\infty$ for k between -0.3 and -0.4.

Section 13.7, page 442

The following exercises require that you solve the problems using a variety of numerical integration techniques (Euler, Runge-Kutta, Adams-Bashforth, Bulirsch-Stoer and implicit Runge-Kutta) first using a fixed step size and then allow the step size to vary according to some convergence criteria. Then we are to compare the solutions graphically. This could be done by repeating the commands for each problem changing only the equation. For example, we could enter

```
$ODEPlotNumber = 0;
ODE[{y' == -50y,y[0]==1/50},y,{t,0,.5},Method->NDSolve,
    StepSize->0.1,PlotSolution->{{t,0,.5},PlotRange->All}];
ODE[{y' == -50y,y[0]==1/50},y,{t,0,.5},Method->Euler,
    VariableStepSize->False,StepSize->0.1,Tolerance->0.001,
    PlotSolution->{{t,0,.5},PlotRange->All}];
Show[GraphicsArray[Graph[{1,2}]]];
```

to obtain the fixed step size graphs and then enter

```
$ODEPlotNumber = 0;
ODE[{y' == -50y,y[0]==1/50},y,{t,0,.5},Method->NDSolve,
    StepSize->0.1,PlotSolution->{{t,0,.5},PlotRange->All}];
ODE[{y' == -50y,y[0]==1/50},y,{t,0,.5},Method->Euler,
    VariableStepSize->True,StepSize->0.1,Tolerance->0.001,
    PlotSolution->{{t,0,.5},PlotRange->All}];
Show[GraphicsArray[Graph[{1,2}]]];
```

to obtain the variable step size graphs. In both cases, we have set the $ODEPlotNumber to zero in order to use the same Show command. These commands are identical except for the value of the VariableStepSize option. If we combined these commands into a single *Mathematica* function, then we become much more efficient. For example, if we collect the necessary steps in a Module and define

```
plt[vary_]:= Module[{},
$ODEPlotNumber = 0;
ODE[{y' == -50y,y[0]==1/50},y,{t,0,.5},Method->NDSolve,
    StepSize->0.1,PlotSolution->{{t,0,.5},PlotRange->All}];
ODE[{y' == -50y,y[0]==1/50},y,{t,0,.5},Method->Euler,
    VariableStepSize->vary,Tolerance->0.001,StepSize->0.1,
    PlotSolution->{{t,0,.5},PlotRange->All}];
Show[GraphicsArray[Graph[{1,2}]]];]
```

then we need only call **plt[False]** to get the first set of graphs and then call **plt[True]** to obtain the second set.

The following command is to be used with each of the answers to Problems 1-6 in Section 13.7. It uses **Module** to combine several individual commands into a single command called **plt[vary]**, whose single argument, which must be **False** or **True**, deactivates or activates the **VariableStepSize** option, respectively. After defining **eq**, first evaluate **plt[False]** and then **plt[True]**.

```
plt[vary_]:= Module[{},
$ODEPlotNumber = 0;
ODE[eq,y,{t,0,.5},Method->NDSolve,StepSize->0.1,
    PlotSolution->{{t,0,.5},PlotRange->All}];
ODE[eq,y,{t,0,.5},VariableStepSize->vary,
    Tolerance->0.001,Method->Euler,StepSize->0.1,
    PlotSolution->{{t,0,.5},PlotRange->All}];
ODE[eq,y,{t,0,.5},VariableStepSize->vary,
    Tolerance->0.001,Method->RungeKutta4,StepSize->0.1,
    PlotSolution->{{t,0,.5},PlotRange->All}];
ODE[eq,y,{t,0,.5},VariableStepSize->vary,
    Tolerance->0.001,Method->AdamsBashforth,StepSize->0.1,
    PlotSolution->{{t,0,.5},PlotRange->All}];
ODE[eq,y,{t,0,.5},VariableStepSize->vary,
    Tolerance->0.001,Method->BulirschStoer,StepSize->0.1,
    PlotSolution->{{t,0,.5},PlotRange->All}];
ODE[eq,y,{t,0,.5},Method->ImplicitRungeKutta,
    StepSize->0.1,Tolerance->0.001,VariableStepSize->vary,
    PlotSolution->{{t,0,.5},PlotRange->All}];
Show[GraphicsArray[{Graph[{1,2}],Graph[{3,4}],
    Graph[{5,6}]}]];]
```

Use *Mathematica* to plot the numerical solutions of the following initial value problems. In each case first use **VariableStepSize->False** and **StepSize->0.1** and integrate over the interval $0 \le t \le 0.5$. Then resolve the problem using **VariableStepSize->True** and **Tolerance->0.001**, again with

StepSize->0.1. Solve each problem using **NDSolve** and compare it with the solutions found by the Euler, Runge-Kutta, Adams-Bashforth, Bulirsch-Stoer and implicit Runge-Kutta methods.

1. $\begin{cases} y' = -50y, \\ y(0) = 1/50 \end{cases}$

Solution.

```
eq = {y' == -50y,y[0] == 1/50};
```

2. $\begin{cases} y' = 10\sin(10y), \\ y(0) = 1 \end{cases}$

Solution.

```
eq = {y' == 10 Sin[10y],y[0] == 1};
```

3. $\begin{cases} y' = \sin(50y)\cos(50y), \\ y(0) = -1 \end{cases}$

Solution.

```
eq = {y' == Sin[50y]Cos[50y],y[0] == -1};
```

4. $\begin{cases} y' = \cos(99y) - \sin(98y), \\ y(0) = 1 \end{cases}$

Solution.

```
eq = {y' == Cos[99y] - Sin[98y],y[0] == 1};
```

5. $\begin{cases} y' = \cos(99y) - \sin(98t), \\ y(0) = 1 \end{cases}$

Solution.

```
eq = {y' == Cos[99y] - Sin[98t],y[0] == 1};
```

6. $\begin{cases} y' = t\cos(100y) - t^5 y^2, \\ y(0) = 1 \end{cases}$

Solution.

```
eq = {y' == t Cos[100y] - t^5 y^2,y[0] == 1};
```

Every **ODE** numerical integration command returns an answer in terms of a list of points representing a discrete approximation to the true solution. For example, if we enter

```
sol = ODE[{y' == y,y[0] == 1/60},y,{t,0,.3},Method->Euler]
```

we would obtain the list of points

```
{{0, 0.0167}, {0.1, 0.0183}, {0.2, 0.0202}, {0.3, 0.0222}}
```

associated with our variable **sol**. If we then enter

```
Last[sol]
```

we get the last point

```
{0.3, 0.0222}
```

and if we enter

```
Last[Last[sol]]
```

we get the ordinate for the last point

```
0.0222
```

In fact, we could simply enter

```
Last[Last[
  ODE[{y' == y,y[0] == 1/60},y,{t,0,.3},Method->Euler]]]
```

and obtain the same result without introducing the variable **sol**. If we then wanted to make the output more attractive, we can use *Mathematica*'s **Print** command. For example,

```
Print["Euler solution at t = .3 is ",Last[Last[
  ODE[{y' == y,y[0] == 1/60},y,{t,0,.3},Method->Euler]]]]
```

will display

```
Euler solution at t = .3 is 0.0222
```

The following command is to be used with each of the answers to Problems 7-12 in Section 13.7. As in the previous exercise, it uses **Module** to combine several individual commands into a single command called **fin[]**, which requires no argument. After defining **eq** and **interval**, evaluate **fin[]** to obtain the solutions. The **ODEDigits** option simply determines the number of digits to display.

```
fin[]:= Module[{},
Print[ODE[eq,y,interval,
    Method->FirstOrderLinear,Form->Equation]];
Print["Exact                    = ",N[ODE[eq,y,interval,
    Method->FirstOrderLinear,Form->Explicit] /.
    t -> Last[interval],4]];
Print["Euler                    = ",Last[Last[ODE[eq,y,interval,
    Method->Euler,StepSize->0.1,ODEDigits->4,
    VariableStepSize->True,Tolerance->0.0001]]]];
Print["Runge-Kutta              = ",Last[Last[ODE[eq,y,interval,
    Method->RungeKutta4,StepSize->0.1,ODEDigits->4,
    VariableStepSize->True,Tolerance->0.0001]]]];
Print["Adams-Bashforth          = ",Last[Last[ODE[eq,y,interval,
    Method->AdamsBashforth,StepSize->0.1,ODEDigits->4,
    VariableStepSize->True,Tolerance->0.0001]]]];
Print["Bulirsch-Stoer           = ",Last[Last[ODE[eq,y,interval,
    Method->BulirschStoer,StepSize->0.1,ODEDigits->4,
    VariableStepSize->True,Tolerance->0.0001]]]];
Print["Implicit Runge-Kutta = ",Last[Last[ODE[eq,y,interval,
    Method->ImplicitRungeKutta,StepSize->0.1,ODEDigits->4,
    VariableStepSize->True,Tolerance->0.0001]]]];]
```

For each of the initial value Problems 7–12, use **ODE** with
VariableStepSize->True,Tolerance->0.0001, and **StepSize->0.1**. Find
the exact value at the indicated point and then obtain a four-decimal approximation at the
indicated point using the Euler, Runge-Kutta, Adams-Bashforth, Bulirsch-Stoer and implicit
Runge-Kutta methods.

7. $y' = y$, $y(0) = 1/60$, $y(1.0)$
Solution.

```
eq = {y' == y,y[0] == 1/60};
interval = {t,0,1};
```

$$y(t) = \frac{e^t}{60}$$

```
Exact                    = 0.0453
Euler                    = 0.04322
Runge-Kutta              = 0.04522
Adams-Bashforth          = 0.0453
Bulirsch-Stoer           = 0.0451
Implicit Runge-Kutta = 0.0453
```

8. $y' = -y/2$, $y(0) = 1$, $y(2.0)$
Solution.

```
eq = {y' == -y/2,y[0] == 1};
interval = {t,0,2};
```

$$y(t) = e^{-t/2}$$

Exact	= 0.3679
Euler	= 0.3643
Runge-Kutta	= 0.3679
Adams-Bashforth	= 0.3679
Bulirsch-Stoer	= 0.368
Implicit Runge-Kutta	= 0.3679

9. $y' = t$, $\quad y(0) = 1$, $\quad y(2.0)$

Solution.

```
eq = {y' == t,y[0] == 1};
interval = {t,0,2};
```

$$y(t) = 1 + \frac{t^2}{2}$$

Exact	= 3.
Euler	= 2.981
Runge-Kutta	= 3.
Adams-Bashforth	= 3.
Bulirsch-Stoer	= 3.
Implicit Runge-Kutta	= 3.

10. $y' = -t^2 + t + 1$, $\quad y(0) = 1$, $\quad y(2.0)$

Solution.

```
eq = {y' == -t^2 + t + 1,y[0] == 1};
interval = {t,0,2};
```

$$y(t) = 1 + t + \frac{t^2}{2} - \frac{t^3}{2}$$

Exact	= 2.333
Euler	= 2.352
Runge-Kutta	= 2.333
Adams-Bashforth	= 2.333
Bulirsch-Stoer	= 2.333
Implicit Runge-Kutta	= 2.333

11. $y' = y - \cos(t) - \sin(t)$, $\quad y(0) = 1$, $\quad y(\pi)$

Solution.

```
eq = {y' == y - Cos[t] - Sin[t],y[0] == 1};
interval = {t,0,Pi};
```

$$y(t) = \cos(t)$$

Exact	= -1.
Euler	= -0.9033
Runge-Kutta	= -0.9979
Adams-Bashforth	= -1.
Bulirsch-Stoer	= -0.9988
Implicit Runge-Kutta	= -0.999

12. $y' - y/t = t$, $\quad y(1) = 2$, $\quad y(5)$

Solution.

```
eq = {y' - y/t == t,y[1] == 2};
interval = {t,1,5};
```

Exact	= 30.
Euler	= 29.75
Runge-Kutta	= 30.
Adams-Bashforth	= 30.
Bulirsch-Stoer	= 30.
Implicit Runge-Kutta	= 30.

$$y(t) = t + t^2$$

Section 13.8, page 447

Use the Numerov method to find approximate solutions to the following initial value problems 1–4. For each problem choose a suitable interval and step size and plot the solution.

1. $$\begin{cases} y'' + t\,y = 0, \\ y(0) = 1,\ y'(0) = 1 \end{cases}$$

Solution.

```
ODE[{y'' + t y == 0,y[0] == 1,y'[0] == 1},y,{t,0,8},
Method->Numerov,PlotSolution->{{t,0,8}}];
```

2. $$\begin{cases} y'' = \dfrac{\cos(t)}{|\cos(t)|} y + t, \\ y(0) = 2, \quad y'(0) = 1 \end{cases}$$

Solution.

```
ODE[{y'' == Cos[t]/Abs[Cos[t]]y + t,y[0] == 2,
y'[0] == 1},y,{t,0,6},
Method->Numerov,PlotSolution->{{t,0,6}}];
```

3. $$\begin{cases} y'' = \sin(t)y + t, \\ y(0) = 0, \quad y'(0) = 1 \end{cases}$$

Solution.

```
ODE[{y'' == Sin[t] y + t,y[0] == 0,y'[0] == 1},y,{t,0,8},
Method->Numerov,PlotSolution->{{t,0,8}}];
```

4. $$\begin{cases} y'' = -\dfrac{y}{t+1} + \tan(5t), \\ y(0) = 2, \quad y'(0) = 1 \end{cases}$$

Solution.

```
ODE[{y'' == -y/(t + 1) + Tan[5t],y[0] == 2,y'[0] == 1},y,
{t,0,10},Method->Numerov,PlotSolution->{{t,0,10}}];
```

The following set of commands is to be used with each of the answers to Problems 5-8 in Section 13.8:

```
f[t_] = ODE[eq,y,t,Method->SecondOrderLinear,Form->Explicit]
TableForm[Table[{1/n,Chop[N[f[1]]],
    Last[Last[ODE[eq,y,{t,0,1},
    Method->Numerov,StepSize->1/n]]]},{n,2,8,2}]]
```

For each of the following Problems 5–8 use **Numerov** to compute an approximate solution at $t = 1.0$. Use step sizes of $h = 1/2, 1/4$, and $1/8$. Then determine the symbolic solution and compare the values of $y(1)$ coming from the symbolic and numerical solutions.

5. $\begin{cases} y'' + 2y = t, \\ y(0) = 1, \quad y'(0) = 0 \end{cases}$

Solution.

```
eq = {y'' + 2y == t,y[0] == 1,y'[0] == 0}
```

$$y(t) = \frac{t}{2} + \cos(\sqrt{2}t) - \frac{\sin(\sqrt{2}t)}{2\sqrt{2}}$$

h	$Y_n(t = 1.0)$	$y(1)$
1/2	0.64	0.307
1/4	0.476	0.307
1/8	0.392	0.307

6. $\begin{cases} y'' = t + 1, \\ y(0) = 1, \quad y'(0) = 0 \end{cases}$

Solution.

```
eq = {y'' == t + y,y[0] == 1,y'[0] == 0}
```

h	$Y_n(t = 1.0)$	$y(1)$
1/2	1.38	1.72
1/4	1.56	1.72
1/8	1.64	1.72

$y(t) = e^t - t$

7. $\begin{cases} y'' = y + t^3, \\ y(0) = 0, \quad y'(0) = 1 \end{cases}$

Solution.

```
eq = {y'' == y + t^3,y[0] == 0,y'[0] == 1}
```

$$y(t) = \frac{7e^t}{2} - \frac{7}{2e^t} - 6t - t^3$$

h	$Y_n(t = 1.0)$	$y(1)$
1/2	1.18	1.23
1/4	1.21	1.23
1/8	1.22	1.23

8. $\begin{cases} y'' = y + e^t \sin(t), \\ y(0) = 1, \quad y'(0) = 0 \end{cases}$

Solution.

```
eq = {y'' == y + Exp[t]Sin[t],y[0] == 1,y'[0] == 0}
```

$$y(t) = \frac{2}{5e^t} + e^t - \frac{2e^t\cos(t)}{5} - \frac{e^t\sin(t)}{5}$$

h	$Y_n(t = 1.0)$	$y(1)$
1/2	1.47	1.82
1/4	1.66	1.82
1/8	1.74	1.82

Chapter 14

Section 14.1, page 455

In Problems 1–4 determine the existence of the improper integral, and find its value if possible.

1. $\displaystyle\int_1^\infty \frac{dt}{t^3}$.

Solution. The exponent in the denominator is larger than one, so that the improper integral exists and can be computed by

$$\int_1^\infty \frac{dt}{t^3} = -\frac{t^{-2}}{2}\bigg|_1^\infty = \frac{1}{2}.$$

2. $\displaystyle\int_1^\infty \frac{dt}{t^5}$.

Solution. The exponent in the denominator is larger than one, so that the improper integral exists and can be computed by

$$\int_1^\infty \frac{dt}{t^5} = -\frac{t^{-4}}{4}\bigg|_1^\infty = \frac{1}{4}.$$

3. $\displaystyle\int_1^\infty \frac{e^t dt}{t^3}$.

Solution. The integrand becomes infinite when $t \longrightarrow \infty$, so that the improper integral does not exist.

4. $\displaystyle\int_1^\infty \frac{\sin(t)dt}{t^3}$

Solution. The integrand is bounded by the function t^{-3}; hence the integral exists. The command

Integrate[Sin[t]/t^3,{t,1,Infinity}]

gives

```
-Pi                       1           1   3        1
---  + HypergeometricPFQ[{-(-)}, {-, -}, -(-)]
 4                        2           2   2        4
```

Hence the integral can be written in terms of a hypergeometric function as $-\pi/4 + {}_{-1/2}F_{1/2,3/2}(-1/4)$. (See Section 21.9 for a discussion of hypergeometric functions.) To find a numerical approximation of the integral we use

```
N[Integrate[Sin[t]/t^3,{t,1,Infinity}]]
```

to get

```
0.37853
```

5. Compute $\mathcal{L}(\cos(b\,t))(s)$ from the definition.

   Solution. The integral can be computed either using integration by parts or with *Mathematica*. The result is

   $$\int_0^\infty e^{-st}\cos(b\,t)\,dt = \frac{e^{-st}}{s^2+b^2}\left(-s\,\cos(b\,t)+b\,\sin(b\,t)\right)\Big|_0^\infty = \frac{s}{s^2+b^2}.$$

6. For which values of the constants a and b does the improper integral

   $$\int_{t_0}^\infty t^9 e^{-at}\sin(b\,t)\,dt \qquad \text{converge?}$$

   Solution. If $a \le 0$, the integrand becomes infinite when $t \longrightarrow \infty$, so that the improper integral does not exist. If $a > 0$, the integrand is bounded by $t^9 e^{-bt}$, which has a convergent improper integral. Hence the integral converges whenever $a > 0$ for all b.

In Problems 7–10 determine which of the following functions is of exponential growth? Find the growth exponent if possible.

7. $f_1(t) = e^{t^2}$

   Solution. The function $f(t) = e^{t^2}$ satisfies

   $$\lim_{t\to\infty}\frac{f(t)}{e^{kt}} = \infty$$

   for any value of k (as can be seen, for example, using l'Hôspital's rule). Hence this function is not of exponential growth.

8. $f_2(t) = t^3 e^{-t}$

   Solution. This function is of exponential growth with exponent -1.

9. $f_3(t) = e^{\sqrt{t}}$

   Solution. This function is of exponential growth with exponent 0.

10. $f_4(t) = \sin(t)\sinh(t)$

Solution. This function is of exponential growth -1.

11. **Theorem 14.4.** *Suppose the Laplace transform of $f(t)$ exists. Then the Laplace transform of $e^{at} f(t)$ also exists, and*

$$\mathcal{L}\big(e^{at} f(t)\big)(s) = \mathcal{L}\big(f(t)\big)(s)\Big|_{s \to s-a}$$

Proof. We use integration by parts with $u = e^{-st}$ and $dv = f'(t)\,dt$:

$$\int_0^M e^{-st} f'(t)\,dt = e^{-st} f(t)\Big|_0^M + \int_0^M s e^{-st} f(t)\,dt. \qquad \text{(S.79)}$$

The first term on the right-hand side of (S.79) equals $e^{-sM} f(M) - f(0)$, which tends to $-f(0)$ when $M \longrightarrow \infty$. The second term tends to $s\mathcal{L}\big(f(t)\big)(s)$. Thus the left-hand side of (S.79) tends to $\mathcal{L}\big(f'(t)\big)(s)$, which completes the proof.

Section 14.2, page 459

Which of the following functions is piecewise continuous?

1. $t - [t]$

Solution. $t - [t]$ is piecewise continuous.

2. $\sin(t)$

Solution. $\sin(t)$ is continuous, in particular piecewise continuous.

3. $t \sin(1/t)$

Solution. $t \sin(1/t)$ is also continuous, since the sine function is bounded by 1, so that

$$\lim_{t \to 0} t \sin(1/t)$$

exists and is zero.

4. $e^{-1/t} \sin(1/t)$

Solution. $e^{-1/t} \sin(1/t)$ is not piecewise continuous, since

$$\lim_{t \to 0} e^{-1/t} \sin(1/t)$$

The function **Ceiling** (corresponding to *Mathematica*'s **Ceiling**) is defined by

$$\textbf{Ceiling}(t) = \text{the smallest integer greater than or equal to } t.$$

5. Show that **Ceiling** is piecewise continuous.

   Solution. Since **Ceiling**$(t) = n + 1$ for $n < t \le n + 1$, the function **Ceiling** is continuous on the interval $n < t \le n + 1$ because it is constant on that interval. It is also easy to check that the left and right limits exist for each n.

6. Plot **Ceiling** for $-1 < t < 5$.
   Solution.

   ```
   Plot[Ceiling[t],{t,-1,5}]
   ```

7. Compute **Ceiling**$(t) -$ **Floor**(t).
   Solution.

   ```
   Plot[Ceiling[t] - Floor[t],{t,-1,5}]
   ```

Section 14.3, page 465

For each of the following initial value problems, find a formula for the Laplace transform $Y(s)$ and then find the solution $y(t)$ for which $Y(s) = \mathcal{L}(y)(s)$.
If we apply the Laplace transform to the general second-order linear equation

$$\begin{cases} y'' + a\,y' + b\,y = f(t), \\ y(0) = y_0, \ y'(0) = y_0' \end{cases}$$

we get

$$Y(s) = \frac{\mathcal{L}(f)(s) + (s + a)y_0 + y_0'}{s^2 + as + b}.$$

This general expression can be used to obtain $Y(s)$ in many of the following problems.

1. $\begin{cases} y'' + 2y' - 15y = 0, \\ y(0) = 2, \ y'(0) = 2 \end{cases}$

   Solution. Using the general solution and partial fraction decomposition, we obtain

   $$Y(s) = \frac{(s+2)2 + 2}{s^2 + 2s - 15} = \frac{2s + 6}{(s+5)(s-3)} = \frac{1}{2}\left(\frac{1}{s+5}\right) + \frac{3}{2}\left(\frac{1}{s-3}\right).$$

The first term is the Laplace transform of $(1/2)e^{-5t}$ and the second term is the Laplace transforms of $(3/2)e^{3t}$. Hence,

$$y(t) = \frac{e^{-5t}}{2} + \frac{3e^{3t}}{2}.$$

The **ODE** command which solves this problem is

```
ODE[{y'' + 2y' - 15y == 0,y[0] == 2,y'[0] == 2},
y,t,Method->Laplace]
```

```
                 3 t
        1      3 E
{{y ->  ----- + -----}}
          5 t    2
        2 E
```

2. $\begin{cases} y'' + 9y = 0, \\ y(0) = 9,\, y'(0) = 3 \end{cases}$

Solution. Using the general solution and partial fraction decomposition, we obtain

$$Y(s) = \frac{(s+0)9 + 3}{s^2 + 9} = 9\left(\frac{s}{s^2 + 3^2}\right) + \frac{3}{s^2 + 3^2}.$$

The first term is the Laplace transform of $9\cos(3t)$ and the second term is the Laplace transforms of $\sin(3t)$. Hence,

$$y(t) = 9\cos(3t) + \sin(3t).$$

The **ODE** command which solves this problem is

```
ODE[{y'' + 9y == 0,y[0] == 9,y'[0] == 3},
y,t,Method->Laplace]
```

```
{{y -> 9 Cos[3 t] + Sin[3 t]}}
```

3. $\begin{cases} y'' + 9y' = 0, \\ y(0) = 0,\, y'(0) = -2 \end{cases}$

Solution. Using the general solution and partial fraction decomposition, we obtain

$$Y(s) = \frac{(s+9)0 - 2}{s^2 + 9s} = -\frac{2}{s(s+9)} = -\frac{2}{9}\left(\frac{1}{s}\right) + \frac{2}{9}\left(\frac{1}{s+9}\right).$$

The first term is the Laplace transform of $-2/9$ and the second term is the Laplace transforms of $(2/9)e^{-9t}$. Hence,

$$y(t) = -\frac{2}{9} + \frac{2e^{-9t}}{9}.$$

The **ODE** command which solves this problem is

```
ODE[{y'' + 9y' == 0,y[0] == 0,y'[0] == -2},
y,t,Method->Laplace]
```

$$\{\{y \,\, -> \,\, -(\frac{2}{9}) \,\, + \,\, \frac{2}{9 \,\, E^{9\,t}}\}\}$$

4. $\begin{cases} y'' - 3y' + 10y = \cos(t), \\ \\ y(0) = 0,\ y'(0) = 0 \end{cases}$

Solution. Using the general solution and partial fraction decomposition, we obtain

$$Y(s) = \frac{s/(s^2 + 1) + (s - 3)0 + 0}{s^2 - 3s + 10} = \frac{s}{(s^2 + 1)((s - 3/2)^2 + (\sqrt{31}/2)^2)}$$

$$= -\frac{1}{30}\left(\frac{1}{s^2 + 1}\right) + \frac{1}{10}\left(\frac{s}{s^2 + 1}\right)$$

$$+ \frac{11}{30\sqrt{31}}\left(\frac{\sqrt{31}/2}{(s - 3/2)^2 + (\sqrt{31}/2)^2}\right) - \frac{1}{10}\left(\frac{s - 3/2}{(s - 3/2)^2 + (\sqrt{31}/2)^2}\right)$$

The first term is the Laplace transform of $-(1/30)\sin(t)$ and the second term is the Laplace transforms of $(1/10)\cos(t)$. The third and forth terms are obtained using Theorem 14.4. For example, since

$$L(\sin(\sqrt{31}/2t))(s) = \frac{\sqrt{31}/2}{s^2 + (\sqrt{31}/2)^2},$$

it follows that

$$L(e^{3t/2}\sin(\sqrt{31}/2t))(s) = \frac{\sqrt{31}/2}{s^2 + (\sqrt{31}/2)^2}\bigg|_{s \to s - 3/2} = \frac{\sqrt{31}/2}{(s - 3/2)^2 + (\sqrt{31}/2)^2}.$$

Similarly,

$$L(e^{3t/2}\cos(\sqrt{31}/2t))(s) = \frac{s - 3/2}{(s - 3/2)^2 + (\sqrt{31}/2)^2}.$$

Hence,

$$y(t) = -\frac{e^{3t/2}}{10}\cos\left(\frac{\sqrt{31}\,t}{2}\right) + \frac{3\cos(t) - \sin(t)}{30} + \frac{11e^{3t/2}}{30\sqrt{31}}\sin\left(\frac{\sqrt{31}\,t}{2}\right).$$

The **ODE** command which solves this problem is

```
ODE[{y'' - 3y' + 10y == Cos[t],y[0] == 0,y'[0] == 0},
y,t,Method->Laplace]
```

```
              (3 t)/2      Sqrt[31] t
             E         Cos[---------]
      Cos[t]                   2            Sin[t]
{{y -> ------ - ----------------------- - ------ +
        10              10                  30

        (3 t)/2      Sqrt[31] t
   11 E        Sin[---------]
                       2
   ------------------------------}}
         30 Sqrt[31]
```

5. $\begin{cases} y'' + cy = \cos(3t), \quad c \neq 9, \\ y(0) = 1,\ y'(0) = 0 \end{cases}$

Solution. Using the general solution and partial fraction decomposition, we obtain

$$Y(s) = \frac{s/(s^2+9) + (s+0)1 + 0}{s^2+c} = \frac{1}{c-9}\left(\frac{s}{s^2+9} + (c-10)\frac{s}{s^2+c}\right).$$

The first term is the Laplace transform of $(1/(c-9))\cos(3t)$ and the second term is the Laplace transforms of $((c-10)/(c-9))\cos(\sqrt{c}\,t)$. Hence,

$$y(t) = \frac{\cos(3t) - 10\cos(\sqrt{c}\,t) + c\cos(\sqrt{c}\,t)}{c-9}.$$

The **ODE** command which solves this problem is

```
ODE[{y'' + c y == Cos[3t],y[0] == 1,y'[0] == 0},
y,t,Method->Laplace]
```

```
       Cos[3 t]    10 Cos[Sqrt[c] t]   c Cos[Sqrt[c] t]
{{y -> -------- - ----------------- + ----------------}}
        -9 + c        -9 + c               -9 + c
```

6. $\begin{cases} y'' + 9y = \cos(3t), \\ y(0) = 1,\ y'(0) = 0 \end{cases}$

Solution. Using the general solution and partial fraction decomposition, we obtain

$$Y(s) = \frac{s/(s^2+9) + (s+0)1 + 0}{s^2+9} = \frac{s}{s^2+9} + \frac{s}{(s^2+9)^2}.$$

The first term is the Laplace transform of $\cos(3t)$ and the second term is

$$-\frac{1}{2}\frac{d}{ds}\left(\frac{1}{s^2+9}\right) = -\frac{1}{6}\frac{d}{ds}\left(\frac{3}{s^2+3^2}\right),$$

and consequently is the Laplace transform of $(t\,\sin(3t))/6$. Hence,

$$y(t) = \cos(3t) + \frac{t\sin(3t)}{6}.$$

The **ODE** command which solves this problem is

```
ODE[{y'' + 9y == Cos[3t],y[0] == 1,y'[0] == 0},
y,t,Method->Laplace]
```

```
                  t Sin[3 t]
{{y -> Cos[3 t] + ----------}}
                      6
```

7. $\begin{cases} y'' + 9y = \cos(3t) + \sin(3t), \\ y(0) = 16,\ y'(0) = 0 \end{cases}$

Solution. When we apply the Laplace transform to both sides of the differential equation we obtain

$$(s^2+9)\mathcal{L}(y)(s) - 16s = \frac{s}{s^2+9} + \frac{3}{s^2+9},$$

which is solved to yield

$$\mathcal{L}(y)(s) = \frac{16s}{s^2+9} + \frac{s+3}{(s^2+9)^2}.$$

The first term is the Laplace transform of $16\cos(3t)$. The second term can be expressed in terms of the Laplace transforms of $t\sin(3t)$ and $t\cos(3t)$ using the method of Example 14.8. Thus

$$\mathcal{L}(t\sin(3\,t))(s) = -\frac{d}{ds}\mathcal{L}(\sin(3\,t))(s)$$

$$= -\frac{d}{ds}\left(\frac{3}{s^2+9}\right) = \frac{6s}{(s^2+9)^2}$$

and

$$L\big(t\cos(3t)\big)(s) = -\frac{d}{ds}L\big(\cos(3t)\big)(s)$$

$$= -\frac{d}{ds}\left(\frac{s}{s^2+9}\right) = \frac{s^2-9}{(s^2+9)^2}.$$

This suggests the partial fraction decomposition

$$\frac{s+3}{(s^2+9)^2} = \frac{s}{(s^2+9)^2} + \frac{1}{6}\left(\frac{s^2+9}{(s^2+9)^2} - \frac{s^2-9}{(s^2+9)^2}\right)$$

$$= \frac{s}{(s^2+9)^2} + \frac{1}{6}\left(\frac{1}{s^2+9} - \frac{s^2-9}{(s^2+9)^2}\right)$$

$$= \frac{1}{6}L(t\sin(3t))(s) + \frac{1}{18}L(\sin(3t))(s) - \frac{1}{6}L(t\cos(3t))(s).$$

Hence the solution is obtained as

$$y(t) = 16\cos(3t) + \frac{1}{18}\sin(3t) + \frac{t\sin(3t)}{6} - \frac{t\cos(3t)}{6}.$$

The **ODE** command which solves this problem is

```
ODE[{y'' + 9y == Cos[3t] + Sin[3t],
y[0] == 16,y'[0] == 0},y,t,Method->Laplace]
```

$$\{\{y \to 16\,\text{Cos}[3\ t] - \frac{t\ \text{Cos}[3\ t]}{6} + \frac{\text{Sin}[3\ t]}{18} + \frac{t\ \text{Sin}[3\ t]}{6}\}\}$$

8. $\begin{cases} y^{(4)} - 16y = 0,\ y(0) = 7, \\ y'(0) = 20,\ y''(0) = -44, \\ y'''(0) = 58 \end{cases}$

Solution. When we apply the Laplace transform to both sides of the differential equation, solve for $Y(s)$ and use partial fraction decomposition, we obtain

$$Y(s) = \frac{53}{16}\left(\frac{1}{s-2}\right) - \frac{85}{16}\left(\frac{1}{s+2}\right) + \frac{11}{8}\left(\frac{2}{s^2+2^2}\right) + 9\left(\frac{s}{s^2+2^2}\right).$$

The first term is the Laplace transform of $(53/16)e^{2t}$, the second term is the Laplace transform of $-(85/16)e^{-2t}$, the third term is the Laplace transform of $(11/8)\sin(2t)$

and the forth term is the Laplace transform of $9\cos(2t)$. Hence,

$$y(t) = -\frac{85e^{-2t}}{16} + \frac{53e^{2t}}{16} + 9\cos(2t) + \frac{11\sin(2t)}{8}.$$

The **ODE** command which solves this problem is

```
ODE[{y'''' - 16y == 0,y[0] == 7,
y'[0] == 20,y''[0] == -44,y'''[0] == 58},
y,t,Method->Laplace]
```

$$\{\{y \to \frac{-85}{16\,E^{2\,t}} + \frac{53\,E^{2\,t}}{16} + 9\,Cos[2\,t] + \frac{11\,Sin[2\,t]}{8}\}\}$$

9. $\begin{cases} y'' - 3y' + 9y = e^{3t}, \\ y(0) = 1, \ y'(0) = 4 \end{cases}$

Solution. When we apply the Laplace transform to both sides of the differential equation, solve for $Y(s)$ and use partial fraction decomposition, we obtain

$$Y(s) = \frac{1}{9}\left(\frac{1}{s-3}\right) + \left(\frac{1}{s^2 - 3s + 9}\right) + \frac{8}{9}\left(\frac{s}{s^2 - 3s + 9}\right).$$

After completing the squares and scaling, we can write

$$Y(s) = \frac{1}{9}\left(\frac{1}{s-3}\right) + \frac{14}{9\sqrt{3}}\left(\frac{3\sqrt{3}/2}{(s-3/2)^2 + (3\sqrt{3}/2)^2}\right)$$

$$+\frac{8}{9}\left(\frac{s-3/2}{(s-3/2)^2 + (3\sqrt{3}/2)^2}\right).$$

The first term is the Laplace transform of $(1/9)e^{3t}$, the second term is the Laplace transform of $(14/(9\sqrt{3}))e^{3t/2}\sin((3/2)\sqrt{3}t)$ and the third term is the Laplace transform of $(8/9)e^{3t/2}\cos((3/2)\sqrt{3}t)$. Hence,

$$y(t) = \frac{e^{3t}}{9} + \frac{8e^{3t/2}\cos((3/2)\sqrt{3}t)}{9} + \frac{14e^{3t/2}\sin((3/2)\sqrt{3}t)}{9\sqrt{3}}.$$

The **ODE** command which solves this problem is

```
ODE[{y'' - 3y' + 9y == E^(3t),
y[0] == 1,y'[0] == 4},
y,t,Method->Laplace]
```

$$\{\{y \rightarrow \frac{E^{3t}}{9} + \frac{8\,E^{(3\,t)/2}\,\cos\!\left[\dfrac{3\,\text{Sqrt}[3]\,t}{2}\right]}{9} + \frac{14\,E^{(3\,t)/2}\,\sin\!\left[\dfrac{3\,\text{Sqrt}[3]\,t}{2}\right]}{9\,\text{Sqrt}[3]} \}\}$$

10. $\begin{cases} y'' - 3y' + 9y = t^4 e^{5t}, \\ y(0) = y'(0) = 0 \end{cases}$

Solution. When we apply the Laplace transform to both sides of the differential equation, solve for $Y(s)$ and use partial fraction decomposition, we obtain

$$Y(s) = \frac{8}{243}\left(\frac{1}{s-3}\right) + \frac{8}{81}\left(\frac{1}{(s-3)^2}\right) + \frac{8}{9}\left(\frac{1}{(s-3)^4}\right) + \frac{8}{3}\left(\frac{1}{(s-3)^5}\right)$$
$$+ \frac{73}{81}\left(\frac{1}{s^2-3s+9}\right) + \frac{251}{243}\left(\frac{s}{s^2-3s+9}\right).$$

After completing the squares and scaling, we can write

$$Y(s) = \frac{8}{243}\left(\frac{1}{s-3}\right) + \frac{8}{81}\left(\frac{1}{(s-3)^2}\right) + \frac{8}{9}\left(\frac{1}{(s-3)^4}\right) + \frac{8}{3}\left(\frac{1}{(s-3)^5}\right)$$
$$+ \frac{397}{243}\left(\frac{3\sqrt{3}/2}{(s-3/2)^2+(3\sqrt{3}/2)^2}\right) + \frac{251}{243}\left(\frac{s-3/2}{(s-3/2)^2+(3\sqrt{3}/2)^2}\right).$$

The first term is the Laplace transform of $(8/243)e^{3t}$, the second term is the Laplace transform of $(8/81)t\,e^{3t}$, the third term is the Laplace transform of $(8/81)t^3\,e^{3t}$ and the forth term is the Laplace transform of $(8/81)t^4\,e^{3t}$. The fifth and sixth terms are obtained using Theorem 14.4 and are the Laplace transforms of $(397/243)e^{3t/2}\sin((3/2)\sqrt{3}t)$ and $(251/243)e^{3t/2}\cos((3/2)\sqrt{3}t)$, respectively. Hence,

$$y(t) = -\frac{8e^{3t}}{243} + \frac{8t\,e^{3t}}{81} - \frac{4t^3e^{3t}}{27} + \frac{t^4e^{3t}}{9} + \frac{251e^{3t/2}\cos((3/2)\sqrt{3}t)}{243}$$
$$+ \frac{397e^{3t/2}\sin((3/2)\sqrt{3}t)}{243\sqrt{3}}.$$

The **ODE** command which solves this problem is

```
ODE[{y'' - 3y' + 9y == t^4 E^(3t),
y[0] == 1,y'[0] == 4},
y,t,Method->Laplace]
```

$$\{\{y \to \frac{-8\,E^{3\,t}}{243} + \frac{8\,E^{3\,t}\,t}{81} - \frac{4\,E^{3\,t}\,t^3}{27} + \frac{E^{3\,t}\,t^4}{9} +$$

$$\frac{251\,E^{(3\,t)/2}\,Cos[\frac{3\,Sqrt[3]\,t}{2}]}{243} +$$

$$\frac{397\,E^{(3\,t)/2}\,Sin[\frac{3\,Sqrt[3]\,t}{2}]}{243\,Sqrt[3]}\}\}$$

11. Prove Corollary 14.12

Solution. We use the method of mathematical induction. The corollary has been proved for $n = 1$. Assuming the truth for the value $n = k$, we differentiate under the integral sign:

$$\frac{d}{ds}\mathcal{L}\left(t^k f(t)\right)(s) = \frac{d}{ds}\int_0^\infty e^{-st}t^k f(t)\,dt = \int_0^\infty \frac{d}{ds}\left(e^{-st}t^k kf(t)\right)dt\right)$$

$$= \int_0^\infty -t^{k+1}e^{-st}f(t)\,dt = -\mathcal{L}(t^{k+1}\,f(t))(s).$$

which proves the statement for the value $n = k + 1$. Therefore, the statement is true for all n.

Section 14.4, page 469

1. Show that
$$\Gamma\left(\frac{3}{2}\right) = \frac{\sqrt{\pi}}{2} \quad \text{and} \quad \Gamma\left(\frac{5}{2}\right) = \frac{3\sqrt{\pi}}{4}.$$

Solution. $\Gamma\left(\frac{5}{2}\right) = \frac{3}{2}\Gamma\left(\frac{3}{2}\right) = \left(\frac{3}{2}\right)\left(\frac{1}{2}\right)\Gamma\left(\frac{1}{2}\right) = \frac{3\sqrt{\pi}}{4}.$

2. Show that $\left(n + \frac{1}{2}\right)! = \frac{\sqrt{\pi}}{2^{2n}}\frac{(2n+1)!}{n!}$

Solution.

$$\left(n+\frac{1}{2}\right)! = \left(n+\frac{1}{2}\right)\left(n-\frac{1}{2}\right)\cdots\left(\frac{1}{2}\right)!$$

$$= \frac{(2n+1)(2n-1)\cdots 1}{2^n}\left(\frac{1}{2}\right)! = \frac{(2n+1)!}{2^{2n}n!}\sqrt{\pi}$$

3. Use *Mathematica* to find numerical approximations to $\Gamma(100)$, $\Gamma(0.01)$ and $\Gamma(i)$.

Solution. We can compute the three values simultaneously with

N[{Gamma[100],Gamma[0.01],Gamma[I]}]

yielding

```
                 155
{9.33262 10    , 99.4326, -0.15495 - 0.498016 I}
```

4. Use *Mathematica*'s command **Plot3D** to plot $\Gamma(u+iv)$ for $-1.5 \le u \le 1.5$ and $-2 \le v \le 2$.

Solution.

Plot3D[Re[Gamma[u + I v]],{u,-3/2,4},{v,-2,2},
PlotPoints->40]

Section 14.5, page 473

Find the following Laplace transforms.

1. $\mathcal{L}(t\cos(bt))$

Solution. Since

$$\mathcal{L}\big(\cos(bt)\big) = F(s) = \frac{s}{s^2+b^2},$$

we use Theorem 14.11 and obtain

$$\mathcal{L}(t\cos(bt)) = -\frac{dF(s)}{ds} = \frac{2s^2}{(s^2+b^2)^2} - \frac{1}{s^2+b^2}.$$

2. $\mathcal{L}(ae^{bt})$

Solution. Since the Laplace transform is linear, we can write

$$\mathcal{L}(ae^{bt}) = a\,\mathcal{L}(e^{bt}) = \frac{a}{s-b}.$$

3. $\mathcal{L}(\sin(at) + \cos(at))$

Solution. Since the Laplace transform is linear, we can write

$$\mathcal{L}(\sin(at) + \cos(at)) = \mathcal{L}(\sin(at)) + \mathcal{L}(\cos(at)) = \frac{a}{s^2 + a^2} + \frac{s}{s^2 + a^2}.$$

4. $\mathcal{L}((\sin(at))^2)$

Solution. Using the linearity of the Laplace transform and the identity

$$(\sin(at))^2 = \frac{1}{2} - \frac{\cos(2at)}{2},$$

we obtain

$$\frac{1}{2s} - \frac{s}{2(s^2 + 4a^2)}.$$

5. $\mathcal{L}(e^t(t - \sin(t)))$

Solution. First, we write the expression as $t\,e^t - e^t \sin(t)$. For the first expression we use Theorem 14.11 to obtain

$$\mathcal{L}(t\,e^t) = -\frac{d}{ds}\left(\frac{1}{s - 1}\right) = \frac{1}{(s - 1)^2}$$

and for the second expression,

$$\mathcal{L}(e^t \sin(t)) = \frac{1}{s^2 + 1}\Big|_{s \to s-1} = \frac{1}{(s - 1)^2 + 1}.$$

Finally, the linearity of the Laplace transform results in $\dfrac{1}{(s - 1)^2} - \dfrac{1}{(s - 1)^2 + 1}$.

6. $\mathcal{L}((t + e^t)^3)$

Solution. After multiplying, we obtain $e^{3t} + 3t\,e^{2t} + 3t^2 e^t + t^3$. The first and last expressions are found in the table of transforms as simply $1/(s - 3)$ and $6/s^4$, respectively. We use Theorem 14.11 to compute the other transforms as follows:

$$\mathcal{L}(3t\,e^{2t}) = -3\frac{d}{ds}\left(\frac{1}{s - 2}\right) = \frac{3}{(s - 2)^2}$$

and

$$\mathcal{L}(3t^2 e^t) = 3\frac{d^2}{ds^2}\left(\frac{1}{s - 1}\right) = \frac{6}{(s - 1)^3}.$$

Hence, the complete solution is $\dfrac{1}{s - 3} + \dfrac{3}{(s - 2)^2} + \dfrac{6}{(s - 1)^3} + \dfrac{6}{s^4}$.

7. $L(t^2 - e^{-9t} + 5)$

Solution. Using the linearity of the Laplace transform and the identity

$$L(t^2 - e^{-9t} + 5) = L(t^2) - L(e^{-9t}) + 5L(1).$$

Hence, the solution is $\dfrac{2}{s^3} - \dfrac{1}{s+9} + \dfrac{5}{s}$.

8. $L(e^{-t}\cosh(k\,t))$

Solution. This is a simple application of Theorem 14.4 by observing that

$$L(e^{-t}\cosh(k\,t)) = L(\cosh(k\,t))\Big|_{s\to s+1} = \frac{s}{s^2 - k^2}\Big|_{s\to s+1}.$$

Hence, the solution is $\dfrac{s+1}{(s+1)^2 - k^2}$.

9. $L(\sin(2t)\cos(2t))$ [Hint: Consider $\sin(4t)$.]

Solution. After writing $\sin(2t)\cos(2t) = (1/2)\sin(4t)$, we obtain

$$L(\sin(2t)\cos(2t)) = \frac{1}{2}L(\sin(4t)) = \frac{1}{2}\left(\frac{4}{s^2+16}\right) = \frac{2}{s^2+16}.$$

10. $L(\sin(t)\cos(2t))$ [Hint: Consider $\sin(t_1 \pm t_2)$.]

Solution. Using the identities $\sin(t_1 + t_2) = \sin(t_1)\cos(t_2) + \cos(t_1)\sin(t_2)$ and $\sin(t_1 - t_2) = \sin(t_1)\cos(t_2) - \cos(t_1)\sin(t_2)$,

we obtain

$$\sin(t_1)\cos(t_2) = \frac{1}{2}(\sin(t_1 + t_2) + \sin(t_1 - t_2)).$$

Therefore,

$$L(\sin(t)\cos(2t)) = \frac{1}{2}L(\sin(3t)) + \frac{1}{2}L(\sin(-t)).$$

Hence, the solution is $\dfrac{3}{2(s^2+9)} - \dfrac{1}{2(s^2+1)}$.

11. $L((\cos(t))^2)$ [Hint: Consider $\cos(2t)$.]

Solution. Using the identity

$$(\cos(t))^2 = \frac{1}{2} + \frac{\cos(2t)}{2},$$

we obtain

$$\mathcal{L}\big((\cos(t))^2\big) = \frac{1}{2}\mathcal{L}(1) + \frac{1}{2}\mathcal{L}\big(\cos(2t)\big) = \frac{1}{2s} + \frac{s}{2(s^2+4)}.$$

12. $\mathcal{L}\big((\sin(t))^3\big)$ [Hint: $(\sin(t))^3 = \sin(t)(\sin(t))^2.$]

   Solution. First, we write $(\sin(t))^3 = \sin(t)(\sin(t))^2 = \sin(t)/2 - \sin(t)\cos(2t)/2$. Then

$$\mathcal{L}\big((\sin(t))^3\big) = \frac{1}{2}\mathcal{L}\big((\sin(t))\big) - \frac{1}{2}\mathcal{L}\big((\sin(t)\cos(2t))\big).$$

Using Exercise 10 we obtain the solution

$$\frac{1}{2(s^2+1)} - \frac{3}{4(s^2+9)} + \frac{1}{4(s^2+1)} = \frac{6}{s^4+10s^2+9}$$

Solve the following initial value problems using Laplace transforms:

13. $\begin{cases} y'(t) + 2y(t) = 0, \\ y(0) = 1 \end{cases}$

   Solution. When we apply the Laplace transform to both sides of the differential equation, incorporate the initial condition and solve for $\mathcal{L}(y(t))$, we obtain

$$\mathcal{L}(y(t)) = \frac{1}{s+2}.$$

Hence, $y(t) = e^{-2t}$.

14. $\begin{cases} y'(t) - y(t) = 2e^t + t\, e^t, \\ y(0) = 2 \end{cases}$

   Solution. When we apply the Laplace transform to both sides of the differential equation, incorporate the initial condition and solve for $\mathcal{L}(y(t))$, we obtain

$$\mathcal{L}(y(t)) = \frac{2s^2 - 2s + 1}{(s-1)^3}.$$

Using partial fraction decomposition, we obtain

$$\frac{2s^2 - 2s + 1}{(s-1)^3} = \frac{2}{s-1} + \frac{2}{(s-1)^2} + \frac{1}{(s-1)^3}$$

We note that

$$\frac{2}{(s-1)^2} = 2\left(\frac{1}{s^2}\Big|_{s\to s-1}\right)$$

and

$$\frac{1}{(s-1)^3} = \frac{1}{2}\left(\frac{2}{s^3}\Big|_{s\to s-1}\right).$$

Hence, from Theorem 14.4, $y(t) = 2e^t + 2t\,e^t + \dfrac{t^2 e^t}{2}$.

15.
$$\begin{cases} y''(t) + y(t) = 2y'(t), \\ y(0) = 1, \quad y'(0) = -1 \end{cases}$$

Solution. When we apply the Laplace transform to both sides of the differential equation, incorporate the initial condition and solve for $\mathcal{L}(y(t))$, we obtain

$$\mathcal{L}(y(t)) = \frac{s-3}{(s-1)^2}.$$

Using partial fraction decomposition, we obtain

$$\frac{s-3}{(s-1)^2} = \frac{1}{s-1} - \frac{2}{(s-1)^2}.$$

Hence, as in Exercise 14, $y(t) = e^t - 2t\,e^t$.

16.
$$\begin{cases} y''(t) + y(t) = \sinh(t), \\ y(0) = 1, \quad y'(0) = 0 \end{cases}$$

Solution. When we apply the Laplace transform to both sides of the differential equation, incorporate the initial condition and solve for $\mathcal{L}(y(t))$, we obtain

$$\mathcal{L}(y(t)) = \frac{s^3 - s + 1}{s^4 - 1}.$$

Using partial fraction decomposition, we obtain

$$\frac{s^3 - s + 1}{s^4 - 1} = \frac{1}{2}\left(\frac{2s-1}{s^2+1}\right) - \frac{1}{4}\left(\frac{1}{s+1}\right) + \frac{1}{4}\left(\frac{1}{s-1}\right).$$

We note that

$$\frac{1}{2}\left(\frac{2s-1}{s^2+1}\right) = 2\left(\frac{1}{s^2+1}\right) - \frac{s}{s^2+1}.$$

Hence, $y(t) = \cos(t) - \dfrac{\sin(t)}{2} - \dfrac{1}{4e^t} + \dfrac{e^t}{4}$.

17.
$$\begin{cases} y''(t) - y'(t) - 6y(t) = 0, \\ y(0) = 2, \quad y'(0) = -1 \end{cases}$$

Solution. When we apply the Laplace transform to both sides of the differential equation, incorporate the initial condition and solve for $\mathcal{L}(y(t))$, we obtain

$$\mathcal{L}(y(t)) = \frac{2s - 3}{s^2 - s - 6}.$$

Using partial fraction decomposition, we obtain

$$\frac{2s - 3}{s^2 - s - 6} = \frac{3}{5}\left(\frac{1}{s - 3}\right) + \frac{7}{5}\left(\frac{1}{s + 2}\right).$$

Hence, $y(t) = \dfrac{3e^{3t}}{5} + \dfrac{7e^{-2t}}{5}$.

18.
$$\begin{cases} y''(t) + y(t) = \sin(2t), \\ y(0) = y'(0) = 0 \end{cases}$$

Solution. When we apply the Laplace transform to both sides of the differential equation, incorporate the initial condition and solve for $\mathcal{L}(y(t))$, we obtain

$$\mathcal{L}(y(t)) = \frac{2}{s^4 + 5s^2 + 4}.$$

Using partial fraction decomposition, we obtain

$$\frac{2}{s^4 + 5s^2 + 4} = \frac{2}{3}\left(\frac{1}{s^2 + 1}\right) - \frac{1}{3}\left(\frac{2}{s^2 + 4}\right).$$

Hence, $y(t) = \dfrac{2}{3}\sin(t) - \dfrac{1}{3}\sin(2t)$.

19.
$$\begin{cases} y''(t) + y(t) = \exp(3t), \\ y(0) = y'(0) = 0 \end{cases}$$

Solution. When we apply the Laplace transform to both sides of the differential equation, incorporate the initial condition and solve for $\mathcal{L}(y(t))$, we obtain

$$\mathcal{L}(y(t)) = \frac{1}{s^3 - 3s^2 + s - 3}.$$

Using partial fraction decomposition, we obtain

$$\frac{1}{s^3 - 3s^2 + s - 3} = \frac{1}{10}\left(\frac{1}{s-3}\right) - \frac{1}{10}\left(\frac{s}{s^2+1}\right) - \frac{3}{10}\left(\frac{1}{s^2+1}\right).$$

Hence, $y(t) = \dfrac{e^{3t} - \cos(t) - 3\sin(t)}{10}$.

20. $\begin{cases} y''(t) + y(t) = t + \exp(t), \\ y(0) = y'(0) = 0 \end{cases}$

Solution. When we apply the Laplace transform to both sides of the differential equation, incorporate the initial condition and solve for $\mathcal{L}(y(t))$, we obtain

$$\mathcal{L}(y(t)) = \frac{s^2 + s - 1}{s^5 - s^4 + s^3 - s^2}.$$

Using partial fraction decomposition, we obtain

$$\frac{s^2 + s - 1}{s^5 - s^4 + s^3 - s^2} = \frac{1}{2}\left(\frac{1}{s-1}\right) + \frac{1}{s^2} - \frac{1}{2}\left(\frac{s}{s^2+1}\right) - \frac{3}{2}\left(\frac{1}{s^2+1}\right).$$

Hence, $y(t) = \dfrac{1}{2}\left(e^t + 2t - \cos(t) - 3\sin(t)\right)$

21. A function $f(t)$ is said to be **periodic** with period $T > 0$ provided that $f(t) = f(t+T)$ for all t. Suppose that if $f(t)$ is piecewise continuous and of exponential growth. If $f(t)$ is periodic with period T, show that

$$\mathcal{L}(f(t)) = \frac{1}{1 - e^{-sT}} \int_0^T e^{-st} f(t)\, dt. \tag{S.80}$$

Solution. First, we write

$$\mathcal{L}(f(t)) = \int_0^T e^{-st} f(t)\, dt + \int_T^\infty e^{-st} f(t)\, dt.$$

Substitution of $t = u + T$ converts the last integral into

$$\int_T^\infty e^{-st} f(t)\, dt = e^{-sT} \int_0^\infty e^{-su} f(u)\, du = e^{-sT} \mathcal{L}(f(t)).$$

Hence,

$$\mathcal{L}(f(t)) = \int_0^T e^{-st} f(t)\, dt + e^{-sT} \mathcal{L}(f(t)).$$

When we solve this equation for $\mathcal{L}(f(t))$ we get (S.80).

22. Show that $L(f(at)) = \dfrac{1}{a}F\left(\dfrac{s}{a}\right)$.

Solution. Using the definition and the transformation $u = at$, we get

$$L(f(at)) = \int_0^\infty e^{-st} f(at)\, dt = \frac{1}{a}\int_0^\infty e^{-su/a} f(u)\, du = \frac{1}{a}F\left(\frac{s}{a}\right).$$

23. If $F(s) = L(f(t))$, show that $L\left(\displaystyle\int_0^t f(\tau)\, d\tau\right) = \dfrac{1}{s}F(s)$.

Solution. Let $G(t) = \displaystyle\int_0^t f(\tau)\, d\tau$. Then $G(0) = 0$ and $G'(t) = f(t)$. Using $L(G'(t)) = sL(G(t)) - 0$, we obtain $L(G'(t)) = F(s) = sL(G(t))$. Now divide by s.

24. If $F(s) = L(f(t))$, show that $L\left(\dfrac{f(t)}{t}\right) = \displaystyle\int_s^\infty F(\tau)\, d\tau$.

Solution. We have

$$F(s) = L(f(t)) = L\left(t\frac{f(t)}{t}\right) = -\frac{d}{ds}L\left(\frac{f(t)}{t}\right).$$

When we integrate this equation with respect to s and use the fact that $\displaystyle\lim_{s\to\infty} F(s) = 0$, we obtain the result.

Section 14.6, page 478

Write the following functions in terms of unit step functions.

14.6. $f(t) = \begin{cases} 0 & \text{if } t < 0, \\ 1 & \text{if } t \ge 0 \end{cases}$

Solution. $f(t) = u_0(t)$

14.6. $f(t) = \begin{cases} 0 & \text{if } t < 0, \\ t & \text{if } t \ge 0 \end{cases}$

Solution. $f(t) = t\, u_0(t)$

14.6. $f(t) = \begin{cases} 0 & \text{if } t < 0, \\ 2 & \text{if } 0 \le t < 1, \\ -1 & \text{if } t \ge 1 \end{cases}$

Solution. $f(t) = 2u_0(t) - 3u_1(t)$

14.6. $f(t) = \begin{cases} 0 & \text{if } t < 0, \\ \sin(t) & \text{if } 0 \le t < \pi, \\ \cos(t) & \text{if } t \ge \pi \end{cases}$

Solution. $f(t) = \sin(t)u_0(t) + \big(\cos(t) - \sin(t)\big)u_\pi(t)$

5. $f(t) = \begin{cases} 1 & \text{if } 0 \le t < 1, \\ e^t & \text{if } 1 \le t < 2, \\ 2 & \text{if } t \ge 2 \end{cases}$

Solution. $f(t) = u_0(t) + (e^t - 1)u_1(t) + (2 - e^t)u_2(t)$

6. $f(t) = \begin{cases} 1 & \text{if } 0 \le t < 1, \\ 4t - t^2 & \text{if } 1 \le t < 2, \\ 1 & \text{if } t \ge 2 \end{cases}$

Solution. $f(t) = u_0(t) - (t^2 - 4t + 1)u_1(t) + (t^2 - 4t + 1)u_2(t)$

Graph the following functions:

7. $f(t) = e^t u_3(t)$
Solution.

```
Plot[E^t UnitStep[t - 3],{t,-1,4}];
```

8. $g(t) = (t - 2)^2 u_2(t)$
Solution.

```
Plot[(t - 2)^2UnitStep[t - 2],{t,-1,4},
PlotRange->All];
```

9. $f(t) = t - (t - 1)u_1(t)$
Solution.

```
Plot[t - (t - 1)UnitStep[t - 1],{t,-1,4},
PlotRange->All];
```

10. $g(t) = (t - 3)u_2(t) - (t - 2)u_3(t) + 2u_5(t)$
Solution.

```
Plot[(t - 3)UnitStep[t - 2] - (t - 2)UnitStep[t - 3],
{t,-1,4},PlotRange->All];
```

Find the Laplace transforms of the following functions:

11. $f(t) = \begin{cases} 3 & \text{if } 0 \leq t < 2, \\ 0 & \text{if } t \geq 2 \end{cases}$

Solution. We first write f in terms of unit step functions as

$$f(t) = 3(u_0(t) - u_2(t)) = 3u_0(t) - 3u_2(t);$$

then we apply Theorem 14.18 and obtain

$$L(f(t))(s) = \frac{3}{s} - \frac{3e^{-2s}}{s} = \frac{3(1 - e^{-2s})}{s}.$$

12. $f(t) = u_\pi(t) \sin(t - \pi)$

Solution. Since f is already in terms of unit step functions, we can apply Theorem 14.18 immediately and obtain

$$L(f(t))(s) = e^{-\pi s} L(\sin(t))(s) = \frac{e^{-\pi s}}{(s^2 + 1)}.$$

13. $f(t) = \begin{cases} 4 & \text{if } 1 \leq t < 3, \\ 0 & \text{if } t < 1 \text{ or } \geq 3 \end{cases}$

Solution. We first write f in terms of unit step functions as

$$f(t) = 4(u_1(t) - u_3(t)) = 4u_1(t) - 4u_3(t);$$

then we apply Theorem 14.18 and obtain

$$L(f(t))(s) = \frac{4e^{-s}}{s} - \frac{4e^{-3s}}{s} = 4\left(\frac{e^{-s}}{s} - \frac{e^{-3s}}{s}\right).$$

14. $f(t) = u_\pi(t) \cos(t)$

Solution. We first write f in terms of unit step functions as

$$f(t) = u_\pi(t) \cos(t) = -u_\pi(t) \cos(t - \pi);$$

then we apply Theorem 14.18 and obtain

$$L(f(t))(s) = -e^{-\pi s} L(\cos(t))(s) = -\frac{se^{-\pi s}}{(s^2 + 1)}.$$

15. $f(t) = u_{\pi/2}(t)\cos(t)$

Solution. We first write f in terms of unit step functions as

$$f(t) = u_{\pi/2}(t)\cos(t) = -u_\pi(t)\sin(t - \pi/2);$$

then we apply Theorem 14.18 and obtain

$$\mathcal{L}(f(t))(s) = -e^{-\pi s/2}\mathcal{L}(\sin(t))(s) = -\frac{e^{-\pi s/2}}{(s^2 + 1)}.$$

16. $f(t) = \begin{cases} |\sin(t)| & \text{if } 0 \le t < 2\pi, \\ 0 & \text{if } t < 0 \text{ or } \ge 2\pi \end{cases}$

Solution. We first write f in terms of unit step functions as

$$f(t) = \sin(t)\big(u_0(t) - u_\pi(t)\big) - \sin(t)\big(u_\pi(t) - u_{2\pi}(t)\big).$$

After expanding and using familiar trigonometric identities, this can be rewritten as

$$f(t) = \sin(t)u_0(t) + 2\sin(t - \pi)u_\pi(t) + \sin(t - 2\pi)u_{2\pi}(t).$$

Now we apply Theorem 14.18 and obtain

$$\mathcal{L}(f(t))(s) = \big(1 + 2e^{-\pi s} + e^{-2\pi s}\big)\mathcal{L}(\sin(t))(s) = \frac{1 + 2e^{-\pi s} + e^{-2\pi s}}{(s^2 + 1)}$$

Find the inverse Laplace transform of the following functions:

17. $\dfrac{e^{-s}}{s}$

Solution. Using Theorem 14.18, we write

$$\mathcal{L}^{-1}\Big(\frac{e^{-s}}{s}\Big)(t) = u_1(t)\mathcal{L}^{-1}\Big(\frac{1}{s}\Big)(t - 1) = u_1(t).$$

18. $\dfrac{e^{-2s}}{s - 3}$

Solution. Using Theorem 14.18, we write

$$\mathcal{L}^{-1}\Big(\frac{e^{-2s}}{s - 3}\Big)(t) = u_2(t)\mathcal{L}^{-1}\Big(\frac{1}{s - 3}\Big)(t - 2) = u_2(t)\left(e^{3t}\Big|_{t \to t-2}\right) = u_2(t)e^{3t-6}.$$

19. $\dfrac{e^{-3s}}{s^2+4}$

Solution. Using Theorem 14.18, we write

$$\mathcal{L}^{-1}\left(\frac{e^{-3s}}{s^2+4}\right)(t) = \frac{u_3(t)}{2}\mathcal{L}^{-1}\left(\frac{2}{s^2+4}\right)(t-3) = \frac{u_3(t)}{2}\left(\sin(2t)\Big|_{t\to t-3}\right)$$

$$= \frac{u_3(t)}{2}\sin(2t-6).$$

20. $\dfrac{e^{-4s}}{s+4}$

Solution. Using Theorem 14.18, we write

$$\mathcal{L}^{-1}\left(\frac{e^{-4s}}{s+4}\right)(t) = u_4(t)\mathcal{L}^{-1}\left(\frac{1}{s+4}\right)(t-4) = u_4(t)\left(e^{-4t}\Big|_{t\to t-4}\right) = u_4(t)e^{16-4t}$$

21. $\dfrac{s+e^{-\pi s}}{a(s+1)^2}$

Solution. Using the linearity of the inverse Laplace transform, we first write

$$\mathcal{L}^{-1}\left(\frac{s+e^{-\pi s}}{a(s+1)^2}\right)(t) = \mathcal{L}^{-1}\left(\frac{s}{a(s+1)^2}\right)(t) + \mathcal{L}^{-1}\left(\frac{e^{-\pi s}}{a(s+1)^2}\right)(t).$$

The first term can be written as

$$\frac{s}{a(s+1)^2} = \frac{1}{a}\left(\frac{1}{s+1} - \frac{1}{(s+1)^2}\right)$$

which inverts to

$$\frac{1}{a}\left(e^{-t} - te^{-t}\right).$$

The second term uses Theorem 14.18 and can be written

$$\frac{u_\pi(t)}{a}\left(te^{-t}\Big|_{t\to t-\pi}\right) = \frac{u_\pi(t)(t-\pi)e^{-t+\pi}}{a}.$$

Therefore, the complete solution is

$$\frac{1}{a}\left(e^{-t} - te^{-t}\right) + \frac{u_\pi(t)(t-\pi)e^{-t+\pi}}{a}.$$

22. $\dfrac{e^{-s} - 2e^{-2s} + 2e^{-3s} - e^{-4s}}{s}$

Solution. Using the linearity of the inverse Laplace transform, we first write

$$\mathcal{L}^{-1}\left(\frac{e^{-s}}{s}\right)(t) - 2\mathcal{L}^{-1}\left(\frac{e^{-2s}}{s}\right)(t) + 2\mathcal{L}^{-1}\left(\frac{e^{-3s}}{s}\right)(t) - \mathcal{L}^{-1}\left(\frac{e^{-4s}}{s}\right)(t).$$

Using Theorem 14.18, we note that

$$\mathcal{L}^{-1}\left(\frac{e^{-as}}{s}\right)(t) = u_a(t),$$

hence the solution becomes $u_1(t) - 2u_2(t) + 2u_3(t) - u_4(t)$.

Section 14.7, page 481

Solve and plot each of the following initial value problems.

1. $\begin{cases} y' = 1 - u_1(t), \\ y(0) = 0 \end{cases}$

Solution. When we apply the Laplace transform to both sides of the differential equation, solve for $Y(s)$ and use partial fraction decomposition, we obtain

$$Y(s) = \frac{1}{s^2} - \frac{e^{-s}}{s^2}.$$

The first term is the Laplace transform of t, the second term is the Laplace transform of $u_1(t)(t - 1)$. Hence,

$$y(t) = t + (1 - t)u_1(t).$$

The **ODE** command which solves this problem is

```
ODE[{y' == 1 - UnitStep[t - 1],y[0] == 0},y,t,
Method->Laplace,PlotSolution->{{t,0,2}}]

{{y -> t + UnitStep[-1 + t] - t UnitStep[-1 + t]}}
```

2. $\begin{cases} y' = 1 - 2u_1(t) + u_2(t), \\ y(0) = 1 \end{cases}$

Solution. When we apply the Laplace transform to both sides of the differential equation, solve for $Y(s)$ and use partial fraction decomposition, we obtain

$$Y(s) = \frac{1}{s} + \frac{1}{s^2} - 2\frac{e^{-s}}{s^2} + \frac{e^{-2s}}{s^2}.$$

The first term is the Laplace transform of 1, the second term is the Laplace transform of t, the third term is the Laplace transform of $2u_1(t)(t-1)$ and the last term is the Laplace transform of $u_2(t)(t-2)$. Hence,

$$y(t) = 1 + t + (2 - 2t)u_1(t) + (t - 2)u_2(t).$$

The **ODE** command which solves this problem is

```
ODE[{y' == 1 - 2UnitStep[t - 1] + UnitStep[t - 2],
y[0] == 1},y,t,Method->Laplace,PlotSolution->{{t,0,3}}]

{{y -> 1 + t - 2 UnitStep[-2 + t] + t UnitStep[-2 + t] +

    2 UnitStep[-1 + t] - 2 t UnitStep[-1 + t]}}
```

3. $$\begin{cases} y' + y = t - u_1(t), \\ y(0) = 1 \end{cases}$$

Solution. When we apply the Laplace transform to both sides of the differential equation, solve for $Y(s)$ and use partial fraction decomposition, we obtain

$$Y(s) = \frac{1}{s^2} - \frac{1}{s} + 2\left(\frac{1}{s+1}\right) - \frac{e^{-s}}{s} + \frac{e^{-s}}{s+1}.$$

The first term is the Laplace transform of t, the second term is the Laplace transform of 1, the third term is the Laplace transform of $2e^{-t}$. The final terms use Theorem 14.18 and become $u_1(t)$ and $u_1(t)e^{1-t}$. Hence,

$$y(t) = 2e^{-t} - 1 + t + (e^{1-t} - 1)u_1(t).$$

The **ODE** command which solves this problem is

```
ODE[{y' + y == t - UnitStep[t - 1],y[0] == 1},y,t,
Method->Laplace,PlotSolution->{{t,0,3}}]

             2
{{y -> -1 + --- + t - UnitStep[-1 + t] +
             t
            E

        1 - t
       -----    UnitStep[-1 + t]}}
        E
```

4. $$\begin{cases} y'' + y = u_3(t), \\ y(0) = 0, \ y'(0) = 1 \end{cases}$$

Solution. When we apply the Laplace transform to both sides of the differential equation, solve for $Y(s)$ and use partial fraction decomposition, we obtain

$$Y(s) = \frac{1}{s^2 + 1} + \frac{e^{-3s}}{s} - \frac{e^{-3s}s}{s^2 + 1}.$$

The first term is the Laplace transform of $\sin(t)$, the second term is the Laplace transform of $u_3(t)$, the third term is the Laplace transform of $u_3(t)\cos(t-3)$. Hence,

$$y(t) = \sin(t) + \left(1 - \cos(t-3)\right)u_3(t).$$

The **ODE** command which solves this problem is

```
ODE[{y'' + y   == UnitStep[t - 3],y[0] == 0,y'[0] == 1},
y,t,Method->Laplace,Form->Explicit,
PostSolution->{ExpandAll[#,Trig->True]&},
PlotSolution->{{t,0,4Pi}}]

Sin[t] + UnitStep[-3 + t] - Cos[3 - t] UnitStep[-3 + t]
```

5.
$$\begin{cases} y'' + y = t - t\,u_1(t), \\ y(0) = y'(0) = 0 \end{cases}$$

Solution. When we apply the Laplace transform to both sides of the differential equation, solve for $Y(s)$ and use partial fraction decomposition, we obtain

$$Y(s) = \frac{1}{s^2} - \frac{1}{s^2 + 1} - \frac{e^{-s}}{s^2} - \frac{e^{-s}}{s} + \frac{e^{-s}}{s^2 + 1} + \frac{e^{-s}s}{s^2 + 1}.$$

The first term is the Laplace transform of t, the second term is the Laplace transform of $\sin(t)$. The final terms use Theorem 14.18 and become $u_1(t)(t-1)$, $u_1(t)$, $u_1(t)\sin(t-1)$ and $u_1(t)\cos(t-1)$ respectively. Hence,

$$y(t) = t - \sin(t) + \left(\cos(1-t) - \sin(1-t) - t\right)u_1(t)$$

The **ODE** command which solves this problem is

```
ODE[{y'' + y == t - t UnitStep[t - 1],y[0] == 0,
y'[0] == 0},y,t,Method->Laplace,PlotSolution->{{t,0,5}}]

{{y -> t - Sin[t] - t UnitStep[-1 + t] +
     Cos[1 - t] UnitStep[-1 + t] -
     Sin[1 - t] UnitStep[-1 + t]}}
```

6.
$$\begin{cases} y'' + y = u_\pi(t) - u_{2\pi}(t), \\ y(0) = 0, \quad y'(0) = 1 \end{cases}$$

Solution. When we apply the Laplace transform to both sides of the differential equation, solve for $Y(s)$ and use partial fraction decomposition, we obtain

$$Y(s) = \frac{1}{s^2+1} + e^{-\pi s}\left(\frac{1}{s} - \frac{s}{s^2+1}\right) - e^{-2\pi s}\left(\frac{1}{s} - \frac{s}{s^2+1}\right).$$

The first term is the Laplace transform of $\sin(t)$, the second term is the Laplace transform of $u_\pi(t)(1 - \cos(t - \pi)) = u_\pi(t)(1 + \cos(t))$ and the third term is the Laplace transform of $u_{2\pi}(t)(1 - \cos(t - 2\pi)) = u_{2\pi}(t)(1 - \cos(t))$. Using half-angle identities, we can write this as

$$y(t) = \sin(t) - 2(\sin(t/2))^2 u_{2\pi}(t) + 2(\cos(t/2))^2 u_\pi(t)$$

The **ODE** command which solves this problem is

```
ODE[{y'' + y == UnitStep[t - Pi] - UnitStep[t - 2Pi],
y[0] == 0,y'[0] == 1},y,t,
Method->Laplace,PlotSolution->{{t,0,10}}]

                     t 2
{{y -> Sin[t] - 2 Sin[-]  UnitStep[-2 Pi + t] +
                     2

        t 2
  2 Cos[-]  UnitStep[-Pi + t]}}
        2
```

7.
$$\begin{cases} y'' + 4y = t - u_{\pi/2}(t)(t - \pi/2), \\ y(0) = y'(0) = 0 \end{cases}$$

Solution. When we apply the Laplace transform to both sides of the differential equation, solve for $Y(s)$ and use partial fraction decomposition, we obtain

$$Y(s) = \frac{1}{4}\left(\frac{1}{s^2}\right) - \frac{1}{8}\left(\frac{2}{s^2+4}\right) - \frac{1}{4}\left(\frac{e^{-\pi s/2}}{s^2}\right) + \frac{1}{8}\left(\frac{e^{-\pi s/2}2}{s^2+4}\right).$$

The first term is the Laplace transform of $t/4$, the second term is the Laplace transform of $\sin(2t)/8$ and the third term is the Laplace transform of $u_{\pi/2}(t)(t - \pi/2)/4$ and the last term is the Laplace transform of $u_{\pi/2}(t)\sin(2t - \pi)/8 = -u_{\pi/2}(t)\sin(2t)/8$. Hence,

$$y(t) = \frac{t}{4} - \frac{\sin(2t)}{8} + \left(\frac{\pi - 2t - \sin(2t)}{8}\right)u_{\pi/2}(t).$$

The **ODE** command which solves this problem is

```
ODE[{y'' + 4y  == t - UnitStep[t - Pi/2](t - Pi/2),
y[0] == 0,y'[0] == 0},y,t,Method->Laplace,
PlotSolution->{{t,0,10}}]
```

$$
\{\{y \to \frac{t}{4} - \frac{\text{Sin}[2\ t]}{8} + \frac{\text{Pi UnitStep}[\dfrac{-\text{Pi}}{2} + t]}{8} -
$$

$$
\frac{t\ \text{UnitStep}[\dfrac{-\text{Pi}}{2} + t]}{4} - \frac{\text{Sin}[2\ t]\ \text{UnitStep}[\dfrac{-\text{Pi}}{2} + t]}{8}\}\}
$$

8. $\begin{cases} y'' + 4y = u_{2\pi}(t)\sin(t), \\ y(0) = 1, \quad y'(0) = 0 \end{cases}$

Solution. When we apply the Laplace transform to both sides of the differential equation, solve for $Y(s)$ and use partial fraction decomposition, we obtain

$$
Y(s) = \frac{s}{s^2+4} + \frac{1}{3}\left(\frac{e^{-2\pi s}}{s^2+1}\right) - \frac{1}{6}\left(\frac{e^{-2\pi s}2}{s^2+4}\right).
$$

The first term is the Laplace transform of $\cos(2t)$, the second term is the Laplace transform of $u_{2\pi}(t)\sin(t-2\pi)/3 = u_{2\pi}(t)\sin(t)/3$ and the last term is the Laplace transform of $u_{2\pi}(t)\sin(2t-4\pi)/6 = u_{2\pi}(t)\sin(2t)/6$. Hence,

$$
y(t) = \cos(2t) + u_{2\pi}(t)\left(\frac{\sin(t)}{3} - \frac{\sin(2t)}{6}\right).
$$

The **ODE** command which solves this problem is

```
ODE[{y'' + 4y == UnitStep[t - 2Pi]Sin[t],
y[0] == 1,y'[0] == 0},y,t,Method->Laplace,
PlotSolution->{{t,0,10}}]
```

$$
\{\{y \to \text{Cos}[2\ t] + \frac{4\ \text{Cos}[\dfrac{t}{2}]\ \text{Sin}[\dfrac{t}{2}]^3\ \text{UnitStep}[-2\ \text{Pi} + t]}{3}\}\}
$$

9. $\begin{cases} y'' + y = u_{\pi}(t) - u_{2\pi}(t), \\ y(0) = 0, \quad y'(0) = 1 \end{cases}$

Solution. When we apply the Laplace transform to both sides of the differential equation, solve for $Y(s)$ and use partial fraction decomposition, we obtain

$$Y(s) = \frac{1}{s^2+1} + \frac{e^{-\pi s}}{s} - \frac{e^{-\pi s}s}{s^2+1} - \frac{e^{-2\pi s}}{s} + \frac{e^{-2\pi s}s}{s^2+1}.$$

The first term is the Laplace transform of $\sin(t)$, the second term is the Laplace transform of $u_\pi(t)$, the third term is the Laplace transform of $u_\pi \cos(t - \pi) = -u_\pi \cos(t)$, the forth term is the Laplace transform of $u_{2\pi}(t)$ and the last term is the Laplace transform of $u_{2\pi} \cos(t - 2\pi) = u_{2\pi} \cos(t)$. Hence,

$$y(t) = \sin(t) + u_\pi(t)\,(1 + \cos(t)) - u_{2\pi}(t)\,(1 - \cos(t)).$$

The **ODE** command which solves this problem is

```
ODE[{y'' + y == UnitStep[t - Pi] - UnitStep[t - 2Pi],
y[0] == 0,y'[0] == 1},y,t,Method->Laplace,
PlotSolution->{{t,0,10}}]

{{y -> Sin[t] - UnitStep[-2 Pi + t] +
    Cos[t] UnitStep[-2 Pi + t] + UnitStep[-Pi + t] +
    Cos[t] UnitStep[-Pi + t]}}
```

10. $\begin{cases} y'' + 4y' + 3y = 1 - u_2(t) - u_4(t) + u_6(t), \\ y(0) = y'(0) = 0 \end{cases}$

Solution. When we apply the Laplace transform to both sides of the differential equation, solve for $Y(s)$ and use partial fraction decomposition, we obtain

$$Y(s) = \frac{1}{3}\left(\frac{1}{s}\right) - \frac{1}{2}\left(\frac{1}{s+1}\right) + \frac{1}{6}\left(\frac{1}{s+3}\right)$$

$$-e^{-2s}\left(\frac{1}{3}\left(\frac{1}{s}\right) - \frac{1}{2}\left(\frac{1}{s+1}\right) + \frac{1}{6}\left(\frac{1}{s+3}\right)\right)$$

$$-e^{-4s}\left(\frac{1}{3}\left(\frac{1}{s}\right) - \frac{1}{2}\left(\frac{1}{s+1}\right) + \frac{1}{6}\left(\frac{1}{s+3}\right)\right)$$

$$+e^{-6s}\left(\frac{1}{3}\left(\frac{1}{s}\right) - \frac{1}{2}\left(\frac{1}{s+1}\right) + \frac{1}{6}\left(\frac{1}{s+3}\right)\right)$$

The first term is the Laplace transform of $1/3$, the second term is the Laplace transform of $e^{-t}/2$ and the third term is the Laplace transform of $e^{-3t}/6$. The other terms are derived from these three using Theorem 14.18. Hence,

$$y(t) = \frac{e^{-3t}}{6}\left((1 - 3e^{2t} + 2e^{3t}) + (3e^{2+2t} - 2e^{3t} - e^{6})u_2(t)\right.$$

$$\left.(3e^{4+2t} - 2e^{3t} - e^{12})u_4(t) - (3e^{6+2t} - 2e^{3t} - e^{18})u_6(t)\right).$$

The **ODE** command which solves this problem is

```
ODE[{y'' + 4y' + 3y  == 1 - UnitStep[t - 2] -
UnitStep[t - 4] + UnitStep[t - 6],y[0] == 0,y'[0] == 0},
y,t,Method->Laplace,PlotSolution->{{t,0,10}}]
```

yielding

$$\{\{y \to \frac{1}{3} + \frac{1}{3 t} - \frac{1}{t} + \frac{\text{UnitStep}[-6 + t]}{3} +$$

$$\frac{E^{18 - 3 t} \quad \text{UnitStep}[-6 + t]}{6} - \frac{E^{6 - t} \quad \text{UnitStep}[-6 + t]}{2} -$$

$$\frac{\text{UnitStep}[-4 + t]}{3} - \frac{E^{12 - 3 t} \quad \text{UnitStep}[-4 + t]}{6} +$$

$$\frac{E^{4 - t} \quad \text{UnitStep}[-4 + t]}{2} - \frac{\text{UnitStep}[-2 + t]}{3} -$$

$$\frac{E^{6 - 3 t} \quad \text{UnitStep}[-2 + t]}{6} + \frac{E^{2 - t} \quad \text{UnitStep}[-2 + t]}{2}\}\}$$

Section 14.8, page 486

Use the Laplace transform to solve the following initial value problems:

1. $\begin{cases} y'(t) + y(t) = \delta(t - 2), \\ y(0) = 2 \end{cases}$

Solution. When we apply the Laplace transform to both sides of the differential equation, incorporate the initial condition and solve for $\mathcal{L}(y(t))$, we obtain

$$\mathcal{L}(y(t)) = \frac{1 + 2e^{2s}}{e^{2s}(s + 1)}.$$

Using partial fraction decomposition, we obtain

$$\frac{1 + 2e^{2s}}{e^{2s}(s+1)} = 2\left(\frac{1}{s+1}\right) + \frac{e^{-2s}}{s+1}.$$

Hence, $y(t) = 2e^{-t} + e^{2-t}u_2(t)$.

2. $\begin{cases} y'' + 2y' + 2y = t\,\delta(t - \pi), \\ y(0) = y'(0) = 0 \end{cases}$

Solution. When we apply the Laplace transform to both sides of the differential equation, incorporate the initial conditions and solve for $\mathcal{L}(y(t))$, we obtain

$$\mathcal{L}(y(t)) = \frac{\pi e^{-\pi s}}{s^2 + 2s + 2}.$$

Using partial fraction decomposition, we obtain

$$\frac{\pi e^{-\pi s}}{s^2 + 2s + 2} = \pi\left(\frac{e^{-\pi s}}{(s+1)^2 + 1}\right).$$

Since

$$\mathcal{L}^{-1}\left(\frac{e^{-\pi s}}{(s+1)^2 + 1}\right) = e^{\pi}e^{-t}\sin(t - \pi)u_\pi(t),$$

we obtain $y(t) = -\pi\,e^{\pi-t}\sin(t)u_\pi(t)$.

3. $\begin{cases} y'(t) - 3y(t) = t\,\delta(t - 2), \\ y(0) = 1 \end{cases}$

Solution. When we apply the Laplace transform to both sides of the differential equation, incorporate the initial condition and solve for $\mathcal{L}(y(t))$, we obtain

$$\mathcal{L}(y(t)) = \frac{2 + e^{2s}}{e^{2s}(s - 3)}.$$

Using partial fraction decomposition, we obtain

$$\frac{2 + e^{2s}}{e^{2s}(s - 3)} = \frac{1}{s - 3} + 2\left(\frac{e^{-2s}}{s - 3}\right).$$

Hence, $y(t) = e^{3t} + 2e^{3t-6}u_2(t)$.

4. $\begin{cases} y''(t) + y(t) = \delta(t - \pi/2) + \delta(t - 3\pi/2), \\ y(0) = y'(0) = 0 \end{cases}$

Solution. When we apply the Laplace transform to both sides of the differential equation, incorporate the initial conditions and solve for $\mathcal{L}(y(t))$, we obtain

$$\mathcal{L}(y(t)) = \frac{1 + e^{\pi s}}{e^{3\pi/2}(s^2 + 1)}.$$

Using partial fraction decomposition, we obtain

$$\frac{1 + e^{\pi s}}{e^{3\pi/2}(s^2 + 1)} = \frac{e^{-3\pi/2s}}{s^2 + 1} + \frac{e^{-\pi/2s}}{s^2 + 1}.$$

Since

$$\mathcal{L}^{-1}\left(\frac{e^{-as}}{s^2 + 1}\right) = \sin(t - a)u_a(t),$$

we obtain $y(t) = \cos(t)u_{3\pi/2}(t) - \cos(t)u_{\pi/2}(t)$.

5. $\begin{cases} y''(t) - 2y'(t) = 1 + \delta(t - 2), \\ y(0) = 0, \ y'(0) = 1 \end{cases}$

Solution. When we apply the Laplace transform to both sides of the differential equation, incorporate the initial conditions and solve for $\mathcal{L}(y(t))$, we obtain

$$\mathcal{L}(y(t)) = \frac{s + e^{2s} + s\,e^{2s}}{e^{2s}s^2(s - 2)}.$$

Using partial fraction decomposition, we obtain

$$\frac{s + e^{2s} + s\,e^{2s}}{e^{2s}s^2(s - 2)} = -\frac{3}{4}\left(\frac{1}{s}\right) + \frac{3}{4}\left(\frac{1}{s - 2}\right) - \frac{1}{2}\left(\frac{1}{s^2}\right) + \frac{1}{2}\left(\frac{e^{-2s}}{s - 2}\right) - \frac{1}{2}\left(\frac{e^{-2s}}{s}\right).$$

Hence, $y(t) = \dfrac{-3 + 3e^{2t}}{4} - \dfrac{t}{2} + \dfrac{(e^{2t-4} - 1)u_2(t)}{2}.$

6. $\begin{cases} y''(t) + 2y'(t) + y = \delta(t - 1), \\ y(0) = y'(0) = 0 \end{cases}$

Solution. When we apply the Laplace transform to both sides of the differential equation, incorporate the initial conditions and solve for $\mathcal{L}(y(t))$, we obtain

$$\mathcal{L}(y(t)) = \frac{e^{-s}}{(s + 1)^2}.$$

Since

$$\mathcal{L}^{-1}\left(\frac{1}{(s+1)^2}\right) = (t-1)e^{1-t},$$

we obtain $y(t) = (t-1)e^{1-t}u_1(t)$.

7. $\begin{cases} y''(t) + 4y'(t) + 13y = \delta(t-\pi) + \delta(t-3\pi), \\ y(0) = 1, \quad y'(0) = 0 \end{cases}$

Solution. When we apply the Laplace transform to both sides of the differential equation, incorporate the initial conditions and solve for $\mathcal{L}(y(t))$, we obtain

$$\mathcal{L}(y(t)) = \frac{1 + e^{2\pi s} + 4e^{3\pi s} + se^{3\pi s}}{e^{3\pi s}(s^2 + 4s + 13)}.$$

Using partial fraction decomposition, we obtain

$$\frac{1 + e^{2\pi s}}{e^{3\pi s}(s^2 + 4s + 13)} = \frac{e^{-3\pi s}}{(s+2)^2 + 9} + \frac{e^{-\pi s}}{(s+2)^2 + 9}$$

$$+ \frac{4}{3}\left(\frac{3}{(s+2)^2 + 9}\right) + \frac{s}{(s+2)^2 + 9}.$$

Since

$$\mathcal{L}^{-1}\left(\frac{1}{(s+2)^2 + 9}\right) = \frac{1}{3}\left(e^{-2t}\sin(3t)\right),$$

we obtain $y(t) = e^{-2t}\cos(3t) + \dfrac{2e^{-2t}\sin(3t)}{3} - \dfrac{e^{6\pi-2t}\sin(3t)u_{3\pi}(t)}{3}$

$$- \frac{e^{2\pi-2t}\sin(3t)u_\pi(t)}{3}.$$

8. $\begin{cases} y''(t) - 7y'(t) + 6y = e^t + \delta(t-2) + \delta(t-4), \\ y(0) = y'(0) = 0 \end{cases}$

Solution. When we apply the Laplace transform to both sides of the differential equation, incorporate the initial conditions and solve for $\mathcal{L}(y(t))$, we obtain

$$\mathcal{L}(y(t)) = \frac{-1 - e^{2s} + e^{4s} + s + se^{2s}}{e^{4s}(s-6)(s-1)^2}.$$

Using partial fraction decomposition, we obtain

$$\frac{-1 - e^{2s} + e^{4s} + s + s\,e^{2s}}{e^{4s}(s-6)(s-1)^2} = \frac{1}{25}\left(\frac{1}{s-6}\right) - \frac{1}{5}\left(\frac{1}{(s-1)^2}\right) - \frac{1}{25}\left(\frac{1}{s-1}\right)$$

$$+ \frac{1}{5}\left(\frac{e^{-2s}}{s-6}\right) - \frac{1}{5}\left(\frac{e^{-2s}}{s-1}\right) + \frac{1}{5}\left(\frac{e^{-4s}}{s-6}\right) - \frac{1}{5}\left(\frac{e^{-4s}}{s-1}\right).$$

Since

$$\mathcal{L}^{-1}\left(\frac{e^{-as}}{s-b}\right) = e^{b(t-a)}u_a(t),$$

we obtain $y(t) = \dfrac{-e^t + e^{6t} - 5t\,e^t}{25} + \dfrac{(e^{6t-12} - e^{t-2})u_2(t)}{5}$

$$+ \frac{(e^{6t-24} - e^{t-4})u_4(t)}{5}.$$

9. Given $y'' + a\,y' + b\,y = f(t)$, where a and b are constants and f is piecewise continuous and of exponential growth, show that the effect of replacing $f(t)$ by $f(t) + c\,\delta(t)$ has the same effect as increasing the initial value of $y'(0)$ by the constant c. A similar result holds for higher-order equations.

Solution. Computing the Laplace transform of the original equation, we get

$$\mathcal{L}(y(t))(s^2 + a\,s + b) - (a+s)y(0) - y'(0) = \mathcal{L}(f(t)).$$

If we replace $f(t)$ with $f(t) + c\,\delta(t)$, we obtain

$$\mathcal{L}(y(t))(s^2 + a\,s + b) - (a+s)y(0) - y'(0) = \mathcal{L}(f(t)) + c,$$

which is equivalent to

$$\mathcal{L}(y(t))(s^2 + a\,s + b) - (a+s)y(0) - (y'(0) + c) = \mathcal{L}(f(t)),$$

effectively replacing $y'(0)$ with $y'(0) + c$. For an nth-order linear equation, the result is clearly equivalent to replacing $y^{(n-1)}(0)$ with $y^{(n-1)}(0) + c$.

10. With A, a, b, c constant and $c > 0$, consider the initial value problem

$$\begin{cases} y'' + y = A\,\delta(t-c), \\ y(0) = a,\ y'(0) = b. \end{cases} \tag{S.81}$$

Under what conditions, if any, is $y(t) = 0$ for $t \geq c$? Equivalently, is it possible to choose the strength and location of the impulse in such way as to cancel the oscillation completely?

Solution. Applying the Laplace transform to (S.81), we obtain

$$\mathcal{L}\big(y(t)\big) = \frac{as}{s^2+1} + \frac{b}{s^2+1} + \frac{Ae^{-sc}}{s^2+1}. \qquad (S.82)$$

When we apply the inverse Laplace transform to (S.82), we get

$$y(t) = a\,\cos(t) + b\,\sin(t) + A\,\sin(t-c)u_c(t).$$

After writing the solution to the homogeneous equation $y''+y = 0$ in amplitude-phase form, we obtain

$$y(t) = R\,\sin(t+\phi) + A\,\sin(t-c)u_c(t),$$

where $R = \sqrt{a^2+b^2}$ and $\phi = \arctan(a/b)$. We note that

$$A\,\sin(t-c)u_c(t) = \begin{cases} 0 & \text{if } t < c, \\ A\,\sin(t-c) & \text{if } t \geq c. \end{cases}$$

Hence, if we choose $A = R$ and $c = \pi - \phi$, we get

$$y(t) = \begin{cases} R\,\sin(t+\phi) & \text{if } t < c, \\ 0 & \text{if } t \geq c. \end{cases}$$

Therefore, with these choices, there is no oscillation beyond $t = c$.

Section 14.9, page 489

Find the convolution $(f_1 * f_2)(t)$ in the following cases:

1. $f_1(t) = 1,\ f_2(t) = t,\ t > 0$
Solution.

$$\int_0^t 1u\,du = \frac{u^2}{2}\Big|_0^t = \frac{t^2}{2}$$

2. $f_1(t) = t^2,\ f_2(t) = t,\ t > 0$
Solution.

$$\int_0^t (t-u)\,u^2\,du = t\int_0^t u^2\,du - \int_0^t u^3\,du = t\frac{u^3}{3} - \frac{u^4}{4}\Big|_0^t = \frac{t^4}{12}$$

3. $f_3(t) = \sin(t)$, $f_2(t) = \cos(t)$, $t > 0$

Solution.

$$\int_0^t \sin(t-u)\cos(u)\,du = \sin(t)\int_0^t (\cos(u))^2\,du - \cos(t)\int_0^t \sin(u)\cos(u)\,du.$$

First, we have

$$\sin(t)\int_0^t (\cos(u))^2\,du = \sin(t)\left(\frac{u}{2} + \frac{\sin(2u)}{4}\right)\Big|_0^t = \frac{t\sin(t)}{2} + \frac{\sin(t)\sin(2t)}{4}.$$

Also,

$$-\cos(t)\int_0^t \sin(u)\cos(u)\,du = -\cos(t)\frac{(\sin(u))^2}{2}\Big|_0^t = -\frac{\cos(t)(\sin(t))^2}{2}.$$

Since $\sin(t)\sin(2t)/4 = \cos(t)(\sin(t))^2/2$, the solution is $\dfrac{t\sin(t)}{2}$.

4. $f_1(t) = e^{at}$, $f_2(t) = e^{bt}$, $t > 0$

Solution.

$$\int_0^t e^{at-au}e^{bu}\,du = e^{at}\int_0^t e^{(b-a)u}\,du = e^{at}\frac{e^{(b-a)u}}{b-a}\Big|_0^t$$

Hence, the solution is $\dfrac{e^{bt}-e^{at}}{b-a}$.

Use the convolution theorem to find the inverse Laplace transform of each of the following functions.

5. $\dfrac{1}{s(s+1)}$

Since

$$\mathcal{L}^{-i}\left(\frac{1}{s}\right) = 1,$$

and

$$\mathcal{L}^{-1}\left(\frac{1}{s+1}\right) = e^{-t},$$

the convolution theorem implies that

$$\mathcal{L}^{-1}\left(\frac{1}{s(s+1)}\right) = \int_0^t 1\,e^{-u}\,du = -e^{-u}\Big|_0^t = 1 - e^{-t}.$$

6. $\dfrac{1}{s(s^2+4)}$

Since

$$\mathcal{L}^{-1}\left(\frac{1}{s}\right) = 1,$$

and

$$\mathcal{L}^{-1}\left(\frac{1}{s^2+4}\right) = \frac{1}{2}\mathcal{L}^{-1}\left(\frac{2}{s^2+4}\right) = \sin(2t),$$

the convolution theorem implies that

$$\mathcal{L}^{-1}\left(\frac{1}{s(s^2+4)}\right) = \frac{1}{2}\int_0^t 1\,\sin(2u)\,du = -\left.\frac{\cos(2u)}{4}\right|_0^t = \frac{1-\cos(2t)}{4}.$$

7. $\dfrac{1}{(s+1)(s-2)}$

Since

$$\mathcal{L}^{-1}\left(\frac{1}{s+1}\right) = e^{-t},$$

and

$$\mathcal{L}^{-1}\left(\frac{1}{s-2}\right) = e^{2t},$$

the convolution theorem implies that

$$\mathcal{L}^{-1}\left(\frac{1}{(s+1)(s-2)}\right) = \int_0^t e^{-t+u}e^{2u}\,du = e^{-t}\left.\frac{e^{3u}}{3}\right|_0^t = \frac{e^{2t}-e^{-t}}{3}.$$

8. $\dfrac{1}{(s+4)^2}$

Since

$$\mathcal{L}^{-1}\left(\frac{1}{s+4}\right) = e^{-4t},$$

the convolution theorem implies that

$$\mathcal{L}^{-1}\left(\frac{1}{(s+4)^2}\right) = \int_0^t e^{-4t+4u}e^{-4u}\,du = e^{-4t}\,u\Big|_0^t = t\,e^{-4t}.$$

9. $\dfrac{s}{(s^2+9)^2}$

Since

$$\mathcal{L}^{-1}\left(\frac{s}{(s^2+9)}\right) = \cos(3t),$$

and

$$\mathcal{L}^{-1}\left(\frac{1}{s^2+9}\right) = \frac{1}{3}\mathcal{L}^{-1}\left(\frac{3}{s^2+9}\right) = \sin(3t),$$

the convolution theorem implies that

$$\mathcal{L}^{-1}\left(\frac{s}{(s^2+9)^2}\right) = \frac{1}{3}\int_0^t \sin\left(3(t-u)\right)\cos(3u)\,du.$$

This integral can be expanded into

$$\frac{\sin(3t)}{3}\int_0^t (\cos(3u))^2\,du - \frac{\cos(3t)}{3}\int_0^t \sin(3t)\cos(3u)\,du.$$

After integrating, we obtain $\dfrac{t\sin(3t)}{6}$

10. $\dfrac{s}{(s^2+4s-5)^2}$

Since

$$\mathcal{L}^{-1}\left(\frac{s}{(s^2+4s-5)}\right) = \mathcal{L}^{-1}\left(\frac{s}{(s+2)^2-9}\right) = \frac{e^t+5e^{-5t}}{6},$$

and

$$\mathcal{L}^{-1}\left(\frac{1}{s^2+4s-5}\right) = \frac{1}{3}\mathcal{L}^{-1}\left(\frac{3}{(s+2)^2-9}\right) = \frac{e^{-2t}\sinh(3t)}{3} = \frac{e^t-e^{-5t}}{6},$$

the convolution theorem implies that

$$\mathcal{L}^{-1}\left(\frac{s}{(s^2+4s-5)^2}\right) = \frac{1}{36}\int_0^t \left(e^{(t-u)}+5e^{-5(t-u)}\right)\left(e^u-e^{-5u}\right)du.$$

This integral can be expanded into

$$\frac{e^t}{36}\int_0^t \left(1-e^{-6u}\right)du + \frac{5e^{-5t}}{36}\int_0^t \left(e^{6u}-1\right)du.$$

After integrating, we obtain $\dfrac{e^t-e^{-5t}}{54} + \dfrac{t\,e^t - 5t\,e^{-5t}}{36}$

Section 14.10, page 492

Use *Mathematica* to find the Laplace transforms of the following functions of t:

1. $t^2 \sin(t)$

Solution.

> `LaplaceTransform[t^2Sin[t],t,s]//Simplify`

$$\frac{6s^2 - 2}{(s^2 + 1)^3}$$

2. $t^2 y'(t)$

Solution.

> `LaplaceTransform[t^2 y'[t],t,s]//Simplify`

$$s \frac{d^2}{ds^2} \mathcal{L}(y)(s) + 2 \frac{d}{ds} \mathcal{L}(y)(s)$$

3. $\cos(m\,t) \sin(n\,t)$

Solution.

> `Integrate[Exp[-s t]Cos[m t]Sin[n t],`
> `{t,0,Infinity}]//Simplify`

$$\frac{\sqrt{(m+n)^2}}{2(s^2 + (m+n)^2)} - \frac{\sqrt{(m-n)^2}}{2(s^2 + (m-n)^2)}$$

4. $y'''(t) + 2y''(t) + y'(t) + 2y(t)$

Solution.

> `Collect[LaplaceTransform[y'''[t] + 2y''[t] +`
> `y'[t] + 2y[t],t,s],`
> `{LaplaceTransform[y[t],t,s],y[0],y'[0]}]`

$$(2 + s + 2s^2 + s^3)\mathcal{L}(y)(s) - (1 + 2s + s^2)y(0) - (2 + s)y'(0) - y''(0)$$

5. $u_\pi(t) - u_{2\pi}(t)$

Solution.

```
LaplaceTransform[UnitStep[t - Pi]
- UnitStep[t - 2Pi],t,s]//Simplify
```

$$\frac{e^{\pi s} - 1}{s e^{2\pi s}}$$

6. $t\cos(t)e^{2t}$

Solution.

```
LaplaceTransform[t Cos[t]E^(2t),t,s]//Simplify
```

$$\frac{s^2 - 4s - 3}{(s^2 - 4s + 5)^2}$$

7. $t - u_{\pi/2}(t)(t - \pi)$

Solution.

```
LaplaceTransform[
t - UnitStep[t - Pi/2](t - Pi),t,s]//Simplify
```

$$\frac{\pi s + 2e^{\pi s/2} - 2}{2s^2 e^{\pi s/2}}$$

8. $\delta(t - 1) + 2\delta(t - 2) + 3\delta(t - 3)$

Solution.

```
LaplaceTransform[DiracDelta[t - 1] +
2DiracDelta[t - 2] + 3DiracDelta[t - 3],t,s]//Simplify
```

$$\frac{3 + 2e^s + e^{2s}}{e^{3s}}$$

Use *Mathematica* to find the inverse Laplace transforms of the following functions of s:

9. $\dfrac{1}{s^2(s + 1)^3}$

Solution.

```
InverseLaplaceTransform[1/(s^2(s + 1)^3),s,t]//Simplify
```

$$\frac{6 - 6e^t + 4t + 2t\,e^t + t^2}{2e^t}$$

10. $\dfrac{1}{s^3(s^2+4)^2}$

Solution.

```
InverseLaplaceTransform[1/(s^3(s^2 + 4)^2),s,t]//Simplify
```

$$\frac{-2+2t^2+2\cos(2t)+t\sin(2t)}{64}$$

11. $s^2 \mathcal{L}(y(t))(s), \quad y(0)=y'(0)=0$

Solution.

```
InverseLaplaceTransform[s^2
LaplaceTransform[y[t],t,s],s,t] /.{y[0]->0,y'[0]->0}
```

$y''(t)$

12. $\dfrac{s}{(s^2+9)^6}$

Solution.

```
InverseLaplaceTransform[s/(s^2 + 9)^6,s,t]//Together
```

$$\frac{-105t^2\cos(3t)+90t^4\cos(3t)+35t\sin(3t)-135t^3\sin(3t)+27t^5\sin(3t)}{25194240}$$

13. $\dfrac{s}{(s^2+4s-5)^2}$

Solution.

```
InverseLaplaceTransform[s/(s^2 + 4s - 5)^2,s,t]//Together
```

$$\frac{-2+2e^{6t}-15t+3t\,e^{6t}}{108e^{5t}}$$

14. $\dfrac{s}{(s^2+4s+5)^2}$

Solution.

```
InverseLaplaceTransform[
s/(s^2 + 4s + 5)^2,s,t]//ComplexExpand//Simplify
```

$$\frac{2t\cos(t) + (t-2)\sin(t)}{2e^{2t}}$$

15. $s + 5s^4$

Solution.

```
InverseLaplaceTransform[s + 5s^4,s,t]
```

$$\delta'(t) + 5\delta^{(4)}(t)$$

16. $\dfrac{e^{-as}}{(s-b)}$

Solution.

```
InverseLaplaceTransform[E^(-a s)/(s - b),s,t]
```

$$e^{b(t-a)}u_a(t)$$

17. Produce a table of familiar Laplace transforms with the command

```
fns = {1,t,t^n,Exp[a t],Sin[a t],Cos[a t],Sinh[a t],
Cosh[a t],UnitStep[t],DiracDelta[t]};
TableForm[Transpose[{fns,LaplaceTransform[fns,t,s]}]]
```

18. Produce a table of familiar inverse Laplace transforms with the command

```
ifns = {1/s,1/(s + a),1/(s + a)^n,a/(s^2 + a^2),
s/(s^2 + a^2),E^-s,E^-s/s^n,s};
TableForm[Transpose[
{ifns,InverseLaplaceTransform[ifns,s,t]}]]//PowerExpand
```

Use *Mathematica* to solve and plot the following initial value problems:

19.
$$\begin{cases} y'' + 6y' + 9y = t^4\sin(3t), \\ y(0) = 1,\ y'(0) = 4 \end{cases}$$

Solution.

```
ODE[{y'' + 6y' + 9y == t^4 Sin[3t],
y[0] == 1,y'[0] == 4},y,t,Method->Laplace,
Form->Explicit,PlotSolution->{{t,0,2Pi}}]
```

$$y(t) = -\frac{2(119 + 849t)}{243} + \frac{10\cos(3t) - 24t\cos(3t) + 36t^3\cos(3t) - 27t^4\cos(3t)}{486}$$

$$+ \frac{t\left(4\sin(3t) - 9t\sin(3t) + 6t^2\sin(3t)\right)}{81}$$

20. $\begin{cases} y'' + 9y = t^2 \sin(3t), \\ y(0) = 1, \, y'(0) = 4 \end{cases}$

Solution.

```
ODE[{y'' + 9y == t^2 Sin[3t],
y[0] == 1,y'[0] == 4},y,t,
Method->Laplace,Form->Explicit]
```

$$y(t) = \cos(3t) + \frac{t\cos(3t)}{108} - \frac{t^3\cos(3t)}{108} + \frac{431\sin(3t)}{324} + \frac{t^2\sin(3t)}{36}$$

21. $\begin{cases} y^{(4)} + 2y'' + y = t^2 \cos(t), \\ y(0) = y'(0) = y''(0) = y'''(0) = 0 \end{cases}$

Solution.

```
ODE[{y'''' + 2    + y == t^2 Cos[t],
y[0] == 0,y'[0] == 0,    [0] == 0,y'''[0] == 0},y,t,
Method->Laplace,Form->Explicit]
```

$$y(t) = \frac{9t^2\cos(t) - t^4\cos(t) - 9t\sin(t) + 4t^3\sin(t)}{48}$$

22. $\begin{cases} y'' + y = u_2(t) + \delta(t - 2), \\ y(0) = y'(0) = 0 \end{cases}$

Solution.

```
ODE[{y'' + y == UnitStep[t - 1] + DiracDelta[t - 2],
y[0] == 0,y'[0] == 0},y,t,
Method->Laplace,Form->Explicit,
PostSolution->{Expand[#,Trig->True]&}]
```

$$-\sin(t - 2)u_{t-2}(t) + \left(1 - \cos(t - 1)\right)u_{t-1}(t)$$

23. $\begin{cases} y'' + y' + y = t\, u_1(t), \\ y(0) = y'(0) = 0 \end{cases}$

Solution.

```
ODE[{y'' + y' + y == t UnitStep[t - 1],
y[0] == 0,y'[0] == 0},y,t,
Method->Laplace,Form->Explicit]
```

$$y(t) = \left(t - 1 + 2\sqrt{\frac{e}{3}} e^{-t/2} \sin\left(\sqrt{\frac{3}{2}}\, (t - 1) \right) \right) u_1(t)$$

24. $\begin{cases} y'' + y = \delta(t - \pi) - \delta(t - 2\pi), \\ y(0) = y'(0) = 0 \end{cases}$

Solution.

```
ODE[{y'' + y == DiracDelta[t - Pi] - DiracDelta[t - 2Pi],
y[0] == 0,y'[0] == 1},y,t,
Method->Laplace,Form->Explicit,
PlotSolution->{{t,0,4Pi}}]
```

$$y(t) = \left(1 + u_\pi(t) - u_{2\pi}(t) \right) \sin(t)$$

Chapter 15

Section 15.1, page 499

Write each of the following linear differential equations as a first-order system of differential equations.

1. $y'' + y' \sin(t) = 4e^t$

   Solution. $\begin{cases} x_1' = x_2, \\ x_2' = -\sin(t)x_2 + 4e^t \end{cases}$

2. $y'' + 3t\, y' - 2y = 0$

   Solution. $\begin{cases} x_1' = x_2, \\ x_2' = -3t\, x_2 + 2x_1 \end{cases}$

3. $y''' - 3y'' + 2y' - t^2 y = e^t$

   Solution. $\begin{cases} x_1' = x_2, \\ x_2' = x_3, \\ x_3' = 3x_3 - 2x_2 + t^2 x_1 + e^t \end{cases}$

4. $\begin{cases} y_1'' + 2y_1' = 3y_1 + 4y_2, \\ y_2'' - 3y_2' = 4y_1 + 3y_2 \end{cases}$

   Solution.

4. $\begin{cases} x_1' = x_2, \\ x_2' = 3x_1 - 2x_2 + 4x_3, \\ x_3' = x_4, \\ x_4' = 4x_1 + 3x_3 + 3x_4 \end{cases}$

For each of the following systems of equations, find a **single differential equation** satisfied by the functions $x_1(t)$ and $x_2(t)$.

5. $\begin{cases} x_1' = 3x_1 - x_2, \\ x_2' = 2x_1 + 5x_2 \end{cases}$

   Solution. $y'' - 8y' + 17y = 0$

6.
$$\begin{cases} x_1' = t x_1 - 3x_2 + 4e^t, \\ x_2' = 3x_1 - t x_2 + 6e^t \end{cases}$$

Solution. $y'' + (t - 1) y' + (9 - t)y = -14e^t + 4t\, e^t$

7.
$$\begin{cases} x_1''' = x_2, \\ x_2''' = -x_1 \end{cases}$$

Solution. $y^{(6)} + y = 0$

8.
$$\begin{cases} x_1' = x_1 - x_2, \\ x_2' = x_1 + x_2 \end{cases}$$

Solution. $y'' - 2y' + 2y = 0$

9.
$$\begin{cases} x_1'' = 3x_1' + 2x_2' + x_1 - 4x_2 + 5, \\ x_2'' = 5x_1' - 3x_2' + 4x_1 + 5x_2 \end{cases}$$

Solution. $y^{(4)} - 25y'' + 9y' + 9y = -25$

10.
$$\begin{cases} x_1' = 3x_1 - 2x_2, \\ x_2' = 7x_1 + 5x_2 \end{cases}$$

Solution. $y'' - 8y' + 29y = 0$

Section 15.2, page 503

1. Prove the superposition principle for homogeneous linear systems.

Solution.

If $x(t)$ and $y(t)$ are both solutions of the same homogeneous linear system with coefficient matrix $P(t)$, then

$$\frac{d}{dt}\big(a\,x(t) + b\,y(t)\big) = a\frac{dx(t)}{dt} + b\frac{dy(t)}{dt}$$

$$= a\,P(t)x(t) + b\,P(t)y(t) = P(t)\big(a\,x(t) + b\,y(t)\big),$$

which was to be proved.

2. Prove the subtraction principle for general linear systems.

Solution.

If $\mathbf{x}(t)$ and $\mathbf{y}(t)$ are both solutions of the same linear system with coefficient matrix $P(t)$ and right side $\mathbf{g}(t)$, then

$$\frac{d}{dt}\big(\mathbf{x}(t) - \mathbf{y}(t)\big) = \frac{d\mathbf{x}(t)}{dt} - \frac{d\mathbf{y}(t)}{dt}$$

$$= \big(P(t)\mathbf{x}(t) + \mathbf{g}(t)\big) - \big(P(t)\mathbf{y}(t) + \mathbf{g}(t)\big) = P(t)\big(\mathbf{x}(t) - \mathbf{y}(t)\big),$$

which was to be proved.

The following exercises provide some guidance in the proof of the global existence theorem (Theorem 15.2) for linear systems.

3. Show that any solution to the first-order linear system

$$\begin{cases} x_1'(t) = \displaystyle\sum_{j=1}^{n} p_{1j}(t)x_j(t) + g_1(t), \\ \vdots \qquad\qquad \vdots \\ x_n'(t) = \displaystyle\sum_{j=1}^{n} p_{nj}(t)x_j(t) + g_n(t), \end{cases} \qquad\qquad \text{(S.83)}$$

with the initial conditions $x_1(t_0) = X_1, \ldots, x_n(t_0) = X_n$ is also a solution to the system of integral equations

$$\begin{cases} x_1(t) = X_1 + \displaystyle\int_{t_0}^{t} \left(\sum_{j=1}^{n} p_{1j}(s)x_j(s) + g_1(s) \right) ds, \\ \vdots \qquad\qquad \vdots \\ x_n(t) = X_n + \displaystyle\int_{t_0}^{t} \left(\sum_{j=1}^{n} p_{nj}(s)x_j(s) + g_n(s) \right) ds. \end{cases} \qquad\qquad \text{(S.84)}$$

Solution.

Applying the fundamental theorem of calculus to (S.83), we have for $1 \le i \le n$ that

$$x_i(t) = x_i(t_0) + \int_{t_0}^{t} x_i'(s)\, ds$$

$$= X_i + \int_{t_0}^{t} \left(\sum_{j=1}^{n} p_{ij}(s)x_j(s) + g_i(s) \right) ds,$$

which was to be proved.

4. Show that any solution to the system of integral equations (15.20) is also a solution to the first-order linear system (15.19) with the initial conditions $x_1(t_0) = X_1, \ldots, x_n(t_0) = X_n$.

Solution.

If $x_i(t)$ is a solution of (15.20), then we can apply the fundamental theorem of calculus to obtain

$$x_i'(t) = \frac{d}{dt}\left(X_i + \int_{t_0}^t \left(\sum_{j=1}^n p_{ij}(s)x_j(s) + g_i(s)\right)ds\right) = \sum_{j=1}^n p_{ij}(t)x_j(t) + g_i(t),$$

which was to be proved

5. If $X_1, \ldots, X_n$ are given numbers, define $x_i^0(t) = X_i$, and for $i = 1, \ldots, n$ and for $m \geq 0$ define

$$x_i^{m+1}(t) = X_i + \int_{t_0}^t \left(g_i(s) + \sum_{j=1}^n p_{ij}(s)x_j^m(s)\right)ds. \qquad (S.85)$$

Show that

$$\left|x_i^1(t) - x_i^0(t)\right| \leq P(t - t_0)\sum_{i=1}^n |x_i^0| + \int_{t_0}^t G(s)\,ds,$$

and that for $m \geq 1$

$$\left|x_i^{m+1}(t) - x_i^m(t)\right| \leq P\sum_{j=1}^n \int_{t_0}^t \left|x_j^m(s) - x_j^{m-1}(s)\right|ds,$$

where P is an upper bound for the numbers $|p_{ij}(s)|$, for $1 \leq i, j \leq n$, $t_0 \leq s \leq t$ and

$$G(t) = \sum_{i=1}^n |g_i(t)|.$$

Solution.

For $i = 1, \ldots, n$ we have

$$\left|x_i^1(t) - x_i^0(t)\right| = \left|\int_{t_0}^t \left(g_i(s) + \sum_{j=1}^n p_{ij}(s)X_j\right)ds\right| \leq \int_{t_0}^t \left(G(s) + P\sum_{j=1}^n |X_j|\right)ds$$

$$= \int_{t_0}^t G(s)\,ds + P(t - t_0)\sum_{j=1}^n |X_j|.$$

For $m \geq 1$, we have

$$\left| x_i^{m+1}(t) - x_i^m(t) \right| = \left| \int_{t_0}^t \sum_{j=1}^n p_{ij}(s) \left(x_j^m(s) - x_j^{m-1}(s) \right) ds \right|$$

$$\leq P \sum_{j=1}^n \int_{t_0}^t \left| x_j^m(s) - x_j^{m-1}(s) \right| ds,$$

which was to be proved.

6. Use the results of Exercise 5 to show that for $m \geq 0$

$$\left| x_i^{m+1}(t) - x_i^m(t) \right| \leq \frac{|P(t-t_0)|^{m+1}}{(m+1)!} \sum_{i=1}^n |x_i^0| + P^m \int_{t_0}^t \frac{(t-s)^m}{m!} G(s)\, ds \quad \text{(S.86)}$$

for $1 \leq i \leq n$.

Solution. We use the method of mathematical induction. For $m = 0$ (S.86) has been proved in Exercise 5. Assuming (S.86) for the value $m-1$, we write

$$\left| x_i^{m+1}(t) - x_i^m(t) \right| \leq P \sum_{j=1}^n \int_{t_0}^t \left| x_j^m(s) - x_j^{m-1}(s) \right| ds$$

$$\leq P \sum_{j=1}^n \int_{t_0}^t \left(\frac{|P(s-t_0)|^m}{m!} \sum_{i=1}^n |x_i^0| + P^{m-1} \int_{t_0}^s \frac{(s-u)^{m-1}}{(m-1)!} G(u)\, du \right) ds$$

$$\leq \frac{P^{m+1}(t-t_0)^{m+1}}{(m+1)!} + P^m \int_{t_0}^t \frac{(t-s)^m}{m!} G(s)\, ds,$$

which was to be proved.

7. Use the result of Exercise 6 to prove Theorem 15.2.

Solution. From the result in Exercise 6, for each i the sequence of functions $x_i^m(t)$ are uniformly convergent, in the sense that

$$\lim_{m,n \to \infty} \max_{t_0 \leq s \leq t} |x_i^m(s) - x_i^n(s)| = 0.$$

Therefore, there is a limiting function

$$x_i(t) = \lim_{m \to \infty} x_i^m(t),$$

and one can take the limit in (S.85) to obtain

$$x_i(t) = \lim_{m\to\infty} x_i^{m+1}(t) = \lim_{m\to\infty}\left(X_i + \int_{t_0}^t \left(g_i(s) + \sum_{j=1}^n p_{ij}(s)x_j^m(s)\right)ds\right)$$

$$= X_i + \int_{t_0}^t \left(g_i(s) + \sum_{j=1}^n p_{ij}(s)x_j(s)\right)ds.$$

Therefore, by Exercise 4, we have proved that the limit $x_i(t)$ satisfies the system of differential equations (S.83), as required.

Section 15.3, page 507

Find the general solution to each of the following upper/lower triangular systems.

1.
$$\begin{cases} x_1' = 4x_1 - 3x_2 + 7, \\ x_2' = 4x_2 + 5 \end{cases}$$

Solution.
$$\begin{cases} x_1(t) = -\dfrac{43}{16} + C_1 t\, e^{4t} + C_2 t\, e^{4t}, \\ x_2(t) = -\dfrac{5}{4} + C_2 e^{4t} \end{cases}$$

2.
$$\begin{cases} x_1' = x_1 + 4x_2, \\ x_2' = 4x_2 - 6 \end{cases}$$

Solution.
$$\begin{cases} x_1(t) = -6 + C_1 e^t + C_2(e^{4t} - e^t), \\ x_2(t) = \dfrac{3}{2} + C_2 e^{4t} \end{cases}$$

3.
$$\begin{cases} x_1' = -x_1 + 2x_2 + x_3, \\ x_2' = x_2 + x_3, \\ x_3' = 2x_3 \end{cases}$$

Solution.
$$\begin{cases} x_1(t) = C_1 e^{-t} + C_2(e^t - e^{-t}) + C_3(e^{2t} - e^t), \\ x_2(t) = C_2 e^t + C_3(e^{2t} - e^t), \\ x_3(t) = C_3 e^{2t} \end{cases}$$

4.
$$\begin{cases} x_1' = 5x_1 - 4x_2, \\ x_2' = x_2 + 1 \end{cases}$$

Solution.
$$\begin{cases} x_1(t) = -\dfrac{4}{5} + C_1 e^{5t} + C_2(e^t - e^{5t}), \\ x_2(t) = -1 + C_2 e^t \end{cases}$$

5.
$$\begin{cases} x_1' = 3x_1 + 4, \\ x_2' = 5x_1 + 4x_2 + 6 \end{cases}$$

Solution.
$$\begin{cases} x_1(t) = -\dfrac{4}{3} + C_1 e^{3t}, \\ x_2(t) = \dfrac{1}{6} + C_1(e^{4t} - e^{3t}) + C_2 e^{4t} \end{cases}$$

6.
$$\begin{cases} x_1' = x_1 + x_2 + x_3, \\ x_2' = x_2 + x_3, \\ x_3' = x_3 \end{cases}$$

Solution.
$$\begin{cases} x_1(t) = C_1 e^t + C_2 t\, e^t + C_3\left(t\, e^t + \dfrac{t^2 e^t}{2}\right), \\ x_2(t) = C_2 e^t + C_3 t\, e^t, \\ x_3(t) = C_3 e^t \end{cases}$$

Section 15.4, page 514

Find a fundamental set of solutions for each of the following systems of differential equations.

1.
$$\begin{cases} x_1' = 0, \\ x_2' = x_1 \end{cases}$$

Solution. $\left\{ \begin{pmatrix} 1 \\ t \end{pmatrix}, \begin{pmatrix} 0 \\ 1 \end{pmatrix} \right\}$

2.
$$\begin{cases} x_1' = 3x_1, \\ x_2' = 4x_2 \end{cases}$$

Solution. $\left\{ \begin{pmatrix} e^{3t} \\ 0 \end{pmatrix}, \begin{pmatrix} 0 \\ e^{4t} \end{pmatrix} \right\}$

3. $\begin{cases} x_1' = 3x_1, \\ x_2' = 4x_1 \end{cases}$

Solution. $\left\{ \begin{pmatrix} 3e^{3t} \\ 4e^{3t} \end{pmatrix}, \begin{pmatrix} 0 \\ 1 \end{pmatrix} \right\}$

4. $\begin{cases} x_1' = 3x_1, \\ x_2' = 4x_1, \\ x_2' = x_1 + x_2 \end{cases}$

Solution. $\left\{ \begin{pmatrix} 9e^{3t} \\ 12e^{3t} \\ 7e^{3t} \end{pmatrix}, \begin{pmatrix} 0 \\ 1 \\ t \end{pmatrix}, \begin{pmatrix} 0 \\ 0 \\ 1 \end{pmatrix} \right\}$

5. $\begin{cases} x_1' = x_1 + x_2, \\ x_2' = 3x_2 \end{cases}$

Solution. $\left\{ \begin{pmatrix} e^t \\ 0 \end{pmatrix}, \begin{pmatrix} e^{3t} \\ 2e^{3t} \end{pmatrix} \right\}$

6. $\begin{cases} x_1' = 2x_1, \\ x_2' = x_1 + 3x_2 \end{cases}$

Solution. $\left\{ \begin{pmatrix} e^{2t} \\ -e^{2t} \end{pmatrix}, \begin{pmatrix} 0 \\ e^{3t} \end{pmatrix} \right\}$

7. $\begin{cases} x_1' = x_2, \\ x_2' = -x_1 \end{cases}$

Solution. $\left\{ \begin{pmatrix} \cos(t) \\ -\sin(t) \end{pmatrix}, \begin{pmatrix} \sin(t) \\ \cos(t) \end{pmatrix} \right\}$

8. $\begin{cases} x_1' = x_2, \\ x_2' = -x_1, \\ x_3' = 2x_3 \end{cases}$

Solution. $\left\{ \begin{pmatrix} \cos(t) \\ -\sin(t) \\ 0 \end{pmatrix}, \begin{pmatrix} \sin(t) \\ \cos(t) \\ 0 \end{pmatrix}, \begin{pmatrix} 0 \\ 0 \\ e^{2t} \end{pmatrix} \right\}$

Compute the Wronskian of each of the following sets of vector functions.

9. $\left\{ \begin{pmatrix} t \\ 1 \end{pmatrix}, \begin{pmatrix} 1 \\ 0 \end{pmatrix} \right\}$

Solution. -1

10. $\left\{ \begin{pmatrix} e^t \\ e^t \end{pmatrix}, \begin{pmatrix} 4e^{3t} \\ 3e^{3t} \end{pmatrix} \right\}$

Solution. $-e^{4t}$

11. $\left\{ \begin{pmatrix} e^{2t} \\ e^{2t} \end{pmatrix}, \begin{pmatrix} t\,e^{2t} \\ t\,e^{2t} \end{pmatrix} \right\}$

Solution. 0

12. $\left\{ \begin{pmatrix} e^t \\ e^t \\ e^t \end{pmatrix}, \begin{pmatrix} e^{2t} \\ 2e^{2t} \\ 4e^{2t} \end{pmatrix}, \begin{pmatrix} e^{3t} \\ 3e^{3t} \\ 9e^{3t} \end{pmatrix} \right\}$

Solution. $-e^{6t}$

13. $\left\{ \begin{pmatrix} \cos(t) \\ \sin(t) \end{pmatrix}, \begin{pmatrix} -\sin(t) \\ \cos(t) \end{pmatrix} \right\}$

Solution. 1

14. $\left\{ \begin{pmatrix} e^t \cos(t) \\ e^t \sin(t) \end{pmatrix}, \begin{pmatrix} e^t \sin(t) \\ -e^t \cos(t) \end{pmatrix} \right\}$

Solution. $-e^{2t}$

15. $\left\{ \begin{pmatrix} e^t \\ 3e^t \end{pmatrix}, \begin{pmatrix} 4e^{5t} \\ 12e^{5t} \end{pmatrix} \right\}$

Solution. 0

16. $\left\{ \begin{pmatrix} e^t \cos(t) \\ e^t \sin(t) \\ e^{2t} \end{pmatrix}, \begin{pmatrix} e^t \sin(t) \\ -e^t \cos(t) \\ e^{2t} \end{pmatrix}, \begin{pmatrix} e^t \\ e^t \\ e^t \end{pmatrix} \right\}$

Solution. $e^{3t}(-1 + 2e^t \sin(t))$

17 Prove the following formula for differentiation of determinants of matrix-valued functions:

$$\frac{d}{dt} \det \begin{pmatrix} x_{11}(t) & \cdots & x_{1n}(t) \\ \vdots & \ddots & \vdots \\ x_{n1}(t) & \cdots & x_{nn}(t) \end{pmatrix} = \det \begin{pmatrix} x'_{11}(t) & \cdots & x'_{1n}(t) \\ \vdots & \ddots & \vdots \\ x_{n1}(t) & \cdots & x_{nn}(t) \end{pmatrix}$$

$$+ \cdots + \det \begin{pmatrix} x_{11}(t) & \cdots & x_{1n}(t) \\ \vdots & \ddots & \vdots \\ x'_{n1}(t) & \cdots & x'_{nn}(t) \end{pmatrix}.$$

Solution. We denote by $D(t)$ the determinant of the n^2 functions $x_{ij}(t)$ and $D_i(t)$ the determinant obtained by replacing the row $x_{i1}(t), \ldots, x_{in}(t)$ by $x'_{i1}(t), \ldots x'_{in}(t)$. Then from the definition of a determinant

$$D(t) = \sum_{\sigma} sgn(\sigma) x_{1\sigma(1)} x_{2\sigma(2)} \cdots x_{n\sigma(n)}$$

where the sum is over all permutations σ of $\{1, 2, \ldots, n\}$. Applying the derivative and using the product rule, we have

$$D'(t) = \sum_{\sigma} sgn(\sigma) \left(x'_{1\sigma(1)} x_{2\sigma(2)} \cdots x_{n\sigma(n)} + \cdots + x_{1\sigma(1)} x_{2\sigma(2)} \cdots x'_{n\sigma(n)} \right)$$

$$= D_1(t) + \cdots + D_n(t)$$

which was to be proved.

18 Apply the result of Problem 17 to complete the details of the proof of Theorem 15.6 for general n.

Solution. We examine each of the terms separately. The first term is written

$$\det \begin{pmatrix} x'_{11}(t) & \cdots & x'_{1n}(t) \\ \vdots & \ddots & \vdots \\ x_{n1}(t) & \cdots & x_{nn}(t) \end{pmatrix} = p_{11}(t) \det \begin{pmatrix} x_{11}(t) & \cdots & x_{1n}(t) \\ \vdots & \ddots & \vdots \\ x_{n1}(t) & \cdots & x_{nn}(t) \end{pmatrix}$$

$$+ p_{12}(t) \det \begin{pmatrix} x_{21}(t) & \cdots & x_{2n}(t) \\ \vdots & \ddots & \vdots \\ x_{n1}(t) & \cdots & x_{nn}(t) \end{pmatrix} + \cdots + p_{1n}(t) \det \begin{pmatrix} x_{n1}(t) & \cdots & x_{nn}(t) \\ \vdots & \ddots & \vdots \\ x_{n1}(t) & \cdots & x_{nn}(t) \end{pmatrix}.$$

Since a determinant with two equal rows is zero, it follows that all of these determinants are zero except for the first one, which equals $p_{11}(t)W(t)$. The second term is treated similarly, resulting in $p_{22}(t)W(t)$ and so forth for the other terms in the sum, which produces the result that $W'(t) = (p_{11}(t) + \cdots + p_{nn}(t))W(t)$. This is a first-order linear equation, whose unique solution is given by

$$W(t) = W(t_0) \exp \int_{t_0}^{t} (p_{11}(s) + \cdots + p_{nn}(s)) \, ds$$

as required.

Section 15.5, page 528

Find a fundamental set of solutions and the fundamental matrix for each of the following two-dimensional homogeneous systems of linear differential equations.

In each of the following answers, a fundamental set of solutions consists of the columns of the fundamental matrix.

1. $\begin{cases} x_1' = 3x_1 + 3x_2, \\ x_2' = x_1 + 5x_2 \end{cases}$

Solution. A fundamental matrix is $\begin{pmatrix} e^{6t} & -3e^{2t} \\ e^{6t} & e^{2t} \end{pmatrix}$.

2. $\begin{cases} x_1' = -6x_1 + 4x_2, \\ x_2' = -4x_1 + 2x_2 \end{cases}$

Solution. A fundamental matrix is $\begin{pmatrix} e^{-2t} & te^{-2t} \\ e^{-2t} & (t+1/4)e^{-2t} \end{pmatrix}$.

3. $\begin{cases} x_1' = 4x_1 + 2x_2, \\ x_2' = -x_1 + x_2 \end{cases}$

Solution. A fundamental matrix is $\begin{pmatrix} -2e^{3t} & -e^{2t} \\ e^{3t} & e^{2t} \end{pmatrix}$.

4. $\begin{cases} x_1' = x_1 - x_2, \\ x_2' = x_1 + x_2 \end{cases}$

Solution. A fundamental matrix is $\begin{pmatrix} -e^{t}\sin(t) & e^{t}\cos(t) \\ e^{t}\cos(t) & e^{t}\sin(t) \end{pmatrix}$.

5. $\begin{cases} x_1' = 3x_1 + 3x_2, \\ x_2' = x_1 \end{cases}$

Solution. A fundamental matrix is

$$\begin{pmatrix} (3 + \sqrt{21}) \exp\left(\dfrac{3 + \sqrt{21}}{2}t\right) & (3 - \sqrt{21}) \exp\left(\dfrac{3 - \sqrt{21}}{2}t\right) \\ 2 \exp\left(\dfrac{3 + \sqrt{21}}{2}t\right) & 2 \exp\left(\dfrac{3 - \sqrt{21}}{2}t\right) \end{pmatrix}.$$

6. $\begin{cases} x_1' = 4x_1 + 4x_2, \\ x_2' = -4x_1 - 4x_2 \end{cases}$

Solution. A fundamental matrix is $\begin{pmatrix} 1 & t + 1/4 \\ -1 & -t \end{pmatrix}$.

Find a fundamental set of solutions and the fundamental matrix for each of the following three-dimensional homogeneous systems of linear differential equations.

7. $\begin{cases} x_1' = x_1, \\ x_2' = 2x_1 + x_2 - 2x_3, \\ x_3' = 3x_1 + 2x_2 + x_3 \end{cases}$

Solution. A fundamental matrix is $\begin{pmatrix} 0 & 0 & 2e^t \\ e^t \cos(2t) & -e^t \sin(2t) & -3e^t \\ e^t \sin(2t) & e^t \cos(2t) & -2e^t \end{pmatrix}$.

8. $\begin{cases} x_1' = 3x_1 + 2x_2 + 4x_3, \\ x_2' = 2x_1 + 2x_3, \\ x_3' = 4x_1 + 2x_2 + 3x_3 \end{cases}$

Solution. A fundamental matrix is $\begin{pmatrix} 2e^{8t} & e^{-t} & 0 \\ e^{8t} & -2e^{-t} & -2e^{-t} \\ 2e^{8t} & 0 & e^{-t} \end{pmatrix}$.

9. $\begin{cases} x_1' = -2x_1 + x_2 + x_3, \\ x_2' = x_1 - 2x_2 + x_3, \\ x_3' = x_1 + x_2 - 2x_3 \end{cases}$

Solution. A fundamental matrix is $\begin{pmatrix} 1 & e^{-3t} & 0 \\ 1 & -e^{-3t} & e^{-3t} \\ 1 & 0 & -e^{-3t} \end{pmatrix}$.

$$
10. \quad \begin{cases} x_1' = x_1 + 2x_2 - 3x_3, \\ x_2' = x_1 + x_2 + 2x_3, \\ x_3' = x_1 - x_2 + 4x_3 \end{cases}
$$

Solution. A fundamental matrix is
$$
\begin{pmatrix} -e^{2t} & (-t+1)e^{2t} & (-t^2/2+t+1)e^{2t} \\ e^{2t} & t\,e^{2t} & (t^2/2+1)e^{2t} \\ e^{2t} & t\,e^{2t} & (t^2/2)e^{2t} \end{pmatrix}
$$

$$
11. \quad \begin{cases} x_1' = x_1 + x_2, \\ x_2' = x_2, \\ x_3' = 3x_3 \end{cases}
$$

Solution. A fundamental matrix is
$$
\begin{pmatrix} e^t & t\,e^t & 0 \\ 0 & e^t & 0 \\ 0 & 0 & e^{3t} \end{pmatrix}
$$

$$
12. \quad \begin{cases} x_1' = x_1 + x_2 - x_3, \\ x_2' = -x_1 - x_2, \\ x_3' = x_3 \end{cases}
$$

Solution. A fundamental matrix is
$$
\begin{pmatrix} -1 & -t-1 & -2e^t \\ 1 & t & e^t \\ 0 & 0 & e^t \end{pmatrix}
$$

Solve the following two-dimensional initial value problems.

$$
13. \quad \begin{cases} x_1' = 3x_1 + 3x_2, & x_1(0) = 3, \\ x_2' = x_1 + 5x_2, & x_2(0) = 5 \end{cases}
$$

Solution.
$$
\begin{cases} x_1(t) = \dfrac{9e^{6t} - 3e^{2t}}{2}, \\ x_2(t) = \dfrac{9e^{6t} + e^{2t}}{2} \end{cases}
$$

$$
14. \quad \begin{cases} x_1' = -6x_1 + 4x_2, & x_1(0) = 1, \\ x_2' = -4x_1 + 2x_2, & x_2(0) = 3 \end{cases}
$$

Solution.
$$
\begin{cases} x_1(t) = e^{-2t} + 8t\,e^{-2t}, \\ x_2(t) = 3e^{-2t} + 8t\,e^{-2t} \end{cases}
$$

15. $$\begin{cases} x_1' = 4x_1 + 2x_2, & x_1(0) = 2, \\ x_2' = -x_1 + x_2, & x_2(0) = -1 \end{cases}$$

Solution. $$\begin{cases} x_1(t) = 2e^{3t}, \\ x_2(t) = -e^{3t} \end{cases}$$

16. $$\begin{cases} x_1' = x_1 - x_2, & x_1(0) = 4, \\ x_2' = x_1 + x_2, & x_2(0) = 5 \end{cases}$$

Solution. $$\begin{cases} x_1(t) = -5e^t \sin(t) + 4e^t \cos(t), \\ x_2(t) = 5e^t \cos(t) + 4e^t \sin(t) \end{cases}$$

17. $$\begin{cases} x_1' = 3x_1 + 3x_2, & x_1(0) = 0 \\ x_2' = x_1, & x_2(0) = 4 \end{cases}$$

Solution. $$\begin{cases} x_1(t) = 2\sqrt{\dfrac{3}{7}} \sinh\left(\dfrac{(3+\sqrt{21})t}{2}\right), \\ x_2(t) = \left(1 + \sqrt{\dfrac{3}{7}}\right) \cosh\left(\dfrac{(3+\sqrt{21})t}{2}\right) \end{cases}$$

18. $$\begin{cases} x_1' = 4x_1 + 4x_2, & x_1(0) = 3 \\ x_2' = -4x_1 - 4x_2, & x_2(0) = 5 \end{cases}$$

Solution. $$\begin{cases} x_1(t) = 3 + 32t, \\ x_2(t) = 5 - 32t \end{cases}$$

Solve the following three-dimensional initial value problems.

15.5.4. $$\begin{cases} x_1' = x_1, \\ x_2' = 2x_1 + x_2 - 2x_3, \\ x_3' = 3x_1 + 2x_2 + x_3, \\ x_1(0) = 4, x_2(0) = 3, x_3(0) = 1 \end{cases}$$

Solution. $$\begin{cases} x_1(t) = 4e^t, \\ x_2(t) = 3e^t \sin(2t) + 9e^t \cos(2t) - 6e^t, \\ x_3(t) = -3e^t \cos(2t) + 9e^t \sin(2t) + 4e^t \end{cases}$$

15.5.4.
$$
\begin{cases}
x_1' = 3x_1 + 2x_2 + 4x_3, \\
x_2' = 2x_1 + 2x_3, \\
x_3' = 4x_1 + 2x_2 + 3x_3, \\
x_1(0) = 2, \ x_2(0) = 5, \ x_3(0) = 6
\end{cases}
$$

Solution.
$$
\begin{cases}
x_1(t) = \dfrac{14}{3}e^{8t} - \dfrac{8}{3}e^{-t}, \\[2mm]
x_2(t) = \dfrac{7}{3}e^{8t} + \dfrac{8}{3}e^{-t}, \\[2mm]
x_3(t) = \dfrac{14}{3}e^{8t} + \dfrac{4}{3}e^{-t}
\end{cases}
$$

21.
$$
\begin{cases}
x_1' = -2x_1 + x_2 + x_3, \\
x_2' = x_1 - 2x_2 + x_3, \\
x_3' = x_1 + x_2 - 2x_3, \\
x_1(0) = 3, \ x_2(0) = 5, \ x_3(0) = 0
\end{cases}
$$

Solution.
$$
\begin{cases}
x_1(t) = \dfrac{8 + e^{-3t}}{3}, \\[2mm]
x_2(t) = \dfrac{8 + 7e^{-3t}}{3}, \\[2mm]
x_3(t) = \dfrac{8 - 8e^{-3t}}{3}
\end{cases}
$$

22.
$$
\begin{cases}
x_1' = x_1 + 2x_2 - 3x_3, \\
x_2' = x_1 + x_2 + 2x_3, \\
x_3' = x_1 - x_2 + 4x_3, \\
x_1(0) = 1, \ x_2(0) = 2, \ x_3(0) = 3
\end{cases}
$$

Solution.
$$
\begin{cases}
x_1(t) = \dfrac{t^2}{2}e^{2t} - 6t\,e^{2t} + e^{2t}, \\[2mm]
x_2(t) = -\dfrac{t^2}{2}e^{2t} + 5t\,e^{2t} + 2e^{2t}, \\[2mm]
x_3(t) = -\dfrac{t^2}{2}e^{2t} + 5t\,e^{2t} + 3e^{2t}
\end{cases}
$$

23.
$$\begin{cases} x_1' = x_1 + x_2, \\ x_2' = x_2, \\ x_3' = 3x_3, \\ x_1(0) = 3, x_2(0) = 2, x_3(0) = 1 \end{cases}$$

Solution.
$$\begin{cases} x_1(t) = 3e^t + 2t\,e^t, \\ x_2(t) = 2e^t, \\ x_3(t) = e^{3t} \end{cases}$$

24.
$$\begin{cases} x_1' = x_1 + x_2 - x_3, \\ x_2' = -x_1 - x_2, \\ x_3' = x_3, \\ x_1(0) = 8, x_2(0) = 4, x_3(0) = 2 \end{cases}$$

Solution.
$$\begin{cases} x_1(t) = 12 + 14t - 4e^t, \\ x_2(t) = 2 - 14t + 2e^t, \\ x_3(t) = 2e^t \end{cases}$$

25. Suppose $\mathbf{x}^1$ and $\mathbf{x}^2$ are complex conjugate solutions to $\mathbf{x}' = A\mathbf{x}$ where A is a real matrix. Let $\mathbf{u}(t) = \mathfrak{Re}(\mathbf{x}^1(t))$ and $\mathbf{v}(t) = \mathfrak{Im}(\mathbf{x}^1(t))$. Show that $\mathbf{u}$ and $\mathbf{v}$ are real solutions of $\mathbf{x}' = A\mathbf{x}$.

Solution. We have $\mathbf{u} = (1/2)(\mathbf{x}^1 + \mathbf{x}^2)$ and $\mathbf{v} = (1/2i)(\mathbf{x}^1 - \mathbf{x}^2)$. By the superposition principle, $\mathbf{u}$ and $\mathbf{v}$ are real solutions of $\mathbf{x}' = A\mathbf{x}$.

26. In Lemma 15.12 it was shown that the matrix exponential e^{tA} is a fundamental matrix for the linear system $d\mathbf{x}/dt = A\mathbf{x}$. Suppose that $\Phi(t)$ is *any* fundamental matrix for this system. Prove that the matrix exponential can be recovered from the formula

$$e^{tA} = \Phi(t)\Phi(0)^{-1}.$$

[Hint: Both sides of the proposed equation satisfy the same first-order linear system of equations; now appeal to the uniqueness theorem (Theorem 15.1) for first-order linear systems.]

Solution.

Let $\Psi(t) = \Phi(t)\Phi(0)^{-1}$. Computing the derivative, we find that

$$\Psi'(t) = \Phi'(t)\Phi(0)^{-1} = A\Phi(t)\Phi(0)^{-1} = A\Psi(t),$$

and furthermore we have the initial condition $\Psi(0) = I$. But the matrix function e^{tA} also satisfies the same system of equations and the same initial condition. Therefore, by the uniqueness theorem (Theorem 15.1) we infer that

$$e^{tA} = \Psi(t) = \Phi(t)\Phi(0)^{-1},$$

as required.

27. Suppose that the matrix A is diagonalizable with eigenvalues $r_1, \ldots, r_n$ and eigenvectors $\xi_1, \ldots, \xi_n$.

   a. Show that a fundamental matrix is obtained in the form

$$\Phi(t) = \left(\xi_1 e^{r_1 t} \quad \cdots \quad \xi_n e^{r_n t} \right),$$

where the columns are the respective solutions $\xi_j e^{r_j t}$.

Solution.

We denote the matrix elements as $\Phi_{ij}(t) = c_{ij} e^{r_j t}$, where c_{ij} is the ith component of the eigenvector ξ_j. Taking the derivative and using the eigenvector equation, we have $\Phi'_{ij}(t) = c_{ij} r_j e^{r_j t}$ and

$$(A\Phi(t))_{ij} = \sum_{k=1}^{n} a_{ik} c_{kj} e^{r_k t} = r_j c_{ij} e^{r_j t},$$

as required. Since the eigenvalues are distinct, the eigenvectors are linearly independent, hence $\Phi(0)$ is an invertible matrix.

   b. Under what condition on the eigenvectors of a 2×2 matrix A can we be assured that the above fundamental matrices satisfy the commutativity property $\Phi(t)\Phi(s) = \Phi(s)\Phi(t)$ for all s and t?

Solution.

We claim that either the eigenvalues are equal or the eigenvectors lie along the coordinate directions. To see this, write the four terms of the matrix equation $\Phi(t)\Phi(s) = \Phi(s)\Phi(t)$ as follows:

$$\begin{cases} e^{r_1 s}(c_{11}^2 e^{r_1 t} + c_{12} c_{21} e^{r_2 t}) = e^{r_1 t}(c_{11}^2 e^{r_1 s} + c_{12} c_{21} e^{r_2 s}), \\ e^{r_2 s}(c_{11} c_{12} e^{r_1 t} + c_{12} c_{22} e^{r_2 t}) = e^{r_2 t}(c_{11} c_{12} e^{r_1 s} + c_{12} c_{22} e^{r_2 s}), \\ e^{r_1 s}(c_{21} c_{11} e^{r_1 t} + c_{22} c_{21} e^{r_2 t}) = e^{r_1 t}(c_{21} c_{11} e^{r_1 s} + c_{22} c_{21} e^{r_2 s}), \\ e^{r_2 s}(c_{21} c_{12} e^{r_1 t} + c_{22}^2 e^{r_2 t}) = e^{r_2 t}(c_{21} c_{12} e^{r_1 s} + c_{22}^2 e^{r_2 s}). \end{cases} \qquad \text{(S.87)}$$

The equations (S.87) can be rewritten as

$$
\begin{cases}
c_{12}c_{21}(e^{r_2 s}e^{r_1 t} - e^{r_1 s}e^{r_2 t}) = 0, \\
c_{11}c_{12}(e^{r_2 s}e^{r_1 t} - e^{r_1 s}e^{r_2 t}) = 0, \\
c_{22}c_{21}(e^{r_1 s}e^{r_2 t} - e^{r_2 s}e^{r_1 t}) = 0, \\
c_{21}c_{12}(e^{r_2 s}e^{r_1 t} - e^{r_2 t}e^{r_1 s}) = 0.
\end{cases}
\tag{S.88}
$$

If $r_1 = r_2$, the equations (S.88) are automatically satisfied. If $r_1 \neq r_2$ (S.88), will be satisfied if and only if the eigenvectors are proportional to the coordinate vectors. Indeed, from the first line of (S.88), either $c_{12} = 0$ or $c_{21} = 0$. In the first case, we must have $c_{22} \neq 0$ since the eigenvector $\xi_2 \neq 0$. From the third line of (S.88), we must also have $c_{21} = 0$, hence the eigenvectors lie along the coordinate directions. Similarly, if $c_{21} = 0$, then $c_{11} = 0$ and $c_{12} = 0$ both follow, hence the eigenvectors lie along the coordinate directions.

c. Show by example that the commutativity property does not hold for the fundamental matrix

$$
\begin{pmatrix}
2e^t & 3e^{-t} \\
e^t & 2e^{-t}
\end{pmatrix}
$$

belonging to the system

$$
\frac{d\mathbf{x}}{dt} = \begin{pmatrix} 7 & -12 \\ 4 & -7 \end{pmatrix} \mathbf{x}.
$$

Solution. From the solution of part b, we see that the eigenvalues are distinct and that the eigenvectors do not lie along the coordinate directions. Therefore, the equation $\Phi(t)\Phi(s) = \Phi(t)\Phi(s)$ is false in general.

d. Suppose that $\Phi(t)$ is *any* fundamental matrix for the linear system $d\mathbf{x}/dt = A\mathbf{x}$. Show that $\Phi(t) = e^{At}\Phi(0)$. [Hint: Apply the uniqueness theorem (Theorem 15.1) to the columns of both sides of the asserted equation after checking that both sides satisfy the linear system with the same initial conditions.]

Solution.

Use the solution of Exercise 26.

28. Let $\Phi(t)$ be any fundamental matrix for the linear system $d\mathbf{x}/dt = A\mathbf{x}$. Show that the matrix exponential can be recovered through the formula

$$
\Phi(t)\Phi(s)^{-1} = e^{(t-s)A}
$$

for any real numbers s and t. [Hint: Use the result of Problem 27.]

Solution.

The matrix function $t \to U(t) = \Phi(t)\Phi(s)^{-1}$ satisfies the equation $U' = AU$ with the initial conditions $U(s) = I$. But the matrix exponential $e^{(t-s)A}$ satisfies these same conditions, hence by the uniqueness theorem (Theorem 15.1) we must have the equation $\Phi(t)\Phi(s)^{-1} = e^{(t-s)A}$.

Section 15.6, page 541

Find a particular solution to each of the following systems by the method of undetermined coefficients.

1. $\begin{cases} x_1' = x_1 + 4x_2 - e^t, \\ x_2' = x_1 + x_2 + 2e^t \end{cases}$

   Solution. $\begin{cases} x_1 = -2e^t, \\ x_2 = e^t/4 \end{cases}$

2. $\begin{cases} x_1' = -2x_1 + 3x_2 - e^t, \\ x_2' = -x_1 + 2x_2 + e^t \end{cases}$

   Solution. $\begin{cases} x_1 = 2t\,e^t, \\ x_2 = 2t\,e^t \end{cases}$

3. $\begin{cases} x_1' = -x_1 + x_2 - e^t, \\ x_2' = -4x_1 + 3x_2 + e^t \end{cases}$

   Solution. $\begin{cases} x_1 = 3t^2e^t/2, \\ x_2 = 3t^2e^t \end{cases}$

4. $\begin{cases} x_1' = -4x_1 + 8x_2 + 1, \\ x_2' = -2x_1 + 4x_2 + 2t \end{cases}$

   Solution. $\begin{cases} x_1 = t - 2t^2 + 8t^3/3, \\ x_2 = 4t^3/3 \end{cases}$

5. $\begin{cases} x_1' = -2x_1 + x_2 + t, \\ x_2' = -5x_1 + 2x_2 - \sin(2t) \end{cases}$

Solution.
$$\begin{cases} x_1 = 1 - 2t + \dfrac{\sin(2t)}{3}, \\ x_2 = -5t + \dfrac{2\cos(2t) + 2\sin(2t)}{3} \end{cases}$$

6.
$$\begin{cases} x_1' = -x_1 + 4x_2 + 2e^{-t}\cos(t), \\ x_2' = -2x_1 + 3x_2 - e^{-t}\sin(t) \end{cases}$$

Solution.
$$\begin{cases} x_1 = \dfrac{-16\cos(t) - 2\sin(t)}{13e^t}, \\ x_2 = \dfrac{-7\cos(t) + 4\sin(t)}{13e^t} \end{cases}$$

7.
$$\begin{cases} x_1' = -3x_1 + 2x_2 + 2e^{-t}, \\ x_2' = 2x_1 - 3x_2 + 4t \end{cases}$$

Solution.
$$\begin{cases} x_1 = \dfrac{-48 + 40t}{25} + \dfrac{1 + 4t}{4e^t}, \\ x_2 = \dfrac{-52 + 60t}{25} + \dfrac{-1 + 4t}{4e^t} \end{cases}$$

8.
$$\begin{cases} x_1' = 2x_1 - 5x_2 - \cos(t), \\ x_2' = x_1 - 2x_2 + \sin(t) \end{cases}$$

Solution.
$$\begin{cases} x_1 = 2t\cos(t) - t\sin(t), \\ x_2 = t\cos(t) \end{cases}$$

Solve the following initial value problems

9.
$$\begin{cases} x_1' = 2x_1 - x_2 + e^t, \quad x_1(0) = 1, \\ x_2' = 3x_1 - 2x_2 - e^t, \quad x_2(0) = 0 \end{cases}$$

Solution.
$$\begin{cases} x_1 = (1 + 2t)e^t, \\ x_2 = 2t\,e^t \end{cases}$$

10.
$$\begin{cases} x_1' = -x_2 + \cos(t), \quad x_1(0) = 0, \\ x_2' = x_1 + \sin(t), \quad x_2(0) = 0 \end{cases}$$

Solution.
$$\begin{cases} x_1 = t\cos(t), \\ x_2 = t\sin(t) \end{cases}$$

11.
$$\begin{cases} x_1' = 7x_1 - 2x_2 + 2e^{3t}, & x_1(0) = 1, \\ x_2' = 2x_1 + 2x_2 + 4e^{3t}, & x_2(0) = -1 \end{cases}$$

Solution.
$$\begin{cases} x_1 = e^{3t}(2t - 1) + 2e^{6t}, \\ x_2 = e^{3t}(4t - 2) + e^{6t} \end{cases}$$

12.
$$\begin{cases} x_1' = x_2 + e^{2t}\cos(t), & x_1(0) = 1, \\ x_2' = -5x_1 + 4x_2, & x_2(0) = 0 \end{cases}$$

Solution.
$$\begin{cases} x_1 = \dfrac{e^{2t}\left(t\cos(t) + \sin(t) - 2t\sin(t)\right)}{2}, \\ x_2 = -\dfrac{5e^{2t}t\sin(t)}{2} \end{cases}$$

13. Show that the method of undetermined coefficients can be developed to solve a system of equations of the form $x'(t) = Ax + vt^m$. Assuming that $\alpha = 0$ is not an eigenvalue of the matrix A, show that a particular solution can be found in the form $x(t) = w_0 t^m + w_1 t^{m-1} + w_2 t^{m-2} + \cdots + t w_{m-1} + w_m$ for suitably chosen vectors $w_1, \ldots, w_m$.

Solution. For this choice of $x(t)$, we have
$$x'(t) = m w_0 t^{m-1} + (m - 1)w_1 t^{m-2} + \cdots + w_{m-1},$$

$$Ax(t) + vt^m = t^m A w_0 + t^{m-1} A w_1 + \cdots + A w_m + vt^m.$$

Therefore, the linear system will be satisfied if and only if
$$\begin{aligned} A w_0 + v &= 0 \\ A w_1 &= m w_0 \\ &\vdots \\ A w_m &= w_{m-1} \end{aligned}$$

Since zero is not an eigenvalue of the matrix A, it follows that A^{-1} exists and that we can solve these equations uniquely by
$$w_0 = -A^{-1}v_0, \ w_1 = mA^{-1}w_0. \ldots, \ w_m = A^{-1}w_{m-1}.$$

14. With reference to Problem 13, show how to modify the particular solution proposed there to find a particular solution to the system $x'(t) = Ax + vt^m$ in the case that $\alpha = 0$ is an eigenvalue of the matrix A.

Solution. If A is diagonalizable, we may find a particular solution in the form

$$x(t) = w_0 t^{m+1} + w_1 t^m + \cdots + t w_m + w_{m+1}$$

The vectors $w_0, \ldots, w_m$ must satisfy the equations $A w_0 = 0$, $A w_1 + v = (m+1) w_0$, $A w_2 = m w_1, \ldots, A w_1 = w_0$. Following the method of the text, these equations can be solved in case A is diagonalizable.

In case A is not diagonalizable, we may find a particular solution in the form

$$x(t) = w_0 t^{m+k} + w_1 t^{m+k-1} + \cdots + t w_{m+k-1} + w_{m+k}$$

where k depends on the deficiency of the eigenvalue $\alpha = 0$ and the vectors $w_0, \ldots, w_{m+k}$ can be obtained form the Jordan canonical form of A.

Section 15.7, page 545

Find a particular solution to each of the following systems by the method of variation of parameters

1.
$$\begin{cases} x_1' = -2x_1 + 3x_2 + t, \\ x_2' = -x_1 + 2x_2 + e^t \end{cases}$$

Solution.
$$\begin{cases} x_1 = \dfrac{3t\, e^t}{2} - \dfrac{3e^t}{4} + 2t - 1, \\ x_2 = \dfrac{3t\, e^t}{2} - \dfrac{e^t}{4} + t \end{cases}$$

2.
$$\begin{cases} x_1' = -2x_1 + x_2 + \sin(t), \\ x_2' = -5x_1 + 2x_2 - \cos(t) \end{cases}$$

Solution.
$$\begin{cases} x_1 = \dfrac{\sin(t)}{5}, \\ x_2 = -\dfrac{\cos(t) + 3\sin(t)}{5} \end{cases}$$

3.
$$\begin{cases} x_1' = -2x_1 + 4x_2 - 2e^t, \\ x_2' = x_1 + x_2 + e^{-2t} \end{cases}$$

Solution.
$$\begin{cases} x_1 = -e^{-2t}, \\ x_2 = \dfrac{e^t}{2} \end{cases}$$

4.
$$\begin{cases} x_1' = x_2 + e^t, \\ x_2' = x_2 + 2e^t, \\ x_3' = x_1 - 3x_2 + 3x_3 + 3e^t \end{cases}$$

Solution.
$$\begin{cases} x_1 = e^t + \dfrac{t^2 e^t}{2}, \\ x_2 = 2t\, e^t + \dfrac{t^2 e^t}{2}, \\ x_3 = 3t\, e^t + \dfrac{t^2 e^t}{2} \end{cases}$$

5.
$$\begin{cases} x_1' = 4x_1 - 2x_2 + t^{-3}, \\ x_2' = 8x_1 - 4x_2 - t^{-2} \end{cases}$$

Solution.
$$\begin{cases} x_1 = -2 - \dfrac{1}{2t^2} + \dfrac{2}{t} - 2\log(t), \\ x_2 = -4 + \dfrac{5}{t} - 4\log(t) \end{cases}$$

6.
$$\begin{cases} x_1' = 4x_1 + 2x_2 + \cot(t), \\ x_2' = -10x_1 + 4x_2 \end{cases}$$

Solution.
$$\begin{cases} x_1 = \dfrac{1}{2} + \left(\sin(2t) + \dfrac{\cos(2t)}{2}\right)\log(\tan(t)), \\ x_2 = -\dfrac{5\sin(2t)\log(\tan(t))}{2} \end{cases}$$

7.
$$\begin{cases} x_1' = 2x_1 - 5x_2 + \csc(t), \\ x_2' = x_1 - 2x_2 + \sec(t) \end{cases}$$

Solution.
$$\begin{cases} x_1 = -2t\cos(t) - 4t\sin(t) - 5\cos(t)\log(\cos(t)) \\ \qquad + (\cos(t) + 2\sin(t))\log(\sin(t)), \\ x_2 = -2t\sin(t) - 2\cos(t)\log(\cos(t)) - \sin(t))\log(\tan(t)) \end{cases}$$

8.
$$\begin{cases} x_1' = -2x_1 + x_2 - 2t^2, \\ x_2' = 2x_2 + x_3 - t^3 - 2, \\ x_3' = x_1 - 2x_2 - 3t^4 \end{cases}$$

Solution.

$$\begin{cases} x_1 = -\dfrac{989 + 1050t + 738t^2 + 468t^3 + 108t^4}{216}, \\[2mm] x_2 = -\dfrac{125 + 618t + 522t^2 + 252t^3 + 108t^4}{216}, \\[2mm] x_3 = \dfrac{8 + 24t + 36t^2 + 36t^3 + 27t^4}{27} \end{cases}$$

Solve the following initial value problems

9. $\begin{cases} x_1' = -2x_1 + e^t, & x_1(0) = 0, \\ x_2' = x_1 - 2x_2 + 3t, & x_2(0) = 0 \end{cases}$

Solution.
$$\begin{cases} x_1 = \dfrac{-1 + e^{3t}}{3e^{2t}}, \\[2mm] x_2 = \dfrac{23 - 12t - 27e^{2t} + 4e^{3t} + 54t\, e^{3t}}{36e^{2t}} \end{cases}$$

10. $\begin{cases} x_1' = x_1 + x_2 + e^{2t}, & x_1(0) = 2, \\ x_2' = x_1 + x_2 + e^{2t}, & x_2(0) = -1 \end{cases}$

Solution.
$$\begin{cases} x_1 = \dfrac{3 + e^{2t} + 2t\, e^{2t}}{2}, \\[2mm] x_2 = \dfrac{-3 + e^{2t} + 2t\, e^{2t}}{2} \end{cases}$$

11. $\begin{cases} x_1' = 2x_1 + 3x_2 + t, & x_1(0) = 0, \\ x_2' = -3x_1 + 2x_2, & x_2(0) = -1 \end{cases}$

Solution.
$$\begin{cases} x_1 = \dfrac{5 - 26t - e^{2t}(5\cos(3t) + 157\sin(3t))}{169}, \\[2mm] x_2 = -\dfrac{12 + 39t + e^{2t}(157\cos(3t) - 5\sin(3t))}{169} \end{cases}$$

12. $\begin{cases} x_1' = -4x_1 + x_2, & x_1(0) = 0, \\ x_2' = 4x_1 - x_2 + e^{-5t}, & x_2(0) = 1 \end{cases}$

Solution.
$$\begin{cases} x_1 = \dfrac{-6 - 5t + 6e^{5t}}{25e^{5t}}, \\[2mm] x_2 = \dfrac{1 + 5t + 24e^{5t}}{25e^{5t}} \end{cases}$$

13. The purpose of this exercise is to give an independent derivation of the variation of parameters formula for a single second-order equation, beginning with the variation of parameters formula for a system.

    a. Write the second-order equation $y'' + p(t)y' + q(t)y = g(t)$ as a system.

    b. Find a particular solution of this system by formula (15.98).

    c. Identify the first component of this solution as formula (9.83).

Solution.

    a. We write $y'' + p(t)y' + q(t)y = g(t)$ as a system by taking $x_1(t) = y(t)$ and $x_2(t) = y'(t)$, with the result

$$\frac{d}{dt}\begin{pmatrix} x_1 \\ x_2 \end{pmatrix} = \begin{pmatrix} 0 & 1 \\ -q(t) & -p(t) \end{pmatrix}\begin{pmatrix} x_1 \\ x_2 \end{pmatrix} + \begin{pmatrix} 0 \\ g(t) \end{pmatrix},$$

which expresses the two scalar equations

$$\begin{cases} x_1' = x_2, \\ x_2' = -q(t)x_1 - p(t)x_2 + g(t). \end{cases} \qquad (S.89)$$

    b. Suppose now that $\{y_1(t), y_2(t)\}$ forms a fundamental set of solutions of the homogeneous equation $y'' + p(t)y' + q(t)y = 0$ with nonzero Wronskian $W(t) = y_1(t)y_2'(t) - y_1'(t)y_2(t)$. Then we obtain a *fundamental matrix* of the linear system in the form

$$\Phi(t) = \begin{pmatrix} y_1(t) & y_2(t) \\ y_1'(t) & y_2'(t) \end{pmatrix}.$$

Each of the columns of this matrix is a solution of (S.89) and the two columns are linearly independent. Hence $\Phi(t)$ has an inverse matrix given by

$$\Phi(t)^{-1} = \frac{1}{W(t)}\begin{pmatrix} y_2'(t) & -y_2(t) \\ -y_1'(t) & y_1(t) \end{pmatrix}.$$

To find a particular solution to the inhomogeneous equation $y'' + p(t)y' + q(t)y = g(t)$ we follow the above method and look for the solution in the form

$$\mathbf{x}(t) = \Phi(t)\mathbf{u}(t), \qquad \text{where} \qquad \Phi(t)\mathbf{u}'(t) = \begin{pmatrix} 0 \\ g(t) \end{pmatrix}. \qquad (S.90)$$

From the second equation of (S.90) we must have

$$\mathbf{u}'(t) = \Phi(t)^{-1}\begin{pmatrix} 0 \\ g(t) \end{pmatrix} = \frac{1}{W(t)}\begin{pmatrix} -y_2(t)g(t) \\ y_1(t)g(t) \end{pmatrix},$$

or

$$u(t) = \begin{pmatrix} -\int \dfrac{y_2(t)g(t)}{W(t)} dt \\ \int \dfrac{y_1(t)g(t)}{W(t)} dt \end{pmatrix}.$$

c. We have

$$x_1(t) = \left(\mathbf{\Phi}(t)u(t)\right)_1 = y_1(t) \int -\frac{y_2(t)g(t)}{W(t)} dt + y_2(t) \int \frac{y_1(t)g(t)}{W(t)}.$$

This is precisely the same result which was discovered in Section 9.4 by a different method.

Section 15.8, page 548

Use Laplace transforms to solve the following systems of differential equations.

1. $\begin{cases} x_1' = 12x_1 + 5x_2, & x_1(0) = 0, \\ x_2' = -6x_1 + x_2, & x_2(0) = 1 \end{cases}$

Solution. $\begin{cases} x_1 = -5e^{6t} + 5e^{7t}, \\ x_2 = 6e^{6t} - 5e^{7t} \end{cases}$

2. $\begin{cases} x_1' = 4x_1 - 2x_2, & x_1(0) = 2, \\ x_2' = 5x_1 + 2x_2, & x_2(0) = -2 \end{cases}$

Solution. $\begin{cases} x_1 = 2e^{3t}\left(\cos(3t) + \sin(3t)\right), \\ x_2 = e^{3t}\left(-2\cos(3t) + 4\sin(3t)\right) \end{cases}$

3. $\begin{cases} x_1' = -2x_1 + x_2, & x_1(0) = 5, \\ x_2' = -9x_1 + 4x_2, & x_2(0) = -3 \end{cases}$

Solution. $\begin{cases} x_1 = e^t(5 - 18t), \\ x_2 = e^t(-3 + 54t) \end{cases}$

4. $\begin{cases} x_1' = -x_1 - x_2, & x_1(0) = 4, \\ x_2' = -x_2, & x_2(0) = 1, \\ x_3' = -2x_3, & x_3(0) = 1 \end{cases}$

$$
\text{Solution.} \quad \begin{cases} x_1 = e^{-t}(4-t), \\ x_2 = e^{-t}, \\ x_3 = e^{-2t} \end{cases}
$$

5. $\begin{cases} x_1' = -6x_1 + 2x_2, \quad x_1(0) = 1, \\ x_2' = -7x_1 + 3x_2, \quad x_2(0) = 0 \end{cases}$

$$
\text{Solution.} \quad \begin{cases} x_1 = -\dfrac{2}{5}e^{-4t} + \dfrac{2}{5}e^{t}, \\ x_2 = -\dfrac{2}{5}e^{-4t} + \dfrac{7}{5}e^{t} \end{cases}
$$

6. $\begin{cases} x_1' = 4x_1 - x_2, \quad x_1(0) = 1, \\ x_2' = 2x_1 + 5x_2, \quad x_2(0) = 0 \end{cases}$

$$
\text{Solution.} \quad \begin{cases} x_1 = e^{9t/2}\left(\cos\left(\dfrac{\sqrt{7}t}{2}\right) - \dfrac{1}{\sqrt{7}}\sin\left(\dfrac{\sqrt{7}t}{2}\right)\right), \\ x_2 = \dfrac{4e^{9t/2}}{\sqrt{7}}\sin\left(\dfrac{\sqrt{7}t}{2}\right) \end{cases}
$$

7. $\begin{cases} x_1' = x_1 + x_2, \quad x_1(0) = 1, \\ x_2' = -4x_1 + x_2, \quad x_2(0) = 1 \end{cases}$

$$
\text{Solution.} \quad \begin{cases} x_1 = e^{t}\left(\cos(2t) + \dfrac{\sin(2t)}{2}\right), \\ x_2 = e^{t}\left(\cos(2t) - 2\sin(2t)\right) \end{cases}
$$

8. $\begin{cases} x_1' = 2x_2, \quad x_1(0) = 1, \\ x_2' = -2x_1, \quad x_2(0) = 1, \\ x_3' = -3x_4, \quad x_3(0) = 1, \\ x_4' = x_3, \quad x_4(0) = 0 \end{cases}$

$$
\text{Solution.} \quad \begin{cases} x_1 = \cos(2t) + \sin(2t), \\ x_2 = \cos(2t) - \sin(2t), \\ x_3 = \cos(3t), \\ x_4 = \sin(3t) \end{cases}
$$

9.
$$\begin{cases} x' + y' = -3x - 2y + e^{-2t}, & x(0) = 0, \\ 2x' + y' = -2x - y + 1, & y(0) = 0 \end{cases}$$

Solution.
$$\begin{cases} x = 2 + e^{-2t} - 3e^{-t} - t\,e^{-t}, \\ y = -3 - 2e^{-2t} + 5e^{-t} + 2t\,e^{-t} \end{cases}$$

10.
$$\begin{cases} x' = x - y - e^{-t}, & x(0) = 1, \\ y' = 2x + 3y + e^{-t}, & y(0) = 0 \end{cases}$$

Solution.
$$\begin{cases} x = \dfrac{3e^{-t}}{10} + \dfrac{e^{2t}\left(7\cos(t) - 11\sin(2t)\right)}{10}, \\ y = -\dfrac{2e^{-t}}{5} + \dfrac{e^{2t}\left(2\cos(t) + 9\sin(2t)\right)}{5} \end{cases}$$

11.
$$\begin{cases} x' + y' = x, & x(0) = 1, \\ y' + z' = x, & y(0) = 1, \\ z' + x' = x, & z(0) = 1 \end{cases}$$

Solution. $x = y = z = e^{t/2}$

12.
$$\begin{cases} x_1' = 3x_1, & x_1(0) = 1 \\ x_2' = x_1 + 3x_2, & x_2(0) = 1 \\ x_3' = 3x_3, & x_3(0) = 1 \\ x_4' = 2x_3 + 3x_4, & x_4(0) = 1 \end{cases}$$

Solution.
$$\begin{cases} x_1 = e^{3t}, \\ x_2 = e^{3t}(t+1), \\ x_3 = e^{3t}, \\ x_4 = e^{3t}(2t+1), \end{cases}$$

Chapter 16

Section 16.1, page 567

In the following exercises draw the phase portraits of the indicated two-dimensional systems, taking care to label correctly the directions of the arrows and the orientation when appropriate.

1. $\begin{cases} x' = y, \\ y' = -29x - 4y \end{cases}$

Solution. attracting spiral with clockwise rotation

2. $\begin{cases} x' = 5x - y, \\ y' = 3x + y \end{cases}$

Solution. repelling proper node

3. $\begin{cases} x' = 2x - y, \\ y' = 3x - 2y \end{cases}$

Solution. saddle point

4. $\begin{cases} x' = x - 4y, \\ y' = 4x - 7y \end{cases}$

Solution. attracting improper node with counterclockwise orientation

5. $\begin{cases} x' = x - 5y, \\ y' = x - y \end{cases}$

Solution. stable center with counterclockwise orientation

6. $\begin{cases} x' = y, \\ y' = -2x + 2y \end{cases}$

Solution. repelling spiral with clockwise orientation

7. $\begin{cases} x' = 3x - 2y, \\ y' = 4x - y \end{cases}$

Solution. repelling spiral with counterclockwise orientation

8. $\begin{cases} x' = 3x - 4y, \\ y' = x - y \end{cases}$

Solution. repelling improper node

9. $\begin{cases} x' = x + 2y, \\ y' = -5x - y \end{cases}$

Solution. stable center with clockwise orientation

10. $\begin{cases} x' = 13x + 4y, \\ y' = 4x + 7y \end{cases}$

Solution. repelling proper node

Section 16.2, page 576

Find the general solution to each of the following systems of differential equations using **ODE**.

1. $\begin{cases} x' = x + 2y, \\ y' = 4x + 3y \end{cases}$

Solution. Using **ODE**, we write

```
ODE[{x' == x + 2y,y' == 4x + 3y},{x,y},t,
Method->Laplace]
```

to obtain

$$\begin{cases} x(t) = \left(\dfrac{2e^{-t}}{3} + \dfrac{e^{5t}}{3}\right)x(0) + \left(-\dfrac{e^{-t}}{3} + \dfrac{e^{5t}}{3}\right)y(0), \\ y(t) = \left(-\dfrac{2e^{-t}}{3} + \dfrac{2e^{5t}}{3}\right)x(0) + \left(\dfrac{e^{-t}}{3} + \dfrac{2e^{5t}}{3}\right)y(0) \end{cases}$$

2. $\begin{cases} x' = y, \\ y' = -2x + 3y \end{cases}$

Solution. Using **ODE**, we write

```
ODE[{x' == y,y' == -2x + 3y},{x,y},t,
Method->Laplace]
```

to obtain

$$\begin{cases} x(t) = \left(2e^t - e^{2t}\right)x(0) + \left(-e^t + e^{2t}\right)y(0), \\ y(t) = \left(2e^t - 2e^{2t}\right)x(0) + \left(-e^t + 2e^{2t}\right)y(0) \end{cases}$$

3. $\begin{cases} x' = 3x + 2y, \\ y' = -5x + y \end{cases}$

Solution. Using **ODE**, we write

```
ODE[{x' == 3x + 2y,y' == -5x + y},{x,y},t,
Method->Laplace]
```

to obtain

$$\begin{cases} x(t) = \left(e^{2t}\cos(3t) + \dfrac{e^{2t}\sin(3t)}{3}\right)x(0) + \dfrac{2e^{2t}\sin(3t)}{3}y(0), \\ y(t) = \dfrac{-5e^{2t}\sin(3t)}{3}x(0) + \left(e^{2t}\cos(3t) - \dfrac{e^{2t}\sin(3t)}{3}\right)y(0) \end{cases}$$

4. $\begin{cases} x' = 4x - 2y, \\ y' = -4x + 4y \end{cases}$

Solution. Using **ODE**, we write

```
ODE[{x' == 4x - 2y,y' == -4x + 4y},{x,y},t,
Method->Laplace]
```

to obtain

$$\begin{cases} x(t) = e^{4t}\cosh(2\sqrt{2}\,t)x(0) + \dfrac{e^{4t}\sinh(2\sqrt{2}\,t)}{\sqrt{2}}y(0), \\ y(t) = -\dfrac{2e^{4t}\sinh(2\sqrt{2}\,t)}{\sqrt{2}}x(0) + e^{4t}\cosh(2\sqrt{2}\,t)y(0) \end{cases}$$

5. $\begin{cases} x' = 2x + y + 3e^{2t}, \\ y' = -4x + 2y + te^{2t} \end{cases}$

Solution. Using **ODE**, we write

```
ODE[{x' ==  2x +  y + 3 Exp[2t],
      y' == -4x + 2y + t Exp[2t]},
   {x,y},t,Method->Laplace]
```

to obtain

$$\begin{cases} x(t) = e^{2t}\cos(2t)x(0) + \dfrac{e^{2t}\sin(2t)}{2}y(0) + \dfrac{e^{2t}(2t+11\sin(2t))}{8}, \\[3mm] y(t) = -2e^{2t}\sin(2t)x(0) + e^{2t}\cos(2t)y(0) + \dfrac{11e^{2t}(-1+\cos(2t))}{4} \end{cases}$$

6. $\begin{cases} x' = x+y+z-t, \\ y' = 2x+2y+2z-4, \\ z' = 3x+3y+3z-\sin(t) \end{cases}$

Solution. Using **ODE**, we write

```
ODE[{x' ==  x +  y +  z - t,
     y' == 2x + 2y + 2z - 4,
     z' == 3x + 3y + 3z - Sin[t]},
{x,y,z},t,Method->Laplace]
```

to obtain

$$\begin{cases} x(t) = \dfrac{5+e^{6t}}{6}x(0) + \dfrac{e^{6t}-1}{6}y(0) + \dfrac{e^{6t}-1}{6}z(0) \\[3mm] \qquad +\dfrac{61+150t-90t^2}{216} - \dfrac{961e^{6t}}{7992} + \dfrac{\sin(t)-6\cos(t)}{37}, \\[4mm] y(t) = \dfrac{e^{6t}-1}{3}x(0) + \dfrac{2+e^{6t}}{3}y(0) + \dfrac{e^{6t}-1}{3}z(0) \\[3mm] \qquad +\dfrac{61-282t+18t^2}{108} - \dfrac{961e^{6t}}{3996} + \dfrac{2\sin(t)-12\cos(t)}{37}, \\[4mm] z(t) = \dfrac{e^{6t}-1}{2}x(0) + \dfrac{e^{6t}-1}{2}y(0) + \dfrac{e^{6t}+1}{2}z(0) \\[3mm] \qquad +\dfrac{-11+150t+18t^2}{72} - \dfrac{961e^{6t}}{2664} + \dfrac{3+19\sin(t)}{37} \end{cases}$$

7. $\begin{cases} x' = -y - 5t, \\ y' = 3x - 4 \end{cases}$

Solution. Using **ODE**, we write

```
ODE[{x' == -y - 5t,
     y' == 3x - 4},
{x,y},t,Method->Laplace]
```

to obtain

$$
\begin{cases}
x(t) = x(0)\cos(\sqrt{3}\,t) - y(0)\dfrac{\sin(\sqrt{3}\,t)}{\sqrt{3}} - \dfrac{1 - \cos(\sqrt{3}\,t)}{3}, \\[2mm]
y(t) = x(0)\sqrt{3}\sin(\sqrt{3}\,t) + y(0)\cos(\sqrt{3}\,t) - 5t + \dfrac{\sin(\sqrt{3}\,t)}{\sqrt{3}}
\end{cases}
$$

8.
$$
\begin{cases}
x' + y' = x + y + z, \\
y' + z' = 2x + 2y + 2z, \\
z' + x' = 3x + 3y + 3z
\end{cases}
$$

Solution. Using ODE, we write

```
ODE[{x' + y' == x + y + z,
      y' + z' == 2x + 2y + 2z,
      z' + x' == 3x + 3y + 3z},
 {x,y,z},t,Method->Laplace]
```

to obtain

$$
\begin{cases}
x(t) = \dfrac{2 + e^{3t}}{3}x(0) + \dfrac{e^{3t} - 1}{3}y(0) + \dfrac{e^{3t} - 1}{3}z(0), \\[2mm]
y(t) = y(0), \\[2mm]
z(t) = \dfrac{2e^{3t} - 2}{3}x(0) + \dfrac{2e^{3t} - 2}{3}y(0) + \dfrac{2e^{3t} + 1}{3}z(0)
\end{cases}
$$

Solve and plot the following initial value problems using ODE.

9.
$$
\begin{cases}
x' = 2x + 4y, & x(0) = 2, \\
y' = -4x + 2y, & y(0) = -2
\end{cases}
$$

Solution. Using ODE, we write

```
ODE[{x' ==  2x + 4y,
      y' == -4x + 2y,
 x[0] == 2,y[0] == -2}, {x,y},t,
 Method->Laplace,
 PlotSolution->{{t,0,3},PlotRange->All,
 PlotStyle->{{AbsoluteDashing[{1,0}]},
             {AbsoluteDashing[{6,6}]}}}]
```

to obtain

$$\begin{cases} x(t) = 2e^{2t}\cos(4t) - 2e^{2t}\sin(4t), \\ y(t) = -2e^{2t}\cos(4t) - 2e^{2t}\sin(4t) \end{cases}$$

and the plot

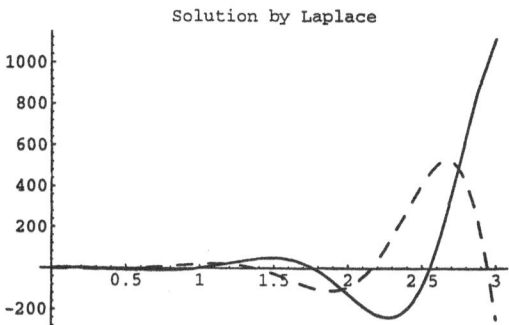

Solution by Laplace

10.
$$\begin{cases} x' = -x + 3y, & x(0) = 0, \\ y' = 3x - y, & y(0) = 4 \end{cases}$$

Solution. Using **ODE**, we write

```
ODE[{x' == -x + 3y,
     y' == 3x -  y,
x[0] == 0,y[0] == 4},{x,y},t,
Method->Laplace,
PlotSolution->{{t,0,1},
PlotStyle->{{AbsoluteDashing[{1,0}]},
            {AbsoluteDashing[{6,6}]}}}]
```

to obtain

$$\begin{cases} x(t) = -2e^{-4t} + 2e^{2t}, \\ y(t) = 2e^{-4t} + 2e^{2t} \end{cases}$$

and the plot

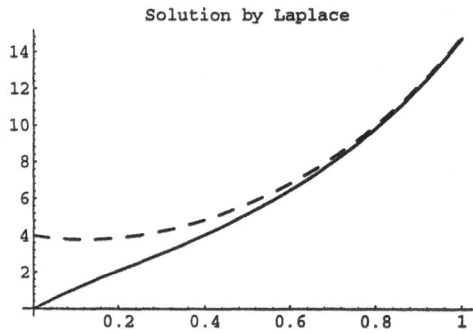

Solution by Laplace

11. $\begin{cases} x' = 4x + 4y, & x(0) = 4, \\ y' = 3x - 4y, & y(0) = 1 \end{cases}$

Solution. Using **ODE**, we write

```
ODE[{x' == 4x + 4y,
     y' == 3x - 4y,
x[0]==4,y[0]==1},{x,y},t,
Method->Laplace,
PlotSolution->{{t,0,1},
PlotStyle->{{AbsoluteDashing[{1,0}]},
            {AbsoluteDashing[{6,6}]}}}}]
```

to obtain

$$\begin{cases} x(t) = 4\cosh(2\sqrt{7}t) + \dfrac{10\sinh(2\sqrt{7}t)}{\sqrt{7}}, \\ y(t) = \cosh(2\sqrt{7}t) + \dfrac{4\sinh(2\sqrt{7}t)}{\sqrt{7}} \end{cases}$$

and the plot

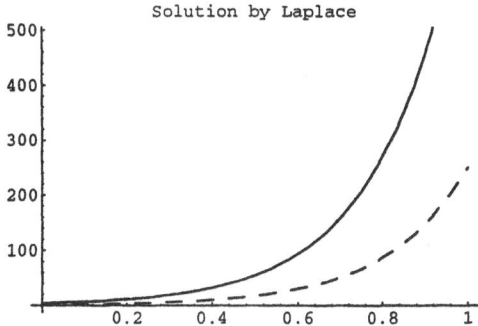

Solution by Laplace

12. $\begin{cases} x' = 5x/2 + 4y, & x(0) = 4, \\ y' = 3x - 4y, & y(0) = 1 \end{cases}$

Solution. Using **ODE**, we write

```
ODE[{x' == 5x/2 + 4y,
      y' == 3x   - 4y,
  x[0] == 4,y[0] == 1}, {x,y},t,
  Method->Laplace,
  PlotSolution->{{t,0,1},
  PlotStyle->{{AbsoluteDashing[{1,0}]},
              {AbsoluteDashing[{6,6}]}}}]
```

to obtain

$$\begin{cases} x(t) = \dfrac{4e^{-11t/2}}{19} + \dfrac{72e^{4t}}{19}, \\ y(t) = -\dfrac{8e^{-11t/2}}{19} + \dfrac{27e^{4t}}{19} \end{cases}$$

and the plot

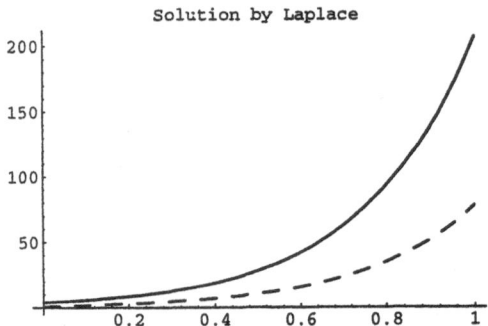

Solution by Laplace

$$
13. \quad
\begin{cases}
x' = 3x + y - z, & x(0) = 1, \\
y' = x + 3y - z, & y(0) = 0, \\
z' = 3x + 3y - z, & z(0) = 0
\end{cases}
$$

Solution. Using **ODE**, we write

```
ODE[{x' == 3x +  y - z,
      y' ==  x + 3y - z,
      z' == 3x + 3y - z,
  x[0] == 1,y[0] == 0,z[0] == 0},{x,y,z},t,
  Method->Laplace,
  PlotSolution->{{t,0,1},
  PlotStyle->{{AbsoluteDashing[{1,0}]},
              {AbsoluteDashing[{6,6}]},
              {AbsoluteDashing[{6,6,2,6}]}}}}]
```

to obtain

$$
\begin{cases}
x(t) = -e^t + 2e^{2t}, \\
y(t) = -e^t + e^{2t}, \\
z(t) = -3e^t + 3e^{2t}
\end{cases}
$$

and the plot

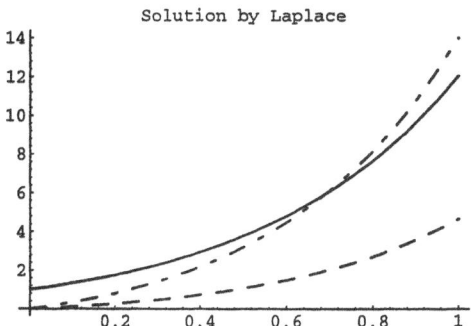

Solution by Laplace

14.
$$\begin{cases} x' = x - y - z, & x(0) = 1, \\ y' = \qquad 2y + 3z, & y(0) = 0, \\ z' = \qquad 3y + z, & z(0) = 0 \end{cases}$$

Solution. Using **ODE**, we write

```
ODE[{x' == x - y -  z,
      y' ==      2y + 3z,
      z' ==      3y +  z,
   x[0] == 1,y[0] == 0,z[0] == 0},{x,y,z},t,
   Method->Laplace,
   PlotSolution->{{t,0,1},
   PlotStyle->{{AbsoluteDashing[{1,0}]},
               {AbsoluteDashing[{6,6}]},
               {AbsoluteDashing[{6,6,2,6}]}}}]
```

to obtain

$$\begin{cases} x(t) = e^t, \\ y(t) = 0, \\ z(t) = 0 \end{cases}$$

and the plot

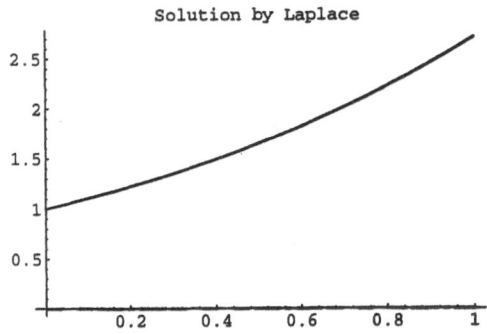

$$\begin{cases} x' = 7x - y + 6z, & x(0) = 1, \\ y' = -10x + 4y - 12z, & y(0) = 1, \\ z' = -2x + y - z, & z(0) = 1 \end{cases}$$

15.

Solution. Using **ODE**, we write

```
ODE[{x' ==    7x -   y +   6z,
       y' == -10x + 4y -  12z,
       z' ==   -2x +  y -   z,
x[0] == 1,y[0] == 1,z[0] == 1},{x,y,z},t,
Method->Laplace,
PlotSolution->{{t,0,1},
PlotStyle->{{AbsoluteDashing[{1,0}]},
             {AbsoluteDashing[{6,6}]},
             {AbsoluteDashing[{6,6,2,6}]}}}}]
```

to obtain

$$\begin{cases} x(t) = 3e^{2t} - 8e^{3t} + 6e^{5t}, \\ y(t) = -3e^{2t} + 16e^{3t} - 12e^{5t}, \\ z(t) = -3e^{2t} + 8e^{3t} - 4e^{5t} \end{cases}$$

and the plot

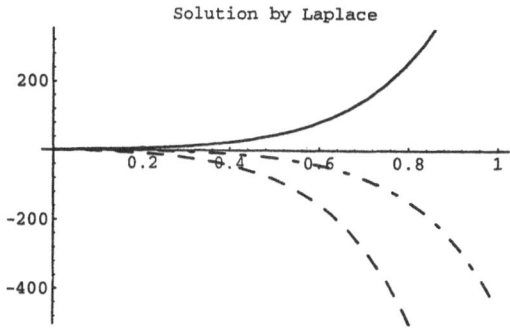

Section 16.3, page 578

Draw the phase portraits of the following linear systems using **ODE**:

1. $$\begin{cases} x'(t) = x, \\ y'(t) = 2y \end{cases}$$

Solution. We enter

```
ODE[{x' == x,y' == 2y,x[0] == a,y[0] == b},
    {x,y},t,Method->LinearSystem,
    Parameters->{{a,-1,1,0.4},{b,-1,1,1}},
    PlotPhase->{{t,-2,2},
    PlotStyle->{{RGBColor[1,0,0],AbsoluteThickness[1]}},
    PlotRange->{{-3,3},{-4,4}}}]
```

and obtain the plot

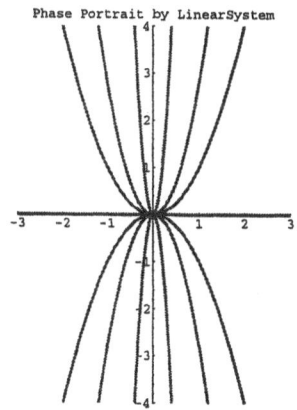

2. $\begin{cases} x'(t) = x, \\ y'(t) = -2y \end{cases}$

Solution. We enter

```
ODE[{x' == x,y' == -2y,x[0] == a,y[0] == b},
    {x,y},t,Method->LinearSystem,
    Parameters->{{a,-1,1,1/2},{b,-1,1,1/2}},
    PlotPhase->{{t,-5,5},
    PlotStyle->{{RGBColor[1,0,0],AbsoluteThickness[1]}},
    PlotRange->{{-2,2},{-2,2}}}]
```

and obtain the plot

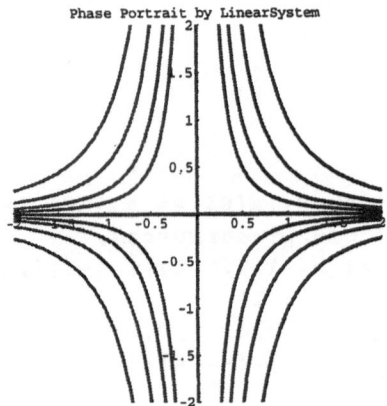

Phase Portrait by LinearSystem

3. $\begin{cases} x'(t) = y/2, \\ y'(t) = -2x \end{cases}$

Solution. we enter

```
ODE[{x' == y/2,y' == -2x,x[0] == a,y[0] == b},
    {x,y},t,Method->LinearSystem,
    Parameters->{{a,-1,1,1/2},{b,-1,1,1/2}},
    PlotPhase->{{t,0,Pi},
    PlotStyle->{{RGBColor[1,0,0],AbsoluteThickness[1]}},
    PlotRange->{{-1.6,1.6},{-2.5,2.5}}}]
```

and obtain the plot

Phase Portrait by LinearSystem

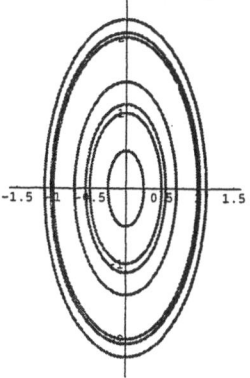

4. $\begin{cases} x'(t) = y, \\ y'(t) = -x \end{cases}$

Solution. We enter

```
ODE[{x' == y,y' == -x,x[0] == a,y[0] == b},
    {x,y},t,Method->LinearSystem,
    Parameters->{{a,-1,1,1/2},{b,-1,1,1/2}},
    PlotPhase->{{t,0,Pi},
    PlotStyle->{{RGBColor[1,0,0],AbsoluteThickness[1]}},
    PlotRange->{{-1.6,1.6},{-1.6,1.6}}}]
```

and obtain the plot

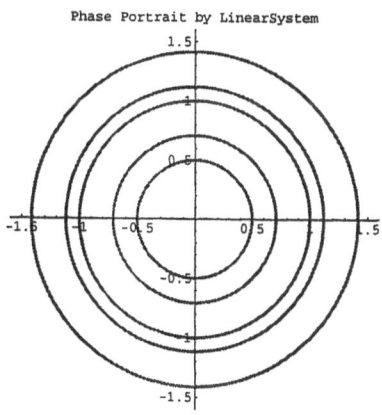

Phase Portrait by LinearSystem

5.
$$\begin{cases} x'(t) = \dfrac{3x + 5y}{2}, \\ y'(t) = \dfrac{-5x + 3y}{2} \end{cases}$$

Solution. We enter

```
ODE[{x' == 3x/2 + 5y/2,y' == -5x/2 + 3y/2,x[0] == a,
    y[0] == b},{x,y},t,Method->LinearSystem,
    Parameters->{{a,-1,1,1/2},{b,-1,1,1/2}},
    PlotPhase->{{t,-10,10},
    PlotStyle->{{RGBColor[1,0,0],AbsoluteThickness[1]}},
    PlotRange->{{-3.2,3.2},{-3.2,3.2}}}]
```

and obtain the plot

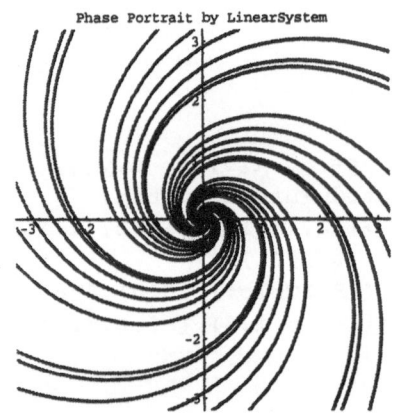

Phase Portrait by LinearSystem

6.
$$\begin{cases} x'(t) = 2x + y, \\ y'(t) = -x + 2y \end{cases}$$

Solution. We enter

```
ODE[{x' == 2x + y,y' == -x + 2y,x[0] == a,
    y[0] == b},{x,y},t,Method->LinearSystem,
    Parameters->{{a,-1,1,1/2},{b,-1,1,1/2}},
    PlotPhase->{{t,-10,10},
    PlotStyle->{{RGBColor[1,0,0],AbsoluteThickness[1]}},
    PlotRange->{{-2,2},{-2,2}}}]
```

and obtain the plot

Phase Portrait by LinearSystem

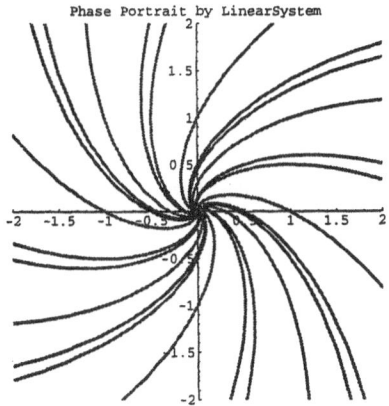

7. $\begin{cases} x'(t) = x, \\ y'(t) = y \end{cases}$

Solution. We enter

```
ODE[{x' == x,y' == y,x[0] == a,y[0] == b},
    {x,y},t,Method->LinearSystem,
    Parameters->{{a,-1,1,1/2},{b,-1,1,1/2}},
    PlotPhase->{{t,-10,10},
    PlotStyle->{{RGBColor[1,0,0],AbsoluteThickness[1]}},
    PlotRange->{{-2,2},{-2,2}}}]
```

and obtain the plot

Phase Portrait by LinearSystem

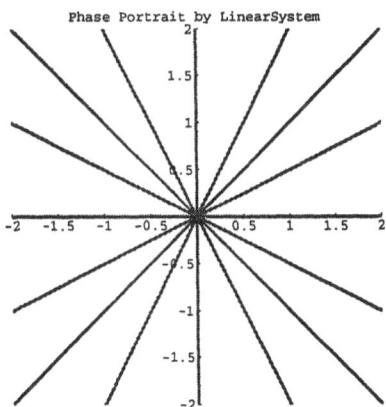

8.
$$\begin{cases} x'(t) = -2x + 5y, \\ y'(t) = -x + y \end{cases}$$

Solution. We enter

```
ODE[{x' == -2x + 5y,y' == -x + y,x[0] == a,
    y[0] == b},{x,y},t,Method->LinearSystem,
    Parameters->{{a,-1,1,1},{b,-1,1,1}},
    PlotPhase->{{t,-10,10},
    PlotStyle->{{RGBColor[1,0,0],AbsoluteThickness[1]}},
    PlotRange->{{-3,3},{-1.5,1.5}}}]
```

and obtain the plot

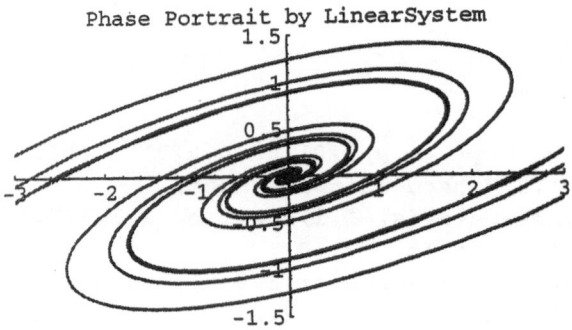

9.
$$\begin{cases} x'(t) = x + y, \\ y'(t) = y \end{cases}$$

Solution. We enter

```
ODE[{x' == x + y,y' == y,x[0] == a,y[0] == b},
    {x,y},t,Method->LinearSystem,
    Parameters->{{a,-1,1,1/2},{b,-1,1,1/2}},
    PlotPhase->{{t,-10,10},
    PlotStyle->{{RGBColor[1,0,0],AbsoluteThickness[1]}},
    PlotRange->{{-2,2},{-2,2}}}]
```

and obtain the plot

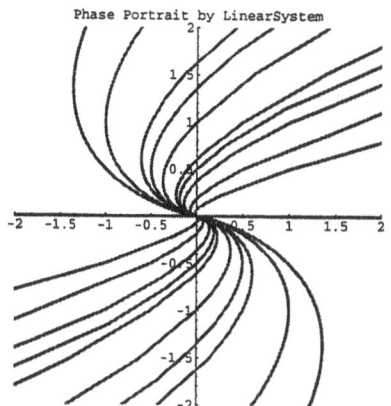

$$10. \quad \begin{cases} x'(t) = \dfrac{3x + 5y}{2}, \\ y'(t) = -\dfrac{5x + 3y}{2} \end{cases}$$

Solution. We enter

```
ODE[{x' == 3x/2 + 5y/2,y' == -5x/2 - 3y/2,x[0] == a,
    y[0] == b},{x,y},t,Method->LinearSystem,
    Parameters->{{a,-1,1,1/2},{b,-1,1,1/2}},
    PlotPhase->{{t,0,2Pi},
    PlotStyle->{{RGBColor[1,0,0],AbsoluteThickness[1]}},
    PlotRange->{{-2.7,2.7},{-2.7,2.7}}}]
```

and obtain the plot

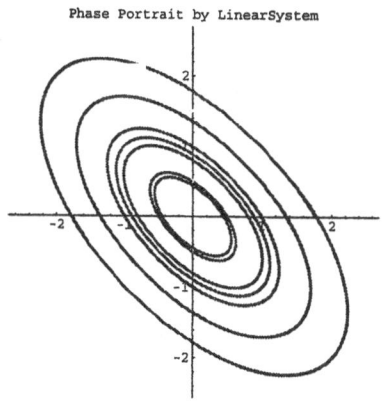

Chapter 17

Section 17.1, page 583

1. Find parametrizations for the ellipse $\dfrac{x^2}{a^2} + \dfrac{y^2}{b^2} - 1 = 0$ and the parabola $x^2 - 4a\, y = 0$.

   Solution. The ellipse is parametrized by the equations $x = a\cos(t)$, $y = b\sin(t)$.
   The parabola is parametrized by the equations $x = t$, $y = t^2/4a$.

2. A **cardioid** is the parametrized curve defined by

$$\text{cardioid}[a](t) = \big(2a\,\cos(t)(1+\cos(t)),\ 2a\,\sin(t)(1+\cos(t))\big).$$

   Find the corresponding implicitly defined curve.

   Solution. We have $\cos(t) + \cos(t)^2 = x/2a$, so that the quadratic formula yields $\cos(t) + (1/2) = \pm\sqrt{(2x+a)/4a}$. On the other hand $x^2 + y^2 = 4a^2(1+\cos(t))^2 = 4a^2(1/2 \pm \sqrt{(2x+a)/4a})^2$. Algebraic simplification leads to

$$\left(\frac{x^2 + y^2 - 2a^2 - 2ax}{4a^2}\right)^2 = \frac{2x+a}{4a}$$

Section 17.2, page 585

1. Show that if $t \longmapsto y(t)$ is a solution of the nth-order equation (17.5), then so is $t \longmapsto y(t - t_0)$.

   Solution. By hypothesis, for any t, we have $y''(t) + F(y^{n-1}(t), \ldots, y(t)) = 0$. Hence we also have $y''(t - t_0) + F(y^{n-1}(t - t_0), \ldots, y(t - t_0)) = 0$. Defining $z(t) = y(t - t_0)$, we see that $z''(t) + F(z^{n-1}(t), \ldots, z(t)) = 0$, which was to be proved.

2. Use **ODE** with **Method->DSolve** and **PlotPhase** to plot several solution curves of the nonautonomous system

$$\begin{cases} x' = t, & x(s) = 1, \\ y' = t^2, & y(s) = 1 \end{cases}$$

   Solution. We enter

```
ODE[{x'==t, y'==t^2,x[s]==1,y[s]==1}, {x,y},t,
Method->LinearSystem,Parameters->{{s,-1,1,0.2}},
PlotPhase->{{t,-1,1}}]
```

and obtain the plot

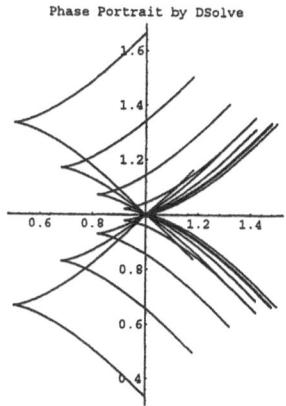

Phase Portrait by DSolve

Section 17.3, page 591

For each of the following systems, find all critical points and the linearized system associated
with each critical point.

1.
$$\begin{cases} x' = x + y^2, \\ y' = x + y \end{cases}$$

Solution.

Critical Point	Linearized System
$(0, 0)$	$\begin{cases} x_1' = x_1, \\ x_2' = x_1 + x_2 \end{cases}$
$(-1, 1)$	$\begin{cases} x_1' = x_1 + 2x_2, \\ x_2' = x_1 + x_2 \end{cases}$

2.
$$\begin{cases} x' = 1 - x\,y, \\ y' = x - y^3 \end{cases}$$

Solution.

Critical Point	Linearized System
$(1, 1)$	$\begin{cases} x_1' = -x_1 - x_2, \\ x_2' = x_1 - 3x_2 \end{cases}$
$(-1, -1)$	$\begin{cases} x_1' = -x_1 + x_2, \\ x_2' = x_1 - 3x_2 \end{cases}$

3. $\begin{cases} x' = x - x^2 - x\,y, \\ y' = 3y - x\,y - 2y^2 \end{cases}$

Solution.

Critical Point	Linearized System
$(0, 0)$	$\begin{cases} x_1' = x_1, \\ x_2' = 3x_2 \end{cases}$
$(0, 3/2)$	$\begin{cases} x_1' = -(1/2), \\ x_2' = -(3/2)x_1 - 3x_2 \end{cases}$
$(1, 0)$	$\begin{cases} x_1' = -x_1 - x_2, \\ x_2' = 2x_2 \end{cases}$
$(-1, 2)$	$\begin{cases} x_1' = x_2, \\ x_2' = -2x_1 - 4x_2 \end{cases}$

4. $\begin{cases} x' = 1 - y, \\ y' = x^2 - y^2 \end{cases}$

Solution.

Critical Point	Linearized System
$(1, 1)$	$\begin{cases} x_1' = -x_2, \\ x_2' = 2x_1 - 2x_2 \end{cases}$
$(-1, 1)$	$\begin{cases} x_1' = -x_2, \\ x_2' = = -2x_1 - 2x_2 \end{cases}$

5.
$$\begin{cases} x' = y, \\ y' = -\sin(x) \end{cases}$$

Solution.

Critical Point	Linearized System
$(n\pi, 0)$ n even	$\begin{cases} x_1' = x_2, \\ x_2' = -x_1 \end{cases}$
$(n\pi, 0)$ n odd	$\begin{cases} x_1' = x_2, \\ x_2' = x_1 \end{cases}$

6.
$$\begin{cases} x' = \sin(y), \\ y' = \cos(x) \end{cases}$$

Solution.

Critical Point	Linearized System
$((m + 1/2)\pi, n\pi)$ m and n even	$\begin{cases} x_1' = x_2, \\ x_2' = -x_1 \end{cases}$
$((m + 1/2)\pi, n\pi)$ m and n odd	$\begin{cases} x_1' = -x_2, \\ x_2' = -x_1 \end{cases}$
$((m + 1/2)\pi, n\pi)$ m even, n odd	$\begin{cases} x_1' = -x_2, \\ x_2' = x_1 \end{cases}$
$((m + 1/2)\pi, n\pi)$ m odd, n even	$\begin{cases} x_1' = x_2, \\ x_2' = x_1 \end{cases}$

7.
$$\begin{cases} x' = (\sin(y))^2, \\ y' = (\cos(x))^2 \end{cases}$$

Solution. The critical points are $((m + (1/2))\pi, n\pi)$, where m and n are integers. In every case the linearized system is $x_1' = 0, x_2' = 0$.

8.
$$\begin{cases} x' = y, \\ y' = -\sin(x) - 4y \end{cases}$$

Solution.

Critical Point	Linearized System
$(m\pi, 0)$ m even	$\begin{cases} x_1' = x_2, \\ x_2' = -x_1 - 4x_2 \end{cases}$
$(m\pi, 0)$ m odd	$\begin{cases} x_1' = x_2, \\ x_2' = x_1 - 4x_2 \end{cases}$

9. $\begin{cases} x' = x(1 - x - y), \\ y' = y(3 - 2y - 2x) \end{cases}$

Solution.

Critical Point	Linearized System
$(0, 0)$	$\begin{cases} x_1' = x_2, \\ x_2' = 3x_2 \end{cases}$
$(0, 3/2)$	$\begin{cases} x_1' = -x_1/2, \\ x_2' = -3x_1 - 3x_2 \end{cases}$
$(1, 0)$	$\begin{cases} x_1' = -x_1 - x_2, \\ x_2' = x_2 \end{cases}$

10. $\begin{cases} x' = x(3 - 2y), \\ y' = y(-4 + 2x) \end{cases}$

Solution.

Critical Point	Linearized System
$(0, 0)$	$\begin{cases} x_1' = 3x_1, \\ x_2' = -4x_2 \end{cases}$
$(2, 3/2)$	$\begin{cases} x_1' = -4x_2, \\ x_2' = 3x_1 \end{cases}$

Section 17.5, page 603

For each of the following nonlinear systems, verify that $(0, 0)$ is a critical point and determine whether or not it is unstable, stable, or asymptotically stable. Discuss spiraling behavior when applicable.

1. $$\begin{cases} x' = x - y + xy, \\ y' = 3x - 2y - xy \end{cases}$$

Solution. $(0, 0)$ is a stable and asymptotically stable critical point for which the solutions spiral.

2. $$\begin{cases} x' = x + x^2 + y^2, \\ y' = y - xy \end{cases}$$

Solution. $(0, 0)$ is an unstable critical point.

3. $$\begin{cases} x' = -2x - y - x(x^2 + y^2), \\ y' = x - y + y(x^2 + y^2) \end{cases}$$

Solution. $(0, 0)$ is a stable and asymptotically stable critical point for which the solutions spiral.

4. $$\begin{cases} x' = y + x(1 - x^2 - y^2), \\ y' = -x + y(1 - x^2 - y^2) \end{cases}$$

Solution. $(0, 0)$ is an unstable critical point.

5. $$\begin{cases} x' = 2x + y + xy^3, \\ y' = x - 2y - xy \end{cases}$$

Solution. $(0, 0)$ is an unstable critical point.

6. $$\begin{cases} x' = x + 2x^2 - y^2, \\ y' = x - 2y + x^3 \end{cases}$$

Solution. $(0, 0)$ is an unstable critical point.

7. $$\begin{cases} x' = y, \\ y' = -x + 4y(1 - x^2) \end{cases}$$

Solution. $(0, 0)$ is an unstable critical point.

8.
$$\begin{cases} x' = 1 + y - e^{-x}, \\ y' = y - \sin(x) \end{cases}$$

Solution. $(0, 0)$ is an unstable critical point.

9.
$$\begin{cases} x' = (1 + x)\sin(y), \\ y' = 1 - x - \cos(y) \end{cases}$$

Solution. $(0, 0)$ is a stable critical point.

10.
$$\begin{cases} x' = 1 - x + y - \cos(x), \\ y' = \sin(x - 3y) \end{cases}$$

Solution. $(0, 0)$ is a stable critical point.

The following exercises illustrate the use of the word *indeterminate.*

11. Show that the system
$$\begin{cases} x' = -y - x(x^2 + y^2), \\ y' = x - y(x^2 + y^2) \end{cases}$$

has a stable critical point at $(0, 0)$. [Hint: Obtain a differential equation for $r^2 = x^2 + y^2$.]

Solution. The function $F = r^2$ satisfies the differential equation $F' = -2F^2$ with the initial condition $F(0) = r_0^2$, for which the solution is $F(t) = r_0^2/(1 + 2t \, r_0^2)$, so that $\lim_{t \to \infty} F(t) = 0$, hence stability.

12. Show that the system
$$\begin{cases} x' = -y + x(x^2 + y^2), \\ y' = x + y(x^2 + y^2) \end{cases}$$

has an unstable critical point at $(0, 0)$. [Hint: Obtain a differential equation for $r^2 = x^2 + y^2$.]

Solution. The function $F = r^2$ satisfies the differential equation $F' = 2F^2$ with the initial condition $F(0) = r_0^2$ for which the solution is $F(t) = r_0^2/(1 - 2t \, r_0^2)$ so that $\lim_{t \to T} F(t) = +\infty$, where $T = 1/2r_0^2$, hence instability.

13. Conclude from Exercises (11) and (12) that a stable center has indeterminate stability type when nonlinear terms are added.

Solution. Both of the nonlinear systems in Exercises (11) and (12) have the same linearized system at $(0, 0)$, which is a stable center. But the nonlinear system in Exercise (11) is asymptotically stable, whereas the nonlinear system in Exercise (12) is unstable. Therefore, the stable center has indeterminate stability type when nonlinear terms are added.

14. Show that the system
$$\begin{cases} x' = -x, \\ y' = y^2 \end{cases}$$
has an unstable critical point at $(0, 0)$.

Solution. The solution of this system is given by $x(t) = x(0)e^{-t}$, $y(t) = y(0)/(1 - ty(0))$. We will disprove the stability condition by taking $\epsilon = 1$. If there were some $\delta > 0$ satisfying stability, then we take $y(0) = \min(1/2, \delta/2)$, $x(0) = 0$. At time $t = 1/y(0) - 2$ we have $y(t) = 1$, which contradicts the requirement that $x(t)^2 + y(t)^2 < \epsilon^2$ for all $t > 0$.

15. Show that the system
$$\begin{cases} x' = -x, \\ y' = -y^3 \end{cases}$$
has a stable critical point at $(0, 0)$.

Solution. The solution of this system is given by $x(t) = x(0)e^{-t}$, $y(t) = y(0)/\sqrt{1 + 2ty(0)^2}$. Given $\epsilon > 0$, we may take $\delta = \epsilon$. Noting that $|x(t)| < |x(0)|$, $|y(t)| < |y(0)|$ for $t > 0$, we see that if $x(0)^2 + y(0)^2 < \delta^2$, then $x(t)^2 + y(t)^2 < \epsilon^2$ for all $t > 0$, as required for stability.

16. Conclude from Exercises 14 and 15 that an attracting center has indeterminate stability type when nonlinear terms are added.

Solution. Both of the nonlinear systems in Exercises 14 and 15 have the same linearized part $x' = -x$, $y' = 0$, which is an attracting center. But the nonlinear system in Exercise 14 is unstable, whereas the nonlinear system in Exercise 15 is stable. Therefore, the attracting center has indeterminate stability type with respect to the addition of nonlinear terms.

Section 17.6, page 614

1. Show that $V(x, y) = x^2 + y^2$ is a Lyapunov function for the system
$$\begin{cases} x' = -x^3 + 2x\, y^2, \\ y' = -2x^2 y - 5y^3 \end{cases}$$

Solution. $V(x, y)$ is clearly positive definite. To compute the time derivative, we write

$$\frac{dV}{dt} = 2x(-x^3 + 2xy^2) + 2y(-2x^2y - 5y^3)$$
$$= -2x^4 - 10y^4$$

which is strictly negative, unless $(x, y) = (0, 0)$. Therefore, $V(x, y)$ satisfies all of the required properties.

2. Find a Lyapunov function of the form $V(x, y) = a x^2 + b y^2$ for the system

$$\begin{cases} x' = -x^3 + x y^2 + x^5, \\ y' = -2x^2y - y^3 - y^5 \end{cases}$$

Solution. $V(x, y)$ is clearly positive definite provided that $a > 0, b > 0$. To proceed further, we write

$$\frac{dV}{dt} = 2ax(-x^3 + x y^2 + x^5,) + 2by(-2x^2y - y^3 - y^5)$$
$$= -2zx^4 + (2a - 4b)x^2y^2 - 2by^4 - 2by^6$$

If in addition we have $0 < a \le 2b$, then $2a - 4b \le 0$ and all of the above terms are non-positive, proving that V is a Lyapunov function.

3. Find a Lyapunov function of the form $V(x, y) = a x^2 + b y^2$ for the system

$$\begin{cases} x' = -x^3 + x y^2 + 3x^7, \\ y' = -2x^2y - 4y^3 - 3y^5 \end{cases}$$

Solution. $V(x, y)$ is clearly positive definite provided that $a > 0, b > 0$. To proceed further, we write

$$\frac{dV}{dt} = 2ax(-x^3 + x y^2 + 3x^7) + 2by(-2x^2y - 4y^3 - 3y^5)$$
$$= -2ax^4 + (2a - 2b)x^2y^2 + 6ax^8 - 8by^4 - 6by^6$$
$$= -2ax^4(1 - 3x^2) + (2a - 2b)x^2y^2 - 8by^4 - 6by^6$$

If in addition we have $0 < a \le b$, then the second term is non-positive. The first term is also non-positive, provided that we restrict x so that $|x| < 1/\sqrt{3}$. Therefore, V is a Lyapunov function in this region.

4. Suppose that we have a linear system with an attracting spiral point corresponding to a 2×2 matrix (a_{ij}). Let

$$V(x_1, x_2) = a_{21}x_1^2 + (a_{22} - a_{11})x_1x_2 - a_{12}x_2^2.$$

a. Show that $V(x_1, x_2) \neq 0$ for $(x_1, x_2) \neq (0, 0)$ and that for any solution of the linear system $dx/dt = Ax$ we have

$$\frac{d}{dt}V(x_1(t), x_2(t)) = (a_{11} + a_{22})V(x_1(t), x_2(t))$$

b. Conclude from a. that all solutions of the linear system satisfy the limiting relation

$$\lim_{t \to \infty} \frac{1}{t} \log|x(t)| = \frac{1}{2}(a_{11} + a_{22}) < 0.$$

Solution. **a.** The eigenvalues $r_{1,2}$ of the matrix (a_{ij}) are found by the quadratic formula as

$$2r_{1,2} = a_{11} + a_{22} \pm \sqrt{(a_{11} - a_{22})^2 + 4a_{12}a_{21}}$$

so that the hypotheses of attracting spiral is equivalent to the inequalities

$$(a_{11} - a_{22})^2 + 4a_{12}a_{21} < 0, \qquad a_{11} + a_{22} < 0,$$

in particular $a_{21}a_{12} < 0$. On the other hand, the function $V(x_1, x_2)$ is defined by a matrix whose determinant is $-a_{12}a_{21} - 4(a_{22} - a_{11})^2 > 0$. Hence $V(x_1, x_2) \neq 0$ for $(x_1, x_2) \neq (0, 0)$(a more detailed estimate follows in part b). Now

$$
\begin{aligned}
dV/dt &= 2a_{21}x_1(a_{11}x_1 + a_{12}x_2) - 2a_{12}x_2(a_{21}x_1 + a_{22}x_2) \\
&+ (a_{22} - a_{11})(x_2(a_{11}x_1 + a_{12}x_2) + x_1(a_{21}x_1 + a_{22}x_2)) \\
&= (a_{21}a_{22} + a_{11}a_{21})x_1^2 + (a_{22}^2 - a_{11}^2)x_1x_2 - (a_{12}a_{22} + a_{12}a_{11})x_2^2 \\
&= (a_{11} + a_{22})\left(a_{21}x_1^2 + (a_{22} - a_{11})x_1x_2 - a_{12}x_2^2\right) \\
&= (a_{11} + a_{22})V(x_1, x_2)
\end{aligned}
$$

b. From the computations in part a), it follows that

$$
\begin{aligned}
\frac{d}{dt}\log|V(x_1(t), x_2(t))| &= (a_{11} + a_{22}), \\
\log|V(x_1(t), x_2(t))| &= \log|V(x_1(0), x_2(0))| + t(a_{11} + a_{22}), \\
\lim_{t \to \infty} \frac{\log|V(x_1(t), x_2(t))|}{t} &= (a_{11} + a_{22}).
\end{aligned}
$$

It remains to show that we can replace $|V(x_1(t), x_2(t))|$ by $|\mathbf{x}(t)|^2$. Since $a_{12}a_{21} < 0$, there is no loss of generality in supposing that $a_{21} > 0 > a_{12}$ (otherwise change V to $-V$). Now for any $s > 0$, $|2x_1x_2| \le sx_1^2 + x_2^2/s$. Applying this to $V(x_1, x_2)$,

$$V(x_1, x_2) \ge x_1^2 \left(a_{21} - s\sqrt{|a_{21}a_{12}|}\right) + x_2^2 \left(a_{12} - \sqrt{|a_{21}a_{12}|/s}\right)$$

We choose $s = \sqrt{|a_{21}/a_{12}|}/2$ and substitute above, to find that

$$V(x_1, x_2) \ge \frac{1}{2}a_{21}x_1^2 - \frac{1}{2}a_{12}x_2^2 \ge m(x_1^2 + x_2^2).$$

where $m = \frac{1}{2}\min(-a_{12}, a_{21})$ On the other hand it is immediate that $|V(x_1, x_2)| \le M(x_1^2 + x_2^2)$, where $M = a_{21} - a_{12} + |a_{11} - a_{22}|$ Hence we have

$$m(x_1^2 + x_2^2) \le V(x_1, x_2) \le M(x_1^2 + x_2^2)$$

$$\log m + 2\log|\mathbf{x}(t)| \le \log|V(x_1(t), x_2(t))| \le \log M + 2\log|\mathbf{x}(t)|$$

Therefore,

$$\lim_{t \to \infty} \frac{\log|\mathbf{x}(t)|}{t} = \lim_{t \to \infty} \frac{\log|V(x_1(t), x_2(t))|}{2t}.$$

5. Suppose that we have the nonlinear system $x' = F(x, y)$, $y' = G(x, y)$ with an isolated critical point at $(0, 0)$. Suppose that the corresponding linearized system defines an attracting spiral point with eigenvalues $r = \lambda \pm i\mu$ with $\lambda < 0$ and $\mu \ne 0$.

   a. Show that the solutions of the nonlinear system satisfy

   $$\lim_{t \to \infty} \frac{1}{t}\log|\mathbf{x}(t)| = \lambda$$

   provided that $x(0)^2 + y(0)^2 < \delta^2$ for a suitable $\delta > 0$. [Hint: Compute the time derivative of $\log V(x_1(t), x_2(t))$ from the previous exercise.]

   b. Apply the result of part a to the damped pendulum system

   $$\begin{cases} x' = y, \\ y' = -\dfrac{g}{l}\sin(x) - cy \end{cases}$$

   with $0 < c < 2\sqrt{g/l}$.

Solution. **a.** The nonlinear system can be written in the form

$$x_1' = a_{11}x_1 + a_{12}x_2 + \epsilon_1(x_1, x_2)|\mathbf{x}|$$
$$x_2' = a_{21}x_1 + a_{22}x_2 + \epsilon_2(x_1, x_2)|\mathbf{x}|$$

where $\epsilon_1(x_1, x_2), \epsilon_2(x_1, x_2) \to 0$ when $(x_1, x_2) \to (0, 0)$. Following the computations in the previous solution, we can write

$$dV/dt = (a_{11} + a_{22})V + \epsilon(x_1, x_2)V(x_1, x_2) \qquad\qquad \text{(S.91)}$$

where $\epsilon(x_1, x_2) \to 0$ when $(x_1, x_2) \to (0, 0)$. By hypothesis $a_{11} + a_{22} < 0$, so we can choose $\delta > 0$ so that $|\epsilon(x, y)| < |a_{11} + a_{22}|/3$ for $x_1^2 + x_2^2 < \delta^2$. Then $(1/V)(dV/dt) < \lambda/2 < 0$ and hence we conclude that $V(x_1(t), x_2(t)) \to 0$ whenever $x_1(0)^2 + x_2(0)^2 < \delta^2$. Returning to (S.91), we see that $\lim_{t\to\infty}(1/V)(dV/dt) = a_{11} + a_{22}$, so by the same steps as in the previous exercise, we conclude that $\lim_{t\to\infty} t^{-1} \log|\mathbf{x}(t)| = (a_{11} + a_{22})/2$, provided that $x_1(0)^2 + x_2(0)^2 < \delta^2$.

b. In the case of the damped pendulum, we have $a_{11} = 0$, $a_{12} = 1$, $a_{21} = -g/l$, $a_{22} = -c < 0$ and the eigenvalues are $2r = -c \pm \sqrt{c^2 - 4g/l}$, which satisfies the hypotheses of part **a.**

Find a suitable Lyapunov function to investigate the stability of the following systems in the neighborhood of $(0, 0)$.

6.
$$\begin{cases} x' = -3y - 2x^3, \\ y' = 2x - 3y^3 \end{cases}$$

Solution. $V(x, y) = 2x^2 + 3y^2$ is a suitable Lyapunov function to prove asymptotic stability.

7.
$$\begin{cases} x' = -x\,y^4, \\ y' = x^4 y \end{cases}$$

Solution. $V(x, y) = x^4 + y^4$ is a suitable Lyapunov function to prove stability.

8.
$$\begin{cases} x' = x + 2x\,y^2, \\ y' = -2y + 4x^2 y \end{cases}$$

Solution. $V(x, y) = x^2 - y^2/2$ proves instability.

9.
$$\begin{cases} x' = -y - x/2 - x^3/4, \\ y' = x - y/2 - y^3/4 \end{cases}$$

Solution. $V(x, y) = x^2 + y^2$ is a suitable Lyapunov function to prove asymptotic stability.

10. $\begin{cases} x' = y + x^3, \\ y' = -x + y^3 \end{cases}$

Solution. $V(x, y) = x^2 + y^2$ is a suitable Lyapunov function to prove instability.

11. $\begin{cases} x' = y + x^2 y^2 - x^5/4, \\ y' = -2x - x^3 y - y^3/2 \end{cases}$

Solution. $V(x, y) = 2x^2 + y^2$ is a suitable Lyapunov function to prove stability.

12. $\begin{cases} x' = x y^4 - 2x^3 - y, \\ y' = 2x^2 y^3 - y^7 + 2x \end{cases}$

Solution. $V(x, y) = x^2 + y^2/2$ is a suitable Lyapunov function to prove stability.

13. $\begin{cases} x' = -2x - 3y, \\ y' = x - y \end{cases}$

Solution. $V(x, y) = x^2 + 3y^2$ is a suitable Lyapunov function to prove asymptotic stability.

14. $\begin{cases} x' = x y - x^3 + y, \\ y' = x^4 - x^2 y - x^3 \end{cases}$

Solution. $V(x, y) = x^4 + 2y^2$ is a suitable Lyapunov function to prove stability.

15. $\begin{cases} x' = -2y - x(x - y)^2, \\ y' = 3x - (3/2)y(x - y)^4 \end{cases}$

Solution. $V(x, y) = 3x^2 + 2y^2$ is a suitable Lyapunov function to prove stability.

16. $\begin{cases} x' = x + x^3, \\ y' = -y - y^3 \end{cases}$

Solution. $V(x, y) = x^2 - y^2$ proves instability.

17. $\begin{cases} x' = x^5 + y^3, \\ y' = x^3 - y^5 \end{cases}$

Solution. $V(x, y) = x^4 - y^4$ proves instability.

18. $\begin{cases} x' = x^3 + 2x\,y^2, \\ y' = x^2 y \end{cases}$

Solution. $V(x, y) = x^4 - y^4$ proves instability.

Chapter 18

Section 18.1, page 627

1. Solve the system (18.7) of coupled oscillators in the case that $m_1 = m_2 = 1$, $k_1 = k_2 = 4$ and $k_{12} = 2.5$ with the initial conditions $x_1(0) = 4$, $x_2(0) = 2$ and $y_1(0) = y_2(0) = 0$. Assume that the equilibrium positions of the two masses are separated by 8 units.

Solution. The angular frequencies are given by

$$\mu_1 = \sqrt{k_1/m_1} = \sqrt{4} = 2 \qquad \text{and} \qquad \mu_2 = \sqrt{(k+2k_{12})/m} = \sqrt{9} = 3.$$

To solve the system with **ODE**, we use

```
prob1=ODE[{x1' == y1,y1' == -6.5x1 + 2.5x2,
           x2' == y2,y2' ==  2.5x1 - 6.5x2,
x1[0] == 4,y1[0] == 0,x2[0] == 2,y2[0] == 0},
    {x1[t],y1[t],x2[t],y2[t]},t,
    Method->LinearSystem]
```

obtaining the output

```
{{x1[t] -> 3. Cos[2. t] + Cos[3. t],
   x2[t] -> 3. Cos[2. t] - 1. Cos[3. t],
   y1[t] -> -6. Sin[2. t] - 3. Sin[3. t],
   y2[t] -> -6. Sin[2. t] + 3. Sin[3. t]}}
```

Thus

$$x_1(t) = 3\cos(2t) + \cos(3t), \qquad x_2(t) = 3\cos(2t) - \cos(3t),$$

$$y_1(t) = -6\sin(2t) - 3\sin(3t), \qquad y_2(t) = -6\sin(2t) = 3\sin(3t).$$

We create the plot using

```
Plot[Evaluate[{x1[t] + 4,x2[t] - 4} /. prob1],{t,0,4Pi}]
```

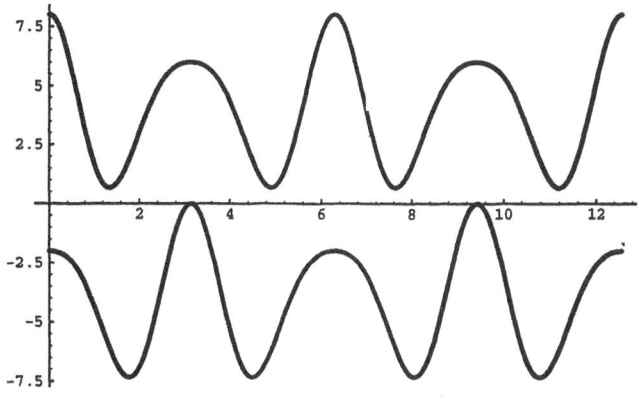

Solution to Exercise 1

2. Solve the system (18.7) of coupled oscillators in the case that $m_1 = m_2 = 1$, $k_1 = k_2 = 4$ and $k_{12} = 6$ with the initial conditions $x_1(0) = 4$, $x_2(0) = 2$ and $y_1(0) = y_2(0) = 0$. Assume that the equilibrium positions of the two masses are separated by 8 units.

Solution. In this case we have

$$\mu_1 = \sqrt{k_1/m_1} = \sqrt{4} = 2 \qquad \text{and} \qquad \mu_2 = \sqrt{(k_1 + 2k_{12})/m_1} = \sqrt{16} = 4.$$

To solve the system with ODE, we use

```
prob2=ODE[{x1' == y1,y1' == -10x1 + 6x2,
            x2' == y2,y2' ==   6x1 - 10x2,
   x1[0] == 4,y1[0] == 0,x2[0] == 2,y2[0] == 0},
        {x1[t],y1[t],x2[t],y2[t]},t,
        Method->LinearSystem]
```

obtaining the output

```
{{x1[t] -> 3 Cos[2 t] + Cos[4 t],x2[t] -> 3 Cos[2 t] - Cos[4 t],
  y1[t] -> -6Sin[2 t] - 4Sin[4 t],y2[t] -> -6Sin[2 t] + 4Sin[4 t]}}
```

Thus

$$x_1(t) = 3\cos(2t) + \cos(4t), \qquad x_2(t) = 3\cos(2t) - \cos(4t),$$

$$y_1(t) = -6\sin(2t) - 4\sin(4t), \qquad y_2(t) = -6\sin(2t) = 4\sin(3t).$$

We create the plot using

```
Plot[Evaluate[{x1[t] + 4,x2[t] - 4} /. prob2],{t,0,4Pi}]
```

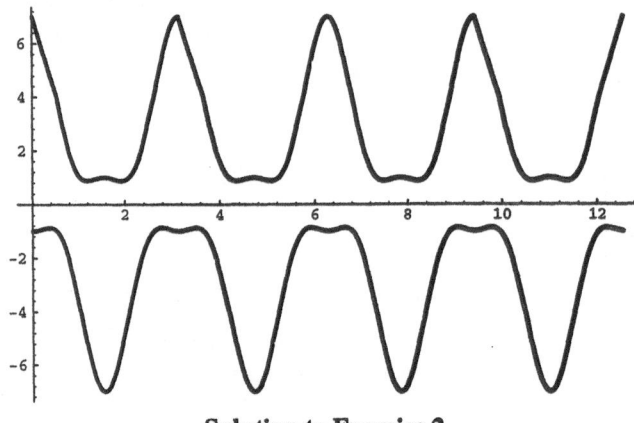

Solution to Exercise 2

3. Show that the solutions Z_a and W_a satisfy the identity $Z_a(t - (\pi/2\mu_1)) = W_a(t)$ for all t. (This is paraphrased as the statement that these two solutions are ninety degrees out of phase with one another.)

Solution. Use the identities $\cos(\alpha - \pi/2) = \sin(\alpha)$ and $\sin(\alpha - \pi/2) = -\cos(\alpha)$.

4. Show that the solutions Z_s and W_s satisfy the identity $Z_s(t - (\pi/2\mu_1)) = W_s(t)$ for all t. (This is paraphrased as the statement that these two solutions are ninety degrees out of phase with one another.)

Solution. Use the identities $\cos(\alpha - \pi/2) = \sin(\alpha)$ and $\sin(\alpha - \pi/2) = -\cos(\alpha)$.

5. Suppose that two equal masses interact in the absence of friction with equal spring constants and equal coupling constants. Let $w_1 = x_1 - x_2$ and $w_2 = x_1 + x_2$. Show that w_1 and w_2 satisfy the *decoupled* system

$$\begin{cases} m\,w_1'' + (k + 2k_{12})w_1 = 0, \\ m\,w_2'' + k\,w_2 = 0. \end{cases} \tag{S.92}$$

where m is the common value of the masses, k is the common value of the spring constants and k_{12} is the common value of the coupling constants. Solve the system (S.92) to obtain w_1 and w_2; then find x_1 and x_2.

Solution. $t \longmapsto (x_1(t), x_2(t))$ is a solution of the system

$$\begin{cases} m\,x_1'' + k\,x_1 = k_{12}(x_2 - x_1), \\ m\,x_2'' + k\,x_2 = -k_{12}(x_2 - x_1). \end{cases} \tag{S.93}$$

When we add the two equations of (S.93) we get the second equation of (S.92), and when we subtract the two equations of (S.93) we get the first equation of (S.92).

The general solutions of each equation of (S.92) can be found separately. Thus

$$\begin{cases} y_1(t) = C_1 \cos(\mu_2 t) + C_2 \sin(\mu_2 t), \\ y_2(t) = C_3 \cos(\mu_1 t) + C_4 \sin(\mu_1 t), \end{cases} \tag{S.94}$$

where the C_j are constants, and

$$\mu_1 = \sqrt{\frac{k}{m}} \qquad \text{and} \qquad \mu_1 = \sqrt{\frac{k + k_{12}}{m}}.$$

6. Suppose that the initial conditions are such that $x_1(0) = -x_2(0)$ and $x_1'(0) = -x_2'(0)$. Show that the resulting solution only contains terms with frequency $\mu_2 = \sqrt{(k + 2k_{12})/m}$.

Solution.

```
ODE[{x1' == y1,
 y1' == -(k + k12)x1/m + k12 x2/m,
 x2' == y2,
 y2' == k12 x1/m - (k + k12)x2/m,
 x1[0] == -a,x2[0] == a, y1[0] == -b,y2[0] == b},
 {x1,x2,y1,y2},t,Method->LinearSystem] //.
  {Sqrt[-k - 2k12]/Sqrt[m] -> I mu2,
   (Sqrt[-k - 2k12] t)/Sqrt[m] -> I mu2 t,
   Sqrt[m]/Sqrt[-k - 2k12] -> -I/mu2}
```

yields

```
                                              I      -I mu2 t
                  -I mu2 t       I mu2 t      - b E
         -(a E          )      a E            2
{{x1 ->  -----------------  -  --------  -  ----------   +
                2                 2            mu2

     I     I mu2 t
    - b E
     2
    ----------   x2 ->
      mu2

                                     I      -I mu2 t
         -I mu2 t       I mu2 t      - b E
       a E           a E            2
       --------  +   --------  +  ----------   -
          2             2           mu2
```

```
  I     I mu2 t
 - b E
  2
 ------------- , y1 ->
     mu2

        -I mu2 t      I mu2 t
 - (b E        )    b E            I     -I mu2 t
 -----------------  -------------  + - a E          mu2 -
        2               2           2

   I     I mu2 t
 - a E           mu2,
   2

           -I mu2 t       I mu2 t
       b E             b E
 y2 ->  ----------  +  ----------  -
           2              2

   I     -I mu2 t      I     I mu2 t
 - a E           mu2 + - a E          mu2}}
   2                   2
```

7. Suppose that the initial conditions are such that $x_1(0) = x_2(0)$, and $x_1'(0) = x_2'(0)$. Show that the resulting solution only contains terms with frequency $\mu_1 = \sqrt{k/m}$.

```
ODE[{x1' == y1,
 y1' == -(k + k12)x1/m + k12 x2/m,
 x2' == y2,
 y2' == k12 x1/m - (k + k12)x2/m,
 x1[0] == a,x2[0] == a, y1[0] == b,y2[0] == b},
 {x1,x2,y1,y2},t,Method->LinearSystem] //.
  {Sqrt[k]/Sqrt[m]  -> mu1,
    (Sqrt[k] t)/Sqrt[m] -> mu1 t,
    Sqrt[m]/Sqrt[k]  -> 1/mu1}
```

yields

```
                              b Sin[mu1 t]
 {{x1 -> a Cos[mu1 t] +  ---------------,
                              mu1

                              b Sin[mu1 t]
    x2 -> a Cos[mu1 t] +  ---------------,
                              mu1

    y1 -> b Cos[mu1 t] - a mu1 Sin[mu1 t],

    y2 -> b Cos[mu1 t] - a mu1 Sin[mu1 t]}}
```

8. Find the eigenvectors of the general matrix which defines the two *different* masses which interact with *different* spring constants and *different* coupling constants.

9. Suppose that two equal masses interact with the same spring constants and the same coupling constants in the presence of identical frictional terms $c\,dx_1/dt$ and $c\,dx_2/dt$. Find a fourth degree polynomial equation to determine the quasiperiod and the relaxation time of this system.

Solution. Substituting into the determinant, we obtain

$$r^2(r+c/m)^2 + 2r(r+c/m)(k+k_{12})/m + \left((k+k_{12})^2 - k_{12}^2\right)/m^2 = 0$$

10. Use **ODE** to solve the initial value problem of Exercise 9 in the case that $m = 1$, $k = 4$, $k_{12} = 6$ and $c = 0.5$, with the initial conditions $x_1(0) = 4$, $x_2(0) = 2$, $y_1(0) = 0$ and $y_2(0) = 0$. Assume that the equilibrium positions of the two masses are separated by 8 units.

Solution. We use

```
prob10=ODE[{x1' == y1,y1' == -10x1 + 6x2 - 0.5y1,
            x2' == y2,y2' ==  6x1 - 10x2 - 0.5y2,
 x1[0] == 4,y1[0] == 0,x2[0] == 2,y2[0] == 0},
 {x1[t],y1[t],x2[t],y2[t]},t,Method->LinearSystem]
```

to get

```
            3. Cos[1.98 t]      Cos[3.99 t]
{{x1[t] ->  --------------  +  -------------  +
               0.25 t             0.25 t
              E                   E

    0.378 Sin[1.98 t]     0.0626 Sin[3.99 t]
    -----------------  +  ------------------ ,
         0.25 t                0.25 t
        E                     E

 x2[t] ->

    3. Cos[1.98 t]     1. Cos[3.99 t]      0.378 Sin[1.98 t]
    --------------  -  --------------  +   ----------------- -
        0.25 t             0.25 t              0.25 t
       E                  E                   E

    0.0626 Sin[3.99 t]
    ------------------ ,
         0.25 t
        E

          -6.05 Sin[1.98 t]     4.01 Sin[3.99 t]
 y1[t] -> -----------------  -  ---------------- ,
               0.25 t               0.25 t
              E                     E

          -6.05 Sin[1.98 t]     4.01 Sin[3.99 t]
 y2[t] -> -----------------  +  ----------------}}
               0.25 t               0.25 t
              E                     E
```

We create the plot using

```
Plot[Evaluate[{x1[t] + 4,x2[t] - 4} /. prob10],{t,0,4Pi}]
```

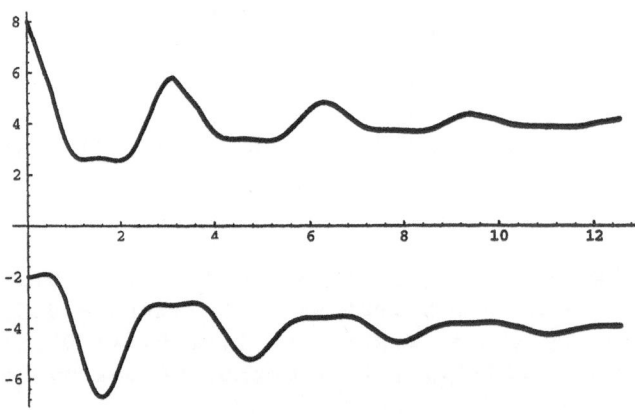

Solution to Exercise 10

Section 18.2, page 631

1. Obtain the system of differential equations for the circuit

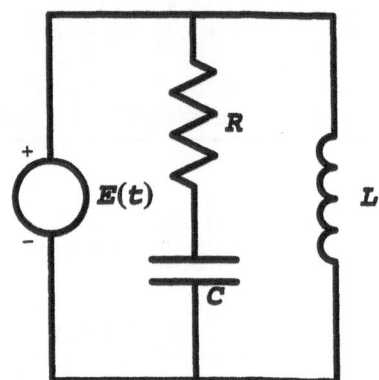

Solution. $\begin{cases} R(I_1' - I_2') + \dfrac{I_1 - I_2}{C} = E'(t), \\ \qquad\qquad L\,I_2' = E(t) \end{cases}$

2. Obtain the system of differential equations for the circuit

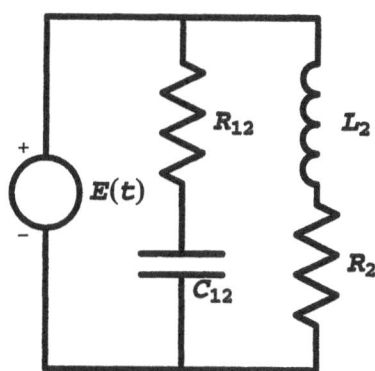

Solution.
$$
\begin{cases}
R(I_1' - I_2') + \dfrac{I_1 - I_2}{C} = E'(t), \\
L I_2' + R I_2 = E(t)
\end{cases}
$$

3. Consider the following circuit.

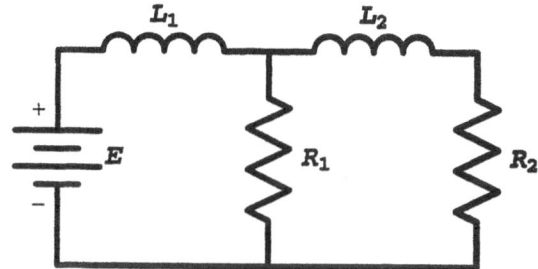

Assume **E** is a constant electromotive force of E volts, **R1** is a resistor of R_1 ohms, **R2** is a resistor of R_2 ohms, **L1** is an inductor of L_1 henrys, **L2** is an inductor of L_2 henrys and the currents are initially zero.

a. Write down the system of differential equations for the currents I_1 and I_2 in each loop.

b. Solve the system explicitly in the case that $E = 100$ volts, $R_1 = 20$ ohms, $R_2 = 40$ ohms, $L_1 = 0.01$ henry and $L_2 = 0.02$ henry.

Solution. **a.** Application of the Kirchhoff voltage law to the left-hand loop yields

$$
L_1 I_1' + R_1 I_{12} = E;
$$

similarly, for the right-hand loop we get

$$
L_2 I_2' + R_2 I_2 - R_1 I_{12} = 0.
$$

Applying the Kirchhoff current law to either junction produces

$$I_1 = I_{12} + I_2.$$

Thus, the system can be rewritten

$$\begin{cases} L_1 I_1' + R_1 I_1 - R_1 I_2 = E \\ L_2 I_2' - R_1 I_1 + (R_1 + R_2) I_2 = 0. \end{cases}$$

b. After substituting the given values into these equations, we obtain the system

$$\begin{cases} 0.01 I_1' + 20 I_1 - 20 I_2 = 100 \\ 0.02 I_2' - 20 I_1 + 60 I_2 = 0, \end{cases} \quad \text{or} \quad \begin{cases} I_1' = -2000 I_1 + 2000 I_2 + 10000 \\ I_2' = 1000 I_1 - 3000 I_2. \end{cases}$$

An **ODE** command which solves this system is

```
ODE[{I1' == -2000 I1 + 2000 I2 + 10000,
     I2' ==  1000 I1 - 3000 I2,
     I1[0] == 0, I2[0] == 0},{I1,I2},t,
     Method->ApproximateLinearSystem]
```

resulting in

```
{{I1 -> 7.5 -  0.833     6.67
                -------- - -------- ,
                4000. t    1000. t
                E          E

  I2 -> 2.5 +  0.833     3.33
                -------- - --------}}
                4000. t    1000. t
                E          E
```

Thus
$$\begin{cases} I_1 \approx 7.5 - 0.833e^{-4000t} - 6.67e^{-1000t}, \\ I_2 \approx 2.5 + 0.833e^{-4000t} - 3.33e^{-1000t}. \end{cases}$$

4. Consider the following circuit.

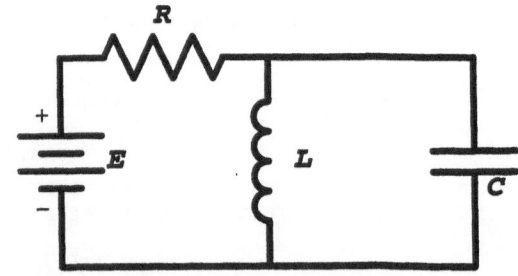

Assume **E** is a constant electromotive force of E volts, **R** is a resistor of R ohms, **L** is an inductor of L henrys, **C** is an inductor of C farads and the currents are initially zero.

    a. Write down the system of differential equations for the currents I_1 and I_2 in each loop.

    b. Solve the system explicitly in the case that $E = 100$ volts, $R = 20$ ohms, $L = 40$ henrys and $C = 0.02$ farad.

Solution. **a.** Application of the Kirchhoff voltage law to the left-hand loop yields

$$R\,I_1 + L\,I'_{12} = E;$$

similarly, for the right-hand loop we get $Q_2/C - L\,I'_{12} = 0$, which when differentiated becomes

$$\frac{Q_2}{C} - L\,I''_{12} = 0.$$

Applying the Kirchhoff current law to either junction produces

$$I_1 = I_{12} + I_2.$$

The system now becomes

$$\begin{cases} L(I'_1 - I'_2) + R\,I_1 = E, \\ -L(I''_1 - I''_2) + \dfrac{I_2}{C} = 0. \end{cases}$$

If we take the derivative of the first equation and then add it to the second equation, we obtain

$$R\,I'_1 + \frac{I_2}{C} = 0,$$

so that the system is finally

$$\begin{cases} L(I'_1 - I'_2) + R\,I_1 = E, \\ R\,I'_1 + \dfrac{I_2}{C} = 0 = 0. \end{cases}$$

After substituting the given values into these equations, we obtain the system

$$\begin{cases} 40(I'_1 - I'_2) + 20 I_1 = 100, \\ 20 I'_1 + 50 I_2 = 0. \end{cases}$$

An **ODE** command which solves this system is

```
ODE[{40(I1' - I2') + 20I1 == 100,20I1' + 50I2 == 0,
I1[0] == 0,I2[0] == 0},{I1,I2},t,Method->DSolve,
Form->Explicit,
PostSolution->{N,Simplify,(N[#,3] /. N[E]->E)&}]
```

yielding

$$\{5. + \frac{3.09}{E^{1.81\ t}} - \frac{8.09}{E^{0.691\ t}}, \frac{2.24}{E^{1.81\ t}} - \frac{2.24}{E^{0.691\ t}}\}$$

Thus
$$\begin{cases} I_1 \approx 5 + 3.09e^{-1.181t} - 8.09e^{0.691t}, \\ I_2 \approx 2.24e^{-1.181t} - 2.24e^{0.691t}. \end{cases}$$

5. Consider the following circuit.

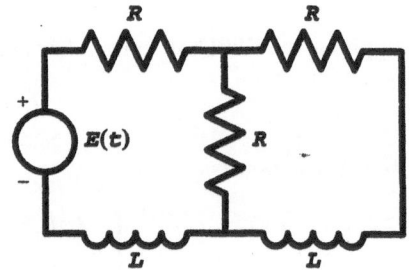

Assume E is an electromotive force with impressed voltage $E(t)$ $110\sin(3t)$ volts, R is a resistor of 10 ohms, L is an inductor of 10 henrys, and the currents are initially zero. Find and plot the currents in each loop.

Solution.

We use

```
ODE[{10 I1' + 10(2I1 - I2) == 0.5Sin[3t],
      10 I2' + 10(2I2 - I1) == 0,I1[0] == 0,I2[0] == 0},
     {I1,I2},t,Method->DSolve,PlotSolution->{{t,0,2Pi}}]
```

to get

$$\{\{I1 \rightarrow \frac{0.00417}{E^{3. t}} + \frac{0.0075}{E^{1. t}} - 0.0117\ Cos[3. t] +$$

$$0.00667\ Sin[3. t],$$

$$-0.00417 \quad 0.0075$$

```
I2 ->  --------  +  --------  -  0.00333 Cos[3. t] -
          3. t        1. t
         E           E

     0.00167 Sin[3. t]}}
```

and the plot

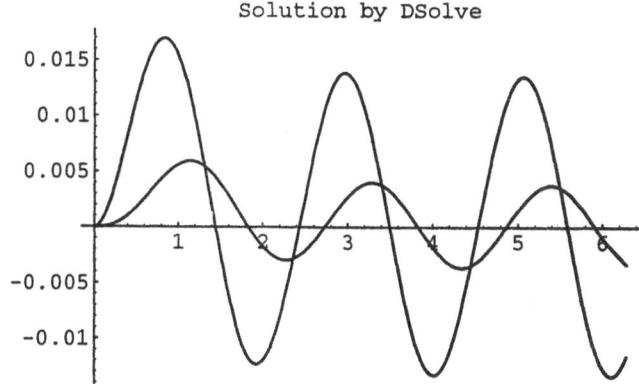

Solution to Exercise 5

$$\text{Thus} \begin{cases} I_1 = 0.00417e^{-3t} + 0.0075e^t - 0.0117\cos(3t) \\ I_2 = -0.00417e^{-3t} + 0.0075e^t - 0.00333\cos(3t). \end{cases}$$

6. After applying the Kirchhoff voltage law to the left-hand loop we get

$$R_1 I_1' + I_{12}/C_1 = E'(t),$$

and the right-hand loop yields

$$R_2 I_2' + I_2/C_2 - I_{12}/C_1 = 0.$$

Applying the Kirchhoff current law to either junction produces

$$I_1 = I_{12} + I_2.$$

The system can now be written

$$\begin{cases} I_1' = -\dfrac{1}{R_1 C_1} I_1 + \dfrac{1}{R_1 C_1} I_2 + \dfrac{E'(t)}{R_1} \\ I_2' = \dfrac{1}{R_2 C_1} I_1 - \dfrac{1}{R_2(C_1 + C_2)} I_2. \end{cases}$$

After substituting the given values into these equations, we obtain the system

$$\begin{cases} I_1' = -500I_1 + 500I_2 + \dfrac{1}{2}\cos(t) \\ I_2' = 1000I_1 - \dfrac{1000}{3}I_2 \end{cases}$$

An **ODE** command which solves this system is

```
ODE[{I1' == -500 I1 + 500 I2 + Cos[t]/2,
     I2' == 1000 I1 - 1000/3 I2,
     I1[0] == 0, I2[0] == 0},{I1,I2},t,
    Method->ApproximateLinearSystem]
```

7. After applying the Kirchhoff voltage law to the left-hand loop, we get

$$L_1 I_1'' + R_1 I_1' + I_{12}/C_1 = E'(t)$$

and the right-hand loop yields

$$L_2 I_2'' + R_2 I_2' + I_2/C_2 - I_{12}/C_1 = 0.$$

Applying the Kirchhoff current law to either junction produces

$$I_1 = I_{12} + I_2.$$

The system can now be written

$$\begin{cases} L_1 I_1'' + R_1 I_1' + I_1/C_1 - I_2/C_1 = E'(t) \\ L_2 I_2'' + R_2 I_2' + I_2/C_2 + I_2/C_1 - I_1/C_1 = 0. \end{cases}$$

After substituting the given values into these equations, we obtain the system

$$\begin{cases} 1.1 I_1'' + 1000 I_1' + 10^6(I_1 - I_2) = 41470\cos(377t) \\ 1.1 I_2'' + 1000 I_2' + 2 10^6 I_2 - 10^6 I_1 = 0. \end{cases}$$

An **ODE** command which solves the system and plots the desired voltage is

```
i2[t_] = Last[ODE[{1.1 I1'' + 1000 I1' + 10^6(I1 - I2)==
        41470 Cos[377. t],1.1 I2'' + 1000 I2' +
        2. 10.^6 I2 - 10^6 I1 == 0,I1[0] == 0,I1'[0]==0,
        I2[0] == 0,I2'[0] == 0},{I1,I2},{t,0,0.05},
        Method->NDSolve,MaxSteps->5000,Form->Explicit]];

Plot[NIntegrate[10^6 i2[s],{s,0,t}],{t,0,.05}];
```

8. After applying the Kirchhoff voltage law to the left-hand loop, we get

$$I_1/C_1 + L_1 I_{12}'' + R_1 I_{12}' = E'(t),$$

and the right-hand loop yields

$$I_2/C_2 + L_2 I_2'' + R_2 I_2' - L_1 I_{12}'' - R_1 I_{12}' = 0.$$

Applying the Kirchhoff current law to either junction produces

$$I_1 = I_{12} + I_2.$$

The system can now be written

$$L_1 I_1'' - L_1 I_2'' + R_1 I_1' - R_1 I_2' + I_1/C_1 = E'(t),$$
$$(L_1 + L_2)I_2'' - L_1 I_1'' + (R_1 + R_2)I_2' - R_1 I_1' + I_2/C_2 = 0.$$

After substituting the given values into these equations, we obtain the system

$$0.1 I_1'' - 0.1 I_2'' + 1000 I_1' - 1000 I_2' + 10^6 I_1 = 41470 \cos(377t)$$
$$0.2 I_2'' - 0.1 I_1'' + 2000 I_2' - 1000 I_1' + 10^6 I_2 = 0.$$

An **ODE** command which solves the system and plots the desired voltage is

```
i2[t_] = Last[ODE[{0.1 I1'' - 0.1 I2'' + 1000 I1' -
         1000 I2' + 10^6 I1 == 41470 Cos[377. t],
         0.2 I2'' - 0.1 I1'' + 2000 I2' - 1000 I1' +
         10^6 I2 == 0,I1[0] == 0,I1'[0] == 0,
         I2[0] == 0,I2'[0] == 0},{I1,I2},{t,0,0.05},
         Method->NDSolve,MaxSteps->5000,Form->Explicit]];
i2p[t_] = i2'[t];
Plot[0.1 i2p[t] + 1000 i2[t],{t,0,0.05}];
```

Section 18.3, page 641

1. The general two-state Markov chain is defined by two parameters $\lambda > 0$, $\mu > 0$ with $q_{11} = -\lambda$, $q_{12} = \lambda$, $q_{21} = \mu$, $q_{22} = -\mu$. Find $P(t)$ and $\lim_{t\to\infty} P(t)$.

Solution. The characteristic equation of Q is

$$0 = \det(r I - Q) = \det\begin{pmatrix} r+\lambda & -\lambda \\ -\mu & r+\mu \end{pmatrix} = (r+\lambda)(r+\mu) - \lambda\mu = r^2 + r(\lambda+\mu),$$

so that the eigenvalues are $r_1 = 0, r_2 = -(\lambda + \mu)$. The eigenvectors are given by $\xi_1 = (1, 1)^T$ and $\xi_2 = (\lambda, -\mu)^T$. The corresponding fundamental matrix is

$$\Phi(t) = \begin{pmatrix} 1 & \lambda e^{-(\lambda+\mu)t} \\ 1 & -\mu e^{-(\lambda+\mu)t} \end{pmatrix} \quad \text{with} \quad \Phi(0)^{-1} = \frac{1}{\lambda+\mu} \begin{pmatrix} \mu & \lambda \\ 1 & -1 \end{pmatrix}.$$

Thus

$$P(t) = \Phi(t)\Phi(0)^{-1} = \frac{1}{\lambda+\mu} \begin{pmatrix} \mu + \lambda e^{-(\lambda+\mu)t} & \lambda - \lambda e^{-(\lambda+\mu)t} \\ \mu - \mu e^{-(\lambda+\mu)t} & \lambda + \mu e^{-(\lambda+\mu)t} \end{pmatrix}$$

$$\lim_{t\to\infty} P(t) = \begin{pmatrix} \mu/(\lambda+\mu) & \lambda/(\lambda+\mu) \\ \mu/(\lambda+\mu) & \lambda/(\lambda+\mu) \end{pmatrix}$$

2. The three-state Ehrenfest Markov chain is defined by the matrix

$$Q = \begin{pmatrix} -2 & 2 & 0 \\ 1 & -2 & 1 \\ 0 & 2 & -2 \end{pmatrix}$$

Find the eigenvalues and eigenvectors of Q and use these to find $P(t) = e^{tQ}$ and $\lim_{t\to\infty} P(t)$.

Solution. The characteristic equation of Q is

$$0 = \det(r\,I - Q) = (r+2)^3 - 4(r+2) = (r+2)(r^2+4r) = r(r+2)(r+4).$$

The eigenvalues are $r_1 = 0, r_2 = -2, r_3 = -4$. The first eigenvector is $\xi_1 = (1, 1, 1)^T$. The second eigenvector is $\xi_2 = (1, 0, -1)^T$. The third eigenvector is $\xi_3 = (1, -1, 1)^T$, leading to the fundamental matrix

$$\Phi(t) = \begin{pmatrix} 1 & e^{-2t} & e^{-4t} \\ 1 & 0 & -e^{-4t} \\ 1 & -e^{-2t} & e^{-4t} \end{pmatrix} \quad \text{with} \quad \Phi(0)^{-1} = \begin{pmatrix} 1/4 & 1/2 & 1/4 \\ 1/2 & 0 & -1/2 \\ 1/4 & -1/2 & 1/4 \end{pmatrix}$$

The matrix $P(t)$ is obtained as the product $P(t) = \Phi(t)\Phi(0)^{-1}$ which leads to the values

$$p_{11}(t) = \frac{1}{4} + \frac{1}{2}e^{-2t} + \frac{1}{4}e^{-4t}$$

$$p_{12}(t) = \frac{1}{2} - \frac{1}{2}e^{-4t}$$

$$p_{13}(t) = \frac{1}{4} - \frac{1}{2}e^{-2t} + \frac{1}{4}e^{-4t}.$$

$$p_{21}(t) = \frac{1}{4} - \frac{1}{4}e^{-4t}$$

$$p_{22}(t) = \frac{1}{2} + \frac{1}{2}e^{-4t}$$

$$p_{23}(t) = \frac{1}{4} - \frac{1}{4}e^{-4t}$$

$$p_{31}(t) = \frac{1}{4} - \frac{1}{2}e^{-2t} + \frac{1}{4}e^{-4t}$$

$$p_{32}(t) = \frac{1}{2} - \frac{1}{2}e^{-4t}$$

$$p_{33}(t) = \frac{1}{4} + \frac{1}{2}e^{-2t} + \frac{1}{4}e^{-4t}$$

The limits are obtained as

$$\lim_{t \to \infty} P(t) = \begin{pmatrix} 1/4 & 1/2 & 1/4 \\ 1/4 & 1/2 & 1/4 \\ 1/4 & 1/2 & 1/4 \end{pmatrix}$$

3. Consider the three-state Markov chain with

$$Q = \begin{pmatrix} -\lambda_0 & \lambda_0 & 0 \\ \lambda_1 & -2\lambda_1 & \lambda_1 \\ 0 & \lambda_0 & -\lambda_0 \end{pmatrix}.$$

Find the eigenvalues and eigenvectors of Q and use these to find $P(t) = e^{tQ}$ and $\lim_{t \to \infty} P(t)$.

Solution. The characteristic equation of Q is

$$0 = \det(r\, I - Q) = (r+\lambda_0)^2(r+2\lambda_1) - \lambda_0(r+\lambda_1) - \lambda_1(r+\lambda_1) = r(r+\lambda_0)(r+\lambda_0+\lambda_1).$$

The eigenvalues are $r_1 = 0, r_2 = -\lambda_0, r_3 = -\lambda_0 - 2\lambda_1$. The first eigenvector is $\xi_1 = (1, 1, 1)^T$. The second eigenvector is $\xi_2 = (1, 0, -1)^T$. The third eigenvector is $\xi_3 = (\lambda_0, -2\lambda_1, \lambda_0)^T$, leading to the fundamental matrix

$$\Phi(t) = \begin{pmatrix} 1 & e^{-\lambda_0 t} & \lambda_0\, e^{-(\lambda_0+2\lambda_1)t} \\ 1 & 0 & -2\lambda_1\, e^{-(\lambda_0+2\lambda_1)t} \\ 1 & -e^{-\lambda_0 t} & \lambda_0\, e^{-(\lambda_0+2\lambda_1)t} \end{pmatrix} \quad \text{with}$$

$$\Phi(0)^{-1} = \frac{1}{2(\lambda_0 + 2\lambda_1)} \begin{pmatrix} 2\lambda_1 & 2\lambda_0 & 2\lambda_1 \\ \lambda_0 + 2\lambda_1 & 0 & -(\lambda_0 + 2\lambda_1) \\ 1 & -2 & 1 \end{pmatrix}.$$

The matrix $P(t)$ is obtained as the product $P(t) = \Phi(t)\Phi(0)^{-1}$ which leads to the values

$$p_{11}(t) = \frac{2\lambda_1 + (\lambda_0 + 2\lambda_1)e^{-\lambda_0 t} + \lambda_0 e^{-(\lambda_0 + 2\lambda_1)t}}{2(\lambda_0 + 2\lambda_1)}$$

$$p_{12}(t) = \frac{2\lambda_0 - 2\lambda_0 e^{-(\lambda_0 + 2\lambda_1)t}}{2(\lambda_0 + 2\lambda_1)}$$

$$p_{13}(t) = \frac{2\lambda_1 - (\lambda_0 + 2\lambda_1)e^{-\lambda_0 t} + \lambda_0 e^{-(\lambda_0 + 2\lambda_1)t}}{2(\lambda_0 + 2\lambda_1)}$$

$$p_{21}(t) = \frac{2\lambda_1 - 2\lambda_1 e^{-(\lambda_0 + 2\lambda_1)t}}{2(\lambda_0 + 2\lambda_1)}$$

$$p_{22}(t) = \frac{2\lambda_0 + 4\lambda_1 e^{-(\lambda_0 + 2\lambda_1)t}}{2(\lambda_0 + 2\lambda_1)}$$

$$p_{23}(t) = \frac{2\lambda_1 - 2\lambda_1 e^{-(\lambda_0 + 2\lambda_1)t}}{2(\lambda_0 + 2\lambda_1)}$$

$$p_{31}(t) = \frac{2\lambda_1 - (\lambda_0 + 2\lambda_1)e^{-\lambda_0 t} + \lambda_0 e^{-(\lambda_0 + 2\lambda_1)t}}{2(\lambda_0 + 2\lambda_1)}$$

$$p_{32}(t) = \frac{2\lambda_0 - 2\lambda_0 e^{-(\lambda_0 + 2\lambda_1)t}}{2(\lambda_0 + 2\lambda_1)}$$

$$p_{33}(t) = \frac{2\lambda_1 + (\lambda_0 + 2\lambda_1)e^{-\lambda_0 t} + \lambda_0 e^{-(\lambda_0 + 2\lambda_1)t}}{2(\lambda_0 + 2\lambda_1)}$$

The limits are obtained as

$$\lim_{t \to \infty} P(t) = \frac{1}{\lambda_0 + 2\lambda_1} \begin{pmatrix} \lambda_1 & \lambda_0 & \lambda_1 \\ \lambda_1 & \lambda_0 & \lambda_1 \\ \lambda_1 & \lambda_0 & \lambda_1 \end{pmatrix}$$

4. Consider the three-state Markov chain with

$$Q = \begin{pmatrix} -\lambda_0 & \lambda_0 & 0 \\ \lambda_1 & -(\lambda_1 + \mu_1) & \mu_1 \\ 0 & \lambda_0 & -\lambda_0 \end{pmatrix}.$$

Find the eigenvalues and eigenvectors of Q and use these to find $P(t) = e^{tQ}$ and $\lim_{t \to \infty} P(t)$.

Solution. The characteristic equation of Q is

$$0 = \det(r\,I - Q) = (r + \lambda_0)^2(r + \lambda_1 + \mu_1) - \lambda_0\mu_1(r + \lambda_0) - \lambda_0\lambda_1(r_+\lambda_0)$$
$$= r(r + \lambda_0)(r + \lambda_0 + \lambda_1 + \mu_1).$$

The eigenvalues are $r_1 = 0$, $r_2 = -\lambda_0$, $r_3 = -\lambda_0 - \lambda_1 - \mu_1$.

The first eigenvector is $\xi_1 = (1, 1, 1)^T$. The second eigenvector is $\xi_2 = (\mu_1, 0, -\lambda_1)^T$. The third eigenvector is $\xi_3 = (-\lambda_0, \mu_1 + \lambda_1, -\lambda_0)^T$, leading to the fundamental matrix

$$\Phi(t) = \begin{pmatrix} 1 & \mu_1 e^{-\lambda_0 t} & -\lambda_0\, e^{-(\lambda_0+\mu_1+\lambda_1)t} \\ 1 & 0 & (\lambda_1 + \mu_1)\, e^{-(\lambda_0+\lambda_1+\mu_1)t} \\ 1 & -\lambda_1 e^{-\lambda_0 t} & -\lambda_0\, e^{-(\lambda_0+\lambda_1+\mu_1)t} \end{pmatrix} \quad \text{with}$$

$$\Phi(0)^{-1} = \zeta \begin{pmatrix} \lambda_1(\lambda_1 + \mu_1) & \lambda_0(\lambda_1 + \mu_1) & \mu_1(\mu_1 + \lambda_1) \\ \lambda_0 + \lambda_1 + \mu_1 & 0 & -(\lambda_0 + \lambda_1 + \mu_1) \\ -\lambda_1 & \mu_1 + \lambda_1 & -\mu_1 \end{pmatrix}$$

where

$$\zeta = \frac{1}{(\mu_1 + \lambda_1)(\mu_1 + \lambda_0 + \lambda_1)}.$$

The matrix $P(t)$ is obtained as the product $P(t) = \Phi(t)\Phi(0)^{-1}$ which leads to the values

$$p_{11}(t) = \frac{\lambda_1(\lambda_1 + \mu_1) + \mu_1(\lambda_0 + \lambda_1 + \mu_1)e^{-\lambda_0 t} + \lambda_0\lambda_1\, e^{-(\lambda_0+\lambda_1+\mu_1)t}}{(\mu_1 + \lambda_1)(\mu_1 + \lambda_1 + \lambda_0)}$$

$$p_{12}(t) = \frac{\lambda_0(\lambda_1 + \mu_1) - \lambda_0(\mu_1 + \lambda_1)\, e^{-(\lambda_0+\lambda_1+\mu_1)t}}{(\mu_1 + \lambda_1)(\mu_1 + \lambda_1 + \lambda_0)}$$

$$p_{13}(t) = \frac{\mu_1(\mu_1 + \lambda_1) - \mu_1(\lambda_0 + \lambda_1 + \mu_1)e^{-\lambda_0 t} + \lambda_0\mu_1\, e^{-(\lambda_0+\lambda_1+\mu_1)t}}{(\mu_1 + \lambda_1)(\mu_1 + \lambda_1 + \lambda_0)}$$

$$p_{21}(t) = \frac{\lambda_1(\mu_1 + \lambda_1) - \lambda_1(\mu_1 + \lambda_1)\, e^{-(\lambda_0+\lambda_1+\mu_1)t}}{(\mu_1 + \lambda_1)(\mu_1 + \lambda_1 + \lambda_0)}$$

$$p_{22}(t) = \frac{\lambda_0(\lambda_1 + \mu_1) + (\lambda_1 + \mu_1)^2\, e^{-(\lambda_0+\lambda_1+\mu_1)t}}{(\mu_1 + \lambda_1)(\mu_1 + \lambda_1 + \lambda_0)}$$

$$p_{23}(t) = \frac{\mu_1(\mu_1 + \lambda_1) - \mu_1(\mu_1 + \lambda_1)\, e^{-(\lambda_0+\lambda_1+\mu_1)t}}{(\mu_1 + \lambda_1)(\mu_1 + \lambda_1 + \lambda_0)}$$

$$p_{31}(t) = \frac{\lambda_1(\lambda_1 + \mu_1) - \lambda_1(\lambda_0 + \lambda_1 + \mu_1)e^{-\lambda_0 t} + \lambda_0\lambda_1\, e^{-(\lambda_0+\lambda_1+\mu_1)t}}{(\mu_1 + \lambda_1)(\mu_1 + \lambda_1 + \lambda_0)}$$

$$p_{32}(t) = \frac{\lambda_0(\lambda_1 + \mu_1) - \lambda_0(\lambda_1 + \mu_1)\, e^{-(\lambda_0+\lambda_1+\mu_1)t}}{(\mu_1 + \lambda_1)(\mu_1 + \lambda_1 + \lambda_0)}$$

$$p_{33}(t) = \frac{\mu_1(\mu_1 + \lambda_1) + \lambda_1(\lambda_0 + \lambda_1 + \mu_1)e^{-\lambda_0 t} + \lambda_0\mu_1 e^{-(\lambda_0 + \lambda_1 + \mu_1)t}}{(\mu_1 + \lambda_1)(\mu_1 + \lambda_1 + \lambda_0)}$$

The limits are obtained as

$$\lim_{t \to \infty} P(t) = \frac{1}{(\mu_1 + \lambda_0 + \lambda_1)} \begin{pmatrix} \lambda_1 & \lambda_0 & \mu_1 \\ \lambda_1 & \lambda_0 & \mu_1 \\ \lambda_1 & \lambda_0 & \mu_1 \end{pmatrix}$$

5. Consider the three-state Markov chain of example 18.10 in case $p = 2/3$, $q = 1/3$. Find the eigenvalues and eigenvectors and use these to find $P(t)$ in terms of real-valued functions.

Solution. We have

$$Q = \begin{pmatrix} -1 & 2/3 & 1/3 \\ 1/3 & -1 & 2/3 \\ 2/3 & 1/3 & -1 \end{pmatrix}$$

From the discussion in the text, the eigenvalues are $r_1 = 0$, $r_2 = -3/2 + i\sqrt{3}/6$, $r_3 = -3/2 - i\sqrt{3}/6$. The corresponding eigenvectors are $\xi_1 = (1, 1, 1)^T$, $\xi_2 = (1, z, z^2)^T$, $\xi_3 = (1, z^2, z)^T$ where $z = e^{2\pi i/3} = -1/2 + i\sqrt{3}/2$. The fundamental matrix is

$$\Phi(t) = \begin{pmatrix} 1 & e^{r_2 t} & e^{r_3 t} \\ 1 & z e^{r_2 t} & z^2 e^{r_3} \\ 1 & z^2 e^{r_2 t} & z e^{r_3 t} \end{pmatrix} \qquad \text{with}$$

$$\Phi(0)^{-1} = \frac{i}{3(z^2 - z)} \begin{pmatrix} z^2 - z & z^2 - z & z^2 - z \\ z^2 - z & z - 1 & 1 - z^2 \\ 1 & 1 - z^2 & z - 1 \end{pmatrix}$$

The matrix $P(t)$ is obtained as the product $P(t) = \Phi(t)\Phi(0)^{-1}$ which leads to the values

$$p_{11}(t) = \frac{1 + e^{r_2 t} + e^{r_3 t}}{3}$$

$$= \frac{1}{3} + \frac{2}{3}e^{-3t/2}\cos(t\sqrt{3}/6)$$

$$p_{12}(t) = \frac{(z^2 - z) + (z - 1)e^{r_1 t} + (1 - z^2)e^{r_3 t}}{3(z^2 - z)}$$

$$= \frac{1}{3} - \frac{1}{3}e^{-3t/2}\cos(t\sqrt{3}/6) + \frac{e^{-3t/2}}{\sqrt{3}}\sin(t\sqrt{3}/6)$$

$$p_{13}(t) = \frac{(z^2 - z) + (1 - z^2)e^{r_1 t} + (z - 1)e^{r_3 t}}{3(z^2 - z)}$$

$$= \frac{1}{3} - \frac{1}{3}e^{-3t/2}\cos(t\sqrt{3}/6) - \frac{e^{-3t/2}}{\sqrt{3}}\sin(t\sqrt{3}/6)$$

The others are obtained from these as follows:

$$p_{22}(t) = p_{33}(t) = p_{11}(t),$$
$$p_{23}(t) = p_{31}(t) = p_{12}(t),$$
$$p_{32}(t) = p_{13}(t) = p_{21}(t).$$

6. Find the eigenvalues and eigenvectors of the random walk model on $n = 4$ points.

Solution. In case $n = 4$ then $z = e^{2\pi i/4} = i$ and the eigenvalues are $r_k = pi^k + qi^{-k} = \{0, (p-q)i - 1, -2, (q-p)i - 1\}$ The eigenvectors are $\xi_1 = (1, 1, 1, 1)^T$, $\xi_2 = (i, -1, -i, 1)$, $\xi_3 = (-1, 1, -1, 1)$, $\xi_4 = (-i, -1, i, 1)$

7. Consider a random walk on n points with $p = q$. Show that the eigenvalues are obtained through the formula

$$r_k = \lambda\big(\cos(2\pi k/n) - 1\big)$$

for $k = 0, 1, \ldots, n - 1$.

Solution. In case $p = q$ formula of example 18.9 simplifies to

$$r_k = \lambda(z^k + z^{-k} - 1 = \lambda\left(\frac{1}{2}e^{2\pi ik/n} + \frac{1}{2}e^{-2\pi ik/n} - 1\right) = \lambda(\cos(2\pi k/n) - 1)$$

8. The purpose of this exercise is to show that the transition matrix $P(t)$ must be a continuous function.

   a. Write the Markov semigroup property in the form (18.23).

   b. Use (18.21) to conclude continuity from the right: $\lim P(t) = P(t_0)$ when t approaches t_0 from the right.

   c. Use (18.21) to show that there exists $\delta > 0$ so that $P(s)$ has an inverse $P(s)^{-1}$ for $0 < s < \delta$, which satisfies $\lim_{s \to 0} P(s)^{-1} = I$.

   d. Establish the equation $P(t_0 - s) = P(t_0)P(s)^{-1}$ for $0 < s < \delta$.

   e. Conclude that $\lim P(t) = P(t_0)$ when t approaches t_0 from the left.

Solution. Part a is immediate from (18.23). Taking $t = t_0$ and letting $s \downarrow 0$ yields part b. From (18.21) we may choose $\delta > 0$ so that $p_{ij}(s) < 1/2$ for $i \neq j$ and $1 - p_{ii}(s) < 1/2$ when $0 < s < \delta$. The inverse of the matrix $P(s)$ can be computed by the geometric series

$$P(s)^{-1} = (I - (I - P(s)))^{-1} = I + \sum_{k=1}^{\infty} (I - P(s))^k.$$

Each of the matrix powers satisfies $(I - P(s))_{ij}^k < 1/2$ for $k = 1, 2, \ldots$ when $0 < s < \delta$. If we replace $1/2$ by an arbitrarily small number $\epsilon > 0$, it is seen that $\lim_{s \downarrow 0} P(s)^{-1} = I.$ which proves part c. To prove part d, we write (18.23) in the form $P(t_0) = P(t_0 - s)P(s)$ and multiply by $P(s)^{-1}$ on the right. Finally, we use part c to let $s \downarrow 0$ to obtain e.

9. The purpose of this exercise is to show that the transition matrix $P(t)$ is a differentiable function.

    a. Integrate the Markov semigroup property (18.23) to obtain the equation

$$\int_t^{t+\delta} P(u)\, du = R\, P(t), \qquad R = \int_0^{\delta} P(u)\, du$$

    b. Argue from the result of Exercise 8 that the matrix R has an inverse and that

$$P(t) = R^{-1} \int_t^{t+\delta} P(u)\, du.$$

    c. Use the fundamental theorem of calculus to deduce the differentiability of the matrix-valued function $t \longrightarrow P(t)$.

Solution. In Exercise 8 it was shown that the Markov semigroup is a continuous function of t. Therefore, it is also integrable and we may integrate (18.23) as follows:

$$\int_0^{\delta} P(t+u)\, du = \int_0^{\delta} P(t)\, P(u)\, du = P(t) \int_0^{\delta} P(u)\, du.$$

Since $|P(u) - I| < 1/2$ for $u < \delta$, it is also true that

$$\int_0^{\delta} (I - P(u))\, du < \delta/2$$

and we argue as in Exercise 8 that the matrix

$$\frac{1}{\delta} \int_0^{\delta} P(u)\, du$$

also has an inverse. Hence R has an inverse and we may write

$$P(t) = R^{-1} \int_t^{t+\delta} P(u)\, du.$$

The right-hand side is the definite integral of a continuous function between the indicated limits, hence is also a differentiable function of t with $P'(t) = R^{-1}P(t + \delta) - P(t)$. ∎

10. This problem discusses the background of the Markov semigroup property, for readers who are familiar with the notions of random variable and conditional probability, written $\mathbf{Prob}(A|B)$. The random variable $X(t)$ represents the state of the systems a time t. The **temporally homogeneous Markov property** is the statement that

$$\mathbf{Prob}\big[X(t_k + s) = j | X(t_0) = i_0, \ldots, X(t_k) = i_k\big] = \mathbf{Prob}\big[X(s) = j | X(0) = i_k\big],$$

where $t_0 < \cdots < t_k$ is any finite set of time points and $i_0, \ldots, i_k, j$ are any members of the set $[1, \ldots, n]$. Suppose that the system of probabilities satisfies the temporally homogeneous Markov property, and let

$$p_{ij}(t) = \mathbf{Prob}\big[X(t) = j | X(0) = i\big].$$

Show that the matrix $P(t) = \big(p_{ij}(t)\big)$ satisfies the Markov semigroup property (18.22) [Hint: Sum over the "intermediate states" which may be assumed by the random variable $X(t)$.]

Solution.

$$
\begin{aligned}
p_{ij}(t + s) &= \mathbf{Prob}\big[X(t + s) = j \,|\, X(0) = i\big] \\
&= \sum_{k=1}^{N} \mathbf{Prob}\big[X(t + s) = j,\, X(t) = k \,|\, X(0) = i\big] \\
&= \sum_{k=1}^{N} \mathbf{Prob}\big[X(t + s) = j \,|\, X(t) = k,\, X(0) = i\big]\mathbf{Prob}\big[X(t) = k | X(0) = i\big] \\
&= \sum_{k=1}^{N} \mathbf{Prob}\big[X(s) = j | X(t) = k\big]\mathbf{Prob}\big[X(t) = k | X(0) = i\big] \\
&= \sum_{k=1}^{N} p_{kj}(s) p_{ik}(t)
\end{aligned}
$$

12. The purpose of this exercise is to show that the nonzero eigenvalues of the Q-matrix lie in the left half-plane.

a. Suppose that $\xi = (c_i)$ is an eigenvector of Q with eigenvalue r. Show that

$$\sum_{j \neq i} q_{ij} c_j = (r + q_{ii}) c_i.$$

b. Deduce from part **a** that

$$|(r + q_{ii}) c_i| \leq |q_{ii}| \max_{1 \leq j \leq n} |c_j|.$$

for $1 \leq i \leq n$.

c. Choose i so that $|c_i| = \max_{1 \leq j \leq n} |c_j|$ to obtain $|r + q_{ii}| \leq |q_{ii}|$ for that value of i.

d. Conclude that either $r = 0$ or the complex number r satisfies $\mathfrak{Re}(r) < 0$.

Solution. The inequality $|r + q_{ii}| \leq |q_{ii}|$ describes a circle in the complex plane which lies wholly in the left half-plane, with the exception of the point $(0, 0)$

13. In this exercise we show that if $q_{ij} > 0$ for all $i \neq j$, then the eigenvalue $r = 0$ has only the eigenvector $\xi = (1, \ldots, 1)^T$. To prove this, let $\xi = (c_1, \ldots, c_n)^T$ be an eigenvector, and write $c_j = \alpha_j + i\beta_j$.

a. Write the eigenvector condition $Q\xi = 0$ in the form

$$\sum_{k \neq j} q_{jk}(c_k - c_j) = 0$$

for $1 \leq j \leq n$.

b. Take the real and imaginary parts of this equation to conclude that

$$\sum_{k \neq j} q_{jk}(\alpha_k - \alpha_j) = 0 \quad \text{and} \quad \sum_{k \neq j} q_{jk}(\beta_k - \beta_j) = 0$$

for $1 \leq j \leq n$.

c. Conclude that α_k and β_k are constants independent of k.

d. Conclude that the eigenvector ξ is a constant multiple of $(1, \ldots, 1)^T$.

Solution. If $\sum_{k \neq j} q_{jk}(\alpha_k - \alpha_j) = 0$, then we pick j so that α_j is equal to its minimum. Then all of the numbers $\alpha_k - \alpha_j$ are nonnegative, as are the coefficients q_{jk}. Since the sum is zero, all of the non-zero terms in the sum must be zero, hence $\alpha_k - \alpha_j$ must be identically zero. A similar argument applied to β_j Shows that $\beta_k - \beta_j$ must be identically zero.

14. In this exercise we show that the eigenvalue $r = 0$ of the Q-matrix has no generalized eigenvectors.

a. Suppose that η is a vector which solves the equation $Q\eta = c(1, \ldots, 1)^T$ for some complex constant c. Show that $P(t)\eta = \eta + t\,c(1, \ldots, 1)^T$. [Hint: Use the power series definition of the matrix exponential e^{tQ}.]

b. Conclude that $c = 0$. [Hint: Take $t \longrightarrow \infty$ in part a and use the inequalities $|p_{ij}(t)| \le 1$.]

c. Generalize to the case $Q^k\eta = c(1, \ldots, 1)^T$ for $k > 1$.

Solution. The matrix exponential is $P(t) = I + tQ + t^2 Q^2/2 + \cdots$. Applied to η, we have $P(t)\eta = \eta + tc(1, \ldots, 1)^T$. But $0 \le p_{ij}(t) \le 1$ for all t, hence we must have $c = 0$.

15. Use the results of Exercises 12, 13 and 14 to show that if $q_{ij} > 0$ for $i \ne j$ then the transition matrix $P(t)$ satisfies $\lim_{t\to\infty} p_{ij}(t) = \pi_j$ for $1 \le i, j \le n$, where π_j is the solution of the equations $\sum_{i=1}^{n} \pi_i q_{ij} = 0$ for $1 \le j \le n$.

Solution. From the Jordan canonical form, we have $P(t) = U\,e^{t\Lambda}\,U^{-1}$ where Λ is the upper-triangular matrix of Jordan blocks. We have already shown that the eigenvalues satisfy $r_1 = 0$ and $\Re(r) < 0$ for $i = 2, \ldots, n$. Therefore

$$\lim_{t\to\infty} e^{t\Lambda} = \begin{pmatrix} 1 & 0 & 0 & \cdots & 0 & 0 \\ 0 & 0 & 0 & \cdots & 0 & 0 \\ \vdots & \vdots & \vdots & \ddots & 0 & 0 \\ 0 & 0 & 0 & \cdots & 0 & 0 \end{pmatrix} =: R$$

$$\lim_{t\to\infty} P(t) = U R U^{-1} := \Pi.$$

The matrix $P(t)$ is nonnegative and has row sums equal to 1, so the same holds true for the limiting matrix Π. To prove the stated properties of the limit, we write the Markov semigroup property in the form

$$P(s)\,P(t) = P(t + s) = P(t)\,P(s)$$

and let $t \longrightarrow \infty$ to obtain

$$P(s)\,\Pi = \Pi = \Pi\,P(s).$$

We now differentiate with respect to s at $s = 0$ to obtain

$$Q\,\Pi = 0 = \Pi\,Q.$$

Therefore, the columns of Π are eigenvectors of Q with eigenvalue zero; hence from Exercise 13, each column must be of the form $\pi_i(1, \ldots, 1)^T$ where $\pi_i \ge 0$. Now

$\Pi Q = 0$ implies that the vector $(\pi_1, \dots, \pi_n)$ must be an eigenvector of Q^T with eigenvalue zero. In detail, for each j

$$0 = \sum_{i=1}^{n} \pi_i q_{ij} = -q_{jj}\pi_j + \sum_{i \neq j} \pi_i q_{ij}.$$

If $\pi_j = 0$ for some j then $0 = \sum_{i \neq j} \pi_i q_{ij}$ which implies that $\pi_j = 0$ for all i, since $q_{ij} > 0$. We conclude that $\pi_j > 0$ and

$$\lim_{t \to \infty} p_{ij}(t) = \pi_j,$$

as required.

16. A Markov chain is said to be **irreducible** if for each pair i, j there exists a set of indices $i_1, \dots, i_N$ so that $q_{ii_1} > 0, q_{i_1 i_2} > 0, \dots, q_{i_N j} > 0$. The problem is to show that an irreducible Markov chain also satisfies Exercises 12,13, 14 and hence also Theorem 18.4.

Solution. Exercises 12 and 14 remain true for any Q. When we re-do Exercise 13 we find that $\alpha_j - \alpha_k = 0$ for all k such that $q_{jk} > 0$. By hypothesis, any two states (i, j) can be joined by a sequence of states $i_1, \dots, i_N$ so that $q_{i_m i_{m+1}} > 0$ along this path. In this way we conclude that $\alpha_k - \alpha_j = 0$ for all j, k, similarly for $\beta_k - \beta_j$. Then we can repeat the reasoning used to solve Exercise 15

In each of the following exercises, find the transition matrix $P(t)$ and $\lim_{t\to\infty} P(t)$ for the given infinitesimal matrix Q. Use the *Mathematica* command `MatrixExp`.

17. $Q = \begin{pmatrix} -1 & 0 & 1 \\ 1 & -2 & 1 \\ 2 & 0 & -2 \end{pmatrix}$

Solution. We must find the exponential of the matrix $t\,Q$, which suggests the *Mathematica* command

`MatrixExp[{{-t,0,t},{t,-2t,t},{2t,0,-2t}}]`

which produces the output

```
   2     1              1     1
{{- - + ----- ,  0,     - - - ----- },
   3     3 t            3      3 t
        3 E                   3 E


   2     1      -2 t    -2 t  1     1
 {- - + ----- - E  ,   E  ,   - - - ----- },
   3     3 t                   3     3 t
        3 E                         3 E
```

$$\{\frac{2}{3} - \frac{2}{3\ E^{3\ t}},\ 0,\ \frac{1}{3} + \frac{2}{3\ E^{3\ t}}\}\}$$

so that the solution is

$$P(t) = \begin{pmatrix} 2/3 + (1/3)e^{-3t} & 0 & 1/3 - (1/3)e^{-3t} \\ 2/3 + (1/3)e^{-3t} - e^{-2t} & e^{-2t} & (1/3) - (1/3)e^{-3t} \\ 2/3 - (2/3)e^{-3t} & 0 & 1/3 + (2/3)e^{-3t} \end{pmatrix}$$

and

$$\lim_{t \to \infty} P(t) = \begin{pmatrix} 2/3 & 0 & 1/3 \\ 2/3 & 0 & 1/3 \\ 2/3 & 0 & 1/3 \end{pmatrix}$$

18. $Q = \begin{pmatrix} -1 & 1 & 0 \\ 1 & -3 & 2 \\ 1 & 1 & -2 \end{pmatrix}$

Solution. We must find the exponential of the matrix $t\ Q$, which suggests the *Mathematica* command

```
MatrixExp[{{-t,t,0},{t,-3t,2t},{t,t,-2t}}]
```

which produces the output

$$\{\{-\frac{1}{2} + \frac{1}{2\ E^{2\ t}},\ \frac{1}{4} - \frac{1}{4\ E^{4\ t}},\ -\frac{1}{4} + \frac{1}{4\ E^{4\ t}} - \frac{1}{2\ E^{2\ t}}\},$$

$$\{-\frac{1}{2} - \frac{1}{2\ E^{2\ t}},\ \frac{1}{4} + \frac{3}{4\ E^{4\ t}},\ -\frac{3}{4} + \frac{3}{4\ E^{4\ t}} + \frac{1}{2\ E^{2\ t}}\},$$

$$\{-\frac{1}{2} - \frac{1}{2\ E^{2\ t}},\ \frac{1}{4} - \frac{1}{4\ E^{4\ t}},\ -\frac{1}{4} + \frac{1}{4\ E^{4\ t}} + \frac{1}{2\ E^{2\ t}}\}\}$$

so that the solution is

$$P(t) = \begin{pmatrix} 1/2 + (1/2)e^{-2t} & 1/4 - (1/4)e^{-4t} & 1/4 + (1/4)e^{-4t} - (1/2)e^{-2t} \\ 1/2 - (1/2)e^{-2t} & 1/4 + (3/4)e^{-4t} & (1/4) - (3/4)e^{-4t} + (1/2)e^{-2t} \\ 1/2 - (1/2)e^{-2t} & 1/4 - (1/4)e^{-4t} & 1/4 + (1/4)e^{-4t} + (1/2)e^{-2t} \end{pmatrix}$$

and

$$\lim_{t \to \infty} P(t) = \begin{pmatrix} 1/2 & 1/4 & 1/4 \\ 1/2 & 1/4 & 1/4 \\ 1/2 & 1/4 & 1/4 \end{pmatrix}$$

19. $Q = \begin{pmatrix} -1 & 1/2 & 1/2 \\ 1 & -2 & 1 \\ 1 & 0 & -1 \end{pmatrix}$

Solution. We must find the exponential of the matrix $t\,Q$, which suggests the *Mathematica* command

MatrixExp[{{-t,t/2,t/2},{t,-2t,t},{t,0,-t}}]

and produces the *Mathematica* output

```
     1    2       1    1     1    1
{{- + ------ , - - ------ , - - ------},
  3    3 t    3    3 t    3    3 t
      3 E          3 E          3 E

  1    1     1    2     1    1
{- - ------ , - + ------ , - - ------},
  3   3 t    3    3 t    3    3 t
      3 E         3 E          3 E

  1    1     1    1     1    2
{- - ------ , - - ------ , - + ------}}
  3   3 t    3    3 t    3    3 t
      3 E         3 E          3 E
```

so that the solution is

$$P(t) = \begin{pmatrix} \dfrac{1}{3} + \dfrac{2e^{-3t}}{3} & \dfrac{1}{3} - \dfrac{e^{-3t}}{3} & \dfrac{1}{3} - \dfrac{e^{-3t}}{3} \\[2ex] \dfrac{1}{3} - \dfrac{e^{-3t}}{3} & \dfrac{1}{3} + \dfrac{2e^{-3t}}{3} & \dfrac{1}{3} - \dfrac{e^{-3t}}{3} \\[2ex] \dfrac{1}{3} - \dfrac{e^{-3t}}{3} & \dfrac{1}{3} - \dfrac{e^{-3t}}{3} & \dfrac{1}{3} + \dfrac{2e^{-3t}}{3} \end{pmatrix}$$

and

$$\lim_{t \to \infty} P(t) = \begin{pmatrix} 1/3 & 1/3 & 1/3 \\ 1/3 & 1/3 & 1/3 \\ 1/3 & 1/3 & 1/3 \end{pmatrix}$$

20. $Q = \begin{pmatrix} -2 & 1 & 1 \\ 2 & -4 & 2 \\ 1 & 1 & -2 \end{pmatrix}$

Solution. We must find the exponential of the matrix $t\,Q$, which suggests the *Mathematica* command

MatrixExp[{{-2t,t,t},{2t,-4t,2t},{t,t,-2t}}]

which produces the *Mathematica* output

```
  2     1        1      1    1
{{- +  ----- +  ----- , - - ----- ,
  5    5 t      3 t     5   5 t
      10 E      2 E         5 E

  2     1        1
  - +  ----- -  ----- },
  5    5 t      3 t
      10 E      2 E

  2     2      1    4     2     2
{- -  ----- , - +  ----- , - - ----- },
 5    5 t     5   5 t     5   5 t
      5 E         5 E         5 E

  2     1        1     1    1
{- +  ----- -  ----- , - - ----- ,
 5    5 t      3 t    5   5 t
      10 E     2 E        5 E

  2     1        1
  - +  ----- +  ----- }}
  5    5 t      3 t
      10 E      2 E
```

so that the solution is

$$P(t) = \begin{pmatrix} \dfrac{2}{5} + \dfrac{e^{-5t}}{10} + \dfrac{e^{-3t}}{2} & \dfrac{1}{5} - \dfrac{e^{-5t}}{5} & \dfrac{2}{5} + \dfrac{e^{-5t}}{10} - \dfrac{e^{-3t}}{2} \\[2ex] \dfrac{2}{5} - \dfrac{2e^{-5t}}{5} & \dfrac{1}{5} + \dfrac{4e^{-5t}}{5} & \dfrac{2}{5} - \dfrac{2e^{-5t}}{5} \\[2ex] \dfrac{2}{5} + \dfrac{e^{-5t}}{10} - \dfrac{e^{-3t}}{2} & \dfrac{1}{5} - \dfrac{e^{-5t}}{5} & \dfrac{2}{5} + \dfrac{e^{-5t}}{10} + \dfrac{e^{-3t}}{2} \end{pmatrix}$$

and

$$\lim_{t \to \infty} P(t) = \begin{pmatrix} 2/5 & 1/5 & 2/5 \\ 2/5 & 1/5 & 2/5 \\ 2/5 & 1/5 & 2/5 \end{pmatrix}$$

Chapter 19

Section 19.1, page 653

The following exercises ask for solution the problems using a variety of numerical integration techniques (Euler, Runge-Kutta, Adams-Bashforth, Bulirsch-Stoer and implicit Runge-Kutta) first using a fixed step size and then allowing the step size to vary according to some convergence criteria. Afterwards are the the graphs of the solutions should be compared.

The following command is to be used with each of the answers to Problems 1-12 in Section . It uses **Module** to combine several individual commands into a single command called **plt [vary]**, whose single argument, which must be **False** or **True**, deactivates or activates the **VariableStepSize** option, respectively. After defining **sys** and **interval**, first evaluate **plt [False]** and then **plt [True]**.

```
plt[vary_]:= Module[{},
ODE[sys,{x,y},interval,Method->NDSolve,
    PlotPhase->{interval,PlotRange->All}];
ODE[sys,{x,y},interval,VariableStepSize->vary,
    Tolerance->0.001,Method->Euler,StepSize->0.1,
    ODEMaxStepSize->0.1,
    PlotPhase->{interval,PlotRange->All}];
ODE[sys,{x,y},interval,VariableStepSize->vary,
    Tolerance->0.001,Method->RungeKutta4,StepSize->0.1,
    ODEMaxStepSize->0.1,
    PlotPhase->{interval,PlotRange->All}];
ODE[sys,{x,y},interval,VariableStepSize->vary,
    Tolerance->0.001,Method->AdamsBashforth,StepSize->0.1,
    ODEMaxStepSize->0.1,
    PlotPhase->{interval,PlotRange->All}];
ODE[sys,{x,y},interval,VariableStepSize->vary,
    Tolerance->0.001,Method->BulirschStoer,StepSize->0.1,
    ODEMaxStepSize->0.1,
    PlotPhase->{interval,PlotRange->All}];
ODE[sys,{x,y},interval,VariableStepSize->vary,
    Method->ImplicitRungeKutta,StepSize->0.1,
    Tolerance->0.001,ODEMaxStepSize->0.1,
    PlotPhase->{interval,PlotRange->All}];]
```

Use *Mathematica* to plot the numerical phase plane solutions of the following initial value problems over the indicated intervals. Use the Euler, Runge-Kutta, Adams-Bashforth, Bulirsch-Stoer, and implicit Runge-Kutta methods. In each case first use the options

VariableStepSize->False and StepSize->0.1 and integrate over the interval indicated in the problem. Then solve the problem again using ODEMaxStepSize->0.1, VariableStepSize->True and Tolerance->0.001, again with the option StepSize->0.1. If possible, find the exact solution.

1.
$$\begin{cases} x' = x - 2y, & x(0) = 0, \\ y' = 2x + y, & y(0) = 2, \\ 0 \le t \le 10 \end{cases}$$

Solution.

```
sys = {x' == x - 2y,x[0] == 0,
       y' == 2x + y,y[0] == 2};
interval = {t,0,10};
```

2.
$$\begin{cases} x' = -2x - 3y, & x(0) = 1, \\ y' = 2x + y, & y(0) = -1, \\ 0 \le t \le 10 \end{cases}$$

Solution.

```
sys = {x' == -2x - 3y,x[0] == 1,
       y' == 2x + y,y[0] == -1};
interval = {t,0,10};
```

3.
$$\begin{cases} x' = 3\sin(x) - 4y, & x(0) = 1, \\ y' = 3x + 2y, & y(0) = 1, \\ 0 \le t \le 2 \end{cases}$$

Solution.

```
sys = {x' == 3Sin[x] - 4y,x[0] == 1,
       y' == 3x + 2y,y[0] == 1};
interval = {t,0,2};
```

4.
$$\begin{cases} x' = -2x - 5\cos(y), & x(0) = -2, \\ y' = 4x - 2y, & y(0) = 3, \\ 0 \le t \le 2 \end{cases}$$

Solution.

```
sys = {x' == -2x - 5Cos[y],x[0]==-2,
       y' == 4x - 2y,y[0]==3};
interval = {t,0,2};
```

5.
$$\begin{cases} x' = 2x - \tan(y), & x(0) = 1, \\ y' = \cos(x) + 2y, & y(0) = 1, \\ 0 \le t \le 1 \end{cases}$$

Solution.

```
sys = {x' == 2x - Tan[y],x[0]==1,
       y' == Cos[x] + 2y,y[0]==1};
interval = {t,0,1};
```

6.
$$\begin{cases} x' = -x + y + y^3, & x(0) = 1, \\ y' = x^2, & y(0) = -1, \\ 0 \le t \le 2 \end{cases}$$

Solution.

```
sys = {x' == -x + y + y^3,x[0]==1,
       y' == x^2,y[0]==-1};
interval = {t,0,5};
```

7.
$$\begin{cases} y'' + \sin(y) = 0, \\ y(0) = 1, \ y'(0) = 0, \\ 0 \le t \le 10 \end{cases}$$

Solution.

```
Clear[x,y]
eq = {y'' + Sin[y] == 0,y[0] == 1,y'[0] == 0};
sys = ODE[eq,y,t,Transformation->ConvertToSystem,
        TransformationVariable->w];
interval = {t,0,10};
{x,y} = {w1,w2};
```

8.
$$\begin{cases} y'' - t \sin(y') + y^2 = 0, \\ y(0) = 0, & y'(0) = 1, \\ 0 \le t \le 4 \end{cases}$$

Solution.

```
Clear[x,y]
eq = {y'' - t Sin[y'] + y^2 == 0,y[0]==0,y'[0]==1};
sys = ODE[eq,y,t,Transformation->ConvertToSystem,
      TransformationVariable->w];
interval = {t,0,4};
{x,y} = {w1,w2};
```

9.
$$\begin{cases} y'' + \sin(y') + y^2 = t, \\ y(0) = 1, \quad y'(0) = 1, \\ 0 \le t \le 10 \end{cases}$$

Solution.

```
Clear[x,y]
eq = {y'' + Sin[y'] + y^2 == t,y[0]==1,y'[0]==1};
sys = ODE[eq,y,t,Transformation->ConvertToSystem,
      TransformationVariable->w];
interval = {t,0,10};
{x,y} = {w1,w2};
```

10.
$$\begin{cases} y'' - \sin(y') - \tan(y) = t^2, \\ y(0) = 1, \quad y'(0) = 1, \\ 0 \le t \le 5 \end{cases}$$

Solution.

```
Clear[x,y]
eq = {y'' - Sin[y'] - Tan[y] == t^2,y[0]==1,y'[0]==1};
sys = ODE[eq,y,t,Transformation->ConvertToSystem,
      TransformationVariable->w];
interval = {t,0,5};
{x,y} = {w1,w2};
```

11.
$$\begin{cases} y'' + y\cos(y') + y^2 = t, \\ y(0) = 0, \quad y'(0) = 0, \\ 0 \le t \le 10 \end{cases}$$

Solution.

```
Clear[x,y]
eq = {y'' + y Cos[y'] + y^2 == t,y[0]==0,y'[0]==0};
sys = ODE[eq,y,t,Transformation->ConvertToSystem,
```

```
            TransformationVariable->w];
    interval = {t,0,10};
    {x,y} = {w1,w2};
```

$$
12. \quad \begin{cases} y'' + y^2 \sin(y') + y = \sqrt{t}, \\ y(0) = 0, \quad y'(0) = 0, \\ 0 \leq t \leq 20 \end{cases}
$$

Solution.

```
    Clear[x,y]
    eq = {y'' + y^2 Sin[y'] + y == Sqrt[t],y[0]==0,y'[0]==0};
    sys = ODE[eq,y,t,Transformation->ConvertToSystem,
            TransformationVariable->w];
    interval = {t,0,20};
    {x,y} = {w1,w2};
```

Section 19.2, page 659

Use **ODE** to plot the solution to the following initial value problems.

$$
1. \quad \begin{cases} x' = 2x(1 - 3y), \quad x(0) = 1, \\ y' = -4y(1 - x), \quad y(0) = b, \\ (0.5 \leq b \leq 4.5) \end{cases}
$$

Solution.

```
    ODE[{x'== 2x(1 - 3y),y' == -4y(1 - x),
    x[0] == 1,y[0] == b}, {x,y},{t,0,6},
    Parameters->{{b,1/2,9/2,1/2}},
    Method->NDSolve,NumericalOutput->None,
    PlotPhase->{{t,0,6},PlotRange->All,
    PlotStyle->{{RGBColor[1,0,0]}}}]
```

$$
2. \quad \begin{cases} x' = x(1 - 1.2y), \quad x(0) = a, \\ y' = -y(1 - 0.9x), \quad y(0) = 3, \\ (1 \leq a \leq 7) \end{cases}
$$

Solution.

```
    ODE[{x'== x(2 - 1.2y),y' == -y(1 - 0.9x),
    x[0] == a,y[0] == 3}, {x,y},{t,0,8},
```

```
Parameters->{{a,1,7,1}},
Method->NDSolve,NumericalOutput->None,
PlotPhase->{{t,0,8},PlotRange->All,
PlotStyle->{{RGBColor[1,0,0]}}}}]
```

3. An extension of the Lotka-Volterra equations (19.5) includes a saturation effect, which is caused by a large number of prey. The more general equations are

$$\begin{cases} x' = ax - \dfrac{bxy}{1+sx}, \\ y' = -cy + \dfrac{fxy}{1+sx}, \end{cases} \tag{S.95}$$

where a, b, c, f, and s are constants. The first four parameters have the same meaning that they had in (19.5), while s measures the saturation. The parameters a, b, c, f are all positive, but there is no restriction on s.

a. Determine the critical solutions of (S.95) and their attraction types.

b. Solve and plot (S.95) in the case that $a = 1, b = 1/90, c = 2, f = 1/50$, and $s = 0.003$.

Solution. a. The critical points can be computed using

```
CriticalPoints[{x'[t] == x[t](a - b y[t]/(1 + s x[t])),
                y'[t] == -y[t](-c +f x[t]/(1 + s x[t]))},
{x[t],y[t]},t]//Simplify
```

yielding

$$\{\{0,\ 0\},\ \{\frac{c}{f-cs},\ \frac{af}{bf-bcs}\}\}$$

Solution. b. We may use

```
ODE[{x' ==  x(1 - (y/90)/(1+0.003x)),
     y' == -y(2 - (x/50)/(1+0.003x)),
x[0] == a,y[0] == 50},{x,y},{t,0,12},
Method->NDSolve,Parameters->{{a,10,50,10}},
PlotPhase->{{t,0,12}}];
```

to obtain the phase-plane plot

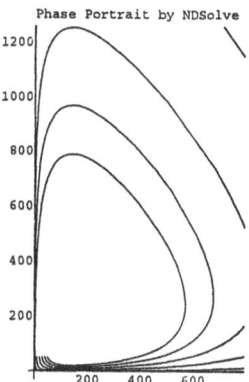

Phase Portrait by NDSolve

4. An improved version for the predator-prey model results when the the differential equation for the prey is modified so that it has the form of a logistic equation in the absence of predators. The new system is

$$\begin{cases} x' = x(a - g\,x - b\,y), \\ y' = y(-c + f\,x), \end{cases} \tag{S.96}$$

where a, b, c, f, and g are positive constants.

    a. Determine the critical solutions of (S.96) and their attraction types.

    b. Solve and plot (S.96) in the case that $a = 1$, $b = 1/90$, $c = 2$, $f = 1/50$, and $g = 1/130$.

Solution. **a.** The critical points can be computed using

```
CriticalPoints[{x'[t] == x[t](a - g x[t] - b y[t]),
                y'[t] == -y[t](-c +f x[t])},
{x[t],y[t]},t]
```

yielding

$$\left\{\{0,\ 0\},\ \left\{\frac{a}{g},\ 0\right\},\ \left\{\frac{c}{f},\ \frac{a}{b} - \frac{c\,g}{b\,f}\right\}\right\}$$

Solution. **b.** We may use

```
ODE[{x' ==  x(1 - x/130 -(y/90)),
     y' == -y(2 - (x/50)),
```

```
x[0] == a,y[0] == 50},{x,y},{t,-0.5,15},
Method->NDSolve,MaxSteps->2000,
Parameters->{{a,5,135,10}},
PlotPhase->{{t,-0.5,15},
PlotRange->{{0,150},{0,90}}}];
```

to obtain the phase-plane plot

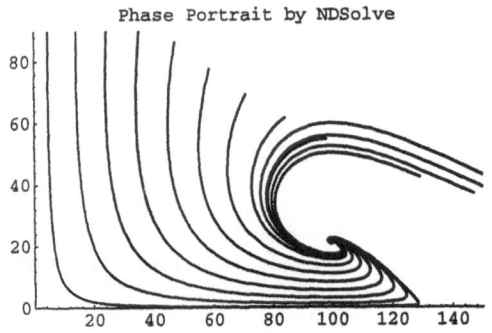

Phase Portrait by NDSolve

5. Show that the solution curves of (19.5) are closed.

Solution. The solution of (19.5) is given by (19.8), which can be rewritten as

$$y^a e^{-by} = q \, x^{-c} e^{fx}$$

where q is a positive constant. Consider the function $F(y) = y^a e^{-by}$. This function has the following properties:

$$\begin{cases} F(0) = 0, \\ \lim_{y \to \infty} F(y) = 0, & \text{(S.97)} \\ F(y) \text{ has a positive maximum for } y = b/a. \end{cases}$$

It can be proved that (S.97) implies that if $K > 0$ and $K \neq F(b/a)$, then the equation

$$F(y) = K$$

has either two positive solutions or no solutions. Similarly, if $G(x) = x^c e^{fx}$, $L > 0$ and $L \neq G(f/c)$, then the equation

$$G(x) = L$$

has either two positive solutions or no solutions. These facts imply that the solution curves of (19.5) are closed.

Section 19.4, page 673

1. A pendulum is displaced through an angle of $\pi/3$ radians. What is its period?

Solution.

$$P(k) = 4K(k)\sqrt{L/g} = 4K\left(\sin((\pi/3)/2)^2\right)\sqrt{L/g} = 6.743\sqrt{L/g} \quad \text{seconds.}$$

2. A *two second pendulum* is one which makes a full swing in two seconds. Using the standard value $g = 32.174$ feet per second^2, find the length of the two second pendulum.

Solution. $L = 4g/\pi^2 = 13.04$ feet.

3. If a two second pendulum is displaced through an angle of $\pi/2$ radians, determine the time required for it to make one complete oscillation.

Solution.

$$P(k) = 4K(k)\sqrt{L/g} = 4K(\sin((\pi/2)/2)^2)\sqrt{13.04/32.174} = 4.72 \quad \text{seconds.}$$

4. Repeat Exercise 3 if the initial displacement is $\pi/6$ radians.

Solution.

$$P(k) = 4K(k)\sqrt{L/g} = 4K(\sin((\pi/6)/2)^2)\sqrt{13.04/32.174} = 4.07 \quad \text{seconds.}$$

5. Use ODE and plot the solution to a two second pendulum with an initial displacement of $\pi/2$. Experiment with the plotting range to obtain an illustrative comparison with Exercise 3.

Solution.

```
ODE[{y'' + (32.174/13.04) Sin[y] == 0,
    y[0]==Pi/2,y'[0]==0},y,t,
    Method->TableLookup,PlotSolution->{{t,0,4.72}}]
```

6. Use ODE to obtain the general solutions to the linear undamped pendulum and the nonlinear undamped pendulum, each with initial conditions $y(0) = Y_0$ and $y'(0) = 0$.

Solution.

```
g /: Im[g] = 0;
L /: Im[L] = 0;
g /: Positive[g] = True;
L /: Positive[L] = True;
ODE[{y'' + (g/L) y == 0,y[0]==y0,y'[0]==0},y,t,
Method->SecondOrderLinear,Form->Equation]
```

$$y == y0 \; Cos[\frac{Sqrt[g]\; t}{Sqrt[L]}]$$

```
ODE[{y'' + (g/L) Sin[y] == 0,y[0]==y0,y'[0]==0},y,t,
Method->TableLookup,Form->Equation]
```

$$y == 2 \; ArcSin[JacobiSN[Sqrt[\frac{g}{L}] \; t + EllipticK[Sin[\frac{y0}{2}]^2],$$

$$Sin[\frac{y0}{2}]^2] \; Sin[\frac{y0}{2}]]$$

7. Use the results of the previous problem and compare the plots of the solutions to a two second pendulum with an initial displacement of $\pi/3$. Experiment with the plotting range to obtain an illustrative comparison.

Solution.

```
ODE[{y'' + (32.174/13.04) y == 0,y[0]==Pi/3,
y'[0]==0},y,t,GraphLabel->1,
Method->SecondOrderLinear,PlotSolution->{{t,0,20}}]
ODE[{y'' + (32.174/13.04) Sin[y] == 0,
y[0]==Pi/3,y'[0]==0},
y,t,Method->TableLookup,GraphLabel->2,
PlotSolution->{{t,0,20},PlotStyle->{{RGBColor[1,0,0]}}}]
Show[{Graph[1],Graph[2]}];
```

8. When we compare the exact solution to the undamped pendulum problem with the linear version obtained through a small angle approximation, initially we see that the solutions disagree more and more as t increases. Assuming a two second pendulum with an initial displacement of $\pi/3$ radians, find the smallest time t for which the two solutions exhibit a maximum difference. That is, a time when their positions are exactly opposite one another.

Solution.

```
linear = ODE[{y'' + (32.174/13.04) y == 0,
    y[0]==Pi/3,y'[0]==0},y,t,
    Method->SecondOrderLinear,Form->Explicit]
nonlinear = ODE[{y'' + (32.174/13.04) Sin[y] == 0,
    y[0]==Pi/3,y'[0]==0},y,t,
    Method->TableLookup,Form->Explicit]
FindRoot[linear == -nonlinear,{t,20}]

{t -> 19.6744}
```

Section 19.5, page 677

1. Plot a curve α whose curvature is given by $k(s) = s^2$.

Solution.

```
cloth2[s_]=ODE[{x' == Cos[theta[s]],y' == Sin[theta[s]],
theta' == s^2,x[0] == 0,y[0] == 0,theta[0] == 0},
{x,y,theta},{s,-5,5},Method->NDSolve,Form->Explicit]

ParametricPlot[Evaluate[Drop[cloth2[s],{1}]],{s,-5,5},
PlotPoints->80,AspectRatio->Automatic,
PlotRange->All]
```

2. Plot a curve α whose curvature is given by $k(s) = s^2 \sin(s)$.

Solution.

```
cloth2[s_]=ODE[{x' == Cos[theta[s]],y' == Sin[theta[s]],
theta' == s^2 Sin[s],x[0] == 0,y[0] == 0,theta[0] == 0},
{x,y,theta},{s,-5,5},Method->NDSolve,Form->Explicit]

ParametricPlot[Evaluate[Drop[cloth2[s],{1}]],{s,-5,5},
PlotPoints->80,AspectRatio->Automatic,
PlotRange->All]
```

Chapter 20

Section 20.1, page 687

Find the first few terms in the Taylor series representation of each of the following functions about the specified point t_0.

1. $f(t) = \sin(t), \quad t_0 = 0$

Solution. $f(t) = t - \dfrac{t^3}{3!} + \dfrac{t^5}{5!} - \cdots$

2. $f(t) = \sin(t), \quad t_0 = \pi/2$

Solution. $f(t) = 1 - \dfrac{1}{2}\left(t - \dfrac{\pi}{2}\right)^2 + \dfrac{1}{4!}\left(t - \dfrac{\pi}{2}\right)^4 - \cdots$

3. $f(t) = e^{3t}, \quad t_0 = 1$

Solution. $f(t) = e^3 + 3e^3(t-1) + \dfrac{9e^3}{2}(t-1)^2 + \cdots$

4. $f(t) = e^{t^2}, \quad t_0 = 0$

Solution. $f(t) = 1 + t^2 + \dfrac{t^4}{2} + \dfrac{t^6}{3!} + \cdots$

5. $f(t) = \dfrac{3}{2-t}, \quad t_0 = 0$

Solution. $f(t) = \dfrac{3}{2} + \dfrac{3}{4}t + \dfrac{3}{8}t^2 + \dfrac{3}{16}t^3 + \cdots$

6. $f(t) = \dfrac{3+t}{3-t}, \quad t_0 = 0$

Solution. $f(t) = 1 + \dfrac{2}{3}t + \dfrac{2}{9}t^2 + \dfrac{2}{27}t^3 + \cdots$

7. $f(t) = t\cos(3t), \quad t_0 = 0$

Solution. $f(t) = t - \dfrac{9}{2}t^3 + \dfrac{27}{8}t^5 - \cdots$

8. $f(t) = (t+2)e^{3t}, \quad t_0 = 0$

Solution. $f(t) = 2 + 7t + 12t^2 + \dfrac{27}{2}t^3 + \cdots$

9. $f(t) = (t+2)e^{3t}$, $t_0 = 1$

Solution. $f(t) = 3e^3 + 10e^3(t-1) + \dfrac{33e^3}{2}(t-1)^2 + \cdots$

10. $f(t) = \cosh(t)$, $t_0 = 0$

Solution. $f(t) = 1 + \dfrac{t^2}{2} + \dfrac{t^4}{4!} + \cdots$

Use the general product rule (Lemma 20.6) to compute the following higher derivatives:

11. $(t^3 e^{4t})''$

Solution. $(16t^3 + 24t^2 + 6t)e^{4t}$

12. $\left(\sin(2t)\cosh(3t)\right)'''$

Solution. $-9\sin(2t)\sinh(3t) + 46\cos(2t)\cosh(3t)$

13. $\left(\cos(5t)e^{2t}\right)^{(4)}$

Solution. $41\cos(5t)e^{2t} + 840\sin(5t)e^{2t}$

14. $\left(t^3 \log(t)\right)''$

Solution. $5t + 6t\log(t)$

Section 20.3, page 697

Solve the following initial value problems by the method of power series at the indicated points with the indicated initial values.

1. $\begin{cases} y'' - 4y = 0, \\ t_0 = 0,\ Y_0 = 1,\ Y_1 = 0 \end{cases}$

Solution. $y_{n+2} = 4y_n$,

$$y(t) = 1 + 2t^2 + \frac{2}{3}t^4 + \frac{4}{45}t^6 + \cdots$$

2. $\begin{cases} y'' - t\,y' - y = 0, \\ t_0 = 0,\ Y_0 = 2,\ Y_1 = 3 \end{cases}$

Solution. $y_{n+2} = (n+1)y_n$,

$$y(t) = 2 + 3t + t^2 + t^3 + \frac{t^4}{4} + \cdots$$

3. $\begin{cases} y'' - t\,y' - y = 0, \\ t_0 = 1,\, Y_0 = 1,\, Y_1 = 3 \end{cases}$

Solution. $y_{n+2} = y_{n+1} + (n+1)y_n,$

$$y(t) = 1 + 3(t-1) + 2(t-1)^2 + \frac{5}{3}(t-1)^3 + \frac{11}{12}(t-1)^4 + \cdots$$

4. $\begin{cases} y'' + 3t\,y' + 4y = 0, \\ t_0 = 0,\, Y_0 = 0,\, Y_1 = 1 \end{cases}$

Solution. $y_{n+2} = -(3n+4)y_n,$

$$y(t) = t - \frac{7}{6}t^3 + \frac{91}{120}t^5 - \cdots$$

5. $\begin{cases} y'' - 2t\,y' + 2y = 0, \\ t_0 = 0,\, Y_0 = 0,\, Y_1 = 1 \end{cases}$

Solution. $y_{n+2} = 2(n-1)y_n, \qquad y(t) = t$

6. $\begin{cases} y'' + 9t^2 y = 0, \\ t_0 = 0,\, Y_0 = 1,\, Y_1 = 3 \end{cases}$

Solution. $y_{n+2} = -9n(n-1)y_{n-2},$

$$y(t) = 1 + 3t - \frac{3}{4}t^4 - \frac{27}{20}t^5 + \cdots$$

7. $\begin{cases} y'' + t\,y' + 4y = 0, \\ t_0 = 0,\, Y_0 = 3,\, Y_1 = 6 \end{cases}$

Solution. $y_{n+2} = -(n+4)y_n,$

$$y(t) = 3 + 6t - 6t^2 - 5t^3 + \cdots$$

8. $\begin{cases} y'' - 3y' + 6y = 0, \\ t_0 = 0,\, Y_0 = 5,\, Y_1 = 3 \end{cases}$

Solution. $y_{n+2} = 3y_{n+1} - 6y_n,$

$$y(t) = 5 + 3t - \frac{21}{2}t^2 - \frac{27}{2}t^3 + \cdots$$

9. $\begin{cases} y'' + 5t\,y' + 6t^2 y = 0, \\ t_0 = 0,\, Y_0 = 1,\, Y_1 = 0 \end{cases}$

Solution. $y_{n+2} = -5n\, y_n - 6n(n-1)y_{n-2}$,

$$y(t) = 1 - \frac{t^4}{2} + \frac{t^6}{3} - \frac{t^8}{8} + \cdots$$

10. $\begin{cases} y'' - 3y' + 6y = 0, \\ t_0 = 0,\ Y_0 = 5,\ Y_1 = 3 \end{cases}$

Solution. $y_{n+2} - 3y_{n+1} + 6y_n = 0$,

$$y(t) = 5 + 3t - \frac{21}{2}t^2 - \frac{27}{2}t^3 + \cdots$$

11. $\begin{cases} y'' + 5t\, y' + 6t^2 y = 0, \\ t_0 = 0,\ Y_0 = 0,\ Y_1 = 1 \end{cases}$

Solution. $y_{n+2} = -5n\, y_n - 6n(n-1)y_{n-2}$,

$$y(t) = t - \frac{5}{6}t^3 + \frac{13}{40}t^5 - \frac{25}{84}t^7 + \cdots$$

12. $\begin{cases} y'' - 2t\, y' + 5y = 0, \\ t_0 = 0,\ Y_0 = 1,\ Y_1 = 0 \end{cases}$

Solution. $y_{n+2} = (2n - 5)y_n$,

$$y(t) = 1 - \frac{5}{2}t^2 + \frac{5}{24}t^4 + \frac{t^6}{48} + \cdots$$

13. $\begin{cases} y'' - 2t\, y' + 5y = 0, \\ t_0 = 0,\ Y_0 = 0,\ Y_1 = 1 \end{cases}$

Solution. $y_{n+2} = (2n - 5)y_n$,

$$y(t) = t - \frac{t^3}{2} - \frac{t^5}{40} - \frac{t^7}{336} - \cdots$$

14. $\begin{cases} y'' - t\, y' = 0,\ t_0 = 1, \\ Y_0 = 1,\ Y_1 = 0 \end{cases}$

Solution. $y_{n+2} = y_{n+1} + n\, y_n, \qquad y(t) = 1$

15. $\begin{cases} y'' + (t + 1)y' + 4y = 0, \\ t_0 = 0,\ Y_0 = 0,\ Y_1 = 1 \end{cases}$

Solution. $y_{n+2} = -y_{n+1} - (n+4)y_n,$

$$y(t) = t - \frac{t^2}{2} - \frac{2}{3}t^3 + \frac{7}{6}t^4 + \cdots$$

16.
$$\begin{cases} y'' + (t+1)y' + 4y = 0, \\ t_0 = -1, \ Y_0 = 0, \ Y_1 = 1 \end{cases}$$

Solution. $y_{n+2} = -(n+4)y_n,$

$$y(t) = (t+1) - \frac{5}{6}(t+1)^3 + \frac{7}{24}(t+1)^5 - \frac{(t+1)^7}{20} + \cdots$$

Section 20.4, page 701

Solve and plot the following Airy initial value problems over the stated ranges.

1.
$$\begin{cases} y'' - ty = 0, \\ y(0) = 3^{-2/3}(\Gamma(2/3))^{-1}, \\ y'(0) = -3^{-1/3}(\Gamma(1/3))^{-1}, \\ -15 < t < 5 \end{cases}$$

Solution. $y(t) = \text{Ai}(t)$

2.
$$\begin{cases} y'' - ty = 0, \\ y(0) = -3^{-1/6}(\Gamma(2/3))^{-1}, \\ y'(0) = -3^{-1/3}(\Gamma(1/3))^{-1}, \\ -15 < t < 2 \end{cases}$$

Solution. $y(t) = \text{Bi}(t)$

3.
$$\begin{cases} y'' - 9ty = 0, \\ y(0) = 0, \ y'(0) = 1, \\ -1 < t < 4 \end{cases}$$

Solution. $y(t) = \dfrac{\Gamma(1/3)}{(2)3^{5/6}}\left(\text{Bi}(3^{2/3}t) - \text{Ai}(3^{2/3}t)\right)$

4.
$$\begin{cases} y'' + 36t y = 0, \\ y(0) = 0, \ y'(0) = 1, \\ -1 < t < 3 \end{cases}$$

Solution. $y(t) = \dfrac{\Gamma(1/3)}{2} \left(\dfrac{\text{Ai}(6^{2/3}t)}{12^{1/3}} - \dfrac{\text{Bi}(6^{2/3}t)}{2^{2/3}\,3^{5/6}} \right)$

5. Use *Mathematica* to give a pseudoproof that the Wronskian of **AiryAi[t]** and **AiryBi[t]** is $1/\pi$. What about a mathematical proof?

Solution.

```
N[Limit[Det[{{AiryAi[t],AiryBi[t]},{AiryAiPrime[t],
AiryBiPrime[t]}}],t->0]] == N[1/Pi]
```

True

6. Use *Mathematica* to show that the products **AiryAi[t]AiryAi[t]**, **AiryAi[t]AiryBi[t]** and **AiryBi[t]AiryBi[t]** are linearly independent solutions to $y''' - 4t\,y' - 2y = 0$.

Solution.

```
N[Limit[Simplify[Wronskian[
{AiryAi[#]^2&,AiryAi[#] AiryBi[#]&,
AiryBi[#]^2&}][t]],t->0]] == N[2/Pi^3]
```

True

Section 20.5, page 705

1. Prove the orthogonality relation

$$\int_{-1}^{1} P_m(t) P_n(t)\, dt = 0, \qquad \text{for } m \neq n. \tag{S.98}$$

Solution. We write the Legendre equation for P_m as

$$[(1 - t^2) P_m']' + m(m + 1) P_m = 0.$$

Multiply by $P_n(t)$ and integrate over $-1 \le t \le 1$ to obtain

$$\int_{-1}^{1} P_n(t)[(1 - t^2) P_m']'\, dt + m(m + 1) \int_{-1}^{1} P_m(t) P_n(t)\, dt = 0.$$

The first term can be integrated by parts, noting that the integrand is zero at $t = \pm 1$, to obtain

$$\int_{-1}^{1} P_m'(t) P_n'(t)(1 - t^2)\, dt + m(m + 1) \int_{-1}^{1} P_m(t) P_n(t)\, dt = 0$$

Interchanging the roles of (m, n), we have

$$\int_{-1}^{1} P_n'(t) P_m'(t)(1 - t^2) \, dt + n(n+1) \int_{-1}^{1} P_n(t) P_m(t) \, dt = 0$$

Since the first terms are equal, we conclude that

$$m(m+1) \int_{-1}^{1} P_n(t) P_m(t) \, dt = n(n+1) \int_{-1}^{1} P_n(t) P_m(t) \, dt.$$

If the integral were non-zero, we would conclude that $m(m+1) = n(n+1)$, which is possible if and only if $m = n$. Therefore, the integral must be non-zero, which proves the required orthgonality relation.

2. Use (20.48) to show that

$$\int_{-1}^{1} Q_n(t) P_n(t) \, dt = 0, \tag{S.99}$$

where $Q_n(t)$ is a polynomial of degree $n - 1$ or less.

Solution. First, we show by mathematical induction, that any polynomial Q_n of degree $n - 1$ can be written as $Q_n(t) = c_0 P_0(t) + \cdots + c_{n-1} P_{n-1}(t)$, for some constants $c_0, \ldots, c_{n-1}$. This is clearly true for $n = 1$. Assuming that it holds for all polynomials of degree $n - 2$, we write $Q_n(t) = c_n t^n + Q_{n-1}(t)$. But the Legendre polynomial $P_{n-1}(t) = d_{n-1} t^{n-1} + R_{n-1}(t)$ where $R_{n-1}(t)$ is a polynomial of degree $n - 2$. Using this together with the induction hypothesis on Q_{n-1}, we have

$$Q_n(t) = c_n \left(\frac{P_{n-1}(t) - R_{n-1}(t)}{d_n} \right) + \sum_{j=0}^{n-2} c_j P_j(t) = \sum_{j=0}^{n-1} c_j' P_j(t)$$

Now finally

$$\int_{-1}^{1} Q_n(t) P_n(t) \, dt = \sum_{j=0}^{n-1} c_j' \int_{-1}^{1} P_n(t) P_j(t) \, dt = 0.$$

3. Prove that if $Q_n(t)$ is a polynomial of degree n such that

$$\int_{-1}^{1} Q_n(t) P_k(t) \, dt = 0$$

for $k = 1, \ldots, n - 1$, then $Q_n(t) = c \, P_n(t)$ for some constant c.

Solution. From the proof of 2, we can write $Q_n(t) = c_0 P_0(t) + \cdots + c_n P_n(t)$. By orthogonality,

$$c_j = \frac{2}{2j+1} \int_{-1}^{1} P_j(t) Q_n(t) \, dt = 0$$

for $j < n$. Therefore, $Q_n(t) = c_n P_n(t)$, which was to be proved.

4. To prove Rodrigues' formula (20.47), follow the following steps:

    a. Define $W_n(t) = \dfrac{d^n}{dt^n}(t^2 - 1)^n$ and show that $W_n(t)$ is a polynomial of degree n with leading coefficient $(2n)!/n!$.

    b. Use integration by parts n times to show that for any polynomial $Q_n(t)$ of degree $n - 1$ or less we have

$$\int_{-1}^{1} W_n(t) Q_n(t) \, dt = 0.$$

Solution. The function $(t^2 - 1)^n$ is clearly a polynomial of degree $2n$ whose leading term is t^{2n}. Differentiating this n times produces a polynomial of degree $2n - n = n$, where the coefficient of t^n is $2n(2n - 1) \cdots (n + 1) = (2n)!/n!$. Now if $Q_{n-1}(t)$ is any polynomial of degree $n - 1$ or less, we may write

$$\int_{-1}^{1} W_n(t) Q_n(t) \, dt = \int_{-1}^{1} Q_n(t) \frac{d^n}{dt^n}(t^2 - 1)^n \, dt = -\int_{-1}^{1} Q'_n(t) \frac{d^{n-1}}{dt^{n-1}}(t^2 - 1)^n \, dt.$$

Each time we integrate by parts, we encounter the term $(t^2 - 1)$ which is zero at the endpoints, together with the integral of a derivative of $Q_n(t)$, multiplied by powers of $t^2 - 1$. But the nth derivative of $Q_n(t)$ is zero, hence repeating the partial integration n times reveals that the original integral is zero.

5. Use Exercise 4 to show that $P_n(t) = c W_n(t)$ for some constant c. Then compare the leading coefficients of $P_n(t)$ and $W_n(t)$ to conclude that $c = 1/(2^n n!)$.

Solution. Since $W_n(t)$ is a polynomial of degree n, it may be written in the form $W_n(t) = \sum_{0 \le k \le n} c_k P_k(t)$. But the orthogonality established above shows that $W_n(t)$ is orthogonal to $P_k(t)$ whenever $k < n$. Therefore $W_n(t) = c_n P_n(t)$. To compute c_n, recall that the coefficient of t^n in $P_n(t)$ is $(2n)!/[2^n (n!)^2]$. On the other hand the coefficient of t^n in $W_n(t)$ was shown above to be $(2n)!/n!$. Thus $(2n)!/n! = c_n (2n)!/[2^n (n!)^2]$ which is solved to obtain $c_n = 2^n n!$, hence $c = 1/(2^n n!)$.

6. The nth Hermite polynomial is defined by the Rodrigues formula

$$H_n(t) = (-1)^n e^{t^2} \frac{d^n}{dt^n} e^{-t^2}.$$

Compute $H_n(t)$ for $0 \le n \le 6$.

Solution. Writing $D = d/dt$, we have

$$
\begin{aligned}
D(e^{-t^2}) &= -2t\,e^{-t^2} \\
D^2(e^{-t^2}) &= [(-2t)^2 - 2]e^{-t^2} = [4t^2 - 2]e^{-t^2} \\
D^3(e^{-t^2}) &= [(-2t)(4t^2 - 2) + 8t]e^{-t^2} = [-8t^3 + 12t]e^{-t^2} \\
D^4(e^{-t^2}) &= [(-2t)(12t - 8t^3) + (12 - 24t^2)]e^{-t^2} = [16t^4 - 48t^2 + 12]e^{-t^2} \\
D^5(e^{-t^2}) &= [(-2t)(16t^4 - 48t^2 + 12) + (64t^3 - 96t)]e^{-t^2} = [-32t^5 + 160t^3 - 120t]e^{-t^2} \\
D^6(e^{-t^2}) &= (-2t)(-32t^5 + 160t^3 - 120t) + (-160t^4 + 480t^2 - 120)e^{-t^2} \\
&= [64t^6 - 480t^4 + 700t^2 - 120]e^{-t^2}
\end{aligned}
$$

Therefore, we have

$$
\begin{aligned}
H_0(t) &= 1 \\
H_1(t) &= 2t \\
H_2(t) &= 4t^2 - 2 \\
H_3(t) &= 8t^3 - 12t \\
H_4(t) &= 16t^4 - 48t^2 + 12 \\
H_5(t) &= 32t^5 - 160t^3 + 120t \\
H_6(t) &= 64t^6 - 480t^4 + 700t^2 - 120
\end{aligned}
$$

7. Show that H_n satisfies the differential equation $y'' - 2t\,y' + 2n\,y = 0$.

Solution. Using the notation $D = d/dt$, we re-write the definition of H_n as $(-1)^n e^{-t^2} H_n(t) = D^n(e^{-t^2})$. Changing n to $n+1$, we can write, on the one hand

$$
\begin{aligned}
(-1)^{n+1} e^{-t^2} H_{n+1} &= D^{n+1}(e^{-t^2}) = D(D^n(e^{-t^2})) = (-1)^n D(e^{-t^2} H_n) \\
&= (-1)^n [-2t e^{-t^2} H_n + e^{-t^2} H_n']
\end{aligned}
$$

so that

$$
H_{n+1}(t) = 2t\,H_n(t) - H_n'(t) \tag{S.100}
$$

On the other hand we can use the product rule for higher derivatives to write

$$
\begin{aligned}
D^{n+1}(e^{-t^2}) &= D^n(D(e^{-t^2})) = D^n(-2te^{-t^2}) = -2t\,D^n(e^{-t^2}) - 2n\,D^{n-1}(e^{-t^2}) \\
&= (-1)^n 2t e^{-t^2} H_n - 2n(-1)^{n-1} e^{-t^2} H_{n-1},
\end{aligned}
$$

so that

$$H_{n+1} = 2t H_n - 2n H_{n-1}. \tag{S.101}$$

Equating the two expressions (S.100) and (S.101) yields the formula $H'_n = 2n H_{n-1}$. Substituting this back into (S.101) we obtain

$$(D - 2t) H_n = -H_{n+1}. \tag{S.102}$$

Applying D to both sides of (S.102) and using (S.100) produces

$$D(D - 2t) H_n = -D H_{n+1} = -2(n+1) H_n$$

which is the required differential equation, since the left side is $H''_n - 2t H'_n - 2 H_n$ and the term $-2 H_n$ cancels from both sides.

8. Use *Mathematica* to give a pseudoproof of Exercise 7 as follows. Define functions **HH** and **HHeq** by

```
HH[n_,t_]:= (-1)^n E^(t^2) D[E^(-t^2),{t,n}]
```

and

```
HHeq[n_,t_]:= Simplify[D[HH[n,t],{t,2}] -
              2t D[HH[n,t],t] + 2n HH[n,t]]
```

Then compute **Table[HHeq[n,t],n,0,6]**

9. It can be shown that

$$H_n(t) = \sum_{k=0}^{[n/2]} \frac{(-1)^k n! (2t)^{n-2k}}{k!(n-2k)!}.$$

Check this formula by hand for $0 \le n \le 6$.

10. Redo Exercise 9 in *Mathematica* as follows. Define a functions **HHH** by

```
HHH[n_,t_]:= Sum[(-1)^k n! (2t)^(n-2k)/(k!(n-2k)!),
             {k,0,Floor[n/2]}]
```

Then compute **Table[HHH[n,t],n,0,6]**

Solution.

```
1
2 t
        2
-2 + 4 t
          3
-12 t + 8 t
        2       4
12 - 48 t  + 16 t
          3       5
120 t - 160 t  + 32 t
          2       4       6
-120 + 720 t  - 480 t  + 64 t
```

11. The nth Chebyshev polynomial is defined by

$$T_n(t) = \cos(n \arccos(t)).$$

Establish the recurrence relation

$$T_n(t) = 2t\,T_{n-1}(t) - T_{n-2}(t)$$

and compute $T_n(t)$ for $0 \le n \le 6$.

Solution. We write $T_n(t) = \cos(n\theta)$ with $\cos\theta = t$. Therefore

$$T_{n+1}(t) + T_{n-1}(t) = \cos((n+1)\theta)) + \cos((n-1)\theta) = 2\cos n\theta \cos\theta = 2t\,T_n(t)$$

To obtain the stated identity we change n to $n-1$. Now clearly $T_0(t) = 1$, $T_1(t) = \cos\theta = t$, so that

$$
\begin{aligned}
T_2(t) &= 2t(t) - 1 = 2t^2 - 1 \\
T_3(t) &= 2t(2t^2 - 1) - t = 4t^3 - 3t \\
T_4(t) &= 2t(4t^3 - 3t) - (2t^2 - 1) = 8t^4 - 8t^2 + 1 \\
T_5(t) &= 2t(8t^4 - 8t^2 + 1) - (4t^3 - 3t) = 16t^5 - 20t^3 + 5t \\
T_6(t) &= 2t(16t^5 - 20t^3 + 5t) - (8t^4 - 8t^2 + 1) = 32t^6 - 48t^4 + 18t^2 - 1
\end{aligned}
$$

12. Show that T_n satisfies the differential equation $(1 - t^2)y'' - t\,y' + n^2 y = 0$.

Solution. Begining with the equations $t = \cos\theta$, $T_n(t) = \cos(n\theta)$, we have $(\sin\theta)d\theta/dt = -1$ and

$$T_n'(t) = \frac{n \sin n\theta}{\sin\theta}$$

$$T_n''(t) = \frac{-n^2 \cos(n\theta) + n \sin n\theta \cos\theta / \sin\theta}{\sin^2\theta}$$

so that by using the formula for $T_n'(t)$ we obtain

$$(1 - t^2)T_n''(t) = \sin^2\theta\, T_n''(t) = -n^2 T_n(t) + t T_n'(t)$$

as required.

13. Use *Mathematica* to give a pseudoproof of Exercise 12 as follows. Define functions **TT** and **TTeq** by

```
TT[0,t_]:=  1
TT[1,t_]:=  t
TT[n_,t_]:= 2t TT[n - 1,t] - TT[n - 2,t]
```

and

```
TTeq[n_,t_]:= Simplify[(1 - t^2)D[TT[n,t],{t,2}] -
              t D[TT[n,t],t] + n^2 TT[n,t]]
```

Then compute **Table[TTeq[n,t],n,0,6]**

Solution.

```
1
t
         2
-1 + 2 t
             3
-3 t + 4 t
       2       4
1 - 8 t  + 8 t
           3        5
5 t - 20 t  + 16 t
        2        4        6
-1 + 18 t  - 48 t  + 32 t
```

Section 20.6, page 711

1. Show directly that the function $z(t) = (1 - t/r)^{-\alpha}$ is a solution of the equation

$$z'' - \frac{P z'}{(1 - t/r)} - \frac{Q z}{(1 - t/r)^2} = 0,$$

provided that α is a solution of the equation $\alpha(\alpha + 1) - P\alpha r - Qr^2 = 0$.

Solution. The derivatives are computed as

$$z'(t) = (\alpha/r)(1 - t/r)^{-\alpha-1}, \qquad z''(t) = \alpha(\alpha + 1)/r^2(1 - t/r)^{-\alpha-2}$$

so that

$$z'' - \frac{P z'}{(1 - t/r)} - \frac{Q z}{(1 - t/r)^2} = (1/r^2)\left(\alpha(\alpha + 1) - P\alpha r - Qr^2\right)(1 - t/r)^{-\alpha-2},$$

from which the conclusion follows.

2. Show how to modify the above proof in order to obtain a power series solution of the equation $y'' + p(t)y' + q(t)y = r(t)$.

Solution. By changing t to t/r, we may assume that $r = 1$. Following the discussion in the text, we may further assume that the expansions of the functions $p(t)$, $q(t)$, $r(t)$ satisfy

$$|p_n| \le Pn!, \qquad |q_n| \le Qn!, \qquad |r_n| \le Rn!$$

for some positive constants P, Q, R. By increasing these we can further assume that $P + Q \ge 2$. Now let $\alpha > 0$ be the solution of the quadratic equation $\alpha(\alpha + 1) - P\alpha - Q = R$, which also satisfies the inequality $\alpha \ge 1$, from the choice of P, Q. The power series $Z(t) = (1 - t)^{-\alpha} = \sum_{n=0}^{\infty} Z_n t^n / n!$ converges for $|t| < 1$ and satisfies the differential equation

$$Z'' - \frac{P}{1-t}Z' - \frac{Q}{(1-t)^2}Z = \frac{R}{(1-t)^{\alpha+2}}.$$

In particular, the coefficients P_n, Q_n, R_n satisfy $|p_n| \le P_n$, $|q_n| \le Q_n$, $|r_n| \le R_n$ and

$$Z_{n+2} - \sum_{k=0}^{n} \binom{n}{k} P_k Z_{n+1-k} - \sum_{k=0}^{n} \binom{n}{k} Q_k Z_{n-k} = R_n, \qquad n = 0, 1, 2, \dots$$

Now define $Y_0=1$, $Y_1 = \alpha$ and for $n \ge 0$,

$$Y_{n+2} + \sum_{k=0}^{n} \binom{n}{k} p_k Y_{n+1-k} + \sum_{k=0}^{n} \binom{n}{k} q_k Y_{n-k} = r_n. \qquad \text{(S.103)}$$

We want to prove that $|Y_n| \le Z_n$ for $n = 0, 1, 2, \dots$. This is true for $n = 0, 1$ by definition. Assuming this inequality for the values $0, \dots, n+1$ from (S.103) we have

$$\begin{aligned}
|Y_{n+2}| &\le |r_n| + \sum_{k=0}^{n} \binom{n}{k} |p_k||Y_{n+1-k}| + \sum_{k=0}^{n} \binom{n}{k} |q_k||Y_{n-k}| \\
&\le R_n + \sum_{k=0}^{n} \binom{n}{k} P_k Z_{n+1-k} + \sum_{k=0}^{n} \binom{n}{k} Q_k Z_{n-k} \\
&= R_n + (Z_{n+2} - R_n) = Z_{n+2}
\end{aligned}$$

Hence the power series for $Y(t)$ also converges for $|t| < 1$, which was to be proved.

Section 20.7, page 712

Use *Mathematica* to do Exercises 1–16 on pages 697–698 of Section 20.3.

1.
$$\begin{cases} y'' - 4y = 0, \\ t_0 = 0, Y_0 = 1, Y_1 = 0 \end{cases}$$

Solution. Using

```
ODE[{y'' - 4 y == 0,y[0] == 1,y'[0] == 0},y,t,
Method->SeriesForm,Degree->11,
Form->Explicit]
```

we obtain

$$1 + 2 t^2 + \frac{2 t^4}{3} + \frac{4 t^6}{45} + \frac{2 t^8}{315} + \frac{4 t^{10}}{14175}$$

2.
$$\begin{cases} y'' - t\, y' - y = 0, \\ t_0 = 0, Y_0 = 2, Y_1 = 3 \end{cases}$$

Solution. Using

```
ODE[{y'' -t y' - y == 0,y[0] == 2,y'[0] == 3},y,t,
Method->SeriesForm,Degree->9,
Form->Explicit]
```

we obtain

$$2 + 3 t + t^2 + t^3 + \frac{t^4}{4} + \frac{t^5}{5} + \frac{t^6}{24} + \frac{t^7}{35} + \frac{t^8}{192} + \frac{t^9}{315}$$

3.
$$\begin{cases} y'' - t\, y' - y = 0, \\ t_0 = 1, Y_0 = 1, Y_1 = 3 \end{cases}$$

Solution. Using

```
ODE[{y'' -t y' - y == 0,y[1] == 1,y'[1] == 3},y,t,
Method->SeriesForm,Degree->6,ExpansionPoint->1,
Form->Explicit]
```

we obtain

$$1 + 3 (-1 + t) + 2 (-1 + t)^2 + \frac{5 (-1 + t)^3}{3} +$$

$$\frac{11 (-1 + t)^4}{12} + \frac{31 (-1 + t)^5}{60} + \frac{43 (-1 + t)^6}{180}$$

4. $\begin{cases} y'' + 3t\,y' + 4y = 0, \\ t_0 = 0,\, Y_0 = 0,\, Y_1 = 1 \end{cases}$

Solution. Using

```
ODE[{y'' + 3t y' +4 y == 0,y[0] == 0,y'[0] == 1},y,t,
Method->SeriesForm,Degree->11,ExpansionPoint->0,
Form->Explicit]
```

we obtain

$$t - \frac{7\,t^3}{6} + \frac{91\,t^5}{120} - \frac{247\,t^7}{720} + \frac{1235\,t^9}{10368} - \frac{7657\,t^{11}}{228096}$$

5. $\begin{cases} y'' - 2t\,y' + 2y = 0, \\ t_0 = 0,\, Y_0 = 0,\, Y_1 = 1 \end{cases}$

Solution. Using

```
ODE[{y'' - 2t y' +2 y == 0,y[0] == 0,y'[0] == 1},y,t,
Method->SeriesForm,Degree->11,ExpansionPoint->0,
Form->Explicit]
```

we obtain

$$t$$

6. $\begin{cases} y'' + 9t^2 y = 0, \\ t_0 = 0,\, Y_0 = 1,\, Y_1 = 3 \end{cases}$

Solution. Using

```
ODE[{y'' +9t^2 y == 0,y[0] == 1,y'[0] == 3},y,t,
Method->SeriesForm,Degree->11,ExpansionPoint->0,
Form->Explicit]
```

we obtain

$$1 + 3\,t - \frac{3\,t^4}{4} - \frac{27\,t^5}{20} + \frac{27\,t^8}{224} + \frac{27\,t^9}{160}$$

7. $\begin{cases} y'' + t\,y' + 4y = 0, \\ t_0 = 0,\, Y_0 = 3,\, Y_1 = 6 \end{cases}$

Solution. Using

```
ODE[{y'' + t y' + 4 y == 0,y[0] == 3,y'[0] == 6},y,t,
Method->SeriesForm,Degree->8,ExpansionPoint->0,
Form->Explicit]
```

we obtain

$$3 + 6\,t - 6\,t^2 - 5\,t^3 + 3\,t^4 + \frac{7\,t^5}{4} - \frac{4\,t^6}{5} - \frac{3\,t^7}{8} + \frac{t^8}{7}$$

8. $\begin{cases} y'' - 3y' + 6y = 0, \\ t_0 = 0,\ Y_0 = 5,\ Y_1 = 3 \end{cases}$

Solution. Using

```
ODE[{y'' -3 y' + 6 y == 0,y[0] == 5,y'[0] == 3},y,t,
Method->SeriesForm,Degree->11,ExpansionPoint->0,
Form->Explicit]
```

we obtain

$$5 + 3\,t - \frac{21\,t^2}{2} - \frac{27\,t^3}{2} - \frac{39\,t^4}{8} + \frac{9\,t^5}{8} + \frac{123\,t^6}{80} +$$

$$\frac{279\,t^7}{560} + \frac{99\,t^8}{4480} - \frac{153\,t^9}{4480} - \frac{3\,t^{10}}{256} - \frac{657\,t^{11}}{492800}$$

9. $\begin{cases} y'' + 5t\,y' + 6t^2 y = 0, \\ t_0 = 0,\ Y_0 = 1,\ Y_1 = 0 \end{cases}$

Solution. Using

```
ODE[{y'' +5t y' + 6t^2 y == 0,y[0] == 1,y'[0] == 0},y,t,
Method->SeriesForm,Degree->11,ExpansionPoint->0,
Form->Explicit]
```

we obtain

$$1 - \frac{t^4}{2} + \frac{t^6}{3} - \frac{t^8}{8} + \frac{t^{10}}{30}$$

10. $\begin{cases} y'' - 3y' + 6y = 0, \\ t_0 = 0,\ Y_0 = 5,\ Y_1 = 3 \end{cases}$

Solution. Using

```
ODE[{y'' -3 y' + 6 y == 0,y[0] == 5,y'[0] == 3},y,t,
Method->SeriesForm,Degree->11,ExpansionPoint->0,
Form->Explicit]
```

we obtain

$$5 + 3t - \frac{21 t^2}{2} - \frac{27 t^3}{2} - \frac{39 t^4}{8} + \frac{9 t^5}{8} + \frac{123 t^6}{80} +$$

$$\frac{279 t^7}{560} + \frac{99 t^8}{4480} - \frac{153 t^9}{4480} - \frac{3 t^{10}}{256} - \frac{657 t^{11}}{492800}$$

11. $\begin{cases} y'' + 5t\,y' + 6t^2 y = 0, \\ t_0 = 0,\ Y_0 = 0,\ Y_1 = 1 \end{cases}$

Solution. Using

```
ODE[{y'' +5t y' + 6t^2 y == 0,y[0] == 0,y'[0] == 1},y,t,
Method->SeriesForm,Degree->11,ExpansionPoint->0,
Form->Explicit]
```

we obtain

$$t - \frac{5 t^3}{6} + \frac{13 t^5}{40} - \frac{25 t^7}{336} + \frac{157 t^9}{17280} + \frac{101 t^{11}}{295680}$$

12. $\begin{cases} y'' - 2t\,y' + 5y = 0, \\ t_0 = 0,\ Y_0 = 1,\ Y_1 = 0 \end{cases}$

Solution. Using

```
ODE[{y'' -2t y' + 5 y == 0,y[0] == 1,y'[0] == 0},y,t,
Method->SeriesForm,Degree->11,ExpansionPoint->0,
Form->Explicit]
```

we obtain

$$1 - \frac{5 t^2}{2} + \frac{5 t^4}{24} + \frac{t^6}{48} + \frac{t^8}{384} + \frac{11 t^{10}}{34560}$$

13. $\begin{cases} y'' - 2t\,y' + 5y = 0, \\ t_0 = 0,\ Y_0 = 0,\ Y_1 = 1 \end{cases}$

Solution. Using

```
ODE[{y'' -2t y' + 5 y == 0,y[0] == 0,y'[0] == 1},y,t,
Method->SeriesForm,Degree->11,ExpansionPoint->0,
Form->Explicit]
```

we obtain

$$t - \frac{t^3}{2} - \frac{t^5}{40} - \frac{t^7}{336} - \frac{t^9}{2688} - \frac{13\,t^{11}}{295680}$$

14. $\begin{cases} y'' - t\,y' = 0, t_0 = 1, \\ Y_0 = 1, Y_1 = 0 \end{cases}$

Solution. Using

```
ODE[{y'' -t y'  == 0,y[1] == 1,y'[1] == 0},y,t,
Method->SeriesForm,Degree->11,ExpansionPoint->1,
Form->Explicit]
```

we obtain

1

15. $\begin{cases} y'' + (t+1)y' + 4y = 0, \\ t_0 = 0, Y_0 = 0, Y_1 = 1 \end{cases}$

Solution. Using

```
ODE[{y'' +(t+1) y' +4y == 0,y[0] == 0,y'[0] == 1},y,t,
Method->SeriesForm,Degree->8,ExpansionPoint->0,
Form->Explicit]
```

we obtain

$$t - \frac{t^2}{2} - \frac{2\,t^3}{3} + \frac{5\,t^4}{12} + \frac{3\,t^5}{20} - \frac{49\,t^6}{360} - \frac{4\,t^7}{315} + \frac{29\,t^8}{1120}$$

16. $\begin{cases} y'' + (t+1)y' + 4y = 0, \\ t_0 = -1, Y_0 = 0, Y_1 = 1 \end{cases}$

Solution. Using

```
ODE[{y'' +(t+1) y' +4y == 0,y[-1] == 0,y'[-1] == 1},y,t
Method->SeriesForm,Degree->11,ExpansionPoint->-1,
Form->Explicit]
```

we obtain

0

Use **ODE** to compute the exact solution and the series solution to each of the following equations, and check that they are equivalent.

17. $y' = y$

Solution. The commands

```
ODE[y' == y,y,t,Method->SeriesForm,Form->Explicit]

Collect[Series[ODE[y' == y,y,t,Method->Separable,
Form->Explicit],{t,0,5}],C[1]]
```

both yield

$$(1 + t + \frac{t^2}{2} + \frac{t^3}{6} + \frac{t^4}{24} + \frac{t^5}{120})\ C[1]$$

18. $(1+t^2)y' - y = 0$

Solution. The commands

```
ODE[(1 + t^2) y' - y == 0,y,t,Method->SeriesForm,
Form->Explicit]

Collect[Series[ODE[(1 + t^2)y' - y == 0,y,t,
Method->Separable,
Form->Explicit],{t,0,5}],C[1]]
```

both yield

$$(1 + t + \frac{t^2}{2} - \frac{t^3}{6} - \frac{7 t^4}{24} + \frac{t^5}{24})\ C[1]$$

19. $t^2 y'' + t\, y' - y = 0$

Solution. The commands

```
Normal[Series[
ODE[t^2 y'' + t y' - y == 0,y,t,Method->CauchyEuler,
Form->Explicit],{t,0,5}]]
```

```
ODE[t^2 y'' + t y' - y == 0,y,t,Method->SeriesForm,
Form->Explicit]
```

are both equivalent to

$$\frac{C[1]}{t} + t\ C[2]$$

20. $t^2 y'' + y' + (t^2 - 1/4)y$

Solution. The commands

```
ODE[t^2 y'' + t y' + (t^2 - 1/4)y == 0,y,t,
Method->SeriesForm,Form->Explicit]

Collect[Series[ODE[t^2 y'' + t y' + (t^2 - 1/4)y == 0,
y,t,Method->TableLookup,Form->Explicit],
{t,0,5}]/Sqrt[2/Pi],{C[1],C[2]}]
```

are both equivalent to

$$(\text{Sqrt}[t] - \frac{t^{5/2}}{6} + \frac{t^{9/2}}{120})\ C[1] + (\frac{1}{\text{Sqrt}[t]} - \frac{t^{3/2}}{2} + \frac{t^{7/2}}{24})\ C[2]$$

21. $y'' - t\,y' + 4y = 0$

Solution. The commands

```
ODE[y'' - t y' + 4 y == 0,y,t,Method->SeriesForm,
Form->Explicit]

Collect[Series[
ODE[y'' - t y' + 4 y == 0,y,t,Method->TableLookup,
Form->Explicit],{t,0,5}],{C[1],C[2]}]
```

are both equivalent to

$$(1 - 2 t^2 + \frac{t^4}{3})\ C[1] + (t - \frac{t^3}{2} + \frac{t^5}{40})\ C[2]$$

22. $(1 - t^2)y'' - 2t\,y' + 12y = 0$

Solution. The commands

```
ODE[(1 - t^2)y'' - 2t y' + 12 y == 0,y,t,
Method->SeriesForm,Form->Explicit]

Collect[Series[
ODE[(1 - t^2)y'' - 2t y' + 12 y == 0,y,t,
Method->TableLookup,
Form->Explicit],{t,0,5}],{C[1],C[2]}]
```

are both equivalent to

$$(1 - 6 t^2 + 3 t^4) \, C[1] + (t - \frac{5 t^3}{3}) \, C[2]$$

Chapter 21

Section 21.2, page 724

Find the general solution of each of the following Cauchy-Euler equations.

1. $t^2 y'' - 7t\, y' + 15y = 0$
    Solution. $y(t) = C_1 t^5 + C_2 t^3$

2. $t^2 y'' - 3t\, y' + 4y = 0$
    Solution. $y(t) = C_1 t^2 + C_2 t^2 \log(t)$

3. $t^2 y'' + t\, y' - 4y = 0$
    Solution. $y(t) = C_1 t^2 + C_2 t^{-2}$

4. $t^2 y'' + t\, y' + 4y = 0$
    Solution. $y(t) = C_1 \cos(2\log(t)) + C_2 \sin\cos(2\log(t))$

5. $(t+1)^2 y'' + (t+1)y' - 9y = 0$
    Solution. $y(t) = C_1(t+1)^3 + C_2(t+1)^{-3}$

6. $(t-2)^2 y'' + 5(t-2)y' + 8y = 0$
    Solution. $y(t) = \dfrac{C_1 \cos(2\log(t-2)) + C_1 \sin(2\log(t-2))}{(t-2)^2}$

Solve the following initial value problems.

7.
$$\begin{cases} t^2 y'' + t\, y' - 9y = 0, \\ y(1) = 1,\ y'(1) = -1 \end{cases}$$

Solution. $y(t) = \dfrac{2t^{-3} + t^3}{3}$

8.
$$\begin{cases} t^2 y'' + ty' - 2.25y = 0, \\ y(2) = 0,\ y'(2) = 3 \end{cases}$$

Solution. $y(t) \approx -5.66t^{-1.5} + 0.707t^{1.5}$

9.
$$\begin{cases} t^2 y'' - 9t\, y' + 25y = 0, \\ y(e) = 1,\ y'(e) = 0 \end{cases}$$

Solution. $y(t) = \dfrac{t^5 (6 - 5\log(t))}{e^5}$

10.
$$\begin{cases} t^2 y'' + 2t\, y' - 132y = 0, \\ y(1) = 0,\ y'(-1) = 2 \end{cases}$$

Solution. $y(t) = \dfrac{-2t^{-12} + 2t^{11}}{23}$

11.
$$\begin{cases} t^3 y''' - t^2 y'' - 2t\, y' + 6y = 0, \\ y(1) = y'(1) = 0,\ y''(1) = 1 \end{cases}$$

Solution. $y(t) = \dfrac{t^{-1} - 4t^2 + 3t^3}{12}$

12.
$$\begin{cases} t^3 y''' - 3t^2 y'' - 12t\, y' + 60y = 0, \\ y(-2) = 1,\ y(-2) = -1,\ y(-2) = 2 \end{cases}$$

Solution. $y(t) = -\dfrac{4t^{-3}}{7} - \dfrac{t^4}{28} - \dfrac{t^5}{64}$

13. Carry out the following steps to show that the Cauchy-Euler equation $t^2 y'' + t\, p\, y' + q\, y = 0$ can be transformed to an equation of constant-coefficients by the change of the independent variable. Let $z = \log(t)$.

a. Show that

$$\frac{dy}{dt} = \frac{1}{t}\frac{dy}{dz} \quad \text{and} \quad \frac{d^2y}{dt^2} = \frac{1}{t^2}\left(\frac{d^2y}{dz^2} - \frac{dy}{dz}\right).$$

Solution. Using the chain rule we get

$$\frac{dy}{dt} = \frac{dy}{dz}\frac{dz}{dt} = \frac{1}{t}\frac{dy}{dz} \quad \text{and} \quad \frac{d^2y}{dt^2} = \frac{d}{dt}\left(\frac{dy}{dt}\right) = \frac{1}{t^2}\left(\frac{d^2y}{dz^2} - \frac{dy}{dz}\right).$$

b. Show that the Cauchy-Euler equation becomes

$$\frac{d^2y}{dz^2} + (p-1)\frac{dy}{dz} + q\,y = 0.$$

Solution. We compute

$$t^2y'' + t\,p\,y' + q\,y = \frac{d^2y}{dz^2} - \frac{dy}{dz} + p\frac{dy}{dz} + q\,y = \frac{d^2y}{dz^2} + (p-1)\frac{dy}{dz} + q\,y.$$

14. The following problem develops a portion of the theory of undetermined coefficients for the inhomogeneous Cauchy-Euler equation. Suppose that p, q and r are constants.

   a. (Nonresonant case) Show that a particular solution of the nonhomogeneous Cauchy-Euler equation

$$t^2y'' + t\,p\,y' + q\,y = A\,t^r \qquad\qquad \text{(S.104)}$$

   can be found in the form $y_p(t) = B\,t^r$ if and only if r *is not* a root of the indicial equation $r(r-1) + pr + q = 0$.

Solution. Substituting $y = B\,t^r$ into (S.104), we get

$$B\left(r^2 + (p-1)r + q\right) = B\left(r(r-1) + pr + q\right) = A.$$

Hence we can solve for B if and only if r *is not* a root of the indicial equation.

   b. (First resonant case) Suppose that r *is* a root of the indicial equation $r(r-1) + pr + q = 0$. Show that a particular solution of (21.22) can be found in the form $y_p(t) = B\,t^r\log(t)$ if and only if the roots of the indicial equation are distinct, that is, $2r - 1 + p \neq 0$.

Solution. Substituting $y = B\,t^r\log(t)$ into (S.104), we get

$$B\left((r^2 + (p-1)r + q)\log(t) + 2r - 1 + p\right) = B\,(2r - 1 + p) = A.$$

Hence we can solve for B if and only if $2r - 1 + p \neq 0$.

c. (Second resonant case) Suppose that r is a root of the indicial equation $r(r-1)+pr+q=0$. Show that a particular solution of (21.22) can be found in the form $y_p(t) = B\,t^r((\log(t))^2)$ if and only if the roots of the indicial equation are coincident, that is, $2r-1+p=0$.

Solution. Substituting $y = B\,t^r(\log(t))^2$ into (S.104), we get $2B = A$, which can be solved for B.

Find the general solution to the following nonhomogeneous Cauchy-Euler equations.

15. $t^2y'' - 5t\,y' + 5y = t^3$

Solution. $y(t) = C_1t + C_2t^5 - \dfrac{t^3}{4}$

16. $t^2y'' - 15t\,y' + 63y = t^7$

Solution. $y(t) = C_1t^7 + C_2t^9 - \dfrac{t^2\log(t)}{3}$

17. $t^2y'' - 9t\,y' + 25y = t^7$

Solution. $y(t) = C_1t^5 + C_2t^5\log(t) + \dfrac{t^7}{4}$

18. $t^2y'' + 11t\,y' + 25y = t^{-5}$

Solution. $y(t) = C_1t^{-5} + C_2t^{-5}\log(t) + \dfrac{t^{-5}(\log(t))^2}{2}$

Section 21.3, page 730

Find the indicial equation and the Frobenius solution of each of the following equations, each of which has a regular singular point at $t = 0$.

1. $t^2y'' + t\,y' - t^2y = 0$

Solution. $r^2 = 0$,

$$y(t) = y_0\left(1 + \frac{t^2}{4} + \frac{t^4}{64} + \frac{t^6}{2304} + \cdots\right)$$

2. $t^2y'' + t\,y' + t\,y = 0$

Solution. $r^2 = 0$,

$$y(t) = y_0\left(1 - t + \frac{t^2}{4} - \frac{t^3}{36} + \frac{t^4}{576} + \cdots\right)$$

3. $t^2 y'' - 7t\, y' + (15 + t^2) y = 0$

Solution. $r^2 - 8r + 15 = 0$,

$$y(t) = y_0 t^5 \left(1 - \frac{t^2}{8} + \frac{t^4}{192} - \frac{t^6}{9216} + \cdots \right)$$

4. $t^2 y'' - 3t\, y' + (4 + 3t^2) y = 0$

Solution. $r^2 - 4r + 4 = 0$,

$$y(t) = y_0 t^2 \left(1 - \frac{3}{4} t^2 + \frac{9}{64} t^4 - \cdots \right)$$

5. $t^2 y'' + t\, y' - 4y = 0$

Solution. $r^2 - 4 = 0$,

$$y(t) = y_0 t^2$$

6. $4t^2 y'' + (1 + t^2) y = 0$

Solution. $r^2 - r + 1/4 = 0$,

$$y(t) = y_0 t^{1/2} \left(1 - \frac{t^2}{16} + \frac{t^4}{1024} - \frac{t^6}{147456} + \cdots \right)$$

7. $t^2 y'' + t\, y' + 3t^2 y = 0$

Solution. $r^2 = 0$,

$$y(t) = y_0 \left(1 - \frac{3}{4} t^2 + \frac{9}{64} t^4 - \frac{3}{256} t^6 + \cdots \right)$$

8. $2t^2 y'' + t\, y' + t^2 y = 0$

Solution. $2r^2 - r = 0$,

$$y(t) = y_0 t^{1/2} \left(1 - \frac{t^2}{6} + \frac{t^4}{168} - \frac{t^6}{11088} + \cdots \right)$$

9. $3t^2 y'' + 2t\, y' + t^2 y = 0$

Solution. $3r^2 - r = 0$,

$$y(t) = y_0 t^{1/3} \left(1 - \frac{t^2}{10} + \frac{t^4}{440} - \cdots \right)$$

10. $t^2 y'' + (t + 3t^2) y = 0$

Solution. $r^2 - r = 0$,

$$y(t) = y_0 t \left(1 - \frac{t}{2} - \frac{5t^2}{12} + \frac{23t^3}{144} + \cdots \right)$$

11. $t y'' + 2t y' + 6e^t y = 0$

Solution. $r^2 - r = 0$,

$$y(t) = y_0 t \left(1 - 4t + \frac{17}{3} t^2 - \frac{47}{12} t^3 + \cdots \right)$$

12. $t^2 y'' - t(2+t)y' + (2+t^2)y = 0$

Solution. $r^2 - 3r + 2 = 0$,

$$y(t) = y_0 t^2 \left(1 + t + \frac{t^2}{3} + \frac{t^3}{36} + \cdots \right)$$

13. $t(t-1)y'' + 6t^2 y' + 3y = 0$

Solution. $r^2 - r = 0$,

$$y(t) = y_0 t \left(1 + \frac{3}{2} t + \frac{9}{4} t^2 + \cdots \right)$$

14. $t y'' + 4y' - t y = 0$

Solution. $r^2 - r = 0$,

$$y(t) = y_0 \left(1 + \frac{t^2}{10} + \frac{t^4}{280} + \cdots \right)$$

Section 21.4, page 733

Establish the following identities:

1. $\frac{d}{dt} J_0(t) = -J_1(t)$

Solution. This is the special case $\lambda = 0$ of Exercise 4, solved below.

2. $\frac{d}{dt} t J_1(t) = t J_0(t)$

Solution. This is the special case $\lambda = 1$ of Exercise 3, solved below.

3. $\frac{d}{dt} t^\lambda J_\lambda(t) = t^\lambda J_{\lambda-1}(t)$

Solution. From the power series definition, we have

$$t^\lambda J_\lambda(t) = \sum_{n=0}^{\infty} \frac{(-1)^n t^{2n+2\lambda}}{2^{2n+\lambda} n!(\lambda+n)!}.$$

We differentiate term-by-term to obtain

$$
\begin{aligned}
\frac{d}{dt} t^\lambda J_\lambda(t) &= \sum_{n=0}^{\infty} \frac{(-1)^n (2n+2\lambda) t^{2n+2\lambda-1}}{2^{2n+\lambda} n!(\lambda+n)!} \\
&= t^\lambda \sum_{n=0}^{\infty} \frac{(-1)^n (2n+2\lambda) t^{2n+(\lambda-1)}}{2^{2n+\lambda-1} n!(n+\lambda-1)!} \\
&= t^\lambda J_{\lambda-1}(t).
\end{aligned}
$$

4. $\dfrac{d}{dt} t^{-\lambda} J_\lambda(t) = -t^{-\lambda} J_{\lambda+1}(t)$

Solution. From the power series definition, we have

$$t^{-\lambda} J_\lambda(t) = \sum_{n=0}^{\infty} \frac{(-1)^n t^{2n}}{2^{2n+\lambda} n!(\lambda+n)!}.$$

When we differentiate term-by-term and let $n = m+1$, we obtain

$$
\begin{aligned}
\frac{d}{dt} t^{-\lambda} J_\lambda(t) &= \sum_{n=0}^{\infty} \frac{(-1)^n (2n) t^{2n-1}}{2^{2n+\lambda} n!(\lambda+n)!} \\
&= \sum_{m=0}^{\infty} \frac{(-1)^{m+1} (2m) t^{2m+(\lambda-1)}}{2^{2m+\lambda} m!(m+\lambda+1)!} \\
&= -t^{-\lambda} J_{\lambda+1}(t).
\end{aligned}
$$

5. $J_{1/2}(t) = \left(\dfrac{2}{\pi t}\right)^{1/2} \sin(t)$

Solution. From the power series definition, we have

$$J_{1/2}(t) = \sum_{n=0}^{\infty} \frac{(-1)^n t^{2n+1/2}}{2^{2n+1/2} n!(n+1/2)!}.$$

Recalling the identity $(n+1/2)! = (2n+1)!\sqrt{\pi}/n! 2^{2n+1}$, we obtain

$$J_{1/2}(t) = \sum_{n=0}^{\infty} \frac{(-1)^n t^{2n+1/2} n! 2^{2n+1}}{2^{2n+1/2} n!(2n+1)!\sqrt{\pi}} = \sqrt{\frac{2}{\pi t}} \sum_{n=0}^{\infty} \frac{(-1)^n t^{2n+1}}{(2n+1)!} = \sqrt{\frac{2}{\pi t}} \sin(t)$$

6. $J_{-1/2}(t) = \left(\dfrac{2}{\pi t}\right)^{1/2} \cos(t)$

Solution. Applying Exercise 3 with $\lambda = 1/2$ to the result of Exercise 5, we have

$$t^{1/2} J_{-1/2}(t) = \frac{d}{dt} t^{1/2} J_{1/2}(t) = \sqrt{\frac{2}{\pi}} \frac{d}{dt} \sin(t) = \sqrt{\frac{2}{\pi}} \cos(t)$$

7. $J_{\lambda-1}(t) + J_{\lambda+1}(t) = \dfrac{2\lambda}{t} J_\lambda(t)$

Solution. Referring to the result of Exercise 9 and Exercise 10, we have

$$t J_{\lambda-1} = t J'_\lambda + \lambda J_\lambda, \qquad -t J_{\lambda+1} = t J'_\lambda - \lambda J_\lambda$$

Subtracting the second from the first produces the required result.

8. $J_{\lambda-1}(t) - J_{\lambda+1}(t) = 2\dfrac{d}{dt} J_\lambda(t)$

Solution. Referring to the result of Exercise 9 and Exercise 10, we have

$$t J_{\lambda-1} = t J'_\lambda + \lambda J_\lambda, \qquad -t J_{\lambda+1} = t J'_\lambda - \lambda J_\lambda$$

Adding the second to the first produces the required result.

9. $t\dfrac{d}{dt} J_\lambda(t) + \lambda\, J_\lambda(t) = t\, J_{\lambda-1}(t)$

Solution. If we differentiate the product in Exercise 3 , we obtain

$$
\begin{aligned}
t^\lambda J_{\lambda-1} &= t^\lambda J'_\lambda + \lambda t^{\lambda-1} J_\lambda, \\
J_{\lambda-1} &= J'_\lambda + t^{\lambda-1} J_\lambda, \\
t J_{\lambda-1} &= t J'_\lambda + \lambda J_\lambda.
\end{aligned}
$$

10. $t\dfrac{d}{dt} J_\lambda(t) - \lambda\, J_\lambda(t) = -t\, J_{\lambda+1}(t)$

Solution. If we differentiate the product in Exercise 4 , we obtain

$$
\begin{aligned}
-t^{-\lambda} J_{\lambda+1} &= t^{-\lambda} J'_\lambda + \lambda t^{-\lambda-1} J_\lambda, \\
-J_{\lambda+1} &= J'_\lambda - (\lambda/t) J_\lambda, \\
-t J_{\lambda+1} &= t J'_\lambda - \lambda J_\lambda.
\end{aligned}
$$

11. $\exp\left(\dfrac{x}{2}\left(t - \dfrac{1}{t}\right)\right) = J_0(x) + \displaystyle\sum_{n=1}^{\infty} J_n(x)\left(t^n + (-t)^{-n}\right)$

Solution. Expanding the left side by the exponential power series and using the binomial theorem, we have

$$
e^{\frac{x}{2}(t-\frac{1}{t})} = \sum_{n=0}^{\infty} \frac{1}{n!} \left(\frac{x}{2}\right)^n \left(t - \frac{1}{t}\right)^n
$$

$$
= \sum_{n=0}^{\infty} \frac{1}{n!} \left(\frac{x}{2}\right)^n \sum_{j=0}^{n} \binom{n}{j} t^j \left(\frac{-1}{t}\right)^{n-j}
$$

$$
= \sum_{j=0}^{\infty} \sum_{n=j}^{\infty} \frac{x^n t^{2j-n}(-1)^{n-j}}{2^n j!(n-j)!}
$$

$$
= \sum_{j=0}^{\infty} \sum_{k=0}^{\infty} \frac{x^{j+k} t^{j-k}(-1)^k}{2^{j+k} j!k!}.
$$

The contribution of the terms with $j - k = 0$ is

$$
\sum_{k=0}^{\infty} \frac{x^{2k}(-1)^k}{2^{2k} k!k!} = J_0(x).
$$

The contribution of the terms with $j - k = m > 0$ is

$$
t^m \sum_{k=0}^{\infty} \frac{x^{m+2k}(-1)^k}{2^{m+2k}(m+k)!k!} = t^m J_m(x)
$$

The contribution of the terms with $k - j = m > 0$ is

$$
t^{-m} \sum_{j=0}^{\infty} \frac{x^{m+2j}(-1)^{m+j}}{2^{m+2j}(m+j)!j!} = (-1)^m t^{-m} J_m(x).
$$

Adding these three contributions gives the result.

12. Let λ be any complex number. Check that $J_\lambda(t)$ defined by (21.38) satisfies the Bessel equation $t^2 y'' + t y' + (t^2 - \lambda^2)y = 0$.

Solution.

We have

$$
y' = J_\lambda'(t) = \sum_{n=0}^{\infty} \frac{(-1)^n (2n+\lambda) t^{2n+\lambda-1}}{2^{2n+\lambda} n!(\lambda+n)!},
$$

$$
y'' = J_\lambda''(t) = \sum_{n=0}^{\infty} \frac{(-1)^n (2n+\lambda)(2n+\lambda-1) t^{2n+\lambda-2}}{2^{2n+\lambda} n!(\lambda+n)!},
$$

from which

$$t^2y'' + ty' - \lambda^2 y = \sum_{n=0}^{\infty} \frac{(-1)^n[(2n+\lambda)^2 - \lambda^2]t^{2n+\lambda}}{2^{2n+\lambda}n!(\lambda+n)!}$$

$$= \sum_{n=0}^{\infty} \frac{(-1)^n[4n(n+\lambda)]t^{2n+\lambda}}{2^{2n+\lambda}n!(\lambda+n)!}$$

$$= -t^2 \sum_{n=1}^{\infty} \frac{(-1)^n t^{2(n-1)+\lambda}}{2^{2n+\lambda-2}(n-1)!(\lambda+n-1)!}$$

$$= -t^2 y.$$

Find the general solutions of the following Bessel equations:

13. $t^2y'' + ty' + \left(t^2 - \dfrac{1}{9}\right)y = 0$

Solution. $y = C_1 J_{1/3}(t) + C_2 J_{-1/3}(t)$

14. $t^2y'' + ty' + \left(t^2 - 5\right)y = 0$

Solution. $y = C_1 J_{\sqrt{5}}(t) + C_2 J_{-\sqrt{5}}(t)$

15. $t^2y'' + ty' + \left(4t^2 - \dfrac{1}{25}\right)y = 0$

Solution. $y = C_1 J_{1/5}(2t) + C_2 J_{-1/5}(2t)$

16. $t^2y'' + ty' - \left(7t^2 + \dfrac{1}{49}\right)y = 0$

Solution. $y = C_1 J_{1/7}(\sqrt{7}\,t) + C_2 J_{-1/7}(\sqrt{7}\,t)$

17. **Show that if** $y(t)$ is any solution of the Bessel equation $t^2y'' + t\,y' + (t^2 - \lambda^2)y = 0$, then $w(z) = z^{-c}y(a z^b)$ is a solution of

$$z^2w'' + (2c+1)z\,w' + \left(a^2b^2z^{2b} + c^2 - \lambda^2b^2\right)w = 0,$$

where a, b and c are arbitrary complex numbers.

Solution. Computing the derivatives directly, we have

$$
\begin{aligned}
w(z) &= z^{-c}y(az^b) \\
w'(z) &= -cz^{-c-1}y(az^b) + abz^{b-1-c}y'(az^b) \\
w''(z) &= c(c+1)z^{-c-2}y(az^b) - abcz^{b-c-2}y'(az^b) \\
&\quad + ab(b-c-1)z^{b-c-2}y'(az^b) + a^2b^2z^{2b-c-2}y''(az^b).
\end{aligned}
$$

From this it follows that

$$z^2 w'' + (2c+1)w' = -c^2 z^{-c} y(az^b) + ab^2 z^{b-c} y'(az^b) + a^2 b^2 z^{2b-c} y''(az^b).$$

But the Bessel equation can be re-written as

$$a^2 z^{2b} y''(az^b) + az^b y'(az^b) + (a^2 z^{2b} - \lambda^2) y(az^b) = 0$$

which can be used to re-write the last two terms, leading to the equation

$$z^2 w''(z) + (2c+1)zw'(z) + \left(c^2 + b^2(a^2 z^{2b} - \lambda^2)\right) w(z) = 0$$

which was to be proved.

18. Use Problem 17 to show that any solution of the Airy equation $y'' + t\, y = 0$ can be written as

$$y(t) = t^{1/2} \left(C_1 J_{1/3} \left(\frac{2t^{3/2}}{3} \right) + C_2 J_{-1/3} \left(\frac{2t^{3/2}}{3} \right) \right).$$

Solution. We choose the values $a = 2/3$, $b = 3/2$, $c = -1/2$, $\lambda = 1/3$ and substitute in Problem 17 to obtain the equation $z^2 w'' + z^3 w = 0$, which is the stated Airy equation with independent variable z.

Section 21.5, page 740

1. Use the method of reduction of order to obtain a second linearly independent solution to each of the differential equations in Exercises 1–14 of Section 21.3.

Solution.

1. $r^2 = 0$,

$$y(t) = y_1 \left(-\frac{t^2}{4} - \frac{3}{128} t^4 - \frac{11}{13824} t^6 - \cdots + \left(1 + \frac{t^2}{4} + \frac{t^4}{64} + \frac{t^6}{2304} + \cdots \right) \log(t) \right)$$

2. $r^2 = 0$,

$$y(t) = y_1 \left(2t - \frac{3}{4} t^2 + \frac{11}{108} t^3 - \cdots + \left(1 - t + \frac{t^2}{4} - \frac{t^3}{36} + \cdots \right) \log(t) \right)$$

3. $r^2 - 8r + 15 = 0$,

$$y(t) = y_1 \left(-\frac{t^3}{2} + \frac{t^5}{16} + \frac{t^7}{64} + \cdots + \left(\frac{t^5}{4} - \frac{t^7}{32} + \frac{t^9}{768} \cdots \right) \log(t) \right)$$

4. $r^2 - 4r + 4 = 0$,

$$y(t) = y_1 \left(\frac{3}{4} t^4 - \frac{27}{128} t^6 + \frac{11}{512} t^8 + \cdots + \left(t^2 - \frac{3}{4} t^4 + \frac{9}{64} t^6 \cdots \right) \log(t) \right)$$

5. $r^2 - 4 = 0$,

$$y(t) = y_1 \left(\frac{1}{t^2} \right)$$

6. $r^2 - r + 1/4 = 0$,

$$y(t) = y_1 \left(\frac{t^{5/2}}{16} - \frac{3}{2048} t^{9/2} + \frac{11}{884736} t^{13/2} + \cdots + \left(t^{1/2} - \frac{t^{5/2}}{16} + \frac{t^{9/2}}{1024} + \cdots \right) \log(t) \right)$$

7. $r^2 = 0$,

$$y(t) = y_1 \left(\frac{3}{4} t^2 - \frac{27}{128} t^4 + \frac{11}{512} t^6 - \cdots + \left(1 - \frac{3}{4} t^2 + \frac{9}{64} t^4 + \cdots \right) \log(t) \right)$$

8. $2r^2 - r = 0$,

$$y(t) = y_1 t^{1/2} \left(1 - \frac{t^2}{10} + \frac{t^4}{360} + \cdots \right)$$

9. $3r^2 - r = 0$,

$$y(t) = y_1 t^{1/3} \left(1 - \frac{t^2}{14} + \frac{t^4}{728} + \cdots \right)$$

10. $r^2 - r = 0$,

$$y(t) = y_1 \left(-1 + \frac{t}{2} + 2t^2 - \frac{17}{72} t^3 - \cdots + \left(t - \frac{t^2}{2} - \frac{5}{12} t^3 + \frac{23}{144} t^4 + \cdots \right) \log(t) \right)$$

11. $r^2 - r = 0$,

$$y(t) = y_1 \left(-1 + 4t + 17t^2 + \cdots + \left(6t - 24t^2 + 34t^3 - \frac{47}{2} t^4 + \cdots \right) \log(t) \right)$$

12. $r^2 - 3r + 2 = 0$,

$$y(t) = y_1 \left(-t - t^2 + \frac{t^3}{2} + \cdots + \left(-t^2 - t^3 - \frac{t^4}{3} + \cdots \right) \log(t) \right)$$

13. $r^2 - r = 0$,

$$y(t) = y_1 \left(-1 - \frac{3}{2} t + 3t^2 + \cdots + \left(-3t - \frac{9}{2} t^2 - \frac{27}{4} t^3 + \cdots \right) \log(t) \right)$$

14. $r^2 + 3r = 0$,

$$y(t) = -\frac{1}{3t^3} + \frac{1}{6t} + \frac{t}{24} + \frac{t^3}{432}$$

2. Suppose that the roots of the indicial equation are $r = 0$ and $r = -1$. Show that the second linearly independent solution of (21.23) contains a logarithmic term if and

only if the coefficients in the equation have the property that $q_1 - p_1 \neq 0$.

Solution. If the roots of the indicial equation are $r = 0$ and $r = -1$, then we must have $r(r-1) + p_0 r + q_0 = r(r+1)$, which leads to $p_0 = 2$ and $q_0 = 0$. Also, the solution corresponding to $r = 0$ is a power series, thus, we seek information about the solution corresponding to $r = -1$. From the equation

$$(2r + p_0)Z_1 + (r\, p_1 + q_1)Z_0 = (q_1 - p_1)Z_0 = 0$$

we see that $Z_0 = 0$ if and only if $q_1 - p_1 \neq 0$. (In addition Z_1 can be chosen arbitrarily.) Therefore, the solution obtained under these conditions becomes

$$y_2 = t^{-1} \sum_{n=1}^{\infty} \frac{Z_n}{n!} t^n = Cy_1,$$

for some constant C. Therefore, a second linearly independent solution must contain the logarithmic term.

3. Apply the result of Exercise 2 to the equation $y'' + (2/t)y' + y = 0$. Find the power series solution for the first solution $Y_1(t)$, and then show that the second solution $Y_2(t)$ is also expressed in terms of a power series, with no logarithmic term.

Solution. The indicial equation is $r^2 + r = 0$ yielding roots $r = 0$ and $r = -1$. Also, since $p_1 = q_1 = 0$ and $p_1 - q_1 = 0$. From Exercise 2 there will be no logarithmic term. In fact, the solution is

$$y(t) = y_0\left(1 - \frac{t^2}{6} + \frac{t^4}{120} - \cdots\right) + y_1\left(-\frac{1}{t} + \frac{t}{2} - \frac{t^3}{24} + \cdots\right),$$

or equivalently

$$y(t) = y_0\left(\frac{e^{it} + e^{-it}}{-2t}\right) + y_1\left(\frac{e^{it} - e^{-it}}{2t\,i}\right).$$

4. Show that the power series solution for $y_1(t)$ in Exercise 3 can be expressed as an elementary trigonometric function. Use this, together with the method of reduction of order, to express $y_2(t)$ as an elementary trigonometric function.

Solution. The power series solution is

$$y_1 = 1 - \frac{t^2}{6} + \frac{t^4}{120} - \cdots = \frac{\sin t}{t}.$$

To obtain the second solution by reduction-of-order, we recall that this requires that the ratio satisfy

$$(y_2/y_1)' = W/y_1^2 = (1/t^2)(t/\sin t)^2 = (1/\sin^2 t) = \csc^2 t$$

But the integral of $\csc^2 t$ is $-\cot t$, hence we have

$$y_2/y_1 = -\cot t + C, \quad y_2(t) = \frac{\sin t}{t}(-\cot t + C) = -\frac{\cos t}{t} + C\frac{\sin t}{t}$$

which is an elementary function.

5. Show how to modify the theory of the present section to find a particular solution of the inhomogeneous differential equation

$$t^2 y'' + t\,\widehat{p}(t)y' + \widehat{q}(t)y = t^r \widehat{r}(t),$$

where $\widehat{p}(t)$, $\widehat{q}(t)$, and $\widehat{r}(t)$ have convergent power series about $t = 0$, and r is larger than any root of the corresponding indicial equation.

Solution. We look for the solution in the form $y(t) = t^r \sum_{k=0}^{\infty} Z_k t^k / k!$. When we form the differentiated power series for $y'(t)$, $y''(t)$, we obtain new series, each of which contains the power t^r, which can therefore be cancelled from every term in the equation to be solved. When we equate the coefficients of the various powers of t, we find linear eqations to solve, where there repeatedly occurs the coefficient $r(r-1) + p_0 r + q_0$, which is non-zero by hypothesis. Therefore, all of these equations can be solved uniquely for the respective coefficients $Z_0, Z_1, \ldots$.

6. Suppose that we have an equation with a regular singular point at $t = 0$ written in the standard form $t^2 y'' + t\,\widehat{p}(t)y' + \widehat{q}(t)y = 0$. Suppose further that there are two solutions $y_1(t) = t^{r_1}$ and $y_2(t) = t^{r_2}$ with $r_1 \neq r_2$. Show that $\widehat{p}(t)$ and $\widehat{q}(t)$ are constants, that is, the equation is a Cauchy-Euler equation.

Solution. Assuming that t^r is a solution and substituting, we obtain

$$t^2 r(r-1)t^{r-2} + t\,\widehat{p}(t)rt^{r-1} + \widehat{q}(t)t^r = 0$$

Cancelling the factor t^r everywhere and applying this to $r = r_1, r = r_2$ results in the simultaneous equations

$$r_1(r_1 - 1) + r_1\widehat{p}(t) + \widehat{q}(t) = 0, \qquad r_2(r_1 - 1) + r_2\widehat{p}(t) + \widehat{q}(t) = 0$$

whose determinant is $r_1 - r_2 \neq 0$, by hypothesis. Therefore, the unique solution is obtained as $\widehat{p}(t) = 1 - (r_1 + r_2)$, $\widehat{q}(t) = r_1 r_2$ which are constants. It is immediately verified that

$$r_1(r_1 - 1) + r_1\widehat{p}(t) + \widehat{q}(t) = r_1(r_1 - 1) + r_1(1 - r_1 - r_2) + r_1 r_2 = 0$$

$$r_2(r_2 - 1) + r_2\widehat{p}(t) + \widehat{q}(t) = r_2(r_2 - 1) + r_2(1 - r_1 - r_2) + r_1 r_2 = 0$$

Section 21.6, page 743

1. The **modified Bessel equation** is

$$t^2 y'' + t y' - (t^2 + \lambda^2)y = 0, \qquad \text{where } \lambda \text{ is a constant.} \qquad \text{(S.105)}$$

The **modified Bessel function of the first kind of order** λ is the function $I_\lambda(t)$ given by

$$I_\lambda(t) = \sum_{n=0}^{\infty} \frac{t^{2n+\lambda}}{2^{2n+\lambda} n!(\lambda + n)!}. \qquad \text{(S.106)}$$

Show that $I_\lambda(t)$ is a solution of (S.105) and that $I_\lambda(t) = i^{-\lambda} J_\lambda(i\, t)$.

Solution. $I_\lambda(t) = i^{-\lambda} J_\lambda(i\, t)$ is obvious from (21.70). Then an easy calculation shows that $I_\lambda(t)$ is a solution of (21.69).

2. Suppose that v is not an integer. The **modified Bessel function of the second kind of order** v is the function $K_v(t)$ given by

$$K_v(t) = \left(\frac{\pi}{2}\right) \frac{I_{-v}(t) - I_v(t)}{\sin(v\,\pi)}.$$

If m is an integer, we define

$$K_m(t) = \left(\frac{\pi}{2}\right) \lim_{v \to m} \frac{I_{-v}(t) - I_v(t)}{\sin(v\,\pi)}.$$

Show that if m is an integer, then $I_{-m}(t) = I_m(t)$ and $K_{-m}(t) = K_m(t)$.

Solution. The proof that $I_{-m}(t) = I_m(t)$ is the same as the proof that $J_{-m}(t) = (-1)^m J_m(t)$. Then

$$K_{-m}(t) = \left(\frac{\pi}{2}\right) \lim_{v \to -m} \frac{I_{-v}(t) - I_v(t)}{\sin(v\,\pi)} = \left(\frac{\pi}{2}\right) \lim_{-v \to m} \frac{I_v(t) - I_{-v}(t)}{\sin(-v\,\pi)} =$$

$$\left(\frac{\pi}{2}\right) \lim_{v \to m} \frac{I_{-v}(t) - I_v(t)}{\sin(v\,\pi)}. = K_m(t).$$

3. Prove the following result.

Theorem . *Let λ be a constant. The general solution to the modified Bessel equation*

$$t^2 y'' + t y' - (t^2 + \lambda^2)y = 0$$

is given by

$$y(t) = C_1 I_\lambda(t) + C_2 K_\lambda(t),$$

where C_1 and C_2 are constants. If λ is not an integer, the general solution can also be written as

$$y(t) = C_3 I_\lambda(t) + C_4 I_{-\lambda}(t),$$

where C_3 and C_4 are constants.

Solution. This follows from Exercise 1 and the theorem at the end of Section 21.6.

4. Show that $I_{1/2}(t) = \left(\dfrac{2}{\pi t}\right)^{1/2} \sinh(t)$ and $J_{-1/2}(t) = \left(\dfrac{2}{\pi t}\right)^{1/2} \cosh(t)$.

Solution. This follows from Exercises 5 and 6 of Section 21.4.

Find the general solutions of the following differential equations:

5. $t^2 y'' + t y' + \left(t^2 - 25\right) y = 0$

Solution. $y(t) = C_1 J_5(t) + C_2 Y_5(t)$

6. $t^2 y'' + t y' + \left(t^2 + 25\right) y = 0$

Solution. $y(t) = C_1 I_{5i}(t) + C_2 I_{-5i}(t)$

7. $t^2 y'' + t y' - \left(t^2 + 25\right) y = 0$

Solution. $y(t) = C_1 I_5(t) + C_2 K_5(t)$

8. $t^2 y'' + t y' - \left(t^2 - 25\right) y = 0$

Solution. $y(t) = C_1 J_{5i}(i\,t) + C_2 J_{-5i}(i\,t)$

9. $t^2 y'' + t y' + \left(2t^2 - 3\right) y = 0$

Solution. $y(t) = C_1 J_{\sqrt{3}}(\sqrt{2}\,t) + C_2 J_{-\sqrt{3}}(\sqrt{2}\,t)$

10. $t^2 y'' + t y' + \left(2t^2 + 3\right) y = 0$

Solution. $y(t) = C_1 I_{i\sqrt{3}}(\sqrt{2}\,t) + C_2 I_{-i\sqrt{3}}(\sqrt{2}\,t)$

11. $t^2 y'' + t y' - \left(2t^2 + 3\right) y = 0$

Solution. $y(t) = C_1 I_{\sqrt{3}}(\sqrt{2}\,t) + C_2 I_{-\sqrt{3}}(\sqrt{2}\,t)$

12. $t^2 y'' + t y' - \left(2t^2 - 3\right) y = 0$

Solution. $y(t) = C_1 J_{i\sqrt{3}}(\sqrt{2}\,t) + C_2 J_{-i\sqrt{3}}(\sqrt{2}\,t)$

Section 21.7, page 748

Solve the following differential equations using **ODE**.

1. $t^2 y'' + t y' + (t^2 - 49)y = 0$.

Solution. The command

```
ODE[t^2y'' + t y' + (t^2 -49)y,y,t,Method->TableLookup]
```

yields

```
{{y -> BesselJ[7, t] C[1] + BesselY[7, t] C[2]}}
```

2. $t^2 y'' + t y' + (16t^2 - 2)y = 0$.

Solution. The command

```
ODE[t^2 y'' + t y' + (16t^2 - 2)y,y,t,
    Method->TableLookup]
```

yields

```
{{y -> BesselJ[Sqrt[2], 4 t] C[1] + BesselJ[-Sqrt[2], 4 t] C[2]}}
```

Section 21.8, page 752

1. Prove Lemma 21.16.

Solution.

It can be shown that J_μ is a bounded for any real μ and that Y_ν is bounded for any noninteger real ν. Furthermore, if n is an integer, the only singularity of $Y_n(s)$ is $s = 0$. Therefore, in order for $u(t)$ to be unbounded we must have that Q_2 is nonzero and $c/(m\eta)$ is an integer. In fact, (21.68) implies that $s^{-n} Y_n(s)$ is bounded for any positive integer n. Hence $c/(m\eta)$ is a nonpositive integer. Since c is nonnegative and m and η are positive, it follows that $c = 0$. Thus, (i) implies (ii). That (ii) implies (i) is a consequence of (21.68). ∎

2. Plot the solution to the following initial value problem which models an aging spring with friction:

$$\begin{cases} u'' + 0.08u' + e^{-0.15t}u = 0, \\ u(0) = 1, u'(0) = 0 \end{cases}$$

Solution. We define

```
AgingSpring[Q1_,Q2_,m_,k_,c_,eta_][t_]:=
((2/eta)Sqrt[(k/m)])^(c/(m eta))Exp[-c t/(2m)](
Q1 BesselJ[c/(m eta),(2/eta)Sqrt[(k/m)]Exp[-eta t/2]] +
Q2 BesselY[c/(m eta),(2/eta)Sqrt[(k/m)]Exp[-eta t/2]])
```

Then

```
Plot[Evaluate[AgingSpring[1,0,1,1,0.08,0.15][t]],
{t,0,60},PlotRange->All,DefaultFont->{"Courier",7}]
```

produces the plot

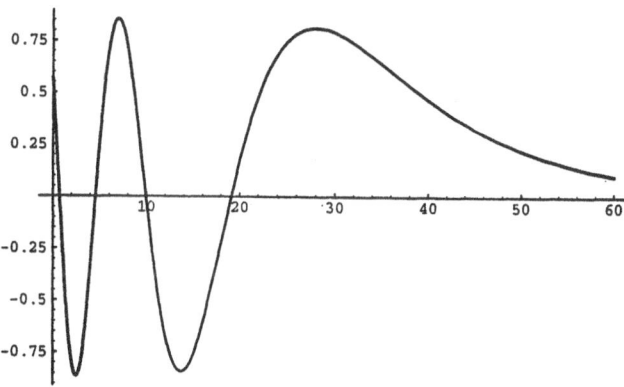

Section 21.9, page 757

1. Show that $F(a; b; c; t) = \sum_{n=0}^{\infty} \dfrac{\Gamma(a+n)\Gamma(b+n)\Gamma(c)t^n}{\Gamma(a)\Gamma(b)\Gamma(c+n)n!}$.

Solution. Since $(a)_n = \dfrac{\Gamma(a+n)}{\Gamma(a)}$, we have

$$F(a; b; c; t) = \sum_{n=0}^{\infty} \frac{(a)_n(b)_n t^n}{n!(c)_n} = \sum_{n=0}^{\infty} \frac{\Gamma(a+n)\Gamma(b+n)\Gamma(c)t^n}{\Gamma(a)\Gamma(b)\Gamma(c+n)n!}.$$

2. Show that if a or b is a negative integer, then $F(a; b; c; t)$ is a polynomial.

Solution. Without loss of generality, let a be a negative integer. Then $(a)_n = 0$ for all $n \geq -a$; hence,

$$\frac{(a)_n(b)_n t^n}{n!(c)_n} = 0$$

for $n \geq -a$. Therefore

$$F(a; b; c; t) = \sum_{n=0}^{\infty} \frac{(a)_n(b)_n t^n}{n!(c)_n} = \sum_{n=0}^{-(a+1)} \frac{\Gamma(a+n)\Gamma(b+n)\Gamma(c)t^n}{\Gamma(a)\Gamma(b)\Gamma(c+n)n!},$$

a polynomial of degree at most $-(a+1)$.

Establish the following identities both by hand and using *Mathematica*:

3. $\log(1+t) = t\, F(1; 1; 2; -t)$

Solution.

```
t Hypergeometric2F1[1,1,2,-t]
```

produces

```
Log[1 + t]
```

4. $\arcsin(t) = t\, F\left(\frac{1}{2}; \frac{1}{2}; \frac{3}{2}; t^2\right)$

Solution.

```
t Hypergeometric2F1[1/2,1/2,3/2,t^2]//PowerExpand
```

produces

```
ArcSin[t]
```

5. $\arctan(t) = t\, F\left(\frac{1}{2}; 1; \frac{3}{2}; -t^2\right)$

Solution.

```
t Hypergeometric2F1[1/2,1,3/2,-t^2]//PowerExpand
```

produces

```
ArcTan[t]
```

6. $e^t = \lim_{t \to \infty} F\left(a; b; b; \dfrac{t}{a}\right)$

Solution.

> `Limit[Hypergeometric2F1[a,b,b,t/a],a->Infinity]`

produces

> E^t

7. $\displaystyle\int_0^{\pi/2} \dfrac{d\phi}{\sqrt{1 - k(\sin(\phi))^2}} = F\left(\dfrac{1}{2}; \dfrac{1}{2}, 1, k\right)$

Solution.

> `(Pi/2)Hypergeometric2F1[1/2,1/2,1,k]`

and

> `Integrate[1/Sqrt[1 - k Sin[phi]^2],phi]`

both produce

> `EllipticK[k]`

8. $\displaystyle\int_0^{\pi/2} \sqrt{1 - k^2(\sin(\phi))^2}\, d\phi = F\left(\dfrac{1}{2}; -\dfrac{1}{2}, 1, k\right)$

Solution.

> `(Pi/2)Hypergeometric2F1[1/2,-1/2,1,k]`

and

> `Integrate[Sqrt[1 - k Sin[phi]^2],phi]`

both produce

> `EllipticE[k]`

Find the general solution to each of the following differential equations.

9. $t(1-t)y'' + \left(\dfrac{1}{2} - 2t\right)y' - \dfrac{y}{4} = 0$

Solution. $y(t) = C_1 \dfrac{1}{\sqrt{1-t}} + C_2 \dfrac{\arcsin(\sqrt{t})}{\sqrt{1-t}}$

10. $t(1-t)y'' + \left(\frac{1}{2} - t\right)y' + \frac{y}{4} = 0$

Solution. $y(t) = C_1\sqrt{1-t} + C_2\sqrt{t}$

11. $2t(1-t)y'' + (3 - 4t)y' + 4y = 0$

Solution. $y(t) = C_1\left(1 - \frac{4t}{3}\right) + C_2\left(\frac{\sqrt{1-t}}{\sqrt{t}} - 4\sqrt{t}\sqrt{1-t}\right)$

12. $t(1-t)y'' + (1 - 3t)y' - y = 0$

Solution. $y(t) = C_1\left(\frac{1}{1-t}\right) + C_2\left(\frac{\log(t)}{1-t}\right)$

Use **ODE** with the option **Method->SecondOrderLinear** to solve each of the following initial value problems:

13. $\begin{cases} y'' = J_0(t), \\ y(0) = y'(0) = 0 \end{cases}$

Solution.

```
ODE[{y'' == BesselJ[0,t],y[0] == 0,y'[0] == 0},
y,t,Method->SecondOrderLinear]
```

14. $\begin{cases} y'' + y = J_0(t), \\ y(0) = y'(0) = 0 \end{cases}$

Solution.

```
ODE[{y'' + y == BesselJ[0,t],y[0] == 0,y'[0] == 0},
y,t,Method->SecondOrderLinear]
```

15. Prove Theorem 21.18 using the following steps.

  a. Assume a Frobenius solution to (21.93) is of the form (21.82). Show that $a_0 r(r + c - 1) = 0$ and

$$a_{n+1} = \frac{(n + r + a)}{(n + r + 1)(n + r + c)} a_n. \qquad \text{(S.107)}$$

  b. If $r = 0$, show that (S.107) implies that

$$a_n = \frac{(a)_n}{n!(c)_n} a_0. \qquad \text{(S.108)}$$

Conclude that $_1F_1(a; c; t)$ is a solution of (21.93).

c. If $r = 1 - c$, show that (S.107) implies that $_1F_1(a + 1 - c; 2 - c; t)$ is a solution of (21.93).

d. If c is not an integer, show that $_1F_1(a; c; t)$ and $_1F_1(a + 1 - c; 2 - c; t)$ are linearly independent.

e. Conclude that $y(t) = C_1 \, _1F_1(a; c; t) + C_2 t^{1-c} \, _1F_1(1 + a - c; 2 - c; t)$ is the general solution of (21.93) in the case that c is not an integer.

Appendix A

Section A.3, page 771

Solve the following systems of linear equations by the method of row reduction. Be careful to distinguish the cases of no solutions, one solution and many solutions.

1. $\begin{cases} x_1 + 3x_2 = 4, \\ x_1 - x_2 = 6 \end{cases}$

Solution. $x_1 = 11/2, x_2 = -1/2$

2. $\begin{cases} x_1 + 4x_2 = 5, \\ 2x_1 + 8x_2 = 11 \end{cases}$

Solution. No solution

3. $\begin{cases} x_1 + x_2 + x_3 = 4, \\ x_1 - 2x_2 + x_3 = 0, \\ 2x_1 - x_2 + 3x_3 = 5 \end{cases}$

Solution. $x_1 = 5/3, x_2 = 4/3, x_3 = 1$

4. $\begin{cases} x_1 + 2x_2 - x_3 = 0, \\ x_1 - x_2 + 2x_3 = 0, \\ 2x_1 + x_2 + x_3 = 0 \end{cases}$

Solution. $x_1 = -s, x_2 = s, x_3 = s$, where s is any real number

5. $\begin{cases} -x_1 + x_2 + 2x_3 = 0, \\ x_1 - x_3 = 0, \\ 3x_1 + x_2 + 2x_3 = 0 \end{cases}$

Solution. $x_1 = 0, x_2 = 0, x_3 = 0$

6. $\begin{cases} x_1 - x_3 = 0, \\ 3x_1 + x_2 + x_3 = 0, \\ -x_1 + x_2 + 2x_3 = 0 \end{cases}$

Solution. $x_1 = 0, x_2 = 0, x_3 = 0$

$$
7. \quad
\begin{cases}
x_1 + x_2 + x_3 + x_4 = 4, \\
x_1 + 2x_2 + 3x_3 + 4x_4 = 3, \\
x_1 - x_2 + 3x_3 - 4x_4 = 0
\end{cases}
$$

Solution. $x_1 = 4 + 11s/6$, $x_2 = 1 - 8s/3$, $x_3 = -1 - s/6$, $x_4 = s$, where s is any real number

$$
8. \quad
\begin{cases}
5x_1 - 2x_2 + 9x_3 + 11x_4 = -29, \\
7x_1 - x_2 + 4x_3 + 5x_4 = -18, \\
6x_1 - 6x_2 + 2x_3 - 11x_4 = -28, \\
x_1 + x_2 + x_3 + x_4 = 0
\end{cases}
$$

Solution. $x_1 = -1$, $x_2 = 3$, $x_3 = -2$, $x_4 = 0$

For each of the following matrices, determine whether it is of Case I (invertible) or Case II (noninvertible). Compute the inverse if the matrix is invertible.

9. $\begin{pmatrix} 1 & 4 \\ 2 & -3 \end{pmatrix}$

Solution. $\begin{pmatrix} 3/11 & 4/11 \\ 2/11 & -1/11 \end{pmatrix}$

10. $\begin{pmatrix} 6 & 2 \\ 3 & -1 \end{pmatrix}$

Solution. $\begin{pmatrix} 1/12 & 1/6 \\ 1/4 & -1/2 \end{pmatrix}$

11. $\begin{pmatrix} 1 & -1 & -1 \\ 3 & -1 & 2 \\ 2 & 2 & 3 \end{pmatrix}$

Solution. $\begin{pmatrix} 7/10 & -1/10 & 3/10 \\ 1/2 & -1/2 & 1/2 \\ -4/5 & 2/5 & -1/5 \end{pmatrix}$

12. $\begin{pmatrix} 1 & 2 & 3 \\ 3 & 5 & 6 \\ 2 & 4 & 5 \end{pmatrix}$

Solution. $\begin{pmatrix} 1 & 2 & -3 \\ -3 & -1 & 3 \\ 2 & 0 & -1 \end{pmatrix}$

13. $\begin{pmatrix} 2 & 3 & 1 \\ -1 & 2 & 1 \\ 4 & -1 & -1 \end{pmatrix}$

Solution. noninvertible

14. $\begin{pmatrix} 0 & 3 & 0 & 0 \\ -4 & 2 & 0 & 0 \\ 0 & -1 & 1 & 0 \\ 0 & 0 & 3 & 1 \end{pmatrix}$

Solution. $\begin{pmatrix} 1/6 & -1/4 & 0 & 0 \\ 1/3 & 0 & 0 & 0 \\ 1/3 & 0 & 1 & 0 \\ -1 & 0 & -3 & 1 \end{pmatrix}$

15. Prove that the identity matrix commutes with all $n \times n$ matrices:
   $A\, I_{n \times n} = I_{n \times n}\, A = A$.

   Solution. Using the definition of matrix multiplication, we have

   $$A\, I_{n \times n} = \left[c_{ij} = \sum_{k=1}^{k=n} a_{ik} i_{kj} \right] = [a_{ij}] = A,$$

   and

   $$I_{n \times n}\, A = \left[c_{ij} = \sum_{k=1}^{k=n} i_{ik} a_{kj} \right] = [a_{ij}] = A.$$

   Therefore, $A\, I_{n \times n} = I_{n \times n}\, A$.

16. Prove that if B is an $n \times n$ matrix that commutes with *every* $n \times n$ matrix A, then
   $B = c\, I_{n \times n}$ for some constant c.

   Solution. If B is an $n \times n$ matrix and $A\,B = B\,A$ for every $n \times n$ matrix A, then

   $$A\,B = \left[\sum_{k=1}^{k=n} a_{ik} b_{kj} \right] = \left[\sum_{k=1}^{k=n} b_{ik} a_{kj} \right] = B\,A.$$

   The i, j entries are equal if and only if $b_{ij} = 0$ for all $i \neq j$ and $b_{ii} = b_{jj}$. Hence,
   $B = c\, I_{n \times n}$ where, for example, $c = b_{11}$.

17. Prove that any two diagonal matrices commute: $\Lambda_1 \Lambda_2 = \Lambda_2 \Lambda_1$.

Solution. If Λ_1 and Λ_2 are diagonal matrices, then

$$\Lambda_1 \Lambda_2 = \left[\sum_{k=1}^{k=n} \lambda_{1ik} \lambda_{2kj} \right] = [\lambda_{1ii} \lambda_{2ij}],$$

and

$$\Lambda_2 \Lambda_1 = \left[\sum_{k=1}^{k=n} \lambda_{2ik} \lambda_{1kj} \right] = [\lambda_{2ii} \lambda_{1ij}].$$

In each case, we get zero whenever $i \neq j$ and $\lambda_{1ii} \lambda_{2ii}$ when $i = j$. Hence, $\Lambda_1 \Lambda_2 = \Lambda_2 \Lambda_1$.

18. **Prove that if B is an $n \times n$ matrix that commutes with *every* diagonal matrix, then B is a diagonal matrix.**

Solution. If B is an $n \times n$ matrix and $D B = B D$ for every $n \times n$ diagonal matrix D, then the i, j entries are $d_i b_{ij}$ and $b_{ij} d_j$. These are equal if and only if $b_{ij} = 0$ for all $i \neq j$. Hence, B is a diagonal matrix.

19. **Prove that the result of interchanging two rows of a matrix can be effected by the left multiplication of the matrix by a certain matrix.**

Solution. Suppose that A is an $m \times n$ matrix. Let E be the matrix obtain from $I_{m \times m}$ by interchanging the ith row with the jth row. Then if we define $E_i = (0, \dots, 1, \dots, 0)$, where the 1 is in the ith position, then

$$E A = \begin{pmatrix} E_1 A^1 & E_1 A^2 & \dots & E_1 A^n \\ \vdots & \vdots & \ddots & \vdots \\ E_j A^1 & E_j A^2 & \dots & E_j A^n \\ \vdots & \vdots & \ddots & \vdots \\ E_i A^1 & E_i A^2 & \dots & E_i A^n \\ \vdots & \vdots & \ddots & \vdots \\ E_n A^1 & E_n A^2 & \dots & E_n A^n \end{pmatrix} = \begin{pmatrix} a_{11} & a_{12} & \dots & a_{1n} \\ \vdots & \vdots & \ddots & \vdots \\ a_{j1} & a_{j2} & \dots & a_{jn} \\ \vdots & \vdots & \ddots & \vdots \\ a_{i1} & a_{i2} & \dots & a_{in} \\ \vdots & \vdots & \ddots & \vdots \\ a_{n1} & a_{m2} & \dots & a_{nn} \end{pmatrix}.$$

This is the matrix obtain from A by interchanging the ith row of A with the jth row of A.

20. **Prove that the result of multiplying a row of a matrix by a fixed constant c can be effected by the left multiplication by a certain matrix.**

Solution. Suppose that A is an $m \times n$ matrix. Let E be the matrix obtain from $I_{m \times m}$ by multiplying the ith row by c. Then if we define $E_i = (0, \ldots, 1, \ldots, 0)$, where the 1 is in the ith position, then

$$E A = \begin{pmatrix} E_1 A^1 & E_1 A^2 & \cdots & E_1 A^n \\ \vdots & \vdots & \ddots & \vdots \\ c E_i A^1 & c E_i A^2 & \cdots & c E_i A^n \\ \vdots & \vdots & \ddots & \vdots \\ E_n A^1 & E_n A^2 & \cdots & E_n A^n \end{pmatrix} = \begin{pmatrix} a_{11} & a_{12} & \cdots & a_{1n} \\ \vdots & \vdots & \ddots & \vdots \\ c a_{i1} & c a_{i2} & \cdots & c a_{in} \\ \vdots & \vdots & \ddots & \vdots \\ a_{n1} & a_{m2} & \cdots & a_{nn} \end{pmatrix}.$$

This is the matrix obtain from A by multiplying the ith row of A by c.

21. Prove that the result of adding a multiple of one row to another row of a matrix can be effected by left multiplication by a certain matrix.

Solution. Suppose that A is an $m \times n$ matrix. Let E be the matrix obtain from $I_{m \times m}$ by adding c times the ith row to the jth row. Then if we define $E_i = (0, \ldots, 1, \ldots, 0)$, where the 1 is in the ith position, then

$$E A = \begin{pmatrix} E_1 A^1 & E_1 A^2 & \cdots & E_1 A^{n-} \\ \vdots & \vdots & \ddots & \vdots \\ (c E_i + E_j)A^1 & (c E_i + E_j)A^2 & \cdots & (c E_i + E_j)A^n \\ \vdots & \vdots & \ddots & \vdots \\ E_n A^1 & E_n A^2 & \cdots & E_n A^n \end{pmatrix}$$

$$= \begin{pmatrix} a_{11} & a_{12} & \cdots & a_{1n} \\ \vdots & \vdots & \ddots & \vdots \\ c a_{i1} + a_{j1} & c a_{i2} + a_{j2} & \cdots & c a_{in} + a_{jn} \\ \vdots & \vdots & \ddots & \vdots \\ a_{n1} & a_{m2} & \cdots & a_{nn} \end{pmatrix}.$$

This is the matrix obtain from A by adding c times the ith row of A to the jth row of A.

22. The **trace** of an $n \times n$ matrix

$$A = \begin{pmatrix} a_{11} & a_{12} & \cdots & a_{1n} \\ a_{21} & a_{22} & \cdots & a_{2n} \\ \vdots & \vdots & \ddots & \vdots \\ a_{n1} & a_{m2} & \cdots & a_{nn} \end{pmatrix}$$

is defined by $\text{tr}\,(A) = a_{11} + \cdots + a_{nn}$. Show that $\text{tr}\,(A\,B) = \text{tr}\,(B\,A)$ for square matrices A and B.

Solution. Suppose that $A\,B = C$, then

$$\text{tr}\,(A\,B) = \text{tr}\,(C) = \left[\sum_i c_{ii}\right] = \left[\sum_i \sum_k a_{ik}b_{ki}\right] = \left[\sum_k \sum_i b_{ik}a_{ki}\right] = \text{tr}\,(B\,A).$$

Section A.4, page 783

Find the eigenvalues and eigenvectors of each of the following 2×2 matrices.

1. $\begin{pmatrix} 3 & 3 \\ 1 & 5 \end{pmatrix}$

Solution.

Eigenvalues	Eigenvectors
$r_1 = 2$	$x_1 = (-3, 1)$
$r_2 = 6$	$x_2 = (1, 1)$

2. $\begin{pmatrix} -6 & 4 \\ -4 & 10 \end{pmatrix}$

Solution.

Eigenvalues	Eigenvectors
$r_1 = 2 + 4\sqrt{3}$	$x_1 = (2 - \sqrt{3}, 1)$
$r_2 = 2 - 4\sqrt{3}$	$x_2 = (2 + \sqrt{3}, 1)$

3. $\begin{pmatrix} 4 & 2 \\ -1 & 1 \end{pmatrix}$

Solution.

Eigenvalues	Eigenvectors
$r_1 = 2$	$x_1 = (-1, 1)$
$r_2 = 3$	$x_2 = (-2, 1)$

4. $\begin{pmatrix} 8 & 6 \\ 4 & 3 \end{pmatrix}$

Solution.

Eigenvalues	Eigenvectors
$r_1 = 0$	$x_1 = (-3, 4)$
$r_2 = 11$	$x_2 = (2, 1)$

5. $\begin{pmatrix} 1 & -1 \\ 1 & 1 \end{pmatrix}$

Solution.

Eigenvalues	Eigenvectors
$r_1 = 1 - i$	$x_1 = (-i, 1)$
$r_2 = 1 + i$	$x_2 = (i, 1)$

6. $\begin{pmatrix} 2 & 2 \\ -9 & 2 \end{pmatrix}$

Solution.

Eigenvalues	Eigenvectors
$r_1 = 2 - 3i\sqrt{2}$	$x_1 = (i\sqrt{2}/3, 1)$
$r_2 = 2 - 3i\sqrt{2}$	$x_2 = (-i\sqrt{2}/3, 1)$

Find the eigenvalues and eigenvectors of each of the following 3×3 matrices.

7. $\begin{pmatrix} 1 & 0 & 0 \\ 2 & 1 & -2 \\ 3 & 2 & 1 \end{pmatrix}$

Solution.

Eigenvalues	Eigenvectors
$r_1 = 1$	$x_1 = (2, -3, 2)$
$r_2 = 1 - 2i$	$x_2 = (0, -i, 1)$
$r_3 = 1 + 2i$	$x_3 = (0, i, 1)$

8. $\begin{pmatrix} 3 & 2 & 4 \\ 2 & 0 & 2 \\ 4 & 2 & 3 \end{pmatrix}$

Solution.

Eigenvalues	Eigenvectors
$r_1 = -1$	$x_1 = (-1, 0, 1)$
$r_2 = -1$	$x_2 = (-1, 2, 0)$
$r_3 = 8$	$x_3 = (2, 1, 2)$

9. $\begin{pmatrix} 2 & 2 & 1 \\ 3 & 4 & 2 \\ 1 & 2 & 1 \end{pmatrix}$

Solution.

Eigenvalues	Eigenvectors
$r_1 = 0$	$x_1 = (0, -1, 2)$
$r_2 = (7 - \sqrt{37})/2$	$x_2 \approx (-0.85, 0.15, 1)$
$r_3 = (7 + \sqrt{37})/2$	$x_3 \approx (1.18, 2.18, 1)$

10. $\begin{pmatrix} 1 & 2 & -3 \\ 1 & 1 & 2 \\ 1 & -1 & 4 \end{pmatrix}$

Solution.

Eigenvalues	Eigenvectors
$r_1 = 2$	$x_1 = (-1, 1, 1)$
$r_2 = 2$	$x_2 = (-1, 1, 1)$
$r_3 = 2$	$x_3 = (-1, 1, 1)$

For each of the following matrices find the eigenvalues, eigenvectors and generalized eigenvectors:

11. $\begin{pmatrix} 3 & 3 \\ 0 & 3 \end{pmatrix}$

Solution.

Eigenvalue	Eigenvector	Generalized Eigenvector
$r_1 = 3$	$x_1 = (1, 0)$	$\eta_{12} = (0, 1/3)$

12. $\begin{pmatrix} 1 & 1 & 0 \\ 0 & 1 & 0 \\ 0 & 0 & 3 \end{pmatrix}$

Solution.

Eigenvalues	Eigenvectors	Generalized Eigenvectors
$r_1 = 1$	$x_1 = (1, 0, 0)$	$\eta_{12} = (0, 1, 0)$
$r_2 = 3$	$x_2 = (0, 0, 1)$	

13. $\begin{pmatrix} 1 & 1 & 1 \\ 1 & 0 & 1 \\ 0 & 1 & 0 \end{pmatrix}$

Solution.

Eigenvalues	Eigenvectors	Generalized Eigenvectors
$r_1 = 1$	$x_1 = (0, -1, 0)$	
$r_2 = 0$	$x_1 = (1, 0, 0)$	
$r_3 = 2$	$x_3 = (3, 2, 1)$	

14. $\begin{pmatrix} 4 & 4 \\ -4 & -4 \end{pmatrix}$

Solution.

Eigenvalue	Eigenvector	Generalized Eigenvector
$r_1 = 3$	$\eta_1 = (1, -1)$	$\eta_{12} = (0, -1/4)$

15. $\begin{pmatrix} 1 & 1 & -1 \\ -1 & -1 & 0 \\ 0 & 0 & 1 \end{pmatrix}$

Solution.

Eigenvalues	Eigenvectors	Generalized Eigenvectors
$r_1 = 0$	$x_1 = (-1, 1, 0)$	$(0, -1, 0)$
$r_2 = 1$	$x_2 = (-2, 1, 1)$	

16. $\begin{pmatrix} 0 & 0 & 0 & 1 \\ 0 & 0 & 1 & 0 \\ 0 & 1 & 0 & 0 \\ 1 & 0 & 0 & 0 \end{pmatrix}$

Solution.

Eigenvalues	Eigenvectors	Generalized Eigenvectors
$r_1 = -1$	$x_1 = (-1, 0, 0, 1)$	
$r_2 = -1$	$x_2 = (0, -1, 1, 0)$	
$r_3 = 1$	$x_3 = (1, 0, 0, 1)$	
$r_4 = 1$	$x_4 = (0, 1, 1, 0)$	

Section A.5, page 787

Compute the exponentials of the following matrices using the definition of the exponential of a matrix.

1. $\begin{pmatrix} 1 & 0 \\ 0 & -1 \end{pmatrix}$

Solution. $\begin{pmatrix} e & 0 \\ 0 & e^{-1} \end{pmatrix}$

2. $\begin{pmatrix} 0 & t \\ -t & 0 \end{pmatrix}$

Solution. $\begin{pmatrix} \cos(t) & -\sin(t) \\ \sin(t) & \cos(t) \end{pmatrix}$

3. $\begin{pmatrix} 0 & 1 & t \\ 0 & 0 & 1 \\ 0 & 0 & 0 \end{pmatrix}$

Solution. $\begin{pmatrix} 1 & 1 & \frac{1}{2}+t \\ 0 & 1 & 1 \\ 0 & 0 & 1 \end{pmatrix}$

4. $\begin{pmatrix} 0 & t \\ t & 0 \end{pmatrix}$

Solution. $\begin{pmatrix} \cosh(t) & \sinh(t) \\ -\sinh(t) & \cosh(t) \end{pmatrix}$

5. $\begin{pmatrix} 0 & t \\ 0 & 0 \end{pmatrix}$

Solution. $\begin{pmatrix} 1 & t \\ 0 & 1 \end{pmatrix}$

6. $\begin{pmatrix} 0 & a & b & c \\ 0 & 0 & d & e \\ 0 & 0 & 0 & f \\ 0 & 0 & 0 & 0 \end{pmatrix}$

Solution. $\begin{pmatrix} 1 & a & b+\dfrac{ad}{2} & c+\dfrac{ae+bf}{2}+\dfrac{adf}{6} \\ 0 & 1 & d & e+\dfrac{df}{2} \\ 0 & 0 & 1 & f \\ 0 & 0 & 0 & 1 \end{pmatrix}$

7. $\begin{pmatrix} 0 & i\pi/3 \\ 0 & -i\pi/3 \end{pmatrix}$

Solution. $\begin{pmatrix} 1 & \dfrac{1+i\sqrt{3}}{2} \\ 0 & \dfrac{1-i\sqrt{3}}{2} \end{pmatrix}$

8. $\begin{pmatrix} a & b \\ a & b \end{pmatrix}$

Solution. $\begin{pmatrix} \dfrac{b+a\,e^{a+b}}{a+b} & \dfrac{b(e^{a+b}-1)}{a+b} \\ \dfrac{a(e^{a+b}-1)}{a+b} & \dfrac{a+b\,e^{a+b}}{a+b} \end{pmatrix}$

Section A.6, page 790

1. Show that the real solutions to the homogeneous system of linear equations

$$\begin{cases} a_{11}x_1 + \cdots + a_{1n}x_n = 0, \\ a_{21}x_1 + \cdots + a_{2n}x_n = 0, \\ \vdots \\ a_{n1}x_1 + \cdots + a_{nn}x_n = 0 \end{cases}$$

form a real vector space and that the complex solutions form a complex vector space. Here solution means an n-dimensional column vector.

2. Show that the set of real solutions of a homogeneous n^{th}-order linear differential equation

$$a_n(t)y^{(n)} + a_{n-1}(t)y^{(n-1)} + \cdots + a_1(t)y' + a_0(t)y = 0$$

form a real vector space and that the complex solutions form a complex vector space.

3. Show that the real solutions to the homogeneous system of linear differential equations

$$\begin{cases} \dfrac{dx_1}{dt} = p_{11}(t)x_1 + \cdots + p_{1n}(t)x_n, \\ \vdots \qquad\qquad \vdots \\ \dfrac{dx_n}{dt} = p_{n1}(t)x_1 + \cdots + p_{nn}(t)x_n. \end{cases}$$

form a real vector space and that the complex solutions form a complex vector space. Here solution means an n-dimensional vector.

4. Show that if x and y are linearly independent, then so are $x + y$ and $x - y$.

Section A.7, page 796

1. Let

$$B = \begin{pmatrix} 0 & 0 & 0 & a \\ 0 & 0 & b & 0 \\ 0 & c & 0 & 0 \\ d & 0 & 0 & 0 \end{pmatrix}$$

Use **MatrixPower** to compute B^k for $k = -1$, $k = 2$ and $k = 7$. In each case use **MatrixForm** to display the matrix. From this information guess the what the matrix B^k is for an arbitrary integer k.

Solution.

$$B^{2k} = \begin{pmatrix} a^{2k}d^{2k} & 0 & 0 & 0 \\ 0 & b^{2k}b^{2k} & 0 & 0 \\ 0 & 0 & b^{2k}b^{2k} & 0 \\ 0 & 0 & 0 & a^{2k}d^{2k} \end{pmatrix}$$

and

$$B^{2k+1} = \begin{pmatrix} 0 & 0 & 0 & a^{2k+1}d^{2k} \\ 0 & 0 & b^{2k+1}c^{2k} & 0 \\ 0 & c^{2k+1}b^{2k} & 0 & 0 \\ d^{2k+1}a^{2k} & 0 & 0 & 0 \end{pmatrix}$$

2. *Mathematica*'s command **Chop** replaces very small numbers by 0. (The default meaning of "very small" is less than 10^{-10}.) Compute the approximate value of the exponential of the matrix **ppp** using **Chop[N[MatrixExp[ppp]]]**.

Solution.

```
Chop[N[MatrixExp[ppp]]]
```

yields

```
{{63.6785, 3.40921, 80.6519},
 {-18.4663, 0.00912114, -23.3275},
 {19.9183, 1.45201, 25.2939}}
```

3. The command **Reverse** when applied to a list creates a new list with the elements reversed. Consequently an $n \times n$ matrix with 1's in the diagonal going from the upper right to lower left can be generated for any integer **n** by the command **Reverse[IdentityMatrix[n]]**. This is the matrix

$$IR_{n \times n} = \begin{pmatrix} 0 & 0 & \cdots & 0 & 1 \\ 0 & 0 & \cdots & 1 & 0 \\ \vdots & \vdots & \ddots & \vdots & \vdots \\ 0 & 1 & \cdots & 0 & 0 \\ 1 & 0 & \cdots & 0 & 0 \end{pmatrix}$$

For example

```
Reverse[IdentityMatrix[5]]//MatrixForm
```

yields

```
0   0   0   0   1
0   0   0   1   0
0   0   1   0   0
0   1   0   0   0
1   0   0   0   0
```

Use **Table** to compute the determinant of **Reverse[IdentityMatrix[n]]** for $1 \le n \le 12$. What is the general formula?

Solution.

```
Table[Det[Reverse[IdentityMatrix[n]]],{n,1,12}]
```

yields

```
{1, -1, -1, 1, 1, -1, -1, 1, 1, -1, -1, 1}
```

Thus, the general formula is $\det(I R_{n \times n}) = (-1)^{n(n+1)/2}$.

Section A.8, page 801

Find the solution sets to the following systems of linear equations using **Solve**.

1.
$$\begin{cases} x_1 + 2x_2 - 3x_3 - 4x_4 = 6, \\ x_1 + 3x_2 + x_3 - 2x_4 = 4, \\ 2x_1 + 5x_2 - 2x_3 - 5x_4 = 10 \end{cases}$$

Solution. Using

```
Solve[{x1 + 2x2 - 3x3 - 4x4 == 6,
        x1 + 3x2 +  x3 - 2x4 == 4,
       2x1 + 5x2 - 2x3 - 5x4 == 10},
 {x1,x2,x3,x4}]
```

we obtain $x_1 = 10 + 11s, x_2 = -2 - 4s, x_3 = s, x_4 = 0$, where s is any real number

2.
$$\begin{cases} 2x_1 + x_2 + 5x_3 + x_4 & = \quad 5, \\ x_1 + x_2 - 3x_3 - 4x_4 & = -1, \\ 3x_1 + 6x_2 - 2x_3 + x_4 & = \quad 8, \\ 2x_1 + 2x_2 + 2x_3 - 3x_4 & = \quad 2 \end{cases}$$

Solution. Using

```
Solve[{2x1 +   x2 + 5x3 +   x4 == 5,
        x1 +   x2 - 3x3 - 4x4 == -1,
       3x1 + 6x2 - 2x3 +   x4 == 8,
       2x1 + 2x2 + 2x3 - 3x4 == 2},
  {x1,x2,x3,x4}]
```

we obtain $x_1 = 2, x_2 = 1/5, x_3 = 0, x_4 = 4/5$

3.
$$\begin{cases} x_1 + x_2 + 2x_3 + x_4 & = 5, \\ 2x_1 + 3x_2 - x_3 - 2x_4 & = 2, \\ 4x_1 + 5x_2 + 3x_3 & = 7 \end{cases}$$

Solution. Using

```
Solve[{x1 +   x2 + 2x3 +   x4 == 5,
       2x1 + 3x2 -   x3 - 2x4 == 2,
       4x1 + 5x2 + 3x3        == 7},
  {x1,x2,x3,x4}]
```

we obtain

```
{}
```

which is *Mathematica*'s notation for the empty set **or no solution.**

4.
$$\begin{cases} x_1 - x_2 & = 1, \\ x_2 - x_3 & = 2, \\ x_3 - x_4 & = 3, \\ x_1 + x_2 + x_3 + x_4 & = 4 \end{cases}$$

Solution. Using

```
Solve[{x1 - x2            == 1,
        x2 - x3            == 2,
        x3 - x4            == 3,
        x1 + x2 + x3 + x4 == 4},
   {x1,x2,x3,x4}]
```

we obtain $x_1 = 7/2, x_2 = -5/2, x_3 = 1/2, x_4 = -5/2$

Redo Exercises 1–4 using **LinearSolve**.

1. $$\begin{cases} x_1 + 2x_2 - 3x_3 - 4x_4 = 6, \\ x_1 + 3x_2 + x_3 - 2x_4 = 4, \\ 2x_1 + 5x_2 - 2x_3 - 5x_4 = 10 \end{cases}$$

Solution. Using

```
nn=
{
{1,2,-3,4},
{1,3,1,-2},
{2,5,-2,-5}
}
LinearSolve[nn,{6,4,10}]
```

we obtain

```
{10, -2, 0, 0}
```

2. $$\begin{cases} 2x_1 + x_2 + 5x_3 + x_4 = 5, \\ x_1 + x_2 - 3x_3 - 4x_4 = -1, \\ 3x_1 + 6x_2 - 2x_3 + x_4 = 8, \\ 2x_1 + 2x_2 + 2x_3 - 3x_4 = 2 \end{cases}$$

Solution. Using

```
nn=
{
{2,1,5,1},
{1,1,-3,-4},
{3,6,-2,1},
{2,2,2,-3}
}
LinearSolve[nn,{5,-1,8,2}]
```

we obtain

$$\{2, \frac{1}{5}, 0, \frac{4}{5}\}$$

3. $\begin{cases} x_1 + x_2 + 2x_3 + x_4 & = 5, \\ 2x_1 + 3x_2 - x_3 - 2x_4 = 2, \\ 4x_1 + 5x_2 + 3x_3 & = 7 \end{cases}$

Solution. Using

```
nn=
{
{1,1,2,1},
{2,3,-1,-2},
{4,5,3,0}
}
LinearSolve[nn,{5,2,7}]
```

we obtain

```
LinearSolve::nosol:
    Linear equation encountered which has no solution.
```

4. $\begin{cases} x_1 - x_2 & = 1, \\ x_2 - x_3 & = 2, \\ x_3 - x_4 & = 3, \\ x_1 + x_2 + x_3 + x_4 = 4 \end{cases}$

Solution. Using

```
nn=
{
{1,-1,0,0},
{0,1,-1,0},
{0,0,1,-1},
{1,1,1,1}
}
LinearSolve[nn,{1,2,3,4}]
```

we obtain

$$\{\frac{7}{2}, \frac{5}{2}, \frac{1}{2}, -(\frac{5}{2})\}$$

Section A.9, page 805

Use **Eigenvalues** and **Eigenvectors** to find the eigenvalues and eigenvectors of the following matrices. Also, compute the exponential of each matrix using **MatrixExp**.

1. $\begin{pmatrix} 3 & 2 \\ 4 & 1 \end{pmatrix}$

Solution.

Eigenvalues	Eigenvectors
$r_1 = -1$	$x_1 = (-1, 2)$
$r_2 = 5$	$x_2 = (1, 1)$

$$\exp(A) = \begin{pmatrix} \dfrac{2e^6 - 1}{3e} & \dfrac{e^6 - 1}{3e} \\[2mm] \dfrac{2e^6 - 2}{3e} & \dfrac{e^6 + 2}{3e} \end{pmatrix}$$

2. $\begin{pmatrix} 6 & -4 \\ 3 & -1 \end{pmatrix}$

Solution.

Eigenvalues	Eigenvectors
$r_1 = 2$	$x_1 = (1, 1)$
$r_2 = 3$	$x_2 = (4, 3)$

$$\exp(A) = \begin{pmatrix} -3e^2 + 4e^3 & 4e^2 - 4e^3 \\ -3e^2 + 3e^3 & 4e^2 - 3e^3 \end{pmatrix}$$

3. $\begin{pmatrix} 3 & -1 \\ 1 & 1 \end{pmatrix}$

Solution.

Eigenvalues	Eigenvectors
$r_1 = 2$	$x_1 = (1, 1)$
$r_2 = 2$	$x_2 = (1, 1)$

$$\exp(A) = \begin{pmatrix} 2e^2 & -e^2 \\ e^2 & 0 \end{pmatrix}$$

4. $\begin{pmatrix} 3 & -8 \\ 2 & 3 \end{pmatrix}$

Solution.

Eigenvalues	Eigenvectors
$r_1 = 3 - 4i$	$x_1 = (-2i, 1)$
$r_2 = 3 + 4i$	$x_2 = (2i, 1)$

$$\exp(A) = \begin{pmatrix} e^3 \cos(4) & -2e^3 \sin(4) \\[2mm] \dfrac{e^3 \sin(4)}{2} & e^3 \cos(4) \end{pmatrix}$$

5. $\begin{pmatrix} 1 & -1 \\ 2 & -3 \end{pmatrix}$

Solution.

Eigenvalues	Eigenvectors
$r_1 = 2 - i$	$x_1 = (-1/2 - i/2, 1)$
$r_2 = 2 + i$	$x_2 = (-1/2 + i/2, 1)$

$$\exp(A) = \begin{pmatrix} e^2(\cos(1) - \sin(1)) & -e^2\sin(1) \\ 2e^2\sin(1) & e^2(\cos(1) + \sin(1)) \end{pmatrix}$$

6. $\begin{pmatrix} 0 & 1 & 0 \\ 0 & 0 & 1 \\ 0 & 0 & 0 \end{pmatrix}$

Solution.

Eigenvalues	Eigenvectors
$r_1 = 0$	$\mathbf{x}_1 = (1, 0, 0)$
$r_2 = 0$	
$r_3 = 0$	

$$\exp(A) = \begin{pmatrix} 1 & 1 & \frac{1}{2} \\ 0 & 1 & 1 \\ 0 & 0 & 1 \end{pmatrix}$$

7. $\begin{pmatrix} 1 & 1 & 1 \\ 0 & 2 & 1 \\ 0 & 0 & 1 \end{pmatrix}$

Solution.

Eigenvalues	Eigenvectors
$r_1 = 1$	$\mathbf{x}_1 = (0, -1, 1)$
$r_2 = 1$	$\mathbf{x}_2 = (1, 0, 0)$
$r_3 = 2$	$\mathbf{x}_3 = (1, 1, 0)$

$$\exp(A) = \begin{pmatrix} e & -e + e^2 & -e + e^2 \\ 0 & e^2 & -e + e^2 \\ 0 & 0 & e \end{pmatrix}$$

8. $\begin{pmatrix} 1 & 2 & 1 \\ 0 & 3 & 1 \\ 0 & 5 & -1 \end{pmatrix}$

Solution.

Eigenvalues	Eigenvectors
$r_1 = -2$	$\mathbf{x}_1 = (-1, -1, 5)$
$r_2 = 1$	$\mathbf{x}_2 = (1, 0, 0)$
$r_3 = 4$	$\mathbf{x}_3 = (1, 1, 1)$

$$\exp(A) = \begin{pmatrix} e & \dfrac{5e^6 - 6e^3 + 1}{6e^2} & \dfrac{e^6 - 1}{6e^2} \\ 0 & \dfrac{5e^6 + 1}{6e^2} & \dfrac{e^6 - 1}{6e^2} \\ 0 & \dfrac{5e^6 - 5}{6e^2} & \dfrac{e^6 + 5}{6e^2} \end{pmatrix}$$

9. $\begin{pmatrix} 4 & -5 & 1 \\ 1 & 0 & -1 \\ 0 & 1 & -1 \end{pmatrix}$

Solution.

Eigenvalues	Eigenvectors
$r_1 = 0$	$x_1 = (1, 1, 1)$
$r_2 = 1$	$x_2 = (3, 2, 1)$
$r_3 = 2$	$x_3 = (7, 3, 1)$

$$\exp(A) = \begin{pmatrix} \dfrac{7e^2 - 6e + 1}{2} & -7e^2 + 9e - 2 & \dfrac{7e^2 - 12e + 5}{2} \\ \dfrac{3e^2 - 4e + 1}{2} & -3e^2 + 6e - 2 & \dfrac{3e^2 - 8e + 5}{2} \\ \dfrac{e^2 - 2e + 1}{2} & -e^2 + 3e - 2 & \dfrac{e^2 - 4e + 5}{2} \end{pmatrix}$$

10. $\begin{pmatrix} -2 & 0 & 1 \\ 1 & 0 & -1 \\ 0 & 1 & -1 \end{pmatrix}$

Solution.

Eigenvalues	Eigenvectors
$r_1 = -1$	$x_1 = (1, 0, 1)$
$r_2 = -1$	
$r_3 = -1$	

$$\exp(A) = \begin{pmatrix} \dfrac{1}{2e} & \dfrac{1}{2e} & \dfrac{1}{2e} \\ \dfrac{1}{e} & \dfrac{2}{e} & -\dfrac{1}{e} \\ \dfrac{1}{2e} & \dfrac{3}{2e} & \dfrac{1}{2e} \end{pmatrix}$$

11. $\begin{pmatrix} 2 & 0 & 0 & 0 \\ 0 & 2 & 0 & 0 \\ 0 & 0 & 3 & 0 \\ 0 & 0 & 0 & 4 \end{pmatrix}$

<table>
<tr><th colspan="2">Eigenvalues</th><th>Eigenvectors</th></tr>
</table>

Eigenvalues	Eigenvectors
$r_1 = 2$	$\mathbf{x}_1 = (1, 0, 0, 0)$
$r_2 = 2$	$\mathbf{x}_2 = (0, 1, 0, 0)$
$r_3 = 3$	$\mathbf{x}_3 = (0, 0, 1, 0)$
$r_4 = 4$	$\mathbf{x}_4 = (0, 0, 0, 1)$

Solution.

$$\exp(A) = \begin{pmatrix} e^2 & 0 & 0 & 0 \\ 0 & e^2 & 0 & 0 \\ 0 & 0 & e^3 & 0 \\ 0 & 0 & 0 & e^4 \end{pmatrix}$$

12.
$$\begin{pmatrix} 3 & 0 & 0 & 0 \\ 4 & 1 & 0 & 0 \\ 0 & 0 & 2 & 1 \\ 0 & 0 & 0 & 2 \end{pmatrix}$$

Solution.

Eigenvalues	Eigenvectors
$r_1 = 1$	$\mathbf{x}_1 = (0, 1, 0, 0)$
$r_2 = 2$	$\mathbf{x}_2 = (0, 0, 1, 0)$
$r_3 = 2$	
$r_4 = 3$	$\mathbf{x}_4 = (1, 2, 0, 0)$

$$\exp(A) = \begin{pmatrix} e^3 & 0 & 0 & 0 \\ 2e^3 - 2e & e & 0 & 0 \\ 0 & 0 & e^2 & e^2 \\ 0 & 0 & 0 & e^2 \end{pmatrix}$$